Organic Transformations in Water

Organic Transformations in Water

Principles and Applications

Edited by Gopinathan Anilkumar, Nissy Ann Harry, and Sankuviruthiyil M. Ujwaldev

WILEY VCH

Editors

Prof. Gopinathan Anilkumar
Mahatma Gandhi University
School of Chemical Sciences
Kottayam
Kerala 686560
India

Dr. Nissy Ann Harry
Catholicate College
Postgraduate and Research Department
of Chemistry
Pathanamthitta
Kerala 689645
India

Dr. Sankuviruthiyil M. Ujwaldev
Sree Kerala Varma College
Postgraduate and Research Department
of Chemistry
Thrissur
Kerala 680011
India

Cover Images: © WhiteJack/Alamy Stock Photo, © Chemical structure was kindly provided by G. Anilkumar, N.A. Harry, S.M. Ujwaldev.

Library of Congress Card No.: applied for

British Library Cataloguing-in-Publication Data
A catalogue record for this book is available from the British Library.

Bibliographic information published by the Deutsche Nationalbibliothek
The Deutsche Nationalbibliothek lists this publication in the Deutsche Nationalbibliografie; detailed bibliographic data are available on the Internet at <http://dnb.d-nb.de>.

Print ISBN: 978-3-527-35377-4
ePDF ISBN: 978-3-527-84683-2
ePub ISBN: 978-3-527-84682-5
oBook ISBN: 978-3-527-84684-9

Typesetting Straive, Chennai, India
Printing CPI Group (UK) Ltd, Croydon CR0 4YY

C9783527353774_251024

Dedicated with profound reverence and gratitude to Professor Dr. Vijay Nair for instilling passion for organic chemistry in the minds of young researchers, nurturing and caring them, moulding them into fine organic chemists, enlightening them, profusely inspiring them through his gentle and distinguished manner, revolutionizing organic synthesis by his unique and unparalleled contributions and for touching the life of one of us (GA) by mentoring him into the fascinating world of organic synthesis.

Contents

About the Editors

Gopinathan Anilkumar was born in Kerala, India, and took his PhD in 1996 from Regional Research Laboratory (renamed as National Institute for Interdisciplinary Science and Technology NIIST-CSIR), Trivandrum, with Dr. Vijay Nair. He did postdoctoral studies at University of Nijmegen, The Netherlands (with Professor Binne Zwanenburg), Osaka University, Japan (with Professor Yasuyuki Kita), Temple University, USA (with Professor Franklin A. Davis), and Leibniz-Institut für Katalyse (LIKAT), Rostock, Germany (with Professor Matthias Beller). He was a senior scientist at AstraZeneca (India). Currently, he is a professor in Organic Chemistry at the School of Chemical Sciences, Mahatma Gandhi University in Kerala, India. His research interests are in the areas of organic synthesis, medicinal chemistry, heterocycles, and catalysis. He has published more than 210 papers in peer-reviewed journals, 7 patents, 10 book chapters, and edited two books entitled "Copper Catalysis in Organic Synthesis" (Wiley-VCH, 2020) and "Green Organic Reactions" (Springer, 2021). He has received Dr. S. Vasudev Award from Govt of Kerala, India, for best research (2016) and Evonik research proposal competition award (second prize 2016). He is a fellow of the Royal Society of Chemistry (UK).

Nissy Ann Harry was born in Kerala, India. She obtained her PhD in organic chemistry from the School of Chemical Sciences, Mahatma Gandhi University, in 2021 under the guidance of Dr. Gopinathan Anilkumar. She has published over 30 papers in peer-reviewed journals and 3 book chapters. She is currently working as an assistant professor in the Postgraduate and Research Department of Chemistry, Catholicate College, Pathanamthitta, Kerala. Her research interests include organic synthesis, catalysis, and green chemistry.

Sankuviruthiyil M. Ujwaldev was born in Kerala, India. He obtained his PhD in organic chemistry from the School of Chemical Sciences, Mahatma Gandhi University, in 2021 under the guidance of Dr. Gopinathan Anilkumar. He has published over 26 papers in peer-reviewed journals and 2 book chapters. He is currently working as an assistant professor in the Postgraduate and Research Department of Chemistry, Sree Kerala Varma College, Thrissur, Kerala.

Preface

Water forms the greenest solvent available for chemical reactions in response to nature's plea for safer chemistry. Solvents comprise a large portion of the overall mass handled in a chemical reaction, thus selecting the right one is crucial. In addition to being inexpensive and benign for the environment, using water as a solvent also produces entirely new reactivity. Water is now employed as an imperative solvent in a wide range of reactions, including oxidation–reduction processes, C–H activations, radical reactions, pericyclic reactions, transition-metal catalysis, and so forth. The switch from conventional organic solvents to green solvents is long overdue. The prime goal of this book is to draw more attention to green organic synthesis among researchers. The chapters are designed to explicate the importance and potency of "nature's reaction solvent" in organic synthesis. This book addresses many aspects of organic synthesis in water, and the 17 chapters demonstrate the complexity and the rapid progress of many areas in this field.

The first chapter comprises a general introduction to organic transformations in water, discussing its historical developments and contributions to green chemistry. The next chapter discusses "on-water reactions" focusing on the mechanistic aspects as well as on a range of reaction types. The third chapter deals with micellar and phase-transfer catalysis that enables the desired transformations to occur in mild aqueous conditions without the need for hazardous organic solvents. In the next chapter, physicochemical aspects and the complex phase behaviors and mechanisms of "on-water" and "in-water" reactions are explained. The fifth chapter focuses on the direct functionalization of inert C—H bonds, in aqueous environments, offering significant advantages in terms of sustainability, safety, and biocompatibility. Performing multicomponent reactions (MCRs) in water offers a more sustainable and environmentally friendly approach to the synthesis of bioactive heterocyclic compounds, which is discussed in the next chapter. The seventh chapter describes the influence of water on the reaction rates, regio- and stereochemical outcomes of pericyclic reactions such as cycloaddition reactions, electrocyclic reactions, sigmatropic rearrangements, and group transfer reactions. Recent progress in olefin metathesis in water is portrayed in Chapter 8, with special reference to water-compatible homogeneous and heterogeneous ruthenium metathesis catalyst systems. Chapter 9 addresses the advances over the past few years in oxidation and reduction reactions using water as the reaction medium. Chapter 10 aims to summarize the recent advances in radical reactions in water, categorizing them according to different catalytic modes. The next chapter primarily delves into synthetic transformations involving carbene intermediates in water. Water as a suitable medium in several nucleophilic additions and substitution reactions in water is showcased in Chapter 12. Chapter 13 explores the advancements and current state of asymmetric synthesis in bulk water using advanced small molecular catalysts, including chiral metal complexes and organocatalysts. The next chapter surveys the recent literature on the synthesis of structurally diverse organic scaffolds utilizing various metal-free organic substances in the aqueous medium. Establishing electroorganic procedures that can be performed in an aqueous solution is the main focus of Chapter 15. Chapter 16 gives an overview of photocatalytic synthetic methods by using water as a reaction medium, by merging the innate green properties of water with the exploitation of a renewable energy source, such as light. Finally, the last chapter illustrates the industrially important organic reactions carried out in the aqueous medium.

Although aqueous organic chemistry has extensive applications and a bright future, this subject is unfocused in current organic textbooks. Leading researchers in the field of organic synthesis from academics across the globe have contributed to this book. This book will be a very valuable reference source for students, postgraduates, research scholars, professors, industrialists, and scientists. We hope our vision to clutch the attention of researchers more toward green organic synthesis will be successful through this book.

Kottayam
September 2024

Gopinathan Anilkumar
Nissy Ann Harry
Sankuviruthiyil M. Ujwaldev

List of Abbreviations

A3	aldehyde-alkyne-amine
ABNO	9-azabicyclo[3.3.1]nonane *N*-oxyl
ADH	alcohol dehydrogenase
API	active pharmaceutical ingredient
ATRP	atom transfer radical polymerization
BIMs	bis(indolyl)methanes
BMIDA	boronic acid *N*-methyliminodiacetic acid
Bpy	bipyridine
BVMs	Baeyer–Villiger monooxygenases
C-H	carbon–hydrogen
C4A4	calix[4]arene tetracarboxylic acid
cBRIDP	di-*t*-butyl(2,2-diphenyl-1-methyl-1-cyclopropyl)phosphine
CDC	cross dehydrogenative coupling
CE	coulombic efficiency
CFP	carbon fiber paper
COMU	1-cyano-2-ethoxy-2-oxoethylidenaminooxy)dimethylamino-morpholinocarbenium hexafluorophosphate
Cryo-TEM	cryogenic transmission electron microscopy
CSTR	continuously stirred tank reactor
CTAB	cetrimonium bromide
CTAC	cetyltrimethylammonium chloride
CTF	covalent triazine framework
β-CD	β-cyclodextrin
CZA	citrazinic acid
CoQ10	coenzyme Q10
D-A	Diels–Alder
DABCO	1,4-diazabicyclo[2.2.2]octane
DBPNA	2,6-dibromo-4-nitroaniline
DBSA	4-dodecylbenzene sulfonic acid
DBU	1,8-diazabicyclo[5.4.0]undec-7-ene
DCGS	diester cationic gemini surfactant
DCILs	dicationic ionic liquids
DCM	dichloromethane

DEAD	diethyl azodicarboxylate
DELs	DNA-encoded libraries
DEPO	diethylphosphine oxide
DFT	density functional theory
DG	directing group
DIPEA	*N*,*N*-di-isopropylethylamine
DLS	dynamic light scattering
DMAD	dimethyl azodicarboxylate
DMC	dimethyl carbonate
DMF	dimethylformamide
DmgH	dimethylglyox
DPBS	Dulbecco's phosphate buffered saline
DPEN	1,2-diphenyl-1,2-ethylenediamine
DTBP	di-*tert*-butyl peroxide
DTMs	dithiomalonates
E factor	environmental factor
EAN	ethyl ammonium nitrate
ECH	electrocatalytic hydrogenation
EDA	electron donor–acceptor
EDA	ethylenediamine
EDC	1-ethyl-3-(3-(dimethylamino)propyl) carbodiimide
EDTA	ethylenediaminetetraaceticacid
EHS	environment health safety
ERED	Ene-reductase
et al.	et alia
EnT	energy transfer
FAD	flavin adenine dinucleotide
FAP	fatty acid photodecarboxylase
FDCA	2,5-furan dicarboxylic acid
FMO	frontier molecular orbital
G1	Grubbs first-generation catalyst
G2	Grubbs second-generation catalyst
GBB	Groebke–Blackburn–Bienayme
HAT	hydrogen atom transfer
HATU	hexafluorophosphate azabenzotriazole tetramethyl uronium
HBD	hydrogen bond donors
HE	Hantzsch ester
HER	hydrogen evolution reaction
HFIP	1,1,1,3,3,3-hexafluoroisopropanol
HG	Hoveyda–Grubbs
HMPC	hydroxypropyl methylcellulose
i-GAL	innovation green aspiration level
I-MCRs	isocyanide-based multicomponent reaction
KREDs	ketoreductases
LUMO	lowest unoccupied molecular orbital

MCCA	medium-chain carboxylic acids
MCR	multicomponent reaction
MNPs	magnetic nanoparticles
MO	molecular orbital
MOF	metal-organic framework
Mpg-CN	mesoporous graphitic carbon nitride
MVK	methyl vinyl ketone
MVS	methyl vinyl sulfone
MWCNTs MNCs	multiwalled carbon nanotubes magnetic nanocomposites
MWI	microwave irradiation
NADPH	nicotinamide adenine dinucleotide phosphate hydrogen
NGO	*N*-doped graphene oxide
NHC	*N*-heterocyclic carbene
Ni-MPA	Ni^{II}-3-mercaptopropionic acid
NMP	*N*-methylpyrrolidone
NMR	nuclear magnetic resonance
NP	nanoparticle
OBBD	B-alkyl-9-oxa-10-borabicyclo[3.3.2]decane
oCB	oxidised carbon black
OEOMA	oligo(ethylene oxide) methyl ether methacrylate
OER	oxygen evolution reaction
OFRs	oscillatory plug flow reactor
o-QMs	*ortho*-quinone methides
ORR	oxygen reduction reaction
PAA	polyacrylic acid
PBAM-ran-PHPAM	poly(*N*-(benzyl acrylamide))-ran-poly(*N*-(2-hydroxypropyl) acrylamide)
PCET	proton-coupled-electron-transfer
PChlide	protochlorophyllide
pDApr	polydiacetylene-proline
PEG	polyethylene glycol
pH	potential of hydrogen
PHU	1-(3,5-bis(trifluoromethyl)phenyl)-3-((R)-phenyl((S)-pyrrolidin-2-yl)methyl)urea
PIDA	poly(diiododiacetylene)
PIL	polymer ionic liquid
PILs	protic ionic liquids
PMI	process mass intensity
ppm	parts per million
PQS	polyethyleneglycol ubiquinol succinate
PTC	phase transfer catalysis
PTS	polyoxyethanyl-α-tocopheryl sebacate
RCM	ring-closing metathesis
RDS	rate determining step
ROMP	ring-opening metathesis polymerization

RT	room temperature
SCF	supercritical fluids
SDOSS	sodium dioctyl sulfosuccinate
SDS	sodium dodecyl sulfate
SERS	surface-enhanced Raman spectroscopy
SET	single electron transfer
SM	Suzuki-Miyaura
SN1	substitution nucleophilic unimolecular
SN2	substitution nucleophilic bimolecular
SNAr	nucleophilic aromatic substitution
SS	stainless-steel
STY	space-time-yield
TBAC	tetrabutylammonium chloride
TBAI	tetra-*n*-butylammonium iodide
TBAOH	tetrabutyl ammonium hydroxide
TBHP	*tert*-butyl hydroperoxide
TEA	triethyl amine
TEM	transmission electron microscopy
TEMPO	(2,2,6,6-tetramethylpiperidin-1-yl)oxyl
TFA	trifluoroacetic acid
TFE	2,2,2-trifluoroethanol
THAC	tetrahexylammonium chloride
THF	tetrahydrofuran
TMEDA	*N*,*N*,*N*′,*N*′-tetramethylethylenediamine
TOF	turnover frequencies
TON	turnover number
TPGS-750-M	DL-α-tocopherol methoxypolyethylene glycol succinate solution
TPP	tetraphenyl phosphonium
TPPMS	sodium diphenylphosphinobenzene-3-sulfonate
TRIP	R-3,3′-Bis(2,4,6-triisopropylphenyl)-1,1′-binaphthyl-2,2′-diyl hydrogenphosphate
Trpy	terpyridine
VFD	vortex fluidic devices
WERSA	water extract of rice straw ash
XPhos	dicyclohexyl[2′,4′,6′-tris(propan-2-yl)[1,1′-biphenyl]-2-yl]phosphane

1

An Introduction to Organic Transformations in Water

Nissy Ann Harry[1] *and Gopinathan Anilkumar*[2,3]

[1] *Catholicate College, Postgraduate and Research Department of Chemistry, Pathanamthitta, Kerala 689645, India*
[2] *Mahatma Gandhi University, School of Chemical Sciences, Priyadarsini Hills PO, Kottayam, Kerala 686560, India*
[3] *Mahatma Gandhi University, Institute for Integrated programs and Research in Basic Sciences (IIRBS), Priyadarsini Hills PO, Kottayam, Kerala 686560, India*

1.1 The Emergence of Water as a Solvent in Organic Chemistry

A world without the plethora of goods produced by industrial organic synthesis is inconceivable in terms of societal welfare. Environmental concerns have been raising their head since the 1940s, coinciding with the expansion of industrial activity [1]. Merely a tiny percentage of the earth's resources are transformed into the intended goods by chemical processes, producing copious amounts of toxic elements and pollutants.

Solvents comprise a significant portion of the total mass needed in a chemical process; therefore, selecting the right one is crucial. Water is the greenest solvent readily accessible and has the least negative environmental effects. The reaction solvent utilized by nature is water. It is an inexhaustible naturally occurring, nonflammable, and abundant solvent. Other intriguing properties of water are the extensive hydrogen bonding, the pH of water can be changed, salts or surfactants can be added to provide salting-in or salting-out effects, and co-solvent or biphasic reaction systems can be used [2].

In 1828, Wohler's urea synthesis was the first organic reaction to employ water as a solvent [3]. The synthesis of indigo by Baeyer and Drewsen in 1882 may be the first instance of real organic synthesis. A sodium hydroxide solution was added to a suspension of *o*-nitrobenzaldehyde in aqueous acetone during their synthesis. The distinctive blue color of indigo appeared instantly, and the product eventually precipitated. But later on, water was practically eliminated from reactions as a result of the development of organometallic chemistry. The availability of conventional solvents for several processes has improved due to advancements in the petrochemical industry. Also, there was a fallacy, "*Corpora non agunt nisi fluida (or liquida) seu*

Organic Transformations in Water: Principles and Applications, First Edition.
Edited by Gopinathan Anilkumar, Nissy Ann Harry, and Sankuviruthiyil M. Ujwaldev.

soluta" ("The substances do not react unless fluid or if dissolved") [4]. This state of concern persisted until Breslow's groundbreaking study in 1980 demonstrated how adding water could accelerate the Diels–Alder process (Scheme 1.1) [5]. Here, water continued to be the most effective medium. Furthermore, even faster rates were produced by the salting-out effect of LiCl, which further reduced the solubility of organic partners in water. In addition, the presence of guanidinium chloride acted to decrease hydrophobic interactions, which resulted in slower reactions. This eliminated the possibility that the dissolution of organic reactants was necessary.

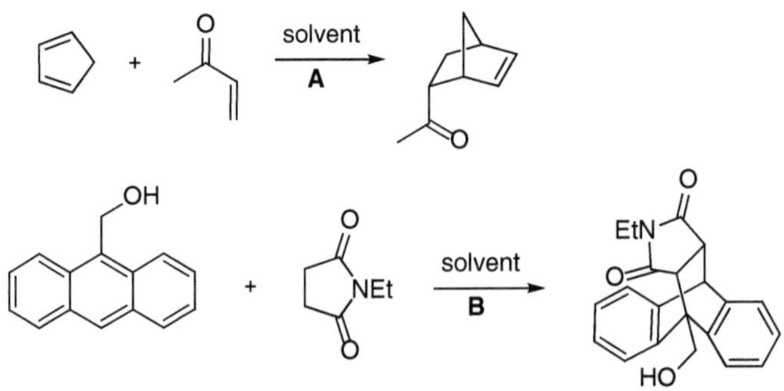

Solvents	A Rate constants $k_2 \times 10^5$	B Rate constants $k_2 \times 10^5$
Water	4400 ± 70	22600 ± 700
Water/β-cyclodextrin	10900	13800
Water/LiCl	10800	Not determined
Water/$C(NH_2^+)_3Cl^-$	4300	Not determined
Acetonitrile	Not determined	107 ± 8
Methanol	75.5	344 ± 25
Isooctane	5.94 ± 0.3	796 ± 71

Scheme 1.1 Acceleration of reaction rate by on-water protocol.

With the release of some masterworks by the Sharpless (Scheme 1.2) and Breslow groups, water has gained popularity as a reaction medium [6, 7]. Sharpless called these water-enhanced reactions between water-insoluble reactants "on-water reactions."

Butler, Coyne, and Fokin defined water-based reactions [8]. If the transition state occurs in bulk water and the substrate solubility is more than 0.01 mol l^{-1}, the reaction is said to occur "in water". Substrates that have a solubility of less than 10^{-5} mol l^{-1} in water and a transition state that takes place on the organic side of the interface are considered to be "on-water" conditions. Lastly, both reaction types are likely to happen simultaneously for reactants with intermediate solubilities [9]. In a most recent review, Kitanosono and Kobayashi suggested classifying all

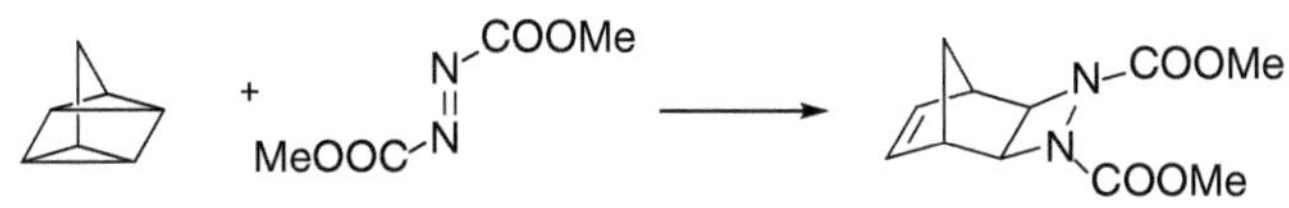

Solvent	Concentration (M)	T (°C)	Time	Yield (%)
Neat	4.53	0	2 h	0
Neat	4.53	23	48 h	85
Toluene	1	80	24 h	74
On-water	4.53	0	1.5 h	93
On-water	4.53	23	10 min	82

Scheme 1.2 Comparison of the reactivity of quadricyclane with dimethyl azodicarboxylate in different solvents.

reactions that employ water as the reaction media as "in water," regardless of the usage of catalysts [10]. Recently, several reports, reviews, and chapters were published showing the importance of water as a green organic solvent [11, 12]. However, the water as a reaction medium is less focused on books [13]. Water is now exclusively employed as a green solvent in several organic transformations, such as cross-coupling reactions, C–H activations, multicomponent reactions [14–16].

1.2 Green Chemistry and Water

Green chemistry, established about two decades ago, has drawn much attention [17]. The design of chemical processes and products to minimize or eliminate the usage or production of hazardous compounds is referred to as "green chemistry," or sustainable chemistry [18]. The choice of green solvents to reduce toxicity and pollution is a significant step toward implementing the twelve principles of green chemistry. Solvents make up between 50% and 80% of the mass in a chemical process. Solvents are an important factor that significantly influence the total toxicity profile of a chemical process.

They typically raise the most concerns regarding process safety issues because of their volatility, flammability, and potential for explosion under specific circumstances [19]. Green alternatives include solventless systems, supercritical fluids (SCF), green solvents like water, and, more recently, ionic liquids.

Among the 12 principles of green chemistry, the use of water as a solvent always increases the greenness and sometimes the efficiency of the reaction (Figure 1.1) [20].

Prevention of waste in a reaction is crucial for sustainable development. According to estimates, solvents make up over 60% of the mass of chemicals in the pharmaceutical and chemical sectors. Reuse of solvents is still rare, and the majority of these solvents are incinerated after use to recover energy, which has a significant negative influence on the environment and waste output [21]. Furthermore, burning

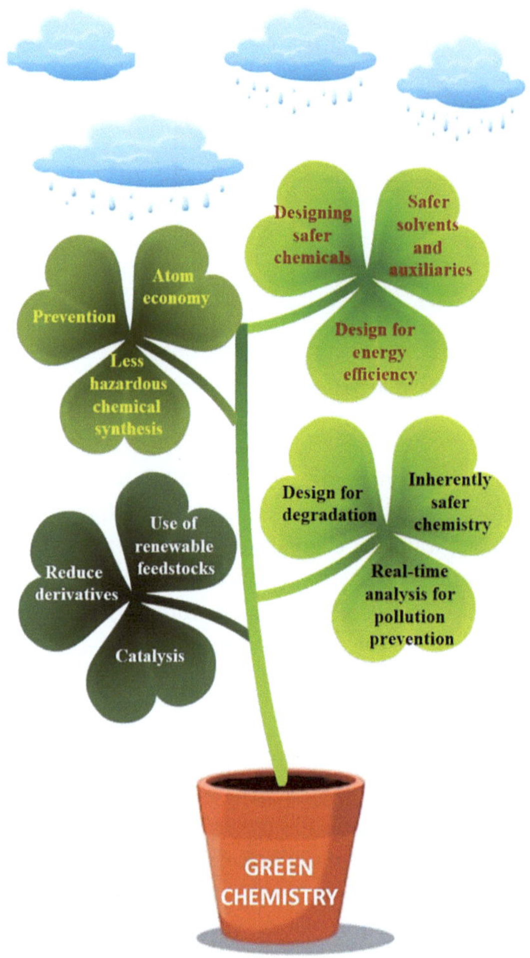

Figure 1.1 Diagrammatic representation of green chemistry principles.

organic solvents produces CO_2, which is largely attributed to the waste output and, to a lesser extent, to climate change. The E factor, positive material identification (PMI), and other typical green chemistry metrics used in reactions conducted in organic solvents are frequently taken into consideration in quantitative studies of waste quantity production [22].

Synthetic methods should be designed in such a way that all reactants that participate in the reaction are included in the final product. Hot-water-promoted, *atom economic* 1,3-rearrangements of cyclic allylic alcohols were reported (Scheme 1.3) [23]. Upon raising the temperature, the pKw of water increases, indicating a greater abundance of the strong base OH^- and strong acid H_3O^+ compared to 25 °C. Raising the temperature also lowers the dielectric constant of water, improving its suitability as a solvent for organic molecules. In contrast to high-temperature water under pressure, hot water (between 60 and 100 °C) is more readily available in both natural and laboratory settings.

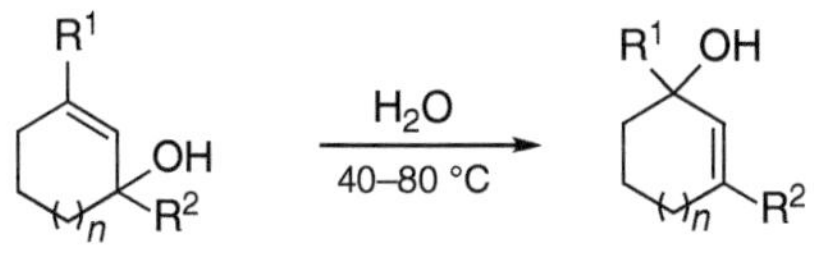

Scheme 1.3 1,3-Rearrangement of cyclic alcohols promoted by hot water.

The effective use of water as a solvent paves the way toward *less hazardous chemical synthesis*. The development of tandem and cascade reaction techniques, which combine as many reactions as feasible to yield the end product in a single operation, is crucial to the creation of greener synthetics. Jamison and coworker synthesized the essential component of the "ladder" polyether marine natural product through a biomimetic cascade cyclization in neutral water (Scheme 1.4) [24].

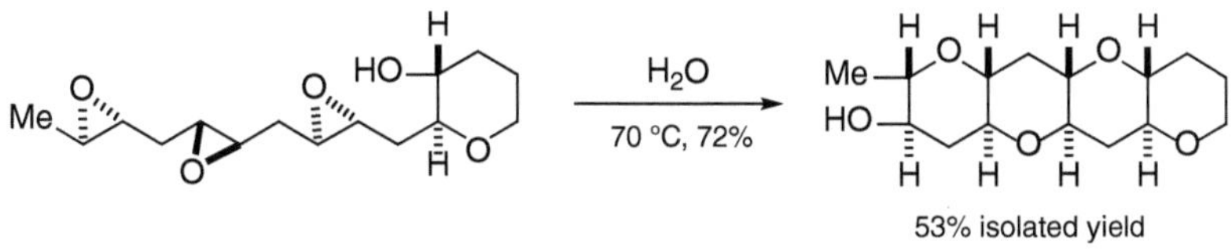

Scheme 1.4 Synthesis of polyether marine natural product in water.

Along with the advent of green chemistry concepts, the phrase "green solvent" was also created and several studies on the *use of safer solvents and auxiliaries* arose. To identify green solvents at an early stage, Capello et al. employed the EHS (E: Environment, H: Health, and S: Safety) assessment approach to rate standard organic and novel solvents (Table 1.1). The approach ranks solvents according to an EHS value between zero and nine using nine effect categories for a solvent, including flammability, risk of explosion, decomposition, and toxicity. An elevated EHS value denotes an elevated danger linked to the solvent. The EHS values of a few chosen solvents

Table 1.1 Selected EHS values of solvents.

Solvent	EHS
Acetone	3.1
Dimethylether	3.9
Methanol	2.7
Acetonritrile	4.5
Ethanol	2.6
THF	3.9
Cyclohexane	4.0
Water	0

Source: Adapted from Capello et al. [25].

used in chemical synthesis are displayed in (Table 1.1) [25]. Water has an EHS value of 0 and is the safest solvent.

Anilkumar and coworkers reported the first iron-catalyzed coupling of aryl iodides with phenols in water as a solvent [26]. Diaryl ethers are produced in good-to-exceptional yields using a 1,2-dimethylethylenediamine (DMEDA) catalytic system and inexpensive and easily accessible $FeCl_3 \cdot 6H_2O$ in this synthesis (Scheme 1.5). Compatibility of this reaction with a broad variety of functional groups further demonstrated its efficiency. Furthermore, the process is made easier because this conversion is done in the presence of air. Therefore, the process offers a simple, affordable, and environmentally responsible method of obtaining diaryl ethers.

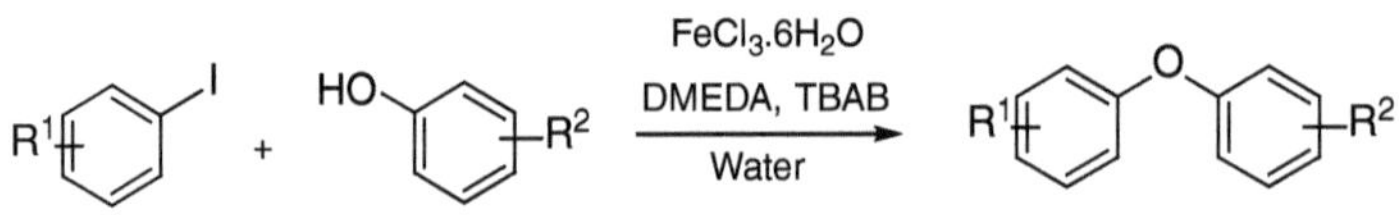

Scheme 1.5 Synthesis of diaryl ethers in water using iron catalysis.

Reducing derivatives is a key step to achieving green chemistry goals. The total synthesis efficiency is increased by the removal of laborious protection-deprotection procedures for specific acidic-hydrogen-containing functional groups when water is used as a solvent. The ability to use water-soluble hydroxyl groups directly without protection or deprotection is a significant benefit of reactions in water. This characteristic is helpful in the chemistry of carbohydrates, where protection and deprotection are frequently used [27]. For example, C-glycosides are potential enzyme inhibitors commonly used in total synthesis. The traditional protocols involve the addition of nucleophiles to a protected sugar while it is anhydrous. Water is now used as a solvent in an effective protocol that eliminates the need for laborious protection and deprotection processes (Scheme 1.6). Using commercially available D-glucose and pentane-2,4-dione as starting materials, a one-step synthesis of 38 β-C glycosidic ketone from an aqueous $NaHCO_3$ solution in nearly quantifiable quantities was achieved [28].

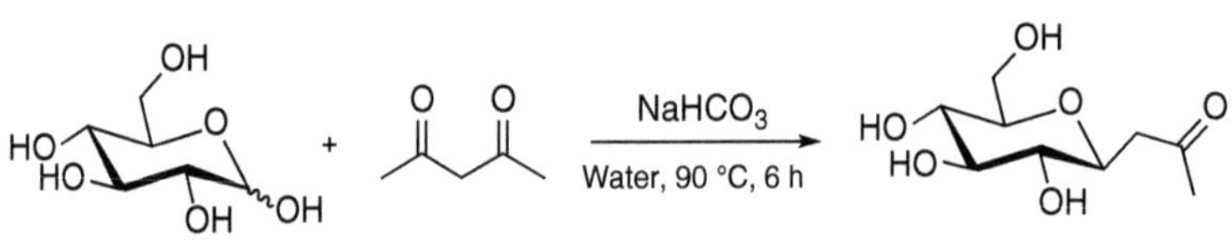

Scheme 1.6 Single-step synthesis of β-C glycosidic ketone in water.

Catalysis via "in water" and "on-water" processes are known approaches [29]. Surfactants and the spontaneous development of micellar aqueous media, however, offer a more widely applicable and practical way to deal with the limited solubility of organic reagents and metal catalysts in water [30]. The hydrophobic effect in water forms aggregates that function as nanoreactors [31]. Higher local concentrations with rate enhancement and particular solvation effects result from this, which favorably influences the chemo-, regio-, and stereo-selectivity of the final product [32].

Amphiphilic surfactant Brij S-100 was employed for oxidative hydroxylation of aromatic boronic acids to phenols using $CuCl_2$ as a catalyst in water without the need for bases or ligands (Scheme 1.7) [33]. The reaction proceeded smoothly with several electron-withdrawing and electron-donating boronic acids to afford the corresponding phenols in moderate-to-excellent yields.

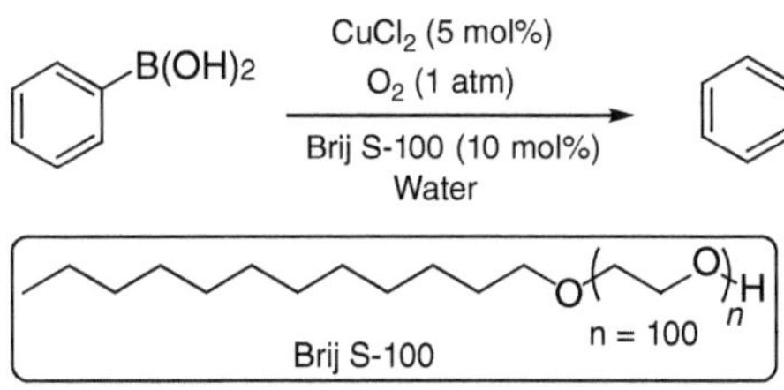

Scheme 1.7 Oxidative hydroxylation of aromatic boronic acids to phenols.

The use of renewable feedstock like bio-based compounds over fossil-based chemicals marks another step toward green chemistry. Research teams have just lately begun to look into the use of biosurfactants in traditional micellar catalysis reactions, such as Pd-catalyzed cross-couplings. High biodegradability and biological synthesis from renewable feedstocks are the two greatest advantages of biosurfactants; these two key benefits will further encourage the use of biosurfactants in micellar catalysis [34].

Design for energy efficiency is a significant step toward green chemistry [35]. The synthesis of 2-substituted 2,3-dihydro-1*H*-perimidines on water using a unique, environmentally friendly, and energy-efficient protocol was reported by Anilkumar and coworkers [36]. Using a catalyst-free on-water technique at room temperature, 1,8-diaminonaphthalene is reacted with various aldehydes to produce the 2,3-dihydro-1*H*-perimidine product in 30 minutes with moderate-to-good yields (Scheme 1.8). A multigram scale reaction is also carried out to guarantee the reaction's scalability.

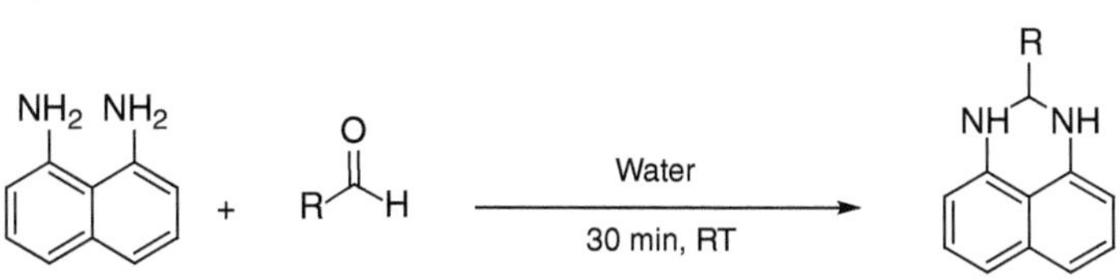

Scheme 1.8 Energy-efficient synthesis of perimidines in water.

Designing for degradation and *designing safer chemicals* that are biodegradable or less harmful to the environment is of utmost importance [37, 38]. Petrochemical compounds are the primary source of surfactants used by the chemical industry. As a result, traditional neutral or anionic surfactants do not meet these criteria. In this sense, emerging designer surfactants mark a significant shift since all newly suggested amphiphilic molecules consist of bio-based units, at least in the hydrophobic portion. For use in the Ullmann type C–S coupling reaction in water mediated by copper salt, an intriguing new surfactant bearing a sugar unit for metal catalysis in water was described (Scheme 1.9). It connects a lactose hydrophilic unit to an aliphatic alkyl chain forming the alkyl lactosamine ALA-14, which is readily synthesized and naturally degradable [39].

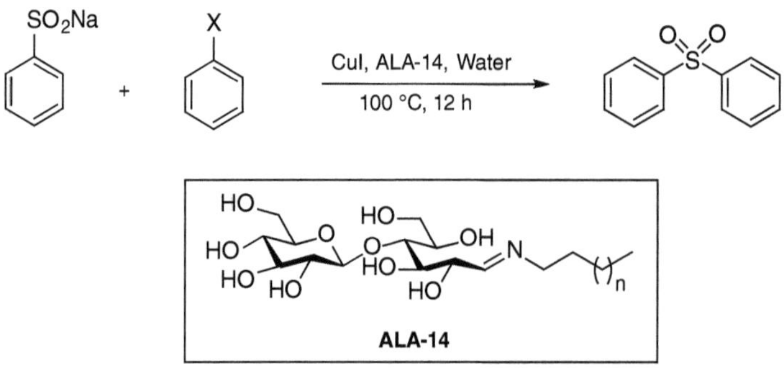

Scheme 1.9 Ullmann C–S coupling using ALA14 in aqueous solution.

Reactions in water thus naturally contribute to *inherently safer chemistry for accident prevention* regarding common dangers, such as fires, explosions, or toxicity on release, in contrast to extensively used catalytic processes in organic solvent systems [40].

Real-time analysis for pollution prevention, in-process monitoring and control before the creation of hazardous chemicals, and analytical methods must be further developed [41]. For reactions carried out in water, such analysis can identify catalyst poisoning, changes in process pH or temperature before a reaction spirals out of control, and other harmful events before a significant catastrophe happens.

1.3 Conclusion

The development of organic reactions in water has received tremendous attention from chemists over the past two decades, and this topic is seeing a significant expansion in study. Lipophilic substrates react in hydrophobic pockets that are generated in water. These pockets serve as reaction vessels that mimic the chemistry found in nature and give us a clear path forward for further research in this field. Employing surfactants in micellar conditions – in which the micelle functions as a nano-reactor is another simplest way to do catalysis in water. The use of water as a solvent for organic transformations has the potential to significantly advance the fields of chemistry, the chemical industry, and environmental preservation. Putting its regulation into action is one of its biggest obstacles. It takes a lot of work to properly screen reaction conditions to facilitate the reactions in water. In the approaching decades, water as a solvent will remain intriguing and beneficial as there will be a lot more reports released in this area. A greater emphasis on large-scale application should be placed on research. More lenient legislation, faster technology transfers between academia and governments, and tax breaks for businesses using such green standards in the workplace are all necessary to enable the adoption of these technologies in the business sector.

References

1 Ritter, S.K. (2001). *Chem. Eng. News* 79: 27–37.

2 Li, C.J. and Anastas, P.T. (2012). *Chem. Soc. Rev.* 41: 1413–1414.

3 Wöhler, F. (1828). *Annalen der Physik und Chemie* 88: 253–256.

4 "The substances do not react unless fluid or if dissolved", quoted by Aristotle (384–322 BCE).

5 Rideout, D.C. and Breslow, R. (1980). *J. Am. Chem. Soc.* 102: 7816–7817.

6 Narayan, S., Muldoon, J., Finn, M.G. et al. (2005). *Angew. Chem. Int. Ed.* 44: 3275–3579.

7 Breslow, R. (2004). *Acc. Chem. Res.* 37: 471–478.

8 (a) Butler, R.N. and Coyne, A.G. (2010). *Chem. Rev.* 110: 6302–6337. (b) Chanda, A. and Fokin, V.V. (2009). *Chem. Rev.* 109: 725–748.

9 (a) Butler, R.N. and Coyne, A.G. (2016). *Org. Biomol. Chem.* 14: 9945–9960. (b) Cortes-Clerget, M., Yu, J., Kincaid, J.R.A. et al. (2021). *Chem. Sci.* 12: 4237–4266.

10 Kitanosono, T. and Kobayashi, S. (2020). *Chem. –Eur. J.* 26: 9408–9429.

11 (a) Harry, N.A., Radhika, S., Neetha, M., and Anilkumar, G. (2019). *Chemistry Select* 4: 12337–12355. (b) Fabris, F., Illner, M., Repke, J.-U. et al. (2023). *Molecules* 28: 4809–4845. (c) Sachdeva, H. and Khaturia, S. (2017). *Biorg. Org. Chem.* 1: 239–243. (d)Kitanosono, T., Masuda, K., Xu, P., and Kobayashi, S. (2018). *Chem. Rev.* 118: 679–746. (e)Li, C.J. and Chen, L. (2006). *Chem. Soc. Rev.* 35: 68–82.

12 Harry, N.A., Rohit, K.R., and Anilkumar, G. (2021). *Green Organic Reactions* (ed. G. Anilkumar and S. Saranya), 33–49. Springer.

13 (a) Lindstrom, U.M. (ed.) (2007). *Organic Reactions in Water: Principles, Strategies and Applications.* Wiley-Blackwell. (b) Li, C.-J. and Chan, T.-H. (ed.) (2006). *Comprehensive Organic Reactions in Aqueous Media.* John Wiley & Sons. (c) Grieco, P.A. (ed.) (1998). *Organic Synthesis in Water.* Springer.

14 Dong, D.-Q., Yang, H., Zhou, M.-Y. et al. (2023). *Curr. Opin. Green Sustain. Chem.* 40: 100778.

15 Dalton, T., Faber, T., and Glorius, F. (2021). *ACS Cent. Sci.* 7: 245–261.

16 Paprocki, D., Madej, A., Koszelewski, D. et al. (2018). *Front. Chem.* 6: 502–523.

17 Anastas, P.T. and Warner, J.C. (ed.) (1988). *Green Chemistry – Theory and Practice.* New York: Oxford University Press.

18 Anastas, P. and Eghbali, N. (2010). *Chem. Soc. Rev.* 39: 301–312.

19 Earle, M.J., Seddon, K.R., Adams, C.J., and Roberts, G. (1998). *Chem. Commun.* 2097–2098.
20 Zhou, F., Hearne, Z., and Li, C.-J. (2019). *Curr. Opin. Green Sustain. Chem.* 18: 118–123.
21 Sheldon, R.A., Bode, M.L., and Akakios, S.G. (2022). *Curr. Opin. Green Sustain. Chem.* 33: 100569–100575.
22 (a) Benison, C.H. and Payne, P.R. (2022). *Curr. Res. Green Sustain. Chem.* 5: 100229–100234. (b) Gonzalez, C. and Lund, C. (2022). *Curr. Opin. Green Sustain. Chem.* 33: 100564.
23 Li, P.-F., Wang, H.-L., and Qu, J. (2014). *J. Org. Chem.* 79: 3955–3962.
24 Vilotijevic, I. and Jamison, T.F. (2007). *Science* 317: 1189–1192.
25 Capello, C., Fischer, U., and Hungerbühler, K. (2007). *Green Chem.* 9: 927–934.
26 Sindhu, K.S., Ujwaldev, S.M., Krishnan, K.K., and Anilkumar, G. (2017). *J. Catal.* 348: 146–150.
27 Guo, J. and Ye, X.-S. (2010). *Molecules* 15: 7235–7265.
28 Rodrigues, F., Canac, Y., and Lubineau, A. (2000). *Chem. Commun.* 2049–2050.
29 Harry, N.A., Shilpa, T., Ujwaldev, S.M., and Anilkumar, G. (2021). *J. Heterocyclic Chem.* 58: 375–381.
30 Cortes-Clerget, M., Kincaid, J.R.A., Akporji, N., and Lipshutz, B.H. (2022). *Supramolecular Catalysis* (ed. P.W.N.M. van Leeuwen and M. Raynal), 467–487. New York: Wiley.
31 Strukul, G., Fabris, F., and Scarso, A. (2022). *Supramolecular Catalysis: New Directions and Developments*, Chapter 31 (ed. P.W.N.M. van Leeuwen and M. Raynal), 451–466. New York: Wiley.
32 Scarso, A. and Strukul, G. (2019). *Green Synthetic Processes and Procedures* (ed. R. Ballini), 268–288. London: Royal Society of Chemistry.
33 Inamoto, K., Nozawa, K., Yonemoto, M., and Kondo, Y. (2011). *Chem. Commun.* 47: 11775–11777.
34 Adamik, R., Herczegh, A.R., Varga, I. et al. (2023). *Green Chem.* 25: 3462–3468.
35 Quadrelli, E.A. (2016). *Green Chem.* 18: 328–330.
36 Harry, N.A., Cherian, R.M., Radhika, S., and Anilkumar, G. (2019). *Tetrahedron Lett.* 60: 150946–150950.
37 Cucciniello, R. and Anastas, P.T. (2021). *Curr. Opin. Green Sustain. Chem.* 31: 100528.
38 DeVito, S.C. (2016). *Green Chem.* 18: 4332–4347.
39 Ge, X., Zhang, S., Chen, X. et al. (2019). *Green Chem.* 21: 2771–2776.
40 Khan, F.I. and Abbasi, S.A. (1998). *J. Loss. Prev. Process Ind.* 11: 361–372.
41 Wang, J. (2002). *Acc. Chem. Res.* 35: 811–816.

2

Chemistry of "On-Water" Reactions

Anthony G. Coyne

University of Cambridge, Yusuf Hamied Department of Chemistry, Lensfield Road, Cambridge CB2 1EW, UK

2.1 Introduction

The use of water as a solvent in organic synthesis can be traced back to Wöhler's synthesis of urea from silver cyanate with ammonium chloride. This is considered one of the foundation reactions in organic synthesis. However, over the next 150 years, the use of water in organic synthesis became less common due to solubility and incompatibility issues with many of the reactions developed. The focus of this chapter is to examine the use of water as a solvent, specifically in what are termed on-water reactions [1]. On-water reactions have been extensively reviewed, and the aim of this chapter is to give an overview of key developments in this area [2–7]. Reactions which have been classed as in-water or using micellar catalysis will be discussed for comparison purposes with on-water reactions; however, these will not be the focus of this chapter [8].

In 1980, Breslow and Rideout reported a kinetic study exploring the Diels–Alder [4+2] cycloaddition reaction of cyclopentadiene and methyl vinyl ketone using water as a solvent [9]. It was observed that the reaction of cyclopentadiene and methyl vinyl ketone as dienophile using was 740 times faster in comparison with that of the organic solvent, isooctane (Scheme 2.1). This was an unexpected result both due to the magnitude of the rate acceleration and that cyclopentadiene is sparingly soluble in water.

Upon changing the dienophile to acrylonitrile, an increase in rate of just over 30 times was measured in comparison with isooctane. These results were surprising for several reasons. First, cycloaddition reactions are generally thought to be "solvent insensitive" with little change in the rate of reaction with a change in the solvent polarity. Second, a highly insoluble diene such as cyclopentadiene underwent a reaction using water as a solvent, and with methyl vinyl ketone as the dienophile, a large rate acceleration was observed. However, upon changing the dienophile to acrylonitrile, the rate increase was not as pronounced.

Breslow and Maitra followed up this kinetic study with a synthetic study by exploring the reactions of cyclopentadiene and a range of dienophiles using water

Organic Transformations in Water: Principles and Applications, First Edition.
Edited by Gopinathan Anilkumar, Nissy Ann Harry, and Sankuviruthiyil M. Ujwaldev.

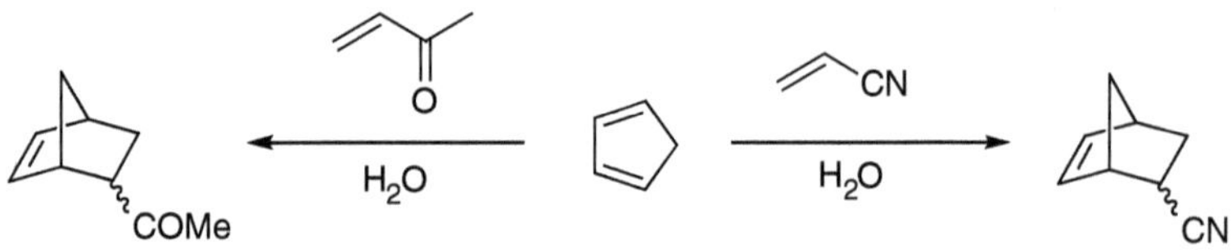

Methyl vinyl ketone		Acrylonitrile	
Solvent	**$k_2 \times 10^5\ M^{-1}s^{-1}$**	**Solvent**	**$k_2 \times 10^5\ M^{-1}s^{-1}$**
Isooctane	5.94	Isooctane	1.90
Methanol	75.5	Methanol	4.0
H_2O	4400	H_2O	59.3

Scheme 2.1 Rates of Diels–Alder 4 + 2 cycloaddition reaction of cyclopentadiene with methyl vinyl ketone and acrylonitrile with various solvents. Source: Adapted from Rideout and Breslow [9].

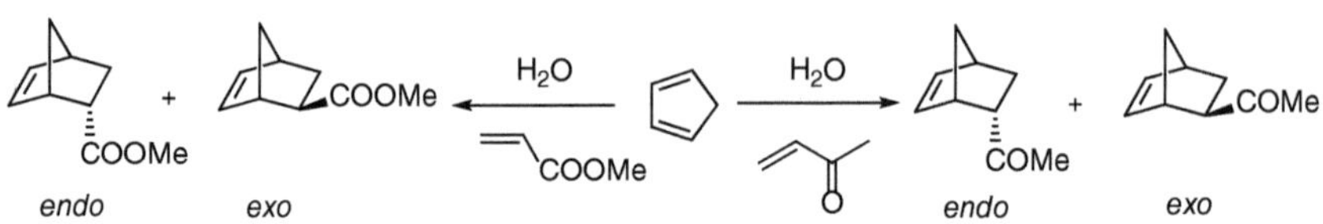

Solvent	Dienophile	*endo/exo*
Cyclopentadiene (excess)	Methyl vinyl ketone	3.85 : 1
	Methyl acrylate	2.9 : 1
Ethanol	Methyl vinyl ketone	8.5 : 1
	Methyl acrylate	5.2 : 1
Water (0.15 M)	Methyl vinyl ketone	21.4 : 1
	Methyl acrylate	9.3 : 1
Water (0.30 M)	Methyl vinyl ketone	18.6 : 1
	Methyl acrylate	5.9 : 1

Scheme 2.2 Diels–Alder 4 + 2 cycloaddition of cyclopentadiene with methyl acrylate and methyl vinyl ketone. Source: Adapted from Breslow and Maitra [10].

as a solvent (Scheme 2.2) [10]. The reaction of cyclopentadiene and methyl vinyl ketone using water as a solvent produced two products, an *endo* and an *exo* isomer (ratio 21.4 : 1). It was observed that the *endo* selectivity of these reactions increased using methyl vinyl ketone as a dienophile, whereas upon changing the dienophile to methyl acrylate, this was not as pronounced (ratio 9.3 : 1). When the reactions were carried out using ethanol as a solvent, an increase in *endo:exo* selectivity was observed; however, this was not to the same extent as that of water as a solvent.

Breslow proposed that the rate increase and the increase in *endo* selectivity using water as a solvent with methyl vinyl ketone were due to the hydrophobic effect. The hydrophobic effect can be explained as in the highly polar environment of the water solvent the hydrophobic organic substrates aggregate together increasing the rate of the reaction. This is however an oversimplification and does not consider the differences in rate observed for methyl vinyl ketone in comparison with acrylonitrile. Subsequent computational studies by Jorgensen and coworkers suggest a more complex mechanistic picture with influences from both the hydrophobic effect as well as a significant contribution from hydrogen bonding effects [11, 12].

In 2005, Sharpless and coworkers reported the $2\sigma + 2\sigma + 2\pi$ reaction of quadricyclane with dimethyl azodicarboxylate (DMAD) using water as a solvent (Scheme 2.3) [1]. When this reaction was carried out using isooctane as the solvent,

(a)

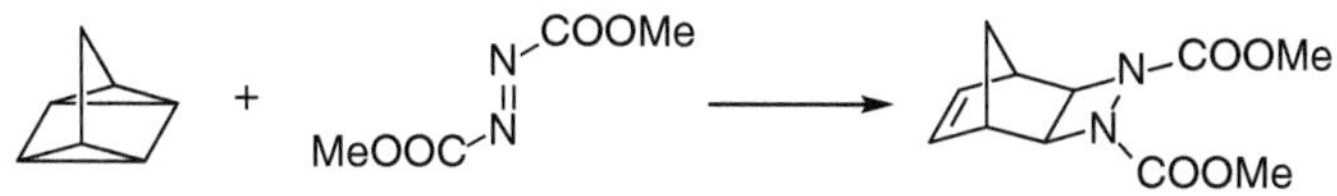

Solvent	Conc (M)	Reaction Time	Yield (%)
Toluene[a]	1	24 h	74
Neat[b]	2	48 h	85
on-H_2O[b]	4.53	10 min	82

(b)

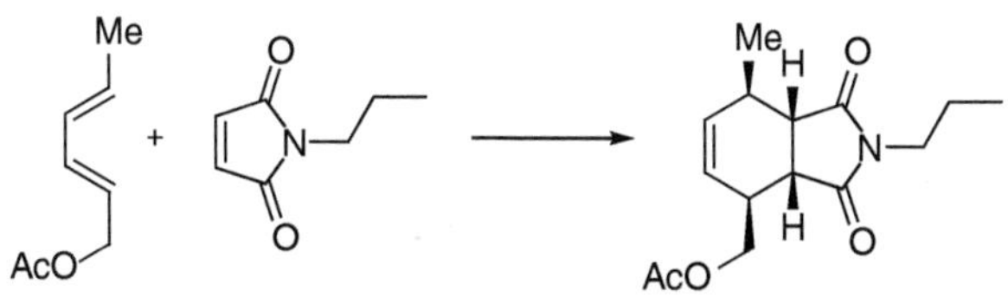

Solvent	Conc (M)	Reaction Time	Yield (%)
Toluene	1	144 h	79
Neat	3.69	10 h	82
on-H_2O	3.69	8 h	81

(c)

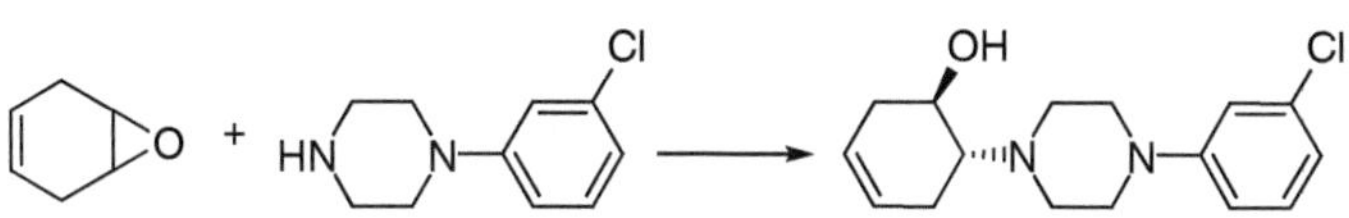

Solvent	Conc (M)	Reaction Time	Yield (%)
Toluene	1	120 h	< 10
Neat	3.88	72 h	76
on-H_2O	3.88	12 h	88

[a] reaction carried out at 80 °C.
[b] reaction carried out at 23 °C.

Scheme 2.3 On-water reactions described by Sharpless and coworkers explore the (a) $2\sigma + 2\sigma + 2\pi$ reaction of quadricyclane with dimethyl azodicarboxylate, (b) Diels–Alder 4 + 2 cycloaddition reaction, and (c) nucleophilic ring opening of epoxides using a range of solvents. Source: Adapted from Narayan et al. [1].

it took >12 hours to complete. Changing to the more polar solvent, methanol, this took 18 hours to go to completion. Surprisingly, when the reaction was carried out using water as a solvent, the reaction was complete within 10 minutes, a significant increase in the rate of reaction. Upon addition of the reactants, it was observed that the highly colored DMAD floated on the water surface, and as the reaction proceeded, this disappeared, leading to the formation of the colorless product.

Sharpless classified these as on-water reactions, as the reactants were seen to float on the water surface. Several other reaction types were also reported by Sharpless and co-workers, including the Diels–Alder cycloaddition reaction and the epoxide ring opening reaction (Scheme 2.1), both of which were also found to be accelerated using water as a solvent. In all three cases reported, the concentration of the reactants was >3 M, which defied the conventional thinking that reactants needed to be fully soluble in the solvent for the reaction to progress. While these reactions show significant rate acceleration, in subsequent studies by other research groups using on-water reactions, the large rate accelerations observed by Sharpless are not always as pronounced or do not occur. This presents one of the major challenges as to how to classify a reaction as either in-water or on-water.

2.1.1 Classification of In-Water and On-Water Reactions

There have been a number of attempts to classify reactions as to whether they are on-water or in-water, and this remains one of the challenges in distinguishing these modes of reaction. Butler and Coyne examined the synthetic and kinetic studies of the 1,3-dipolar cycloaddition reaction of the phthalazinium-2-dicyanomethanide 1,3-dipole with a range of dipolarophiles using water as a solvent [13, 14]. The 1,3-dipole was found to be highly insoluble in water; however, the reactions proceeded as normal using water as a solvent, where the products either crashed out of solution and could be filtered off or floated as an oil on the surface of the water layer. Using the computationally measured solubilities of the reactants, they proposed a general transition from an in-water to an on-water mechanism-dependent upon the solubilities. This approach presents some challenges where the calculated solubilities may not be a true representation of the experimental solubilities. However, as there are a limited number of experimental solubility measurements, the computationally calculated solubilities can be used as a guide to classify these reactions.

Kitanosono and Kobayashi reported a study where they classified reactions depending on their component's interactions with water [15]. These could be divided into three different types: Type I, Type II, and Type III. These types can be further divided into sub-groups. While this allows the classification of reactions into several different reaction types, it presents a challenge to truly understand what is happening on the molecular level using water as a solvent. Reactions, for example, with two components, could be seen as perhaps more straightforward to classify in comparison with a reaction, for example, a Suzuki coupling reaction with up to five or more components. This presents the challenge of classifying reactions as on-water or in-water reactions. Perhaps each reaction needs to be judged on a case-by-case basis considering all the components present rather than on a broader classification of reaction type.

2.1.2 Mechanistic Aspects

One of the major challenges with on-water reactions has been to explain how these reactions occur and, why in some cases, large rate accelerations are observed, this is still not well understood. While there have been a significant number of reactions classified as on-water, the large rate accelerations observed in the initial reports by Sharpless and coworkers are not always observed. Jung and Marcus were one of the first to examine the reactions reported by Sharpless and coworkers using computational methods and proposed that the critical aspect of these lies at the water-oil interface, where OH_{free} groups from the water protrude into the oil phase and act in a catalytic manner [16]. However, understanding the rate acceleration of on-water reactions may lie in the unique chemistry between water and reactants that occurs at an oil droplet-water phase boundary, and while computational approaches can propose mechanisms for these, experimentally, these are highly challenging to study [17].

Beattie, McErlean, and Phippen examined the Diels–Alder 4+2 cycloaddition of cyclopentadiene and dimethyl fumarate using water as a solvent [18]. The reactants, both of which are not highly soluble in water, were found to undergo an on-water reaction where the product was isolated in a yield of 75%. However, the stirring rate of the reaction was shown to be crucial to ensure that the reaction went via an on-water reaction mechanism. When the reaction was carried out using a four-stage piston homogenizer, it was observed that the rate increased significantly, suggesting that the droplet size has a significant influence on the rate. Interestingly, a H/D effect was observed where the yield on-D_2O was 56% different to that of the on-water reaction at 75%.

Kitanosono and Kobayashi proposed a new water model, where three layers of water, with different orientations, can be found at the interface [15]. This model considers catalysts, whether located in the aqueous or lipophilic phases. The first layer, where water molecules orient protons toward the hydrophobic phase (defined as the inner Helmholtz layer), is surrounded by a second layer (defined as the outer Helmholtz layer), and then followed by the bulk layer. The partial polarization of water at the interface would help with the formation of weak interactions with highly oriented transition states. However, as with the previous models, experimental evidence is further needed to explore this (Scheme 2.4).

2.1.3 Influence of Stirring Rate

Organic reactions are generally stirred to ensure the dissolution of the reactants in the solvent; however, little consideration is given to the stirring rate. In the initial reports of the on-water reactions by Sharpless and coworkers, the reaction of quadricyclane and DMAD was vigorously stirred on-water, which enabled the reaction to be carried out in minutes. Ensuring the reaction is vigorously stirred is a recurring theme in many reports where on-water reaction conditions have been

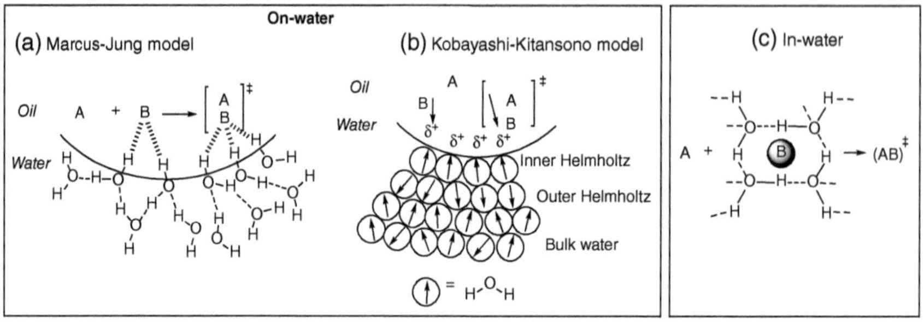

Scheme 2.4 On-water models proposed by (a) Marcus and Jung show H-atoms protruding from the water layer into the oil layer; (b) Kobayashi and Kitansono acidic surfaces with an ordered orientation model at the water-oil interface; and (c) neat and homogeneous reactions. Source: Cortes-Clerget et al. [7]/Royal Society of Chemistry.

explored. Butler et al. examined the reaction of phthalazinium-2-dicyanomethanide with *p*-chlorobenzylidineacetone. When the reaction was carried out at ambient temperature (20 °C) <1% yield of the products was observed after 48 hours. When the reaction temperature was increased to 75 °C, the products (a mixture of diastereoisomers 6.2 : 1) were observed to have an 86% yield. Interestingly, this reaction only occurred upon liquefaction of the *p*-chlorobenzylidine acetone dipolarophile with vigorous stirring (Scheme 2.5) [19, 20]. A similar outcome was

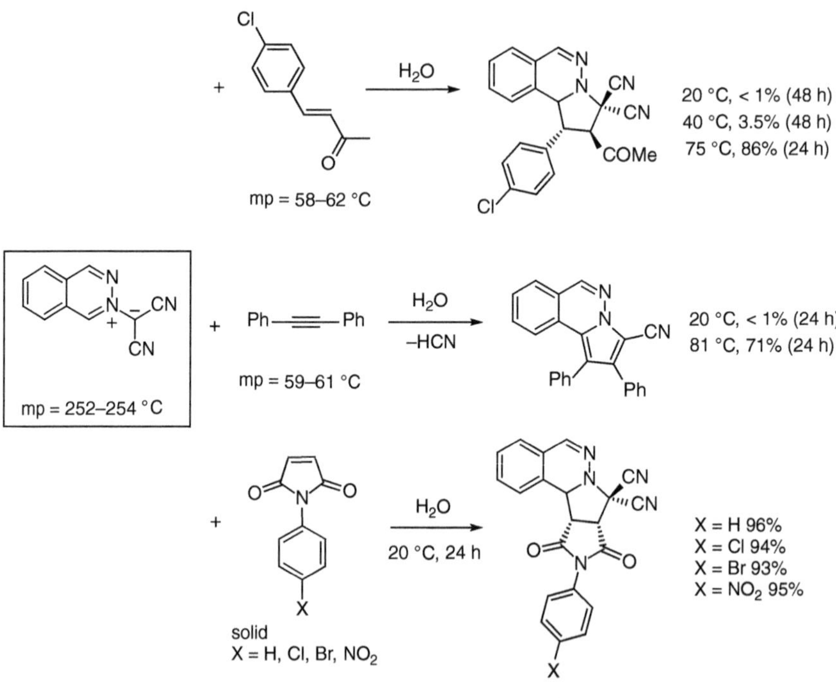

Scheme 2.5 3+2 cycloaddition reaction of phthalazinium-2-dicyanomethanide with a range of solid dipolarophiles on-water.

observed with the reaction of phthalaziuium-2-dicyanomethanide with diphenylacetylene, where the reaction occurred on-water with vigorous stirring above the melting point of the dipolarophile in a yield of 71%. In both cases, liquefaction of the dipolarophile was needed for the reaction to proceed. When the dipolarophile was changed to *para*-substituted N-phenyl maleimides, it was observed that the yellow suspension at the start of the reaction was replaced with a white suspension of the product at the reaction end. This reaction was carried out at 20 °C with vigorous stirring; however, in this instance, no liquefaction of the dipolarophile was needed for the reaction to complete.

Kobayashi and coworkers examined the asymmetric aldol-catalyzed Lewis acid-surfactant (LASC) reaction using water as a solvent [21]. This surfactant catalyst has a long chain anionic surfactant that has a metal cation, and these can form colloidal dispersions of organic molecules in water. When the reaction of benzaldehyde and a silyl enolvinyl ether using 10 mol% of the catalyst ($Sc(DS)_3$) was examined, this led to the formation of a monolayer of the surfactants with the organic molecules encapsulated in them. Upon examination of this reaction at varying stirring rates (0–1400 rpm), a significant difference in the rate of the reaction was observed. With vigorous stirring, the reaction was complete in four hours; however, when the reaction was stopped after four hours with no stirring, the reaction was significantly slower. The higher the stirring rate, the smaller the colloidal dispersions in the aqueous phase, with an increased rate in the reaction comparable to that of the on-water mechanism.

Li and coworkers examined the reaction of indoles with 1,4-benzoquinone in a range of different solvents (Scheme 2.6) [22]. The reaction carried out over 10 hours in organic solvents gave at best a 38% conversion (EtOH). When the solvent was changed to water, it was found that the reactants formed an aqueous suspension that needed to be stirred vigorously, and the product was observed to have an 82% conversion after 10 hours.

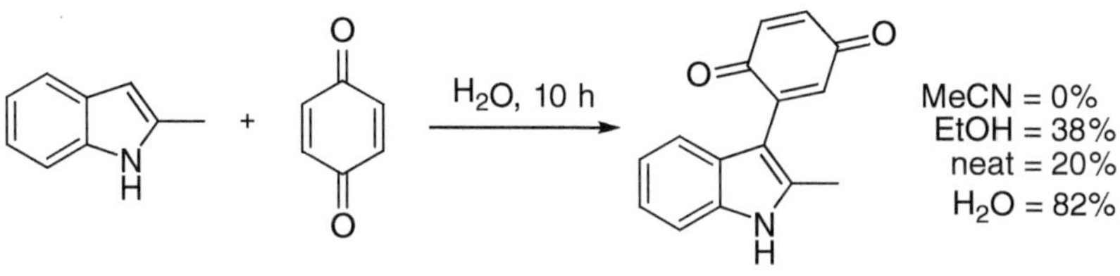

Scheme 2.6 Reaction of 2-methylindole with benzoquinone using various solvents.

A interesting reaction was reported by Capriati and coworkers, who showed that the nucleophilic addition of organolithium reagents to imines could be carried out on-water (Scheme 2.7) [23]. The reaction of aldimine with *n*-BuLi was used as a model reaction, where this was examined using water as a solvent and a variation of the order of addition was also assessed. When the reaction was carried out under vigorous stirring which generated an emulsion, the *n*-BuLi was added to the aldimine, and the yield of the reaction was observed to be 96%. However, changing the stirring

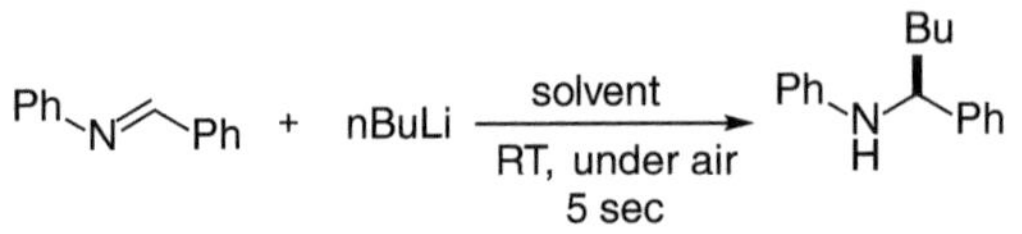

Solvent	Stirring	Yield (%)
H_2O	vortex	96
H_2O	Gentle stirring	82
D_2O	vortex	57
MeOH	vortex	15
H_2O	vortex	20[a]

[a] *n*-BuLi and imine were sequentially added to 1 mL of water

Scheme 2.7 Addition of *n*-butyl lithium to imines using on-water conditions with variation in the stirring speed.

speed to a lower speed under the same reaction conditions gave a yield of 66%. Changing the reaction solvent to D_2O, a drop in yield to 57% was observed. When the reaction was carried out using methanol as a solvent, this generated a product with a 15% yield. Capriati and coworkers proposed that the significant D/H isotopic effect observed on changing from H_2O to D_2O suggested that the on-water catalysis arises from the proton transfer across the water–organic interface. However, more in-depth mechanistic studies are needed to further understand these observations.

The change in the stirring rate of a reaction, while not a factor that is generally considered for organic reactions, could have major implications for reactions, which occur on-water. The formation of organic microdroplets in the aqueous environment could provide the mechanism by which these reactions can offer favorable results over conventional organic solvents. The surface area and the nature of the water–organic interface in these heterogeneous processes are of importance to explaining the observed reactivity changes, and further work to understand this is crucial to the understanding of the mechanism of on-water reactions.

2.1.4 On-Droplet Effects

The understanding of the nature of the surface area and that of the water–organic interface in these heterogeneous processes is challenging. Several groups have explored the use of the microdroplet methodologies to try and understand this. This methodology have been used in a wide range of applications in chemistry and biology, and the applicability of this to on-water reactions has been highlighted [24]. The surface area of an on-water reaction can be maximized by agitation (see Section 2.1.3); however, this can also be achieved by using microdroplets. When the microdroplets are coupled with mass spectroscopy detection, it offers a powerful platform to study these reactions.

Fokin and Eremin examined the on-water chemoselective reaction switch in the synthesis of 1,4-disubstituted 1,2,3-triazoles from organic azides and bromoethenesulfonyl fluoride (BrESF) [25]. The reaction was examined in water microdroplets

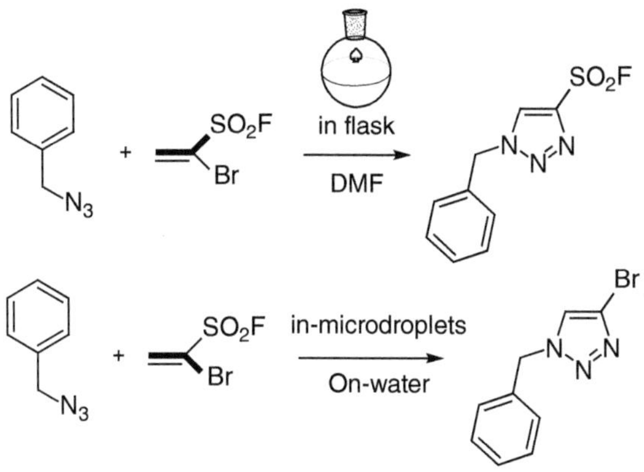

Scheme 2.8 Comparison of the reactions of benzyl azide and BrESF in-flask reaction and in-microdroplet/on-water showing the change in chemoselectivity of the reaction. Source: Eremin and Fokin [25]/American Chemical Society.

at the intersection area between two nebulizer plumes in the electrospray ionization chamber (Scheme 2.8). The reaction was initially examined using four different solvents: acetonitrile, toluene, DCM, and DMF, with varying amounts of water droplets generated by a second nebulizer. When mixture of the benzyl azide and the BrESF in a ratio of 1 : 2 in these four solvents was sprayed into the MS instrument, no signal was observed for the product. However, when water was introduced at a flow rate of 5 μl/min^{-1} by the formation of the product was observed by the accurate mass and isotopic distribution of the product. The initial cycloaddition product was not observed, instead this undergoes rapid elimination-aromatization. The rate of the reaction in microdroplets is significantly higher than in solution and favors an alternate reaction pathway compared to conventional synthesis in a reaction flask. The nature of the water surface and developed H-bonding network are critical for this reaction, and interestingly, when the reaction was carried out on-D_2O droplets, this reaction exhibits a negative solvent isotope effect as great as 8.8.

Zare and coworkers examined the "on-droplet" chemistry of the reaction of quadricyclane and DMAD, which was one of the original reactions reported by Sharpless and coworkers, identifying the on-water reaction [26]. When the reaction was examined in aqueous microdroplets (c. 5 μm), the reaction was accelerated by a factor of 10^2 greater than that of the on-water reaction reported by Sharpless and coworkers. Under these reaction conditions, it was observed that DMAD reacts with itself to form hydrazine carboxylates, which were detected using mass spectroscopy. Zare and coworkers have suggested that the "on-droplet" chemistry is like that of on-water chemistry.

The use of microdroplet methodologies coupled to mass spectroscopy is an active area, and research groups such as those of Richard Zare and Graham Cooks have developed organic reactions that seem to challenge the reactivity that is observed in bulk reactions [27, 28]. The application of this methodology to on-water reactions could offer a method of studying the water–organic interface, which to date has proved challenging using conventional analytical techniques.

2.2 Broadening the Scope of On-Water Reactions

There are a significant number of reviews, which have discussed the scope of on-water reactions, and the emergence of new areas of research has enabled the on-water concept to expand its applicability. This section highlights several areas where the on-water methodology is giving important insights. However, given the large number of reports in the literature using on-water, this is not a comprehensive discussion of these research areas, and reference should be made to the already published reviews in this area.

2.2.1 Light-Mediated Reactions

Radical reactions are synthetically useful reactions in organic synthesis; however, there has been little exploration of using water as a solvent for these reactions [29]. With the emergence of photoredox catalysis in organic synthesis, this has become a highly active area of research [30]. However, there are only a number of reports of on-water reactions using light-mediated conditions. König and coworkers examined a photochemical reaction at the oil-water interface, where they examined as a model the reaction of indole and isoquinoline-1-carbonitrile, under 365 nm light irradiation at 25 °C (Scheme 2.9) [31]. Using water as a reaction solvent, the NH arylation product was obtained in 92% yield. After exploring a range of solvents, it was observed that water gave the best yields for the reaction. When the solvent was changed to methanol, the yield was reduced to 15%, and even with a 1 : 1 mixture of water : methanol, a yield of only 20% was observed. A wide range of substrates were examined, including applying this methodology as a key step for the synthesis of known drug molecules, and in almost all cases, the yields were good to excellent.

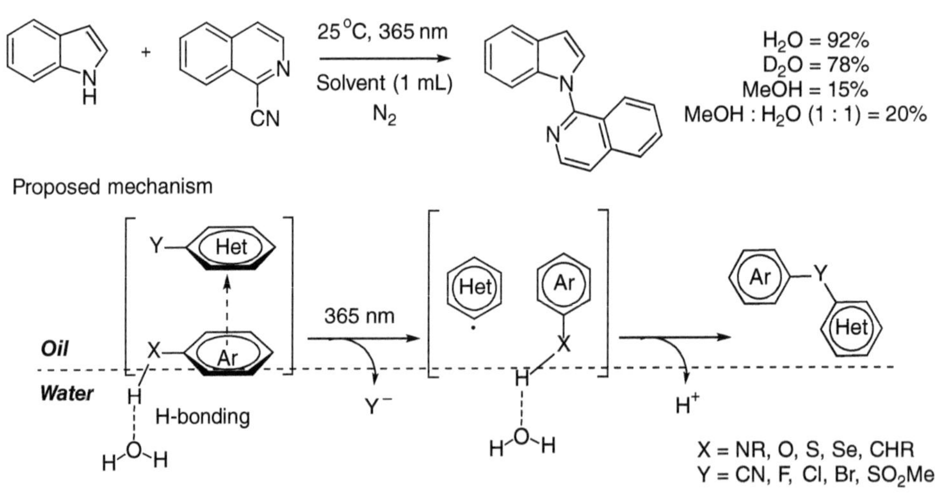

Scheme 2.9 Light-mediated reaction of indole and isoquinoline-1-carbonitrile using on-water methodology.

When the indole and isoquinoline-1-carbonitrile, which are both solids, were combined in the presence of water, yellow oil droplets were observed. Melting point measurements indicated that stoichiometric amounts of these compounds form an eutectic mixture melting at room temperature. Upon melting, the hydrophobic effect of water forces the two reactants to form an electron-donor acceptor (EDA) complex on the water surface, where this forms H-bonds with water, upon irradiation with light at 365 nm this leads to the formation of a radical ion pair, which ultimately leads to the product. A range of substitutions and heterocyclic rings allow this methodology to be applied to a diverse range of different heterocyclic scaffolds.

2.2.2 Electrochemical Reactions

Electrochemical reactions have experienced a surge in interest in the last decade, with the research group of Phil Baran leading the re-emergence of this methodology [32]. Electrochemical organic synthesis using water has been shown to be an ideal solvent for these reactions [33].

Compton and coworkers examined the electrochemical reaction of water immiscible liquid aliphatic acids leading to decarboxylation to afford a hydrocarbon. This can be achieved using an emulsion formed by sonication, so that the organic phase constantly extracts the products [34]. Power ultrasound was used to emulsify hexanoic acid in an aqueous NaOH biphasic system, and after the passage of an electrical charge, this led to the formation of *n*-decane (a Kolbe dimer).

Diniz and coworkers examined the metal-free Reformatsky reaction of ethyl 2-bromo-2-methylpropanoate and benzaldehyde [35]. These were electrolyzed in a divided cell with a potential of −1.2 V versus Ag/AgCl, and several types of carbon cathodes and platinum anodes were used. The product of the reaction was obtained in yields of up to 46% with varying amounts of side products. It was shown that a three-phase system (aqueous, organic, and solid) was essential to optimizing the yield of the coupling product and suppressing side reactions (Scheme 2.10a).

Tanaka and coworkers examined the electroorganic synthesis of an oil-in-water nanoemulsion TEMPO-mediated electrooxidation of amphiphilic alcohols using water (Scheme 2.10b) [36]. Electrooxidation of polyethylene glycol monomethyl ethers gave the corresponding carboxylic acids in good yields. When water soluble alcohols were used, this did not effectively give the corresponding carboxylic acids. In cases of water insoluble alcohols, the electrooxidation did not proceed smoothly. A plausible mechanism for the electrooxidation is that in aqueous solutions of the alcohol, this would interact lipophilically with TEMPO and hydrophobically with water to form a nanoemulsion. This would then disperse in water and bring the TEMPO radical to the anode by diffusion. Electron transfer from the TEMPO radical to the anode would occur in the nanoemulsion to afford the N-oxoammonium ion, which would oxidize the surrounding alcohol. In the case of the water insoluble alcohols, the TEMPO radical would be incorporated into the oil droplets (much larger than the microemulsion) and be separated from the anode so that the oxidation of the TEMPO did not occur readily.

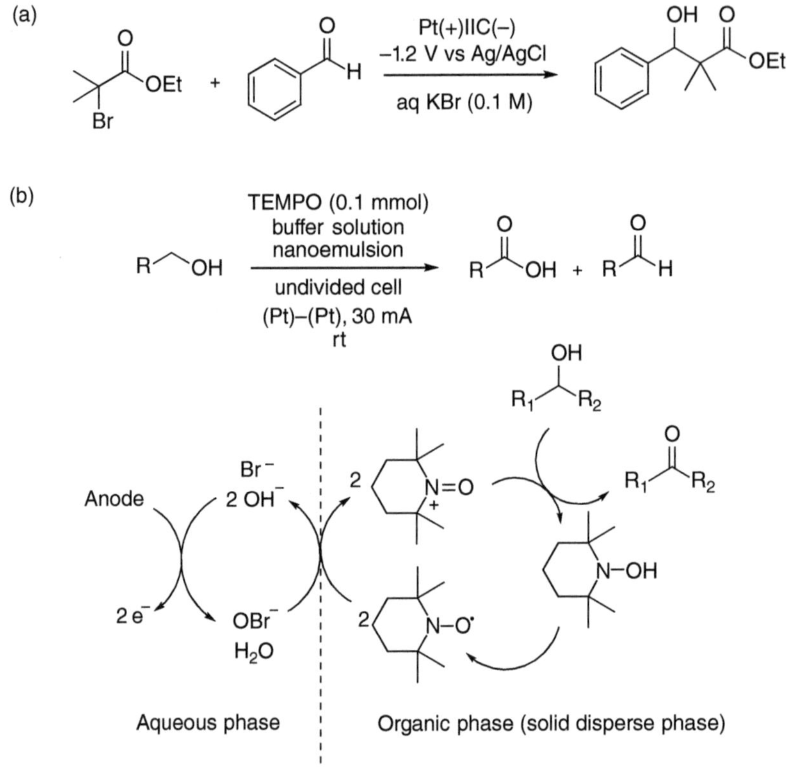

Scheme 2.10 (a) Electrochemical synthesis of a metal-free Reformatsky reaction of ethyl 2-bromo-2-methylpropanoate and benzaldehyde and (b) nanoemulsion TEMPO-mediated electrooxidation of amphiphilic alcohols in water.

Electrochemical organic synthesis, both in-water and on-water, is still in the early stages of exploration in comparison with those in organic solvents; however, there is scope to improve this methodology to include aqueous environments.

2.2.3 Flow Chemistry

Flow chemistry has emerged as a method to synthesize molecules in a continuous fashion in comparison with the more traditional methods employed with batch chemistry in a round-bottomed flask. This methodology has been adapted by a wide range of applications, from the synthesis of active pharmaceutical ingredients (API's) to the synthesis of complex natural products [37, 38]. Using the on-water methodology in flow chemistry presents challenges, especially where reactants and products are suspensions in the aqueous phase, which can prove problematic for flow reactors.

Raston and coworkers developed vortex fluidic devices (VFD), which enhance the on-water effects and catalysis using in-flow [39]. In this reactor, reagents are added to a rapidly spinning tube and react in a thin film. In the confined mode, the mixture remains in the tube; however, this can be converted to a continuous mode where products exit the top. It is hypothesized that the thin film processing could potentially enhance the on-water effect. As a proof of concept, an azide–alkyne cycloaddition reaction was explored. The uncatalyzed reaction (copper-free) was examined between benzyl azide and the disubstituted alkyne, where conversion to triazole was higher than in MeCN or the neat reaction. This was also higher than the magnetic stirring batch reaction. When moved to continuous mode, the conversion of the reactions with a range of substituted azides and terminal alkynes (copper catalyzed from the reactor at a final concentration of 9 ppb) was up to 99%, showing the applicability of this reaction methodology (Scheme 2.11).

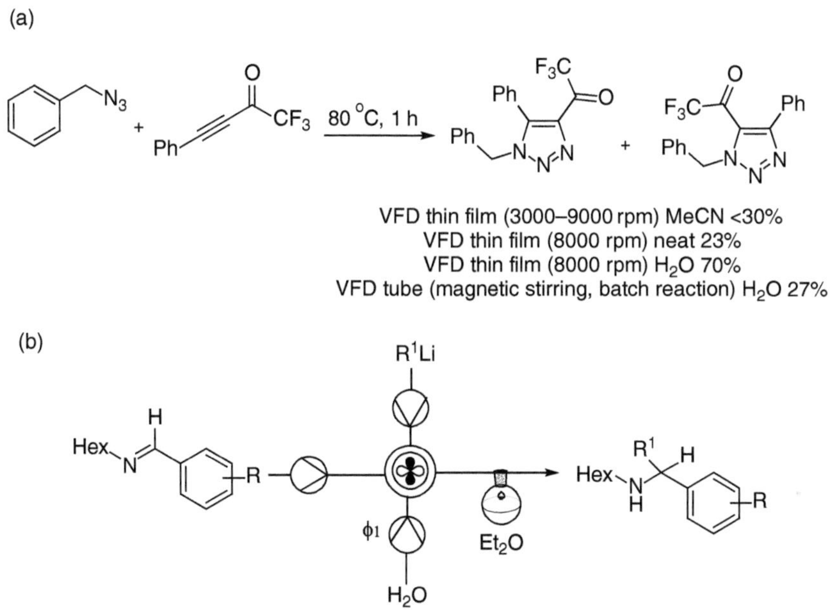

R	R^1	ϕ_1 (mL min^{-1})	Yield (%)
4-Cl	Hex	0.75	93
4-Cl	Hex	1.5	51
4-OMe	*n*-Bu	0.75	80
4-OMe	*n*-Bu	1.5	51

Scheme 2.11 (a) Thin film processing in the VFD enhances the on-water effect for azide and alkyne cycloaddition (non-catalyzed). Conversions are determined by ^{1}H NMR spectroscopy. (b) *In continuo* organolithium addition to imines.

Benaglia and coworkers reported the *in continuo* organolithium, in addition to imines, where the reaction can be carried out in three minutes using an on-water methodology [40]. This methodology builds upon the results reported by Capriati and coworkers [23]. The on-water continuous-flow addition of organolithium reagents to imines was studied in a continuously stirred tank reactor (CSTR) with very fast reaction times of 10–20 seconds, leading to a safer process in comparison with the batch reaction. Using the setup as described in Scheme 2.11b for the organolithium addition to imines, it was found that yields may be improved simply by changing the flow rate of water and adjusting the amount of water. By using a 2 : 1 imine/water flow rate ratio, amine and product were obtained in 93% and 80% isolated yields, respectively, by using about 6 mol/equiv water (Scheme 2.11b), entries 1 and 3. In a single 2.5 ml CSTR, 5 g amine can be synthesized in only three minutes, which is a very significant improvement on the batch-type process.

2.2.4 Prebiotic Chemistry

The origins of life on earth have been debated for some time, and yet there is no universal agreement as to how this arose. However, what is agreed is that water was essential for the appearance of key reactions in prebiotic chemistry, which led to the formation of molecules that were essential for the emergence of life [41, 42]. The majority of on-water reactions reported are unlikely to be relevant for prebiotic chemistry [43]. However, there are a number of reactions, which have been reported using prebiotically plausible molecules that could have arisen in the emergence of life on Earth.

Shapiro and Vigalok explored a Passerini multicomponent reaction, examining both the on-water and in-water reactions by modulating the solubilities of the reactants (Scheme 2.12a) [44]. The two reactants, aldehyde and isocyanide, were added to the reaction mixture, and the aldehyde underwent *in situ* oxidation to

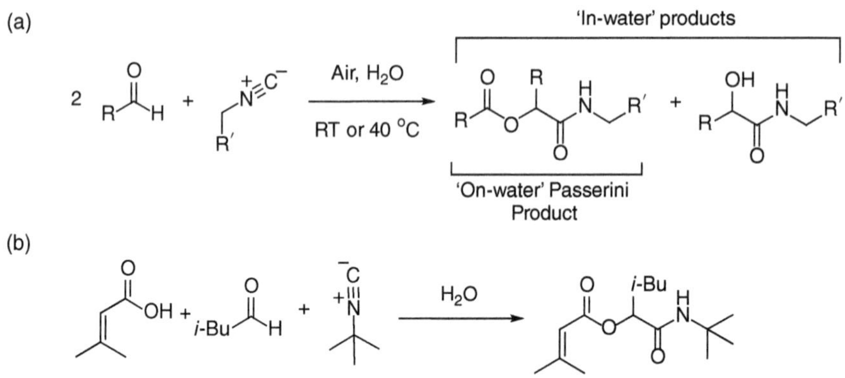

Solvent	Time (h)	Temp (°C)	Yield	Rate (M^{-2} s^{-1})
H_2O	5	50	91	0.10
H_2O	3.5	25	95	0.18
H_2O	2	4	94	1.29
CH_2Cl_2	18	25	45	0.01
MeOH	24	25	15[a]	-

[a]conversion

Scheme 2.12 (a) Passerini reaction explored by Shapiro and Vigalok and dependent upon the nature of the aldehydes (soluble/insoluble), this determined the outcome of the reaction as on-water or in-water. (b) Passerini three component reaction studied by Pirrung and Sarma, where highly insoluble starting materials gave the product in high yields.

form an acid, which then underwent reaction in a Passerini reaction. The reaction was explored with a range of aldehydes and water-insoluble isocyanides. In all cases, the water-insoluble aldehydes gave the three-component Passerini product in high yields, while partially soluble aldehydes gave a mixture of both products. An interesting observation is that the reaction in-water was slower than the reaction on-water.

Pirrung and Sarma examined the Passerini three-component reaction with three different components, and when the reaction was carried out in dichloromethane, a yield of 45% was obtained over 18 hours (Scheme 2.12b) [45]. When the solvent was changed to water, the reaction was complete in 3.5 hours and became heterogeneous because the product is poorly soluble. The three components of the reaction are also highly hydrophobic. The comparison of the rates of reaction shows that the rate using water as a solvent was 18 times faster than that of dichloromethane. When the reaction temperature was lowered to 4 °C, the reaction went to completion in 94% yield and with an even more significant increase in the rate of the reaction.

What is interesting with this reaction is that aldehydes and isocyanides are plausible prebiotic molecules, and the three-component reaction enables molecular complexity to be built up quickly. Sutherland and coworkers examined a prebiotically plausible Passerini multicomponent reaction of phosphates [46, 47]. However, under the reaction conditions, excess aldehyde and isonitrile were required, and the transfer of the phosphate to the newly formed hydroxyl group did not take place. However, despite these limitations of the reaction, it is possible that biology evolved to use these compounds to build up molecules of life.

2.3 Conclusions

The use of water as a solvent continues to be applied to a greater number of reaction types. Interestingly, reactions such as those that use organolithium reagents have been shown to be possible using water as a solvent, showing that conventional thinking for these reactions do not need to be adhered to. While the broadening of the reaction type has expanded, one major challenge that needs further exploration is the mechanistic aspects. An understanding of these reactions could offer insight as to how they occur and why, in some cases, large rate accelerations are observed. The emergence of microdroplet platforms could offer a solution to studying these reactions, but examining them at the molecular level will be key to understanding them. The application of this methodology continues to be applied to a diverse array of reactions, and this is likely to broaden even further into areas such as electrochemistry, flow chemistry, and prebiotic chemistry.

References

1 Narayan, S., Muldoon, J., Finn, M.G. et al. (2005). *Angew. Chem. Int. Ed.* 44: 3275–3279.

2 Harry, N.A., Radhika, S., Neetha, M., and Anilkumar, G. (2019). *ChemistrySelect* 4: 12337–12355.

3 Kitanosono, T. and Kobayashi, S. (2021). *ACS Cent. Sci.* 7: 739–747.
4 Chanda, A. and Fokin, V.V. (2009). *Chem. Rev.* 109: 725–748.
5 Butler, R.N. and Coyne, A.G. (2010). *Chem. Rev.* 110: 6302–6337.
6 Butler, R.N. and Coyne, A.G. (2016). *Org. Biomol. Chem.* 14: 9945–9960.
7 Cortes-Clerget, M., Yu, J., Kincaid, J.R.A. et al. (2021). *Chem. Sci.* 12: 4237–4266.
8 Lipshutz, B.H. (2024). *Green Chem.* 26: 739–752.
9 Rideout, D.C. and Breslow, R. (1980). *J. Am. Chem. Soc.* 102: 7816–7817.
10 Breslow, R. and Maitra, U. (1984). *Tetrahedron Lett.* 25: 1239–1240.
11 Blake, J.F. and Jorgensen, W.L. (1991). *J. Am. Chem. Soc.* 113: 7430–7432.
12 Jorgensen, W.L., Lim, D., and Blake, J.F. (1993). *J. Am. Chem. Soc.* 115: 2936–2942.
13 Butler, R.N., Coyne, A.G., Cunningham, W.J., and Moloney, E.M. (2013). *J. Org. Chem.* 78: 3276–3291.
14 Butler, R.N. and Coyne, A.G. (2015). *J. Org. Chem.* 80: 1809–1817.
15 Kitanosono, T. and Kobayashi, S. (2020). *Chem. – Eur. J.* 26: 9408–9429.
16 Jung, Y. and Marcus, R.A. (2007). *J. Am. Chem. Soc.* 129: 5492–5502.
17 Nguyen, D., Casillas, S., Vang, H. et al. (2021). *J. Phys. Chem. Lett.* 12: 3026–3030.
18 Beattie, J.K., McErlean, C.S.P., and Phippen, C.B.W. (2010). *Chem. – Eur. J.* 16: 8972–8974.
19 Butler, R.N., Coyne, A.G., and Moloney, E.M. (2007). *Tetrahedron Lett.* 48: 3501–3503.
20 Butler, R.N., Coyne, A.G., Cunningham, W.J. et al. (2005). *Helv. Chim. Acta* 88: 1611–1629.
21 Manabe, K., Mori, Y., Wakabayashi, T. et al. (2000). *J. Am. Chem. Soc.* 122: 7202–7207.
22 Zhang, H., Liu, L., Chen, Y. et al. (2006). *Eur. J. Org. Chem.* 2006: 869–873.
23 Dilauro, G., Dell'Aera, M., Vitale, P. et al. (2017). *Angew. Chem. Int. Ed.* 56: 10200–10203.
24 Theberge, A.B., Courtois, F., Schaerli, Y. et al. (2010). *Angew. Chem. Int. Ed.* 49: 5846–5868.
25 Eremin, D.B. and Fokin, V.V. (2021). *J. Am. Chem. Soc.* 143: 18374–18379.
26 Bain, R.M., Sathyamoorthi, S., and Zare, R.N. (2017). *Angew. Chem. Int. Ed.* 56: 15083–15087.
27 Wei, Z., Li, Y., Cooks, R.G., and Yan, X. (2020). *Annu. Rev. Phys. Chem.* 71: 31–51.
28 Gao, D., Jin, F., Lee, J.K., and Zare, R.N. (2019). *Chem. Sci.* 10: 10974–10978.
29 Chatgilialoglu, C., Barata-Vallejo, S., and Gimisis, T. (2024). *Molecules* 29: 569.
30 Shaw, M.H., Twilton, J., and MacMillan, D.W.C. (2016). *J. Org. Chem.* 81: 6898–6926.
31 Tian, Y.-M., Silva, S., Gschwind, R.M., and König, B. (2024). *Science* 383: 750–756.
32 Yan, M., Kawamata, Y., and Baran, P.S. (2017). *Chem. Rev.* 117: 13230–13319.
33 Li, M. and Cheng, X. (2023). *Isr. J. Chem.* 64: e202300067.
34 Wadhawan, J.D., Marken, F., Compton, R.G. et al. (2001). *Chem. Commun.* 1: 87–88.

35 Areias, M.C.C., Bieber, L.W., Navarro, M., and Diniz, F.B. (2003). *J. Electroanal. Chem.* 558: 125–130.

36 Yoshida, T., Kuroboshi, M., Oshitani, J. et al. (2007). *Synlett* 2007: 2691–2694.

37 Porta, R., Benaglia, M., and Puglisi, A. (2016). *Org. Process Res. Dev.* 20: 2–25.

38 Pastre, J.C., Browne, D.L., and Ley, S.V. (2013). *Chem. Soc. Rev.* 42: 8849.

39 Oksdath-Mansilla, G., Kucera, R.L., Chalker, J.M., and Raston, C.L. (2021). *Chem. Commun.* 57: 659–662.

40 Brucoli, J., Puglisi, A., Rossi, S. et al. (2024). *Cell Rep. Phys. Sci.* 5: 101838.

41 Kolb, V.M. (2019). *Handbook of Astrobiology*, vol. 2018 (ed. V.M. Kolb), 331–340. Boca Raton, FL: CRC Press.

42 Ruiz-Lopez, M.F., Francisco, J.S., Martins-Costa, M.T.C., and Anglada, J.M. (2020). *Nat. Rev. Chem.* 4: 459–475.

43 Tran, Q.P., Adam, Z.R., and Fahrenbach, A.C. (2020). *Life* 10: 352.

44 Shapiro, N. and Vigalok, A. (2008). *Angew. Chem. Int. Ed.* 47: 2849–2852.

45 Pirrung, M.C. and Sarma, K.D. (2004). *J. Am. Chem. Soc.* 126: 444–445.

46 Sutherland, J., Mullen, L., and Buchet, F. (2008). *Synlett* 2008: 2161–2163.

47 Mullen, L.B. and Sutherland, J.D. (2007). *Angew. Chem. Int. Ed.* 46: 8063–8066.

3

Chemistry of "In-Water" Reactions: Spotlights on Micellar and Phase-Transfer Catalysis

Tharique N. Ansari[1,2], Gaganpreet Kaur[3], Thomas J. Colacot[2], and Sachin Handa[3]

[1] *University of Louisville, Department of Chemistry, 2320 S Brook Street, Louisville, KY 40292, USA*
[2] *Sigma-Aldrich Co. LLC, 6000 N. Teutonia Ave., Milwaukee, WI 53209, USA*
[3] *University of Missouri, Department of Chemistry, 601 S College Ave, Columbia, MO 65211, USA*

3.1 Introduction

Chemical synthesis has come a long way since Wohler's groundbreaking urea synthesis in 1828 [1, 2]. Since then, synthetic chemistry has become a very powerful area for creating complex molecules with exceptional control over the regio- and stereo-chemistry [2, 3]. This progress has been instrumental in improving the quality of life, such as by producing life-saving drugs, agricultural products to reduce hunger, and components for electronic devices for entertainment and communications. However, these developments come with a cost, where "sustainability" has been largely neglected, thereby increasing pollution and negatively impacting health and safety. The over-reliance on non-sustainable petroleum-derived solvents and the inefficiencies of synthetic practices led to poor atom- and step-economy [4–7]. In pharmaceutical and chemical industries, producing one kilogram of product typically generates an additional 5–100 times of chemical waste, highlighting the inefficiencies during chemical synthesis [8]. Therefore, it is critical to address these challenges by preserving our finite resources with a focus on minimizing waste generation. These initiatives are important to have a healthy society.

With the broader meaning of sustainability, John Werner and Paul Anastas established 12 principles of green chemistry, which served as a guiding platform to reform synthetic practices toward sustainability [7, 9, 10]. By early 2000, various green chemistry tools and metrics were developed to evaluate chemical processes and identify improvement opportunities. These tools included E-factor [11], atom-economy [12], process-mass intensity [13], i-GAL [14], and DOZN™ [15]. Analysis using these tools revealed that organic solvents, while not directly involved in chemical transformations, contribute significantly to waste generation and greenhouse gas emissions [16, 17]. In active pharmaceutical ingredient (API) synthesis, solvents can account for up to 88% of total waste [17]. Process chemists

Organic Transformations in Water: Principles and Applications, First Edition.
Edited by Gopinathan Anilkumar, Nissy Ann Harry, and Sankuviruthiyil M. Ujwaldev.

can minimize the carbon footprint by selecting sustainable solvents, enhancing sustainability, and improving process economics. A shift toward more sustainable organic chemistry practices requires a fundamental change in selecting the proper solvent in the correct quantity [16, 17].

To promote greener practices, chemists have explored alternative solvents [4]. To reduce the carbon footprint associated with petroleum-derived solvents, researchers focus on utilizing solvents derived from bio-renewable resources, such as cyrene [18] and 2-methyl THF [19]. Additionally, the use of water as a reaction medium is considered sustainable because it is safe, noncombustible, affordable, abundant, and has a lower carbon footprint even after use [20–23]. However, despite initial reports, the insolubility of reactants and the sensitivity of catalysts and reagents have impeded the widespread adoption of water as a solvent. To overcome these limitations, introducing amphiphilic molecules as additives has emerged as a strategy to enhance the solubility of organic reactants in the aqueous phase, thereby facilitating chemical reactions [20, 24, 25]. This innovation has revitalized the potential of water as the bulk solvent for enabling organic synthesis. This approach substantially differs from the traditional "on-water" approach as the operating principles differ.

Thus, due to the utmost importance of micellar catalysis and phase-transfer catalysis (PTC), we concisely highlight the fundamental principles governing micellar and PTC without any duplication, as several reviews are available on this topic, as mentioned in Section 3.3. In this chapter, we also highlight the industrial applications of these technologies in the context of the synthesis of APIs and agrochemical products. This chapter offers valuable insights into the potential future directions of these technologies and their associated challenges.

3.2 Understanding the Concept of Micellar Catalysis

Amphiphiles, also called surfactants, have balanced hydrophilic (water-attracting) and hydrophobic (water-repelling) regions within their structure. The term "amphiphilic" comes from the Greek words "amphi," meaning both, and "philos," meaning loving. These molecules play pivotal roles in diverse biological, chemical, and physical processes, especially in forming supramolecular structures, such as micelles, bilayers, and membranes. Amphiphiles typically consist of two distinct parts: a hydrophilic head and a hydrophobic tail. The hydrophilic head contains polar or charged groups that readily interact with water molecules, while the hydrophobic tail is often composed of nonpolar hydrocarbon chains or other water-insoluble groups. Surfactants are classified into four categories based on the charge of the hydrophilic part: cationic, anionic, non-ionic, and zwitterionic. The cationic surfactant has a positively charged hydrophilic head, whereas the anionic surfactant possesses a negatively charged head (generally, carboxylate, sulfonate, or sulfate). In the case of non-ionic surfactants, hydrophilic heads are not electrically charged. The zwitterionic amphiphile has both positively and negatively charged hydrophilic heads. In the context of micellar catalysis, we will be primarily discussing the non-ionic surfactants (Scheme 3.1).

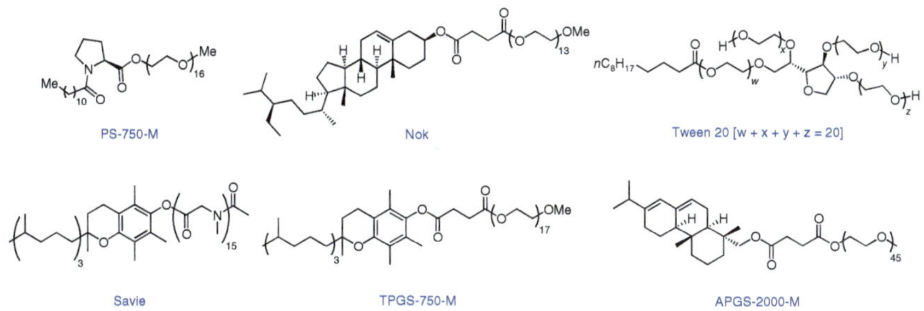

Scheme 3.1 Commonly used surfactants.

When surfactant molecules are introduced into a solution, they organize to minimize their exposure to the surrounding solvent. For example, in a polar solvent such as water, surfactants will align their hydrophilic heads toward the water molecules while their hydrophobic tails aggregate in the core region, away from water. This arrangement results in the formation of micelles [20] and the phenomenon is called micellization [26]. Several factors influence micellization, such as (i) surfactant concentration [27], (ii) temperature [28], (iii) nature of the surfactant [29], and pH of the solution [30]. There are various experimental techniques to understand the micellization process, such as dynamic light scattering, which measures the average size distribution of micelles [31], while cryogenic transmission electron microscopy (cryo-TEM) provides information about the shape and morphology of micelles through direct imaging [32].

These dynamic micellar structures facilitate and enhance chemical reactions under aqueous conditions. The hydrophobic core of the dynamic micelles can serve as a microchamber with distinct solubility properties compared to the bulk solution. This hydrophobic core allows the solubilization of non-polar molecules because of the hydrophobic effect, a phenomenon that arises when nonpolar or hydrophobic molecules clump together to minimize contact with water [24, 33]. The distinct environment within the micelles has the potential to affect the reactivity and selectivity of specific reactions.

3.3 Basic Principles of Phase-Transfer Catalysis

In certain chemical reactions, if one or more of the reactants are insoluble or poorly soluble in the reaction medium, it can significantly affect the reaction outcome. This can be attributed to the ineffective mass transfer, leading to a decreased rate of effective collisions between the reactants. To address these concerns, biphasic reactions can be employed, which involve two phases that are not miscible, typically a liquid–liquid system. This approach allows reactants with poor solubility in a single solvent to be distributed between two phases, enhancing their contact and improving the efficiency of the reaction [34–36]. Typically, one phase acts as the reaction medium, while the other serves as a reservoir, supplying the insoluble reactant. However, due to the immiscible nature of the biphasic solvents, the reactants cannot easily come into contact. In such instances, a phase-transfer catalyst (PTC) is utilized. These catalysts facilitate the transfer of one reactant to the other immiscible solvent phase, thereby enabling the desired transformation to occur more effectively. Technically, catalytic amounts of phase transfer agents promote the interphase transfer of species/reagents, thus enabling reactions between reagents in two immiscible phases. The phase transfer catalyst possesses both functional sites to become soluble in both systems, enabling the transfer of chemicals from one non-miscible, heterogeneous system to another by a chemical process [37–41].

A simple example of PTC is the Williamson ether synthesis [42]. This reaction involves the S_N2 attack of the alkoxide nucleophile on the alkyl halide electrophile. Alkali metal alkoxides/hydroxides are common sources of alkoxide/hydroxide ions, which have a high solubility in aqueous solutions. Organic compounds, such as alkyl halides, are typically insoluble or partially soluble in water. A PTC, such as a quaternary ammonium salt, is usually employed to overcome this solubility issue because it helps transfer alkoxide/hydroxide ions from the aqueous phase to the organic phase, thereby facilitating their reaction with the alkyl halide to form the desired ether.

PTCs are usually organic compounds with hydrophobic (nonpolar alkyl chain) and hydrophilic (polar quaternary ammonium or phosphonium cation) moieties [38–41]. Many PTCs carry a positive charge on their hydrophilic heads [40]. This positive charge helps the PTC to interact with negatively charged species in the aqueous phase, such as anionic reactants or counterions [38]. Some PTCs may have additional functional groups in their structure that can either participate in the reaction or provide specific coordination sites. These functional groups can enhance the catalytic activity or selectivity of the PTC. Sometimes, PTCs include bridging groups that link the hydrophobic tail and the hydrophilic head. These groups can affect the geometry and flexibility of the PTC molecule, which can change its ability to interact with reactants in different phases. Examples of PTCs include (i) quaternary ammonium salts (tetra-butylammonium bromide), (ii) phosphonium salts (tetraphenylphosphonium bromide), and (iii) cationic surfactants.

3.4 Highlights of Micellar Catalysis

Over the last 15 years, a significant literature has emerged, highlighting the potential of aqueous micellar technology as a safer alternative to hazardous organic solvents [20, 24, 33, 43, 44]. Significant contributions from Lipshutz [45–48], Handa [49–53], Gallou [54–56], Kobayashi [35, 57], and others [48, 58–60] have made notable strides in advancing this field, demonstrating this approach's remarkable versatility and efficacy. Its ability to facilitate valuable chemical transformations with improved selectivity under sustainable reaction conditions is particularly notable [53, 61, 62]. The application of micellar technology has profoundly influenced various domains of organic synthesis, including nanocatalysis [63–65], palladium-catalysis [43, 45, 48, 50, 66], amide couplings [52, 67–69], biocatalysis [70–72], and traditional nucleophilic aromatic substitution reactions (S_NAr) [73–75]. Furthermore, the scalability of these reactions has been successfully demonstrated in a myriad of industrial processes [55, 56, 70, 76–78], primarily in the context of multi-kilogram scale synthesis of APIs [55, 56, 70, 76, 77].

3.4.1 Micelle-Enabled Cross-Coupling Reactions

The judicious incorporation of micellar technology in conjunction with palladium catalysis did address some of the contemporary sustainability challenges associated with cross-coupling chemistry. In the last 15 years, numerous reports in this direction have extensively addressed the concerns over (i) catalyst loading and their poor recyclability, (ii) higher reaction temperature, (iii) dependency on hazardous organic solvents, and (iv) stoichiometry of coupling partners or bases. This section will discuss such key reports in the context of various cross-coupling reactions.

3.4.1.1 Suzuki–Miyaura Couplings Enabled by Micellar Catalysis

This name reaction involves the C—C bond formation between alkyl/aryl (pseudo) halides with organoboron compounds in the presence of a base and palladium catalyst, where the easy accessibility and excellent stability of organoboron compounds with the generation of non toxic byproducts make this technology very attractive to the synthetic chemists. As highlighted by Brown and Bostrom [79], this reaction has emerged as the most prevalent palladium-catalyzed cross-coupling reaction in the field of medicinal chemistry over the past three decades. As a result, considerable efforts have been focused on developing environmentally friendly variants of these protocols, given their extensive use in various industrial processes [80, 81].

The first report in this direction appeared in 2008, where Lipshutz and coworkers employed aqueous nano-micelles of non-ionic amphiphile PTS as the reaction medium and (dtbpf)$PdCl_2$ as a catalyst developed for cross-couplings by Colacot and coworker [82] with Et_3N as the base to enable the challenging couplings of aryl halides (iodides, bromides, and chlorides) and boronic acids under milder conditions in water [83]. Further, with the introduction of a newer Pd precatalyst ($AmPhosPdCl_2$) [84], this approach was modified to house the coupling of various heterocyclic substrates, including pyridine, pyrimidines thiophenes, and oxazolines.

Subsequently, significant reports have appeared to tackle numerous challenges in this area [43], including the development of tailored ligands and/or precatalysts [45, 85–87], the utilization of nanoparticle (NP) catalysts instead of traditional homogeneous systems [48], as well as the implementation of flow-reaction processes [88, 89], and other innovative methodologies [58, 65, 90]. Among these advancements, only some of the significantly relevant reports will be discussed in this section.

Lipshutz Lab contributed substantially to developing novel ligands/pre-catalysts to enable this reaction under milder aqueous conditions (Scheme 3.2a) [45, 85, 86, 91]. The first report in this direction appeared in 2015 with the development of the HandaPhos ligand [45]. The ligand was designed to harness the complete micellar effect by increasing the hydrophobicity of the ligand, ensuring its distribution in the hydrophobic micellar inner core. Along with 1000 ppm of $Pd(OAc)_2$, this ligand enabled cross-coupling of a broad range of electrophiles with boronic acids in the

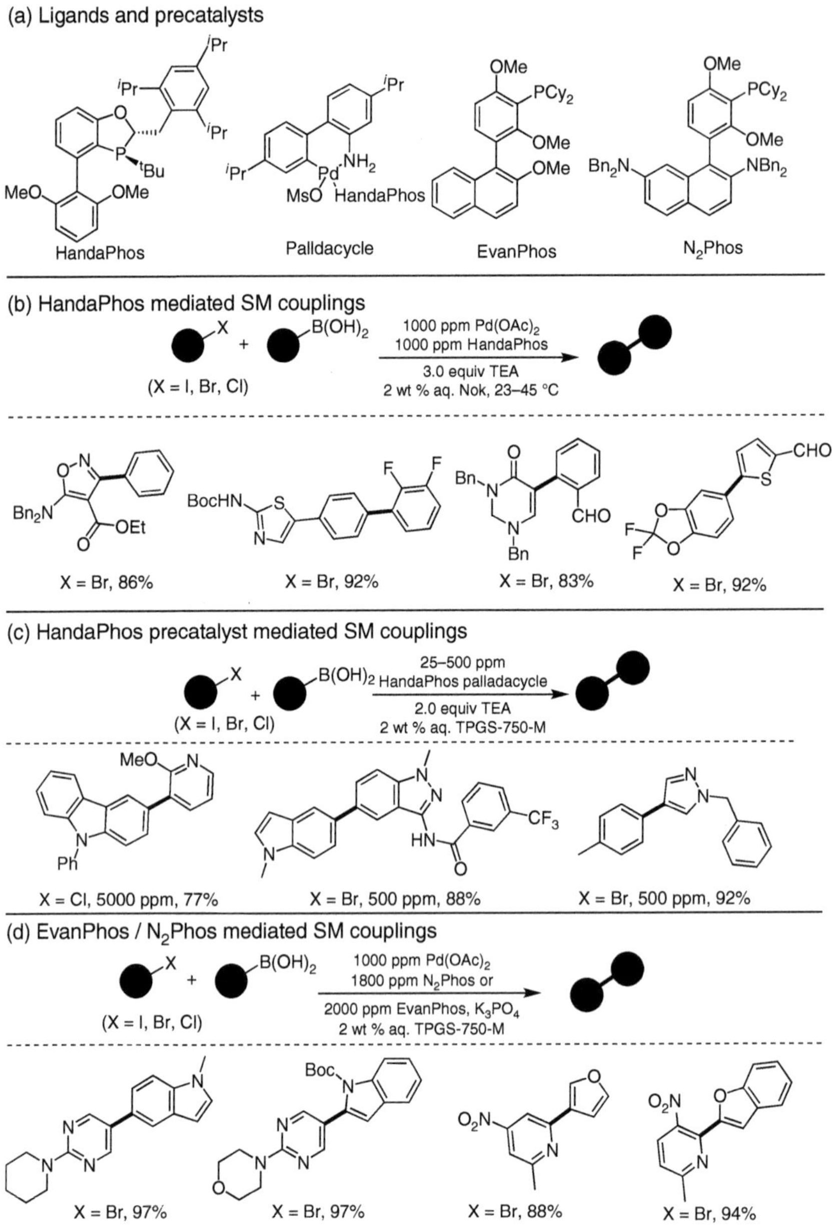

Scheme 3.2 Ligands/precatalysts for micellar Suzuki–Miyaura couplings.

aqueous micelles of Nok (Scheme 3.2b). Notably, the catalyst and reaction medium were recyclable for up to four additional cycles without losing significant activity.

By judicious design of the palladacycle backbone in conjunction with the HandaPhos, the palladium loading can be further reduced to as low as 25 ppm [91]. Takale and Lipshutz introduced this strategy in 2019, showcasing successful couplings of diverse electrophiles and nucleophiles under aqueous conditions (Scheme 3.2c). The lipophilicity of the palladacycle was enhanced by introducing the diisopropyl substitution on the biphenyl group. Separately, the intense synthesis of HandaPhos ligand has raised concern from a sustainability perspective as it substantially affects the overall carbon footprint. In addressing these concerns, Lipshutz developed two ligands, EvanPhos [86] and N_2Phos [85], which can be synthesized easily from readily available starting materials. These ligands enabled the Suzuki–Miyaura (SM) couplings under mild aqueous conditions with ppm levels of Pd loadings (Scheme 3.2d).

In 2018, a novel approach to utilizing a PS-750-M-based micellar strategy was introduced by Handa et al., to overcome the challenges associated with the couplings of (iso)quinolines [87]. Traditionally, these substrates lead to catalyst deactivation due to their binding of the nitrogen with the Pd center, impeding their functionalization via cross-couplings. To address this issue, micellar conditions were employed, using Colacot's highly active π-allyl palladium catalyst (PCy_3Pd(crotyl)Cl). This approach demonstrated successful couplings on a wide range of electrophiles (Br, Cl, and OTf) and nucleophiles (BMIDA, BF_3K, and boronic acids). Notably, the expanded substrate scope accommodated various protective and functional groups, including free OH and NH_2 functional groups (Scheme 3.3a).

The scope of SM coupling is often limited when it comes to sp^2–sp^3 couplings due to the possible deteriorating β-hydride elimination pathway in the catalytic cycle. In 2018, Lipshutz and coworkers presented an elegant solution to this challenge by introducing alkyl OBBD (B-alkyl-9-oxa-10-borabicyclo[3.3.2]decane) reagents as an sp^3 source [92]. These reagents showcased exceptional stability against moisture and air, making them well-suited for implementation in micellar processes. Again, Colacot's (dtbpf)$PdCl_2$ served as the optimal catalyst. By employing mild micellar conditions, this approach successfully enabled a multitude of challenging sp^2–sp^3 couplings (Scheme 3.3b).

Often, the cross-coupling reactions are sensitive to the presence of dissolved oxygen. Proper purging of the micellar medium with argon or nitrogen is typically necessary prior to catalysis to mitigate this challenge. Generally, the dissolved oxygen hinders the regular catalytic cycle by promoting homo-couplings or deborylation of nucleophiles. Micellar catalysis offers an innovative solution by providing micellar inner cores where oxygen is poorly soluble. This enables cross-coupling reactions to proceed under open-air conditions. This approach was reported from Beverina's lab, where the authors demonstrated using BASF's emulsifier Kolliphor EL for SM coupling under atmospheric conditions [58]. The emulsifier creates an oxygen-free hydrophobic inner core, facilitating couplings in open-air environments. The reaction employs (dtbpf)$PdCl_2$ as the catalyst and triethylamine as the base. The versatility of these reaction conditions is showcased through various examples, recycling studies, and the synthesis of several molecular semiconductor cores (Scheme 3.3c).

In recent studies, the micellar conditions have been utilized to develop DNA-encoded libraries (DELs), wherein functionalized DNA fragments undergo arylation using diverse boronic acids [90]. This methodology involves employing

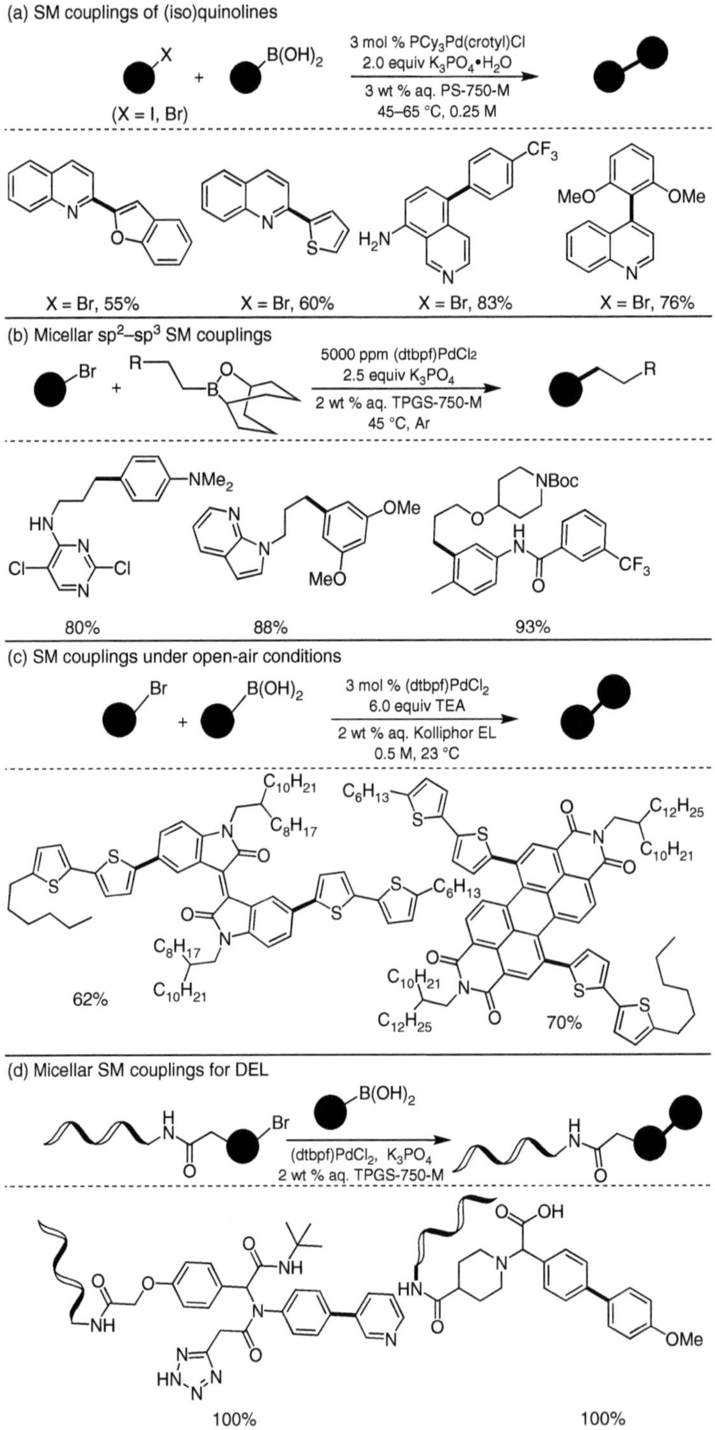

Scheme 3.3 Micellar sp^2–sp^2 and sp^2–sp^3 Suzuki–Miyaura couplings and DEL functionalization.

(dtbpf)$PdCl_2$ or $Pd(PPh_3)_4$ as catalysts, with K_3PO_4 serving as the base within the micelles of TPGS-750-M. This innovative application of micellar catalysis establishes a promising platform for expanding DELs, subsequently facilitating the development of novel therapeutic agents (Scheme 3.3d).

Lipshutz and coworkers explored the integration of flow technology with micellar catalysis to enable sustainable SM couplings [88]. They employed Fe/ppm Pd NPs (800 ppm) as the catalyst under continuous slurry plug flow conditions. Although the reaction yield ranged between 88% and 97%, the substrate scope was limited (no nitrogen heterocycles).

However, in the subsequent paper, a stirred-tank reactor flow setup was utilized to accommodate the coupling of heteroaromatic substrates under milder micellar conditions [89]. The continuous flow procedure exhibited compatibility with a wide range of substrates and delivered excellent reaction yields (Scheme 3.4).

3.4.1.2 C–N Cross-Couplings Enabled by Micellar Catalysis

The cross-couplings of amines with (hetero)aryl (pseudo)halides have surfaced as an efficient method for accessing functionalized aromatic amines and heterocycles [93]. This process has significant implications across various industrial sectors, as these structural motifs play crucial roles in natural products, APIs, organic materials, and catalysts [93, 94]. Over the past 25 years, this approach has evolved into

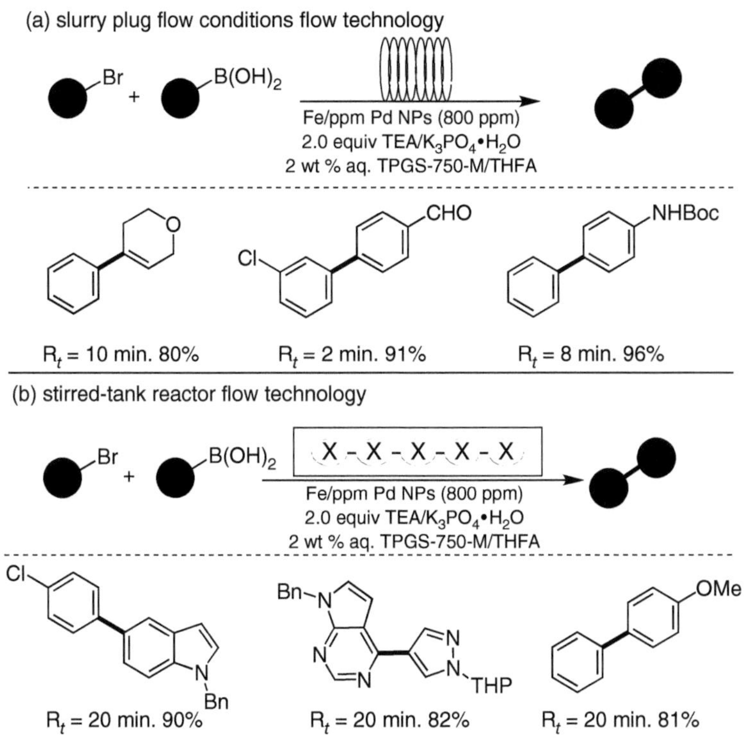

Scheme 3.4 Micellar Suzuki–Miyaura couplings under flow conditions.

a fundamental tool for constructing N—C (sp^2) bonds. The rapid progress in this field is attributed to the development of novel pre-catalysts and ligands, guided by an advanced understanding of the intricate aspects of the reaction mechanism [93, 95, 96]. Compared to the state-of-the-art reports, the micellar variations of these couplings enable the reaction under milder conditions without relying on toxic or hazardous solvents and expensive ligands/precatalysts.

The initial report from Lipshutz's lab employed Takasago's Di-tert-butyl(1-methyl-2,2-diphenylcyclopropyl)phosphine (cBRIDP) as a ligand along with $[(\pi\text{-allyl})PdCl]_2$ as a precursor to effectively catalyze the amination of aryl bromides and chlorides under mild aqueous conditions using nano micelles of PTS [97]. Building upon this work, Schmitt and coworkers further expanded this approach to encompass heterocyclic substrates by increasing the temperature and catalyst loading and replacing the PTS surfactant with TPGS-750-M [98]. Notably, this methodology accommodated the *N*-arylation of amides, carbamates, and urea derivatives. Further, superior performance in these C–N cross-couplings was observed when the cBRIDP ligand was replaced with Buchwald's biaryl phosphine (*t*-BuXPhos) (Scheme 3.5) [99].

In a 2019 report, Lipshutz and coworkers employed Colacot's [*t*-BuXPhosPd(cinnamyl)]OTf as a pre-catalyst for the amination of aryl bromides and chlorides under mild aqueous conditions with ppm levels of [Pd] loadings [100]. This study showcased a diverse substrate scope while minimizing the carried-over palladium to the final product. The scalability of this approach was demonstrated in the organic solvent-free synthesis of 1-(4-((4-fluorophenyl)amino)phenyl)ethan-1-one, where the pure product was easily obtained via filtration and water washing (to remove surfactant) (Scheme 3.6).

In C–N cross-couplings, catalysts are often non-recyclable due to the strong binding of amines with the metal center. Handa and coworkers introduced a novel heterogeneous nanocatalytic approach using a charcoal-immobilized bimetallic Pd–Cu system to address this limitation [65].

The ligand cBRIDP was pivotal in stabilizing the catalytic palladium species, facilitating the desired reaction under mild aqueous conditions. The scope of this approach was broad and accommodated the couplings of numerous heterocycles, including indoles, piperidine, morpholine, etc. Notably, both the catalyst and reaction medium showed recyclability for up to three additional cycles without significant loss of activity (Scheme 3.6). Considering the mild

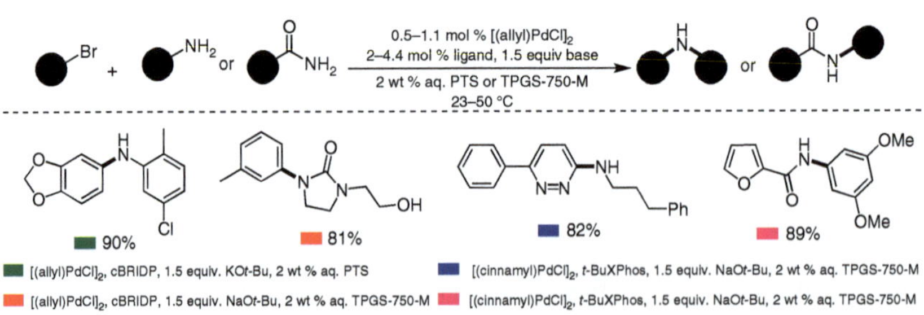

Scheme 3.5 Initial reports on micellar C–N couplings.

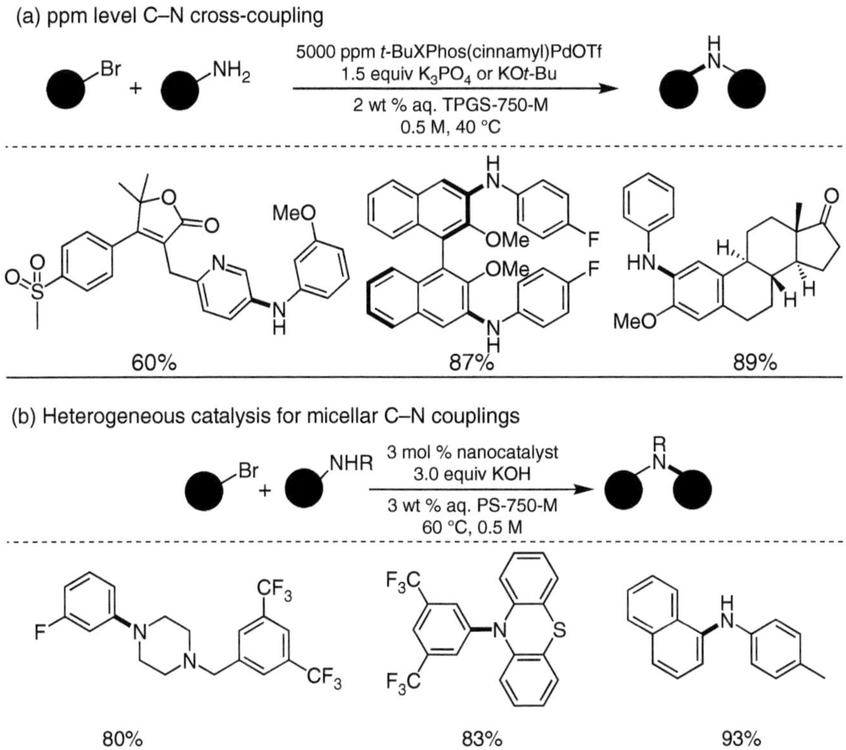

Scheme 3.6 Micellar C–N couplings employing (a) ppm levels of [Pd] and (b) heterogeneous nanocatalysis.

nature of micellar aminations, this can be leveraged to functionalize temperature-sensitive biomolecules directly. In a recent study, Waring and coworkers utilized micellar aminations to generate DELs [101]. This approach employed the cBRIDP:[(cinnamyl)PdCl]$_2$ catalytic system under TPGS-750-M micellar conditions to investigate the *N*-arylation of polynucleotides with various substitutions.

3.4.1.3 Sonogashira Couplings Enabled by Micellar Catalysis

Sonogashira couplings employ terminal olefins as nucleophiles to enable C—C (sp–sp^2) bond formation under Pd or Pd–Cu catalysis conditions. In the context of sustainability, the general reaction conditions are far from ideal. With the introduction of micellar catalysis, most of these challenges were addressed. In the first report of micelle-enabled Sonogashira coupling, copper was not employed, leading to the rediscovery of the pioneering "Heck–Cassar" type alkyne arylation conditions [102]. The catalytic system involved Pd(OAc)$_2$:cBRIDP or (CH$_3$CN)$_2$PdCl$_2$:XPhos, facilitating the formation of sp^2–sp bonds under PTS micellar conditions at room temperature (Scheme 3.7a).

Subsequently, Lipshutz introduced a practical and versatile method for this reaction utilizing HandaPhos ligand and ppm levels of Pd under mild aqueous conditions [103]. This technology used Et$_3$N base, a 2 wt % aqueous solution of TPGS-750-M

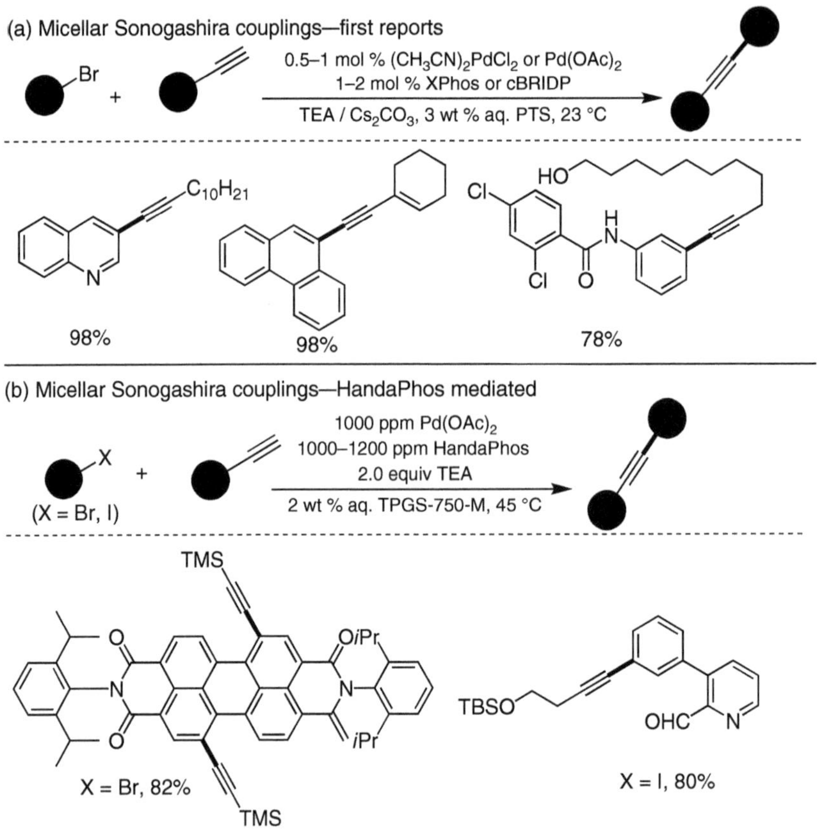

Scheme 3.7 Micellar Sonogashira couplings using (a) initial reports [Pd] and (b) HandaPhos technology.

reaction medium, and a 1 : 1.0–1.5 ratio of [Pd] and ligand. This condition was amenable to the coupling of a broad range of electrophiles and nucleophiles, including various heterocycles, as well as sensitive protective and other functional groups (Scheme 3.7b).

Consequently, from a homogeneous catalysis perspective, the cBRIDP-bound Pd and the Pd–G3 palladacycles of CataCXiumA were identified as effective catalysts for these reactions at ppm levels [54, 104]. The reaction conditions included aqueous nanomicelles of TPGS-750-M and $K_3PO_4 \cdot H_2O$ or Et_3N base.

The cBRIDP-catalyzed couplings displayed smooth and efficient reactions across a wide range of substrates [104], with excellent catalyst recyclability. By employing the G3-palladacycle of CataCXiumA [54], the couplings of challenging heterocyclic compounds with free OH groups were achieved with moderate to excellent yields (Scheme 3.8). Notably, Gallou and Ackermann recently provided a comprehensive review highlighting the remarkable advancements enabled by micellar catalysis in Sonogashira–Hagihara couplings. The NP variants of these reactions are more intriguing and will be discussed in the later section.

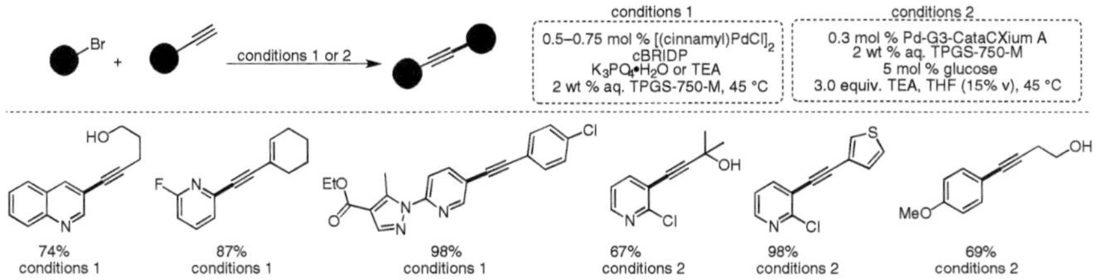

Scheme 3.8 Micellar Sonogashira couplings using cBRIDP/Pd(OAc)$_2$ or Pd-G3-CataCXium.

3.4.1.4 Palladium Catalyzed α-Arylation Reactions Enabled by Micellar Catalysis

The [Pd] catalyzed α-arylation of carbonyls or activating groups offers an efficient method for constructing sp^2–sp^3 C—C bonds [105, 106]. Generally, the reaction involves the intermediacy of enolates or carbanion-type species; due to this reason, chemists believed that such reactions necessitated strictly anhydrous conditions and were incompatible with aqueous environments. However, micellar catalysis provides a solution by enabling these reactions under aqueous conditions through the shielding effect of micelles. The hydrophobic inner cores of micelles act as a protective shield, preventing the undesired quenching of carbanion species by bulk-water molecules. This was first demonstrated by Handa and coworkers in 2017 [31]. The authors employed Colacot's [(*t*-BuXPhos)Pd(allyl)]OTf as the catalyst to enable the α-arylation of nitroalkanes with the chiral surfactant PS-750-M. The enolates from nitroalkanes were generated and consumed *in situ* by leveraging the shielding of micelles. The scope of this approach was evaluated over 20 substrates (Scheme 3.9).

Inspired by these studies, in 2020, Handa and coworkers further expanded the scope of this understanding by developing a novel nanocatalytic technology for the α-arylation of phenyl acetonitriles [50]. Here, the reaction conditions employed the aqueous micelles of PS-750-M and KOH as the base. The nanocatalyst was also synthesized from Colacot's crotyl Pd catalysts (XPhosPd(crotyl)Cl). This study comprehensively evaluated the potential of this approach in enabling this reaction for a broad range of electrophiles (40 + substrates). The authors explored the reaction mechanism employing ^{31}P NMR and successfully detected a carbanion-type

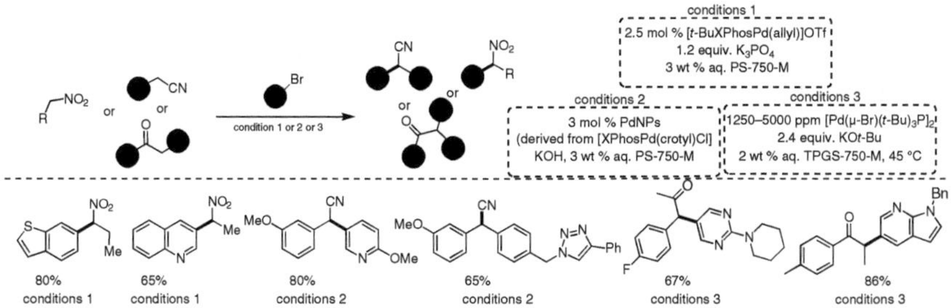

Scheme 3.9 Micellar α-arylation reactions.

intermediate. Further, the scalability was demonstrated with a 50 g scale example of diphenozylate intermediate (Scheme 3.9).

In a recent study, Lipshutz and coworkers presented a method for the α-arylation of ketones using $[Pd(\mu\text{-Br})(t\text{-Bu})_3P]_2$ as the catalyst (1250–5000 ppm), KO*t*Bu as the base, and nano-micelles of TPGS-750-M as the reaction medium (Scheme 3.9) [107]. The reactions proceeded efficiently at 45 °C, allowing for the α-arylation of various ketones and (hetero)aryl bromides. The resulting α-arylated products have been successfully applied in various one-pot reactions, highlighting the versatility and usefulness of this micellar technology.

3.4.1.5 Micellar Nanocatalysis in C–C Couplings

Nanocatalysis has gained significant attention from the synthetic community as a promising alternative to traditional homogeneous/heterogeneous catalysis [108–111]. This approach offers attractive features, including catalyst recyclability, enhanced catalytic activity, improved reaction selectivity, and operationally simple procedures. The consolidation of micellar catalysis with nanocatalysis, known as "micellar nanocatalysis," presents even greener and more sustainable solutions than conventional nanocatalysis [63, 64]. After initial reports in the 2000s discussing the potential of nanocatalysts in surfactant chemistry and cross-coupling reactions [112, 113], the field remained relatively inactive until pioneering studies by Lipshutz and Handa highlighting the effectiveness of Fe–ppm Pd NPs for SM couplings [48]. Since that report, the academic interest in this domain has substantially grown, and various value-added transformations have been demonstrated by leveraging the basic principles of micellar catalysis [64, 114–116].

The seminal 2015 report demonstrated that the Pd impurity within the $FeCl_3$ salt could enable SM couplings to yield functionalized biaryls (Scheme 3.10a,b) [48]. With extensive optimizations, the authors concluded that Fe/ppm Pd NPs can be synthesized from 320 ppm of $Pd(OAc)_2$ as the dopant, SPhos as the ligand, $FeCl_3$ as the support, and MeMgCl as the reductant. These NPs could catalyze the carbon–carbon bond formation between highly functionalized aryl halides and boronic acids under mild aqueous micellar conditions enabled by the designer surfactant TPGS-750-M. Notably, these NPs are highly recyclable, and the catalytic activity was retained even after several rounds of recycling along with the reaction medium. Considering the toxicity of carried-over palladium, this technology was highly attractive from a pharmaceutical industry perspective. Later, Novartis studied the scope of this technology in the context of micellar-flow chemistry [88, 89].

Further, in 2019, the scope of this NP technology was extended to the Cu-free Sonogashira couplings [114]. In this context, the Fe/ppm Pd NPs were modified by replacing the biaryl ligand SPhos with XPhos and increasing the Pd loading to 500 ppm (Scheme 3.10c). These new NPs exhibited excellent activity toward this micellar sp^2–sp type cross-coupling with a broad substrate scope. The optimized reaction conditions employed the aqueous solution of custom-designed tocopherol-derived TPGS-750-M as the reaction medium and triethyl amine (TEA) as a base, and the reaction proceeded smoothly under milder conditions (temperature rt-45 °C). These NPs were also recyclable, with an overall E-factor of 4.14.

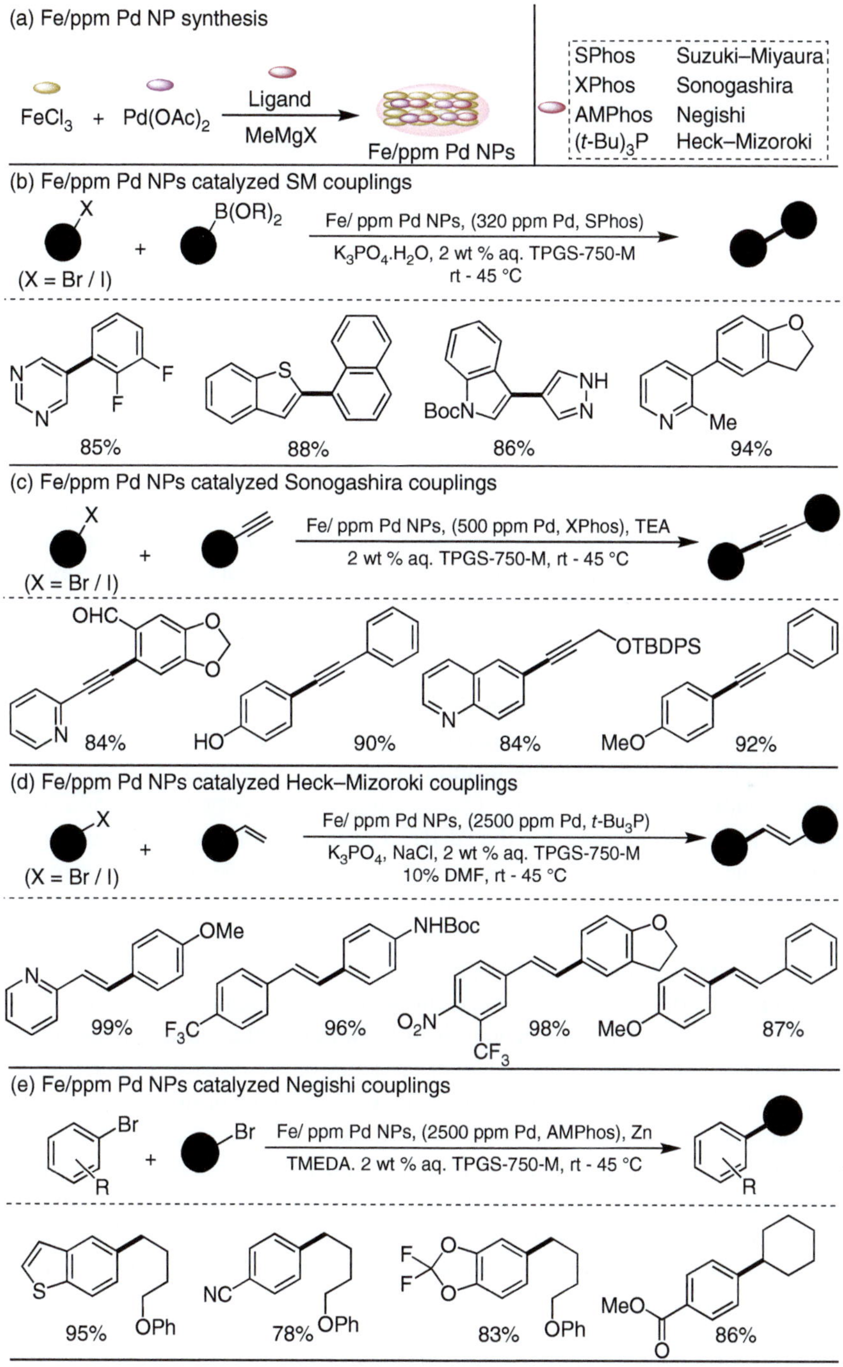

Scheme 3.10 Micellar nanocatalysis in the context of Fe/ppm Pd NPs.

In 2021, Lipshutz and coworkers modified the Fe/ppm Pd NPs recipe by introducing inexpensive *t*-Bu_3P as the ligand and increasing the Pd loading to 2500 ppm [115]. These new NPs were highly active toward challenging Heck–Mizoroki couplings under mild micellar conditions. Here, the micelles of TPGS-750-M played a crucial role in determining the right shape and geometry of these NPs, making them effective for the desired transformations. Notably, the reaction demanded 10% of the reprotoxic solvent *N,N*-dimethylformamide (DMF) for improved reaction yields. This was rationalized by the role of DMF in stabilizing the active NP catalyst (Scheme 3.10d).

Very recently, Lipshutz and coworkers explored the scope of these NPs as the catalysts for the challenging Negishi-type couplings under mild aqueous conditions [116]. The optimal catalytic activity was achieved by employing AmPhos as the ligand and increasing the Pd loadings up to 2500 ppm. The scope of this approach was broad, enabling the sp^2–sp^3-type couplings between highly functionalized alkyl halides and aryl halides. Along with the catalytic amounts of NPs, the optimal conditions employed are four equivalents of zinc powder and five equivalents of TMEDA (as the stabilizer for *in situ*-generated Negishi-type reagents) and 2 wt % aqueous solution of TPGS-750-M. Notably, the catalytic activity significantly dropped after the initial cycles, and additional fresh NP addition was required to achieve desirable conversions (Scheme 3.10e).

From a sustainability perspective, it is essential to note that the carbon footprint of these ligands is often considered heavy. It will be highly regarded if one can perform catalysis without relying on these ligands. In this direction, Handa and coworkers demonstrated the implication of custom-designed proline-derived amphiphile PS-750-M to promote ligand-free micellar nanocatalysis under mild aqueous conditions [51, 117–119]. The seminal work reported by Handa and coworkers in 2021 showcased that the Pd(II) NPs can be stabilized by the carbonyls of amphiphile PS-750-M and can perform oxidative-Heck-type couplings at room temperature (Scheme 3.11a) [51]. The interaction between Pd and the carbonyls of amphiphile was studied using IR and HRMS studies. Further, the scalability of this approach was demonstrated on a 100 g scale example.

Later, in 2021, they further expanded the scope of this approach toward Pd(0) catalysis. Notably, the Pd(0) nanocatalysis in the absence of a phosphine ligand is highly challenging, considering the higher tendency of the Pd(0) surface to aggregate and form catalytically inactive Pd-black [117]. By judiciously leveraging the structural features of the amphiphile PS-750-M, the authors demonstrated the catalytic activity of the ligand-free Pd(0) NPs under mild aqueous conditions. Here, the metal–micelle interaction was studied using various spectroscopic tools, including, IR, ^{13}C NMR, and surface-enhanced Raman spectroscopy (SERS). The NPs effectively enable cross-coupling of aryl-boronic acid with water-sensitive acid chlorides or aryl iodides under mild aqueous conditions (Scheme 3.11b).

The extensive spectroscopic studies in these reports laid the foundation for the basic understanding of metal–micelle cooperativity, which became a key concept in the operational understanding of micellar catalysis [119]. Further, in 2022, this approach was extended toward stabilizing $Pd_2(dba)_3$-derived Pd(0) NPs under ligand-free conditions. Similar to previous reports, the role of PS-750-M in

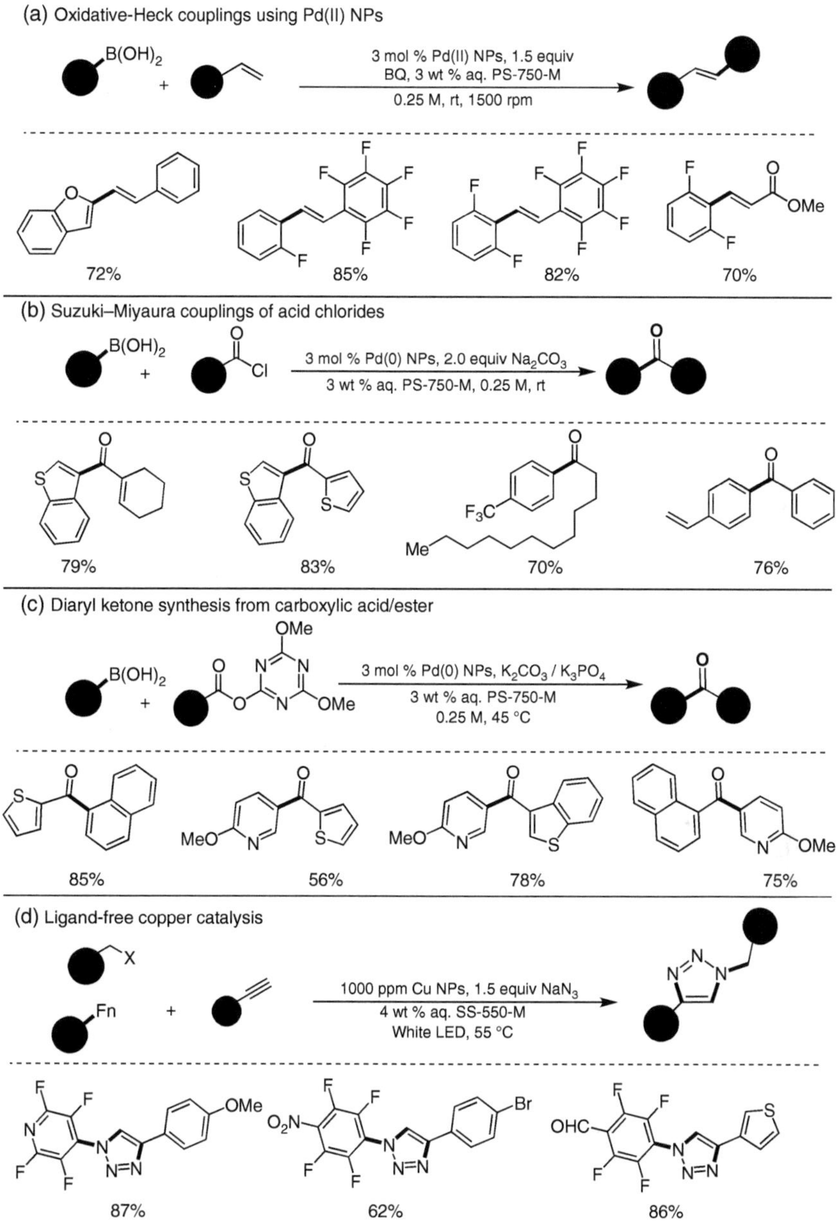

Scheme 3.11 Micellar nanocatalysis enabling ligand-free catalysis.

stabilizing Pd(0) NPs was found to be crucial. In this case, the catalytic activity was demonstrated toward the ketone synthesis employing arylboronic acid and activated-carboxylic acid adducts (Scheme 3.11c).

Later, in 2022, Handa and coworkers demonstrated the implications of this "metal–micelle cooperativity concept" in copper catalysis [118]. The authors reported bench-stable photoactive aqueous nano aggregates of Cu(II) for ppm-level

Cu(I) catalysis in water. The authors developed a modified version of PS-750-M with an optimal PEG-chain length to enable click chemistry under micellar conditions. The reaction proceeded smoothly under photo-assisted condition, with one-pot azidation of electrophiles (benzyl bromides/polyfluoroarenes) followed by the [3+2] cycloaddition to yield the desired triazoles in good yields (Scheme 3.11d).

3.4.2 S_NAr Reactions Employing Micellar Catalysis

S_NAr reactions involve the displacement of leaving groups (F, Cl, Br, etc.) by strong nucleophiles through the intermediacy of Meisenheimer-type complexes [120]. These reactions are extensively employed in drug discovery programs to functionalize the aromatic systems by means of C—N, C—C, and C—O bond formations [121–123]. The S_NAr reactions are attractive due to their higher atom economy and ability to proceed without needing any metals or catalysts. However, from a sustainability perspective, these reactions require significant re-engineering due to their reliance on hazardous dipolar aprotic solvents [124, 125].

To address these challenges, modern micellar catalysis strategies have been employed. Notably, Lipshutz and coworkers developed a general approach employing TPGS-750-M micelles, enabling the incorporation of oxygen, nitrogen, and sulfur nucleophiles at ambient or slightly elevated temperatures [126]. This methodology demonstrates a broad scope, accommodating various heterocyclic systems and sensitive functional groups. Moreover, the in-flask recyclability of the reaction medium was demonstrated, achieving a low E factor value of 4.6–4.8, indicating its efficiency and minimal waste generation (Scheme 3.12a).

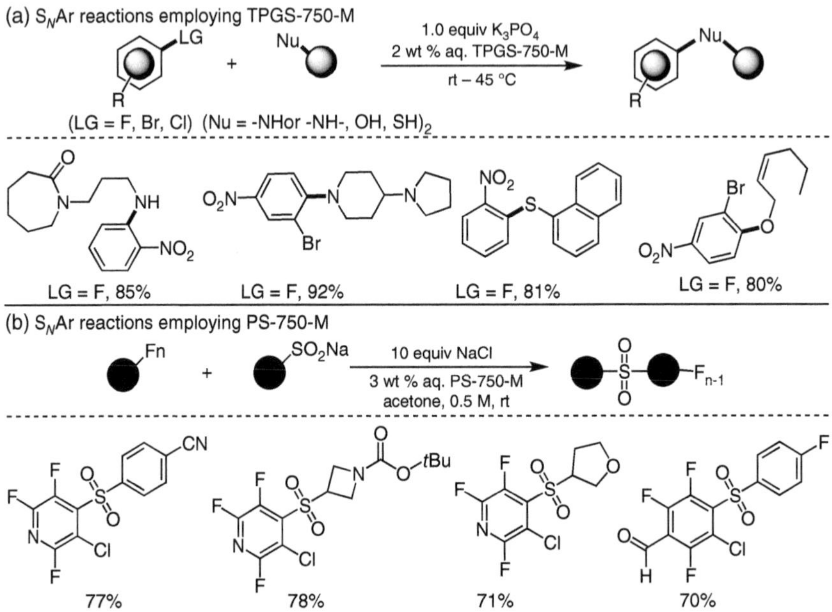

Scheme 3.12 Micellar catalysis in the context of S_NAr reactions.

Despite the broad scope, this report only allowed S_NAr reactions involving neutral nucleophiles, whereas the charged nucleophiles remained incompatible. This can be rationalized by the inability of hydrophobic TPGS-750-M micelles to accommodate charged nucleophiles inside the micelles. To address this challenge, Handa and coworkers used micelles of PS-750-M, which have a polar inner core, for the sulfonylations of polyfluoroarenes using sulfinate salts [74]. The addition of 10 equivalents of NaCl helped in dissolving the sulfinate salt inside the micelles. The optimal conditions also included 10% acetone for fruitful conversion. With the optimal conditions, the researchers evaluated the substrate scope expansion with 24 + substrates. Lithium or sodium salts of aryl and alkyl sulfinic acids showed excellent reactivity. This method was compatible with a wide range of heterocycles, including pyridine, thiophene, furan, and lactam. This approach was scalable, and the product mainly precipitated from the reaction medium, which allowed environmentally responsible recycling of the reaction medium with an E-factor closer to zero (Scheme 3.12b).

A recent study by Abbvie and Handa demonstrated that the hydrophobic pockets derived from the aqueous solution of cellulose derivative (hydroxypropyl methylcellulose (HPMC)) can facilitate scalable S_NAr reactions under mild aqueous conditions [73]. Notably, this is one of the most effective methods for installing a C—N bond in a functionalized aromatic ring. However, since this report does not fall under the category of micellar catalysis, this chapter will not provide a detailed discussion on this topic.

3.4.3 Amide Coupling Employing Micellar Catalysis

Amide coupling reactions hold massive significance in medicinal chemistry, mainly due to the market demand for peptide-based therapeutic agents, which exceeds 25.4 billion USD globally [127]. Traditionally, these transformations in the solution phase involve coupling the carboxylic acid with an amine using stoichiometric coupling or activating reagents in polar aprotic solvents like DMF or dichloromethane (DCM) [128]. However, the use of such solvents poses significant health risks and causes irreparable environmental damage, resulting in substantial investments in process safety and waste disposal measures [124, 128]. The introduction of micellar technology presents an innovative approach to enable water-based amide and peptide synthesis under milder conditions [52, 68, 69]. In 2015, Lipshutz and coworkers reported the first micellar strategy employing the aqueous nanomicelles of TPGS-750-M as the reaction medium, (1-cyano-2-ethoxy-2-oxoethylidenaminooxy)dimethylamino-morpholinocarbenium hexafluorophosphate (COMU) as the coupling reagent, and 2,6-lutidine as the base. This protocol enabled the amidation of various protected amino acids at room temperature (Scheme 3.13a) [68]. The scope of this approach was limited by the low reactivity of aromatic amines. Furthermore, the utility of this method was explored for peptide synthesis in water, focusing on sequential deprotection and coupling at the *N*-terminus in a one-pot fashion. The report successfully synthesized peptides of up to 10 amino acid units with satisfactory overall yields. In another report from Lipshutz's lab, the authors subsequently designed a surfactant molecule,

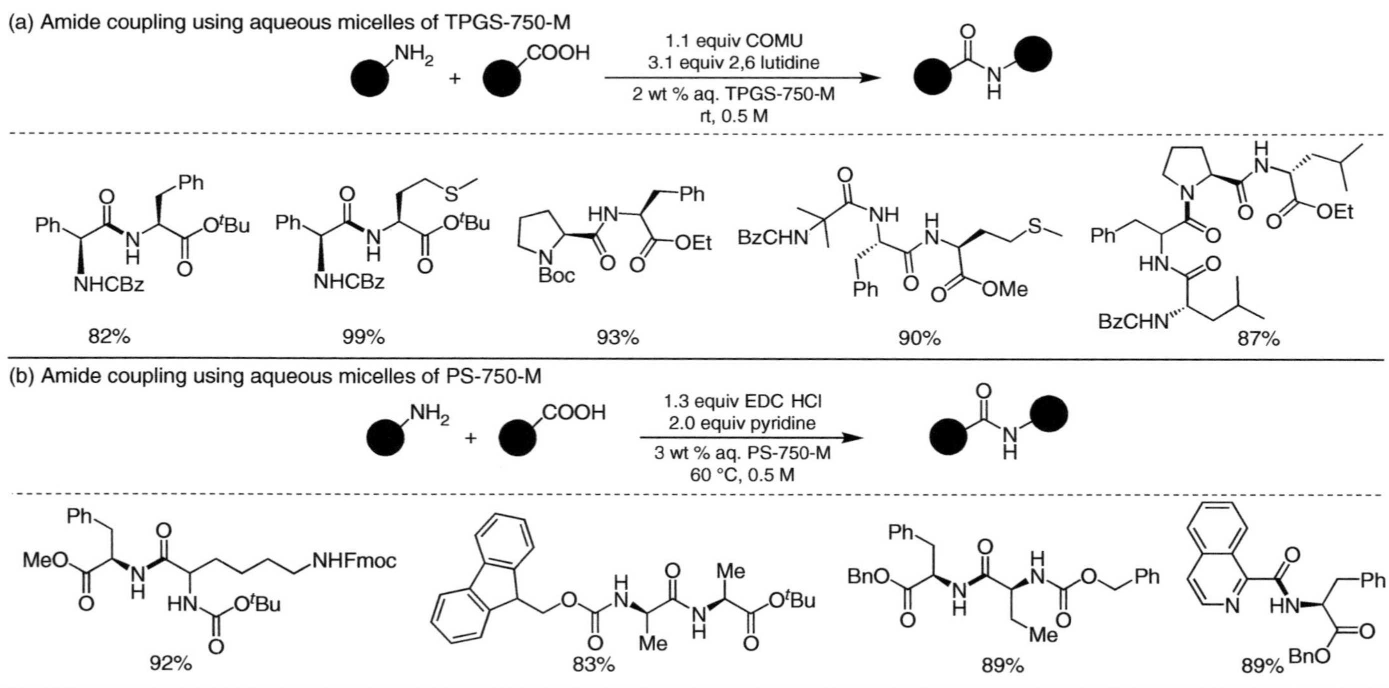

Scheme 3.13 Micellar catalysis in the context amide couplings.

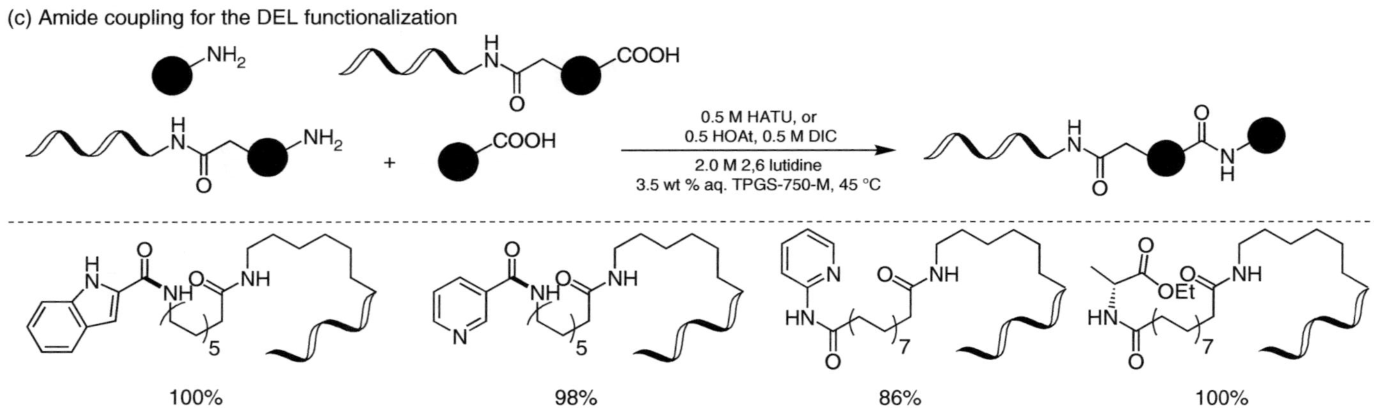

Scheme 3.13 *(Continued)*

MC-1.A, engineered to mimic the essential solvation properties of the polar aprotic solvent DMSO [69]. This amphiphile facilitated the successful amidation of various protected amino acids with moderate to excellent yields. Although these procedures effectively employed water as the primary reaction medium, organic solvents were still required for product extraction and purification.

A recent report by Handa and coworkers established a completely organic solvent-free method using the proline-derived PS-750-M amphiphile [52, 67]. The authors leveraged the structural features of the custom-designed amphiphile to drive the reaction under micellar conditions. Here, the micellar inner core of PS-750-derived aqueous micelles has an unusual polarity for their micellar inner core due to the presence of amide bonds in this region; this helps to solubilize polar reactants like amino acids or peptides (Scheme 3.13b). Utilizing 1-ethyl-3-(3-(dimethylamino)propyl) carbodiimide (EDC) as the coupling reagent and pyridine as the base, the aqueous micelles of PS-750-M micelles demonstrated faster amide coupling of amino acids compared to their organic solvent counterparts (10–40 minutes). Notably, product isolation required only filtration, as the product precipitates upon completion of the reaction. This organic solvent-free approach brings organic chemists one step closer to the ideal amide/peptide synthesis. The generality of this method was confirmed by examining 24 substrates, focusing on real-life examples, while scalability was demonstrated with 11 gram-scale reactions. An epimerization study ensured the retention of stereochemistry, addressing a common issue associated with traditional methods. HPLC analysis validated the complete retention of stereochemistry, overcoming another challenging aspect of these reactions. Very recently, authors expanded the scope of this approach toward the synthesis of two key drug intermediates in a scalable fashion [67].

Furthermore, Wearing and colleagues explored the efficacy of these micellar amidations for the functionalization of DNA molecules, broadening the scope of DELs [129]. Nanomicelles derived from TPGS-750-M were utilized as the reaction medium to successfully enable the coupling of various heterocyclic or alkyl amines/acids with functionalized DNA molecules. Carboxylic acid activation was achieved using HATU, or the combination of *N,N*-diisopropylcarbodiimide (DIC) and HOAt. This approach exhibited a broad scope, accommodating various coupling partners, including amino acids, alkyl amines, and heterocyclic amines, along with their acid variants (Scheme 3.13c).

3.4.4 Biocatalysis Employing Micellar Catalysis

Biocatalysis presents synthetic chemists with a unique opportunity to enhance regio-, chemo-, and stereo-selectivity in chemical synthesis [130–132]. By employing specific enzymes for targeted transformations under mild conditions, biocatalysis presents an exciting alternative to traditional chemo-catalysis. This regime predominantly utilizes water as a solvent, replacing the reliance on toxic/hazardous organic solvents while providing superior selectivity and improved reaction parameters under gentle conditions [130, 133]. Due to these reasons, many industrial processes are adopting this technology to develop greener and efficient chemical processes [134–136].

The integration of micellar technology with biocatalysis holds enormous potential and brings organic synthesis closer to ultimate sustainability. This combination expands substrate compatibility in biocatalysis and enables sequential reaction pathways involving both chemo- and biocatalysts, laying the foundation for next-generation organic synthesis [71, 72, 133, 137, 138]. A recent contribution by Lipshutz and colleagues demonstrates the use of alcohol dehydrogenase (ADH) for stereoselective reduction of functionalized ketones under micellar conditions [137]. Nano-micelles derived from TPGS-750-M facilitate efficient enzyme catalysis by modulating the degree of enzyme saturation through the solubilization of lipophilic reactants in hydrophobic pockets. The study also illustrates the sequential merging of chemo- and biocatalysis, starting with Pd, Cu, Fe, Rh, and Au-catalyzed alkyne/olefin functionalization, followed by ADH-catalyzed ketone reduction (Scheme 3.14a).

The scalability of this approach is showcased by the multi-kilogram-scale synthesis of a chiral 4-piperidinol derivative. Similar TPGS-750-M-based strategies have been developed for the asymmetric reduction of α,β-unsaturated compounds using ene-reductase (ERED) [71] and esterification reactions utilizing lipase enzymes [72] (Scheme 3.14b,c). Notably, no cofactor is required in lipase-mediated esterification, and the reaction proceeds smoothly with a 1 : 1 stoichiometry of acids and alcohols. These methodologies enable multiple sequential reactions by combining chemo- and bio-catalysis [138]. Although this branch of micellar catalysis is still in its early stages, further research is necessary to establish its underlying principles and advance its applications.

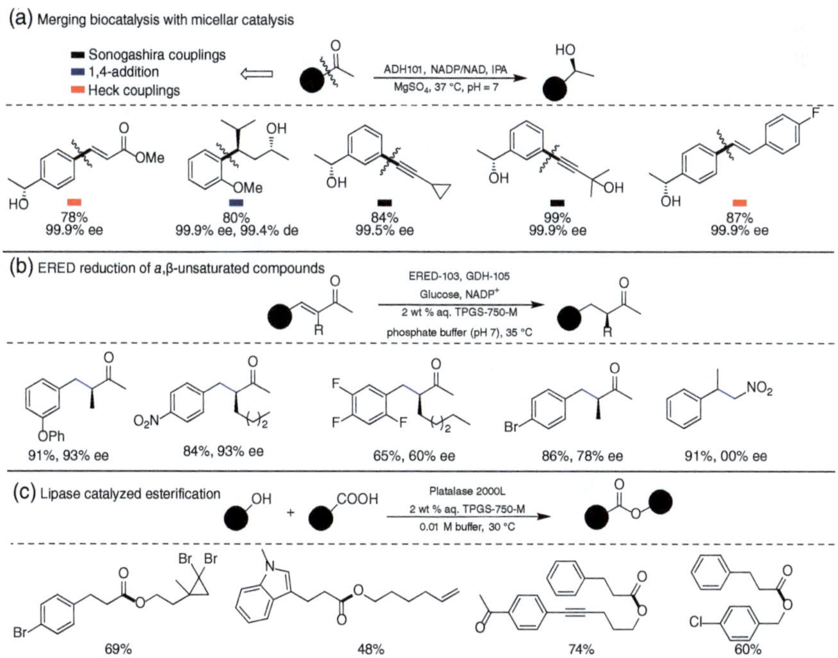

Scheme 3.14 Micellar catalysis in the context of biocatalysis.

3.5 Highlights of PTC

3.5.1 Asymmetric Hydroxylation

PTC has been applied to the asymmetric synthesis of α-aryl-α-hydroxy-δ-lactam derivatives containing a chiral quaternary center (Scheme 3.15). Numerous natural products and bioactive molecules contain lactams and cyclic amines with chiral quaternary centers. α-Aryl-α-hydroxy-δ-lactam derivatives are a prominent class of such compounds [139]. This method involves using a chiral quaternary ammonium salt as a PTC catalyst, KOH as a base, triethyl phosphite as a reducing agent in water, and toluene under an oxygen atmosphere. Excellent yield and enantioselectivity were achieved with only 1 mol % catalyst. The corresponding hydroxylated products with high enantioselectivities were achieved in moderate to high yields using a range of α-aryl-δ-lactams bearing *p*- or *m*-substituents. Moreover, heteroaryl rings such as pyridine and thiophene also gave the hydroxylated product good yield and enantioselectivity under optimized conditions (Scheme 3.15).

3.5.2 Asymmetric S_NAr Reactions

Phase transfer-catalyzed asymmetric S_NAr reactions of carbonyl compounds have been well explored over the last two decades. However, it was only applicable to the electrophiles having strong electron-withdrawing arenes. To broaden the scope, Maruoka and coworkers developed a chiral phase-transfer catalyst modified with binaphthyl, which efficiently facilitated the S_NAr reaction between α-amino acid derivatives and chromium complexes derived from electron-rich fluoroarenes, resulting in the formation of highly enantioselective α-α-disubstituted α-amino acids [140]. In addition to simple alkyl-substituted α-amino acid derivatives, such as phenylalanine and allylglycine, other amino acid derivatives were also well tolerated under reaction conditions. Furthermore, high enantioselectivity was also observed with methionine, serine, and glutamic acid derivatives (Scheme 3.16).

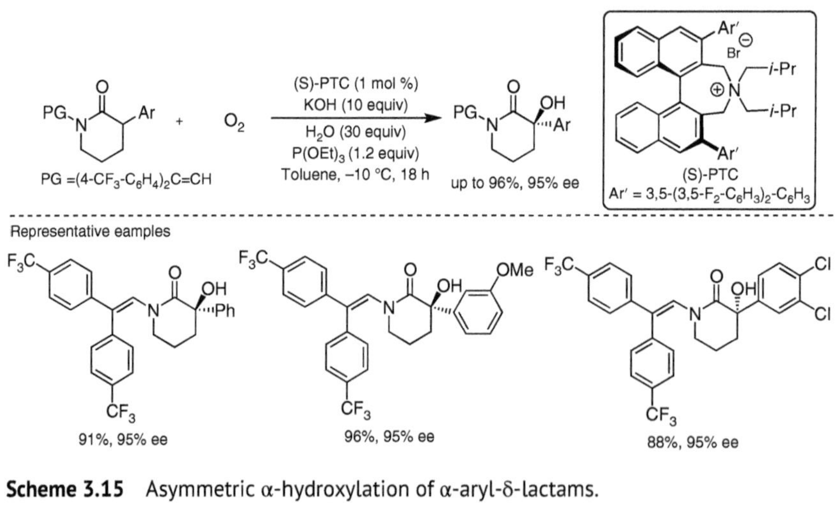

Scheme 3.15 Asymmetric α-hydroxylation of α-aryl-δ-lactams.

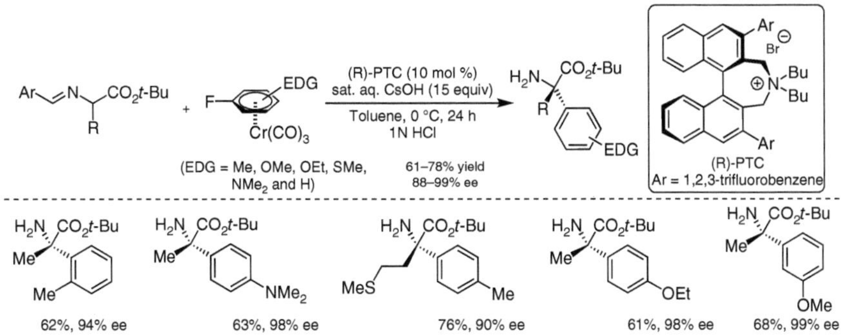

Scheme 3.16 Asymmetric nucleophilic aromatic substitution of α-amino acid derivatives.

3.5.3 Alkylation of 3-Arylpiperidin-2-ones

The chiral quaternary carbon centers are a common feature in nitrogen heterocycles found in natural products, playing a significant and beneficial role as essential building blocks in organic synthesis. For example, 3-alkyl-3-aryl piperidine is the one frequently found in biologically active compounds. Nevertheless, there have been limited successful reports on the catalytic asymmetric synthesis of 3-alkyl-3-aryl-lactams to date, such as those involving Pd-catalyzed allylation and conjugate addition under phase-transfer conditions. In 2020, the Maruoka group developed a catalytic asymmetric alkylation of 3-arylpiperidin-2-ones under phase-transfer conditions [141]. In the presence of 1 mol % chiral quaternary ammonium salt as a catalyst, the reactions of *N*-2,2-diarylvinyl-protected 3-aryl-lactams with diverse alkyl bromides provided the corresponding product in moderate to good yields and high enantioselectivities (Scheme 3.17).

3.5.4 Csp^3–Csp^3 Bond Formation Through Synergistic Effect of Palladium and PTC

While the stereodivergent synthesis of Csp^3—Csp^3 bonds with vicinal stereocenters has been extensively studied through synergistic catalysis, the challenge persists in

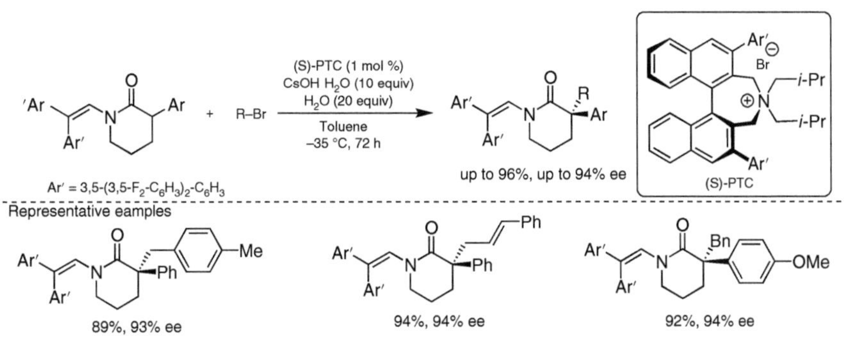

Scheme 3.17 Catalytic asymmetric alkylation of 3-arylpiperidin-2-ones using PTC.

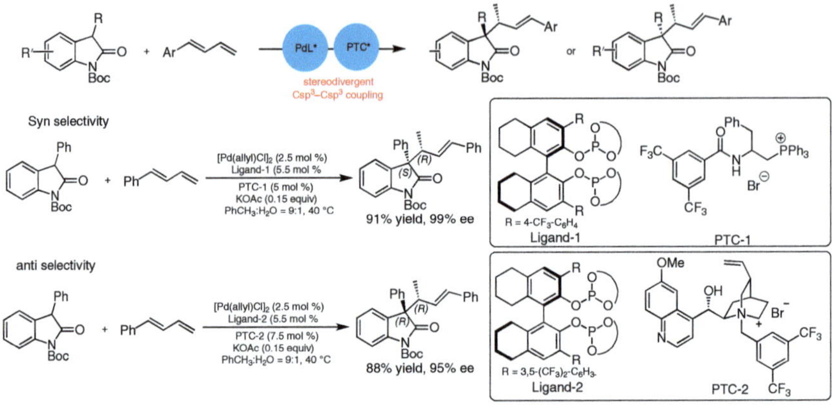

Scheme 3.18 Csp^3-Csp^3 bond formation through the synergistic effect of Palladium and PTC.

Csp^3—Csp^3 coupling with stabilized nucleophiles, primarily due to the competition from a single-catalysis pathway. To solve this issue, Zi and coworkers developed a new method for stereodivergent construction of Csp^3—Csp^3 bonds having vicinal stereocenters by synergistic palladium and PTC. (Scheme 3.18) [142]. The reaction involves 1,3-dienes and oxindoles as coupling partners, quaternary phosphonium salts as a PTC, and bidentate bisphosphite as a ligand. In the course of this reaction, one can achieve the selective production of both *syn* and *anti*-coupling products with good yields and excellent enantio- and diastereoselectivities. This study marks a groundbreaking achievement as the first successful application of metal/PTC synergistic catalysis in the realm of stereodivergent synthesis. The effectiveness of the catalyst is dependent on merging the non-covalent interaction of a stabilized nucleophile with an allylic substitution of a π-allyl-Pd electrophile in a biphasic system.

3.5.5 Enantioselective α-Benzoxylation

The first enantioselective α-benzoxylation of malonic diesters promoted by a phase-transfer catalyst was developed by the Itoh group in 2016 [143]. When benzoyl peroxide and *N*-(9-anthracenylmethyl) cinchonium chloride were added to the α-monosubstituted malonate, the reaction produced the corresponding α,α-disubstituted compounds with high enantioselectivity and excellent yields. This technology was also applied to synthesize the receptor antagonist mineralocorticoid (Scheme 3.19).

3.5.6 Ionic Liquids (ILs) in Phase-Transfer Catalysis

Ionic liquids are a special type of organic salt that remain liquid at low temperatures, typically less than 100 °C. Unlike regular salts with high melting points, ILs can stay in a liquid state over a broad range of temperatures, making them appealing for different applications.

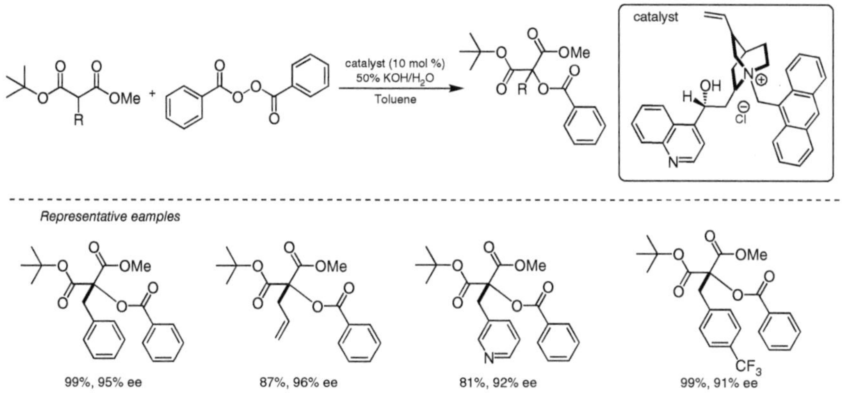

Scheme 3.19 PTC-mediated enantioselective α-benzoxylation.

The primary characteristic of ILs is their ionic nature. They consist of negatively charged ions (anions) and positively charged ions (cations), which are held together by electrostatic forces. The combination of various cations and anions can lead to a wide range of ILs, each with unique properties.

ILs are one of the characteristic types of PTC, composed entirely of ions, and have exceptional physical and chemical properties, including low vapor pressure and high thermal stability. Among them, room temperature ILs specifically have become a widely explored category of ILs that hold promise as an eco-friendly substitute for traditional organic solvents [144]. Moreover, they present a green alternative to the utilization of conventional organic solvents. A subtype known as protic ionic liquids has also gained popularity in the scientific community [145]. The molecular framework of these compounds, featuring protic hydrogens, establishes an extensive network of hydrogen bonds that prove advantageous across various applications.

In 1914, Paul Walden reported on ILs, specifically the protic ionic liquid ethyl ammonium nitrate. However, this groundbreaking discovery remained overlooked mainly by chemists until the late 1970s, when John Nardi and his team rediscovered the potential of ILs [146]. Cammarate and coworkers demonstrated that ILs interact with water through hydrogen bonding [147]. Moreover, the amphiphilic nature of imidazolium-based ILs has been discovered. Lourenco and Afonso were investigating the catalytic potential of ILs, using 1-*N*-butyl-3-methylimidazolium hexafluorophosphate as a reaction medium [148]. Later, polymer-supported ILs were developed and applied for nucleophilic substitutions, such as fluorination [149]. In 2017, Dong Wook Kim's team introduced dicationic ionic liquids and applied them as phase transfer catalysts in a sequence of fluorination reactions where β-elimination byproduct formation was also studied, a common challenge in conventional PTC systems [150]. Notably, the use of hexa-EG-DHIM as a catalyst exhibited excellent chemoselectivity and higher yields. ILs have been applied for a variety of reactions, such as hydrogenation, etherification, esterification, oxidation, epoxidation, polymerization, and organometallic coupling reactions.

In 2017, Sepúlveda et al. presented an innovative strategy for biphasic hydrogenation using ILs [151]. The approach involved using isopropanol as the sole hydrogen

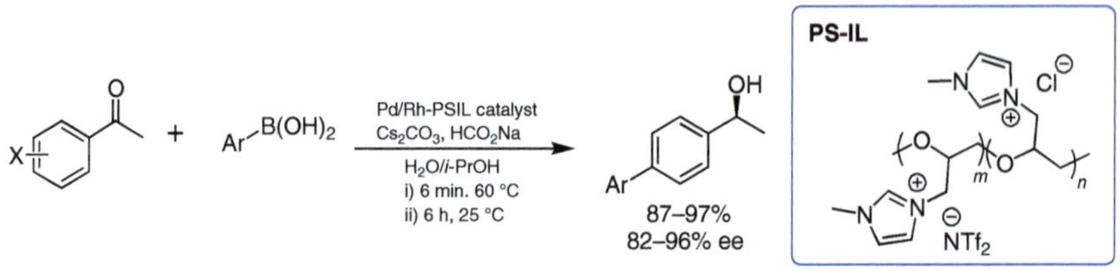

Scheme 3.20 PTC-mediated SM couplings.

donor without bases and generating the Ru catalytic species *in situ*. This method eliminates the need for specialized equipment for high-pressure hydrogen gas and avoids the use of strong bases, making it both environmentally benign and operationally simple.

Subsequently, Li and colleagues created a bifunctional catalyst by anchoring $Pd(OAc)_2$ onto the surface of the polymer [152]. This catalyst demonstrated efficient performance on Suzuki couplings and asymmetric hydrogenation in water through the formation of nano micelles. Notably, it selectively catalyzed the Suzuki couplings at high temperatures, while reduction was observed at lower temperatures. This methodology enabled a one-pot, two-step synthetic route for arylation and asymmetric reduction, utilizing the enantioselective bifunctional catalyst (Scheme 3.20).

3.6 Conclusion

In this chapter, we delve into the advancements in performing organic synthesis using safe solvents under micellar catalysis and PTC. Although both methods allow reactions to occur in aqueous conditions, they each have unique ways of facilitating the desired transformations. Micellar catalysis holds great promise as an alternative to using hazardous solvents such as DMF, *N*-methyl-2-pyrrolidone (NMP), and dioxane, offering a sustainable and eco-friendly option. The industry is actively exploring its potential in chemical process development, recognizing its many benefits in reducing the carbon footprint associated with chemical synthesis. In a similar vein, PTC has significantly contributed to advancing asymmetric organic synthesis, providing practical solutions for challenging synthetic problems. The technologies involving micellar catalysis and PTC are clearly paving the way to make synthetic chemistry more sustainable and greener.

References

1 Wöhler, F. (1828). *Ann. Phys.* 88: 253–256.
2 Nicolaou, K.C. (2018). *Isr. J. Chem.* 58: 104–113.
3 Nicolaou, K.C. (2009). *J. Org. Chem.* 74: 951–972.
4 Dutta, S. (2022). *Ind. Eng. Chem. Res.* 61: 12884–12904.
5 Clark, J.H., Farmer, T.J., Hunt, A.J., and Sherwood, J. (2015). *Int. J. Mol. Sci.* 16: 17101–17159.

6 Trost, B.M. (1991). *Science* 254: 1471–1477.
7 Sheldon, R.A., Bode, M.L., and Akakios, S.G. (2022). *Curr. Opin. Green Sustain. Chem.* 33: 100569.
8 Sheldon, R.A. (1994). *Chemtech* 3: 38–47.
9 Anastas, P.T. and Warner, J.C. (1998). *Green Chemistry Theory and Practice*, 133–190. New York: Oxford University Press.
10 Sheldon, R.A. (2018). *ACS Sustain. Chem. Eng.* 6: 32–48.
11 Sheldon, R.A. (2023). *Green Chem.* 25: 1704–1728.
12 Trost, B.M. (1995). *Angew. Chem. Int. Ed.* 34: 259–281.
13 Jimenez-Gonzalez, C., Ponder, C.S., Broxterman, Q.B., and Manley, J.B. (2011). *Org. Process Res. Dev.* 15: 912–917.
14 Roschangar, F., Li, J., Zhou, Y. et al. (2022). *ACS Sustain. Chem. Eng.* 10: 5148–5162.
15 Sharma, P., Ponnusamy, E., Ghorai, S., and Colacot, T.J. (2022). *J. Organomet. Chem.* 970–971: 122367.
16 Sheldon, R.A. (2019). *Curr. Opin. Green Sustain. Chem.* 18: 13–19.
17 Sheldon, R. (2005). *Green Chem.* 7: 267–278.
18 Stini, N.A., Gkizis, P.L., and Kokotos, C.G. (2022). *Green Chem.* 24: 6435–6449.
19 Jad, Y.E., Acosta, G.A., Govender, T. et al. (2016). *ACS Sustain. Chem. Eng.* 4: 6809–6814.
20 Cortes-Clerget, M., Yu, J., Kincaid, J.R.A. et al. (2021). *Chem. Sci.* 12: 4237–4266.
21 Ansari, T.N., Handa, S., and Gallou, F. (2020). Cross-couplings in water – a better way to assemble new bonds. In: *Organometallic Chemistry in Industry: A Practical Approach* (ed. T.J. Colacot and C.C.C.J. Seechurn), 203–235. Wiley.
22 Chandler, D. (2005). *Nature* 437: 640–647.
23 Chanda, A. and Fokin, V.V. (2009). *Chem. Rev.* 109: 725–748.
24 Lipshutz, B.H., Ghorai, S., and Cortes-Clerget, M. (2018). *Chem. - A Eur. J.* 24: 6672–6695.
25 Dwars, T., Paetzold, E., and Oehme, G. (2005). *Angew. Chem. Int. Ed.* 44: 7174–7199.
26 Floriano, M.A., Caponetti, E., and Panagiotopoulos, A.Z. (1999). *Langmuir* 15: 3143–3151.
27 Zdziennicka, A., Szymczyk, K., Krawczyk, J., and Jańczuk, B. (2012). *Fluid Phase Equilib.* 322–323: 126–134.
28 Rharbi, Y., Li, M., Winnik, M.A., and Hahn, K.G. (2000). *J. Am. Chem. Soc.* 122: 6242–6251.
29 Kondo, K., Klosterman, J.K., and Yoshizawa, M. (2017). *Chem. Eur. J.* 23: 16710–16721.
30 Afzal, S., Lone, M.S., Nazir, N., and Dar, A.A. (2021). *ACS Omega* 6: 14985–15000.
31 Brals, J., Smith, J.D., Ibrahim, F. et al. (2017). *ACS Catal.* 7: 7245–7250.
32 Lam, Y.M., Grigorieff, N., and Goldbeck-Wood, G. (1999). *Phys. Chem. Chem. Phys.* 1: 3331–3334.

33 Hedouin, G., Ogulu, D., Kaur, G., and Handa, S. (2023). *Chem. Commun.* 59: 2842–2853.

34 Makosza, M. (2000). *Pure Appl. Chem.* 72: 1399–1403.

35 Weston, J.O., Miyamura, H., Yasukawa, T. et al. (2017). *Catal. Sci. Technol.* 7: 3985–3998.

36 Starks, C.M. (1971). *J. Am. Chem. Soc.* 93: 195–199.

37 Phipps, R.J., Hamilton, G.L., and Toste, F.D. (2012). *Nat. Chem.* 4: 603–614.

38 Mąkosza, M. and Fedoryński, M. (2003). *Catal. Rev.* 45: 321–367.

39 Zong, L. and Tan, C.-H. (2017). *Acc. Chem. Res.* 50: 842–856.

40 Mąkosza, M. and Fedoryński, M. (1987). *Catalysis in Two-Phase Systems: Phase Transfer and Related Phenomena*, vol. 35 (ed. D.D. Eley, H. Pines, and P.B. Weisz), 375–422. Academic Press.

41 Shirakawa, S. and Maruoka, K. (2013). *Angew. Chem. Int. Ed.* 52: 4312–4348.

42 Hill, J.W. and Corredor, J. (1980). *J. Chem. Educ.* 57: 822. https://doi.org/10.1021/ed057p822.

43 Ansari, T.N., Gallou, F., and Handa, S. (2023). *Coord. Chem. Rev.* 488: 215158.

44 Fleck, N., Roschangar, F., and Haydl, A.M. (2023). *Org. Process Res. Dev.* 27: 822–830.

45 Handa, S., Andersson, M.P., Gallou, F. et al. (2016). *Angew. Chem. Int. Ed.* 55: 4914–4918.

46 Kincaid, J.R.A., Wong, M.J., Akporji, N. et al. (2023). *J. Am. Chem. Soc.* 145: 4266–4278.

47 Lipshutz, B.H., Ghorai, S., Abela, A.R. et al. (2011). *J. Org. Chem.* 76: 4379–4391.

48 Handa, S., Wang, Y., Gallou, F., and Lipshutz, B.H. (2015). *Science* 349: 1087.

49 Kaur, G., Kaur, K., and Handa, S. (2022). *Curr. Opin. Green Sustain. Chem.* 38: 100690.

50 Bihani, M., Ansari, T.N., Finck, L. et al. (2020). *ACS Catal.* 10: 6816–6821.

51 Ansari, T.N., Jasinski, J.B., Leahy, D.K., and Handa, S. (2021). *JACS Au* 1: 308–315.

52 Sharma, S., Buchbinder, N.W., Braje, W.M., and Handa, S. (2020). *Org. Lett.* 22: 5737–5740.

53 Bihani, M., Bora, P.P., Nachtegaal, M. et al. (2019). *ACS Catal.* 9: 7520–7526.

54 Jakobi, M., Gallou, F., Sparr, C., and Parmentier, M. (2019). *Helv. Chim. Acta* 102: e1900024.

55 Parmentier, M., Wagner, M., Wickendick, R. et al. (2020). *Org. Process Res. Dev.* 24: 1536–1542.

56 Gallou, F., Isley, N.A., Ganic, A. et al. (2016). *Green Chem.* 18: 14–19.

57 Kitanosono, T., Miyo, M., and Kobayashi, S. (2016). *ACS Sustain. Chem. Eng.* 4: 6101–6106.

58 Mattiello, S., Rooney, M., Sanzone, A. et al. (2017). *Org. Lett.* 19: 654–657.

59 Ceriani, C., Ghiglietti, E., Sassi, M. et al. (2020). *Org. Process Res. Dev.* 24: 2604–2610.

60 Peacock, H. and Blum, S.A. (2023). *J. Am. Chem. Soc.* 145: 7648–7658.

61 Mattiello, S., Ghiglietti, E., Zucchi, A., and Beverina, L. (2023). *Curr. Opin. Colloid Interface Sci.* 64: 101681.
62 Ogulu, D., Bora, P.P., Bihani, M. et al. (2022). *ACS Appl. Mater. Interfaces* 14: 6754–6761.
63 Wu, B., Miraghaee, S., Handa, S., and Gallou, F. (2022). *Curr. Opin. Green Sustain. Chem.* 38: 100691.
64 Lipshutz, B.H., Caravez, J.C., and Iyer, K.S. (2022). *Curr. Opin. Green Sustain. Chem.* 38: 100686.
65 Ansari, T.N., Taussat, A., Clark, A.H. et al. (2019). *ACS Catal.* 9: 10389–10397.
66 Ansari, T.N., Sharma, S., Hazra, S. et al. (2022). *ACS Catal.* 12: 15686–15695.
67 Sharma, S., Kaur, G., and Handa, S. (2021). *Org. Process Res. Dev.* 25: 1960–1965.
68 Gabriel, C.M., Keener, M., Gallou, F., and Lipshutz, B.H. (2015). *Org. Lett.* 17: 3968–3971.
69 Cortes-Clerget, M., Spink, S.E., Gallagher, G.P. et al. (2019). *Green Chem.* 21: 2610–2614.
70 Wu, B., Zhang, S., Hong, T. et al. (2020). *Org. Process Res. Dev.* 24: 2780–2788.
71 Akporji, N., Singhania, V., Dussart-Gautheret, J. et al. (2021). *Chem. Commun.* 57: 11847–11850.
72 Singhania, V., Cortes-Clerget, M., Dussart-Gautheret, J. et al. (2022). *Chem. Sci.* 13: 1440–1445.
73 Borlinghaus, N., Ansari, T.N., Braje, L.H. et al. (2021). *Green Chem.* 23: 3955–3962.
74 Smith, J.D., Ansari, T.N., Andersson, M.P. et al. (2018). *Green Chem.* 20: 1784–1790.
75 Lee, N.R., Gallou, F., and Lipshutz, B.H. (2017). *Org. Process Res. Dev.* 21: 218–221.
76 Gallou, F. and Lipshutz, B.H. (2018). Organometallic processes in water. In: *Topics in Organometallic Chemistry*, 1–18. Berlin, Heidelberg: Springer.
77 Bailey, J.D., Helbling, E., Mankar, A. et al. (2021). *Green Chem.* 23: 788–795.
78 Compagno, N., Profeta, R., and Scarso, A. (2023). *Curr. Opin. Green Sustain. Chem.* 39: 100729.
79 Brown, D.G. and Boström, J. (2016). *J. Med. Chem.* 59: 4443–4458.
80 Farhang, M., Akbarzadeh, A.R., Rabbani, M., and Ghadiri, A.M. (2022). *Polyhedron* 227: 116124.
81 Ashworth, I.W., Campbell, A.D., Cherryman, J.H. et al. (2018). *Org. Process Res. Dev.* 22: 1801–1808.
82 Grasa, G.A. and Colacot, T.J. (2007). *Org. Lett.* 9: 5489–5492.
83 Lipshutz, B.H., Petersen, T.B., and Abela, A.R. (2008). *Org. Lett.* 10: 1333–1336.
84 Lipshutz, B.H. and Abela, A.R. (2008). *Org. Lett.* 10: 5329–5332.
85 Akporji, N., Thakore, R.R., Cortes-Clerget, M. et al. (2020). *Chem. Sci.* 11: 5205–5212.
86 Landstrom, E.B., Handa, S., Aue, D.H. et al. (2018). *Green Chem.* 20: 3436–3443.
87 Handa, S., Ibrahim, F., Ansari, T.N., and Gallou, F. (2018). *ChemCatChem* 10: 4229–4233.

88 Wood, A.B., Nandiwale, K.Y., Mo, Y. et al. (2020). *Green Chem.* 22: 3441–3444.
89 Wood, A.B., Plummer, S., Robinson, R.I. et al. (2021). *Green Chem.* 23: 7724–7730.
90 Hunter, J.H., Prendergast, L., Valente, L.F. et al. (2020). *Bioconjug. Chem.* 31: 149–155.
91 Takale, B.S., Thakore, R.R., Handa, S. et al. (2019). *Chem. Sci.* 10: 8825–8831.
92 Lee, N.R., Linstadt, R.T.H., Gloisten, D.J. et al. (2018). *Org. Lett.* 20: 2902–2905.
93 Ruiz-Castillo, P. and Buchwald, S.L. (2016). *Chem. Rev.* 116: 12564–12649.
94 Schlummer, B. and Scholz, U. (2004). *Adv. Synth. Catal.* 346: 1599–1626.
95 Surry, D.S. and Buchwald, S.L. (2011). *Chem. Sci.* 2: 27–50.
96 Hartwig, J.F. (2008). *Nature* 455: 314–322.
97 Lipshutz, B.H., Chung, D.W., and Rich, B. (2009). *Adv. Synth. Catal.* 351: 1717–1721.
98 Salomé, C., Wagner, P., Bollenbach, M. et al. (2014). *Tetrahedron* 70: 3413–3421.
99 Wagner, P., Bollenbach, M., Doebelin, C. et al. (2014). *Green Chem.* 16: 4170–4178.
100 Zhang, Y., Takale, B.S., Gallou, F. et al. (2019). *Chem. Sci.* 10: 10556–10561.
101 Graham, J.S., Hunter, J.H., and Waring, M.J. (2021). *J. Org. Chem.* 86: 17257–17264.
102 Lipshutz, B.H., Chung, D.W., and Rich, B. (2008). *Org. Lett.* 10: 3793–3796.
103 Handa, S., Smith, J.D., Zhang, Y. et al. (2018). *Org. Lett.* 20: 542–545.
104 Jin, B., Gallou, F., Reilly, J., and Lipshutz, B.H. (2019). *Chem. Sci.* 10: 3481–3485.
105 Johansson, C.C.C. and Colacot, T.J. (2010). *Angew. Chem. Int. Ed.* 49: 676–707.
106 Sivanandan, S.T., Shaji, A., Ibnusaud, I. et al. (2015). *Eur. J. Org. Chem.* 2015: 38–49.
107 Wood, A.B., Roa, D.E., Gallou, F., and Lipshutz, B.H. (2021). *Green Chem.* 23: 4858–4865.
108 Dandia, A., Parihar, S., Sharma, R. et al. (2020). Nanocatalysis in Green Organic Synthesis. In: *Green Sustainable Process for Chemical and Environmental Engineering and Science* (ed. Inamuddin, R. Boddula, and A.M. Asiri), 71–103. Elsevier.
109 Gellé, A., Jin, T., de la Garza, L. et al. (2020). *Chem. Rev.* 120: 986–1041.
110 Chng, L.L., Erathodiyil, N., and Ying, J.Y. (2013). *Acc. Chem. Res.* 46: 1825–1837.
111 Albonetti, S., Mazzoni, R., and Cavani, F. (2014). Homogeneous, Heterogeneous and Nanocatalysis. In: *Transition Metal Catalysis in Aerobic Alcohol Oxidation* (ed. F. Cardona and C. Parmeggiani), 197–216. The Royal Society of Chemistry.
112 Arcadi, A., Cerichelli, G., Chiarini, M. et al. (2003). *Eur. J. Org. Chem.* 2003: 4080–4086.
113 Li, Y., Hong, X.M., Collard, D.M., and El-Sayed, M.A. (2000). *Org. Lett.* 2: 2385–2388.
114 Handa, S., Jin, B., Bora, P.P. et al. (2019). *ACS Catal.* 9: 2423–2431.
115 Pang, H., Hu, Y., Yu, J. et al. (2021). *J. Am. Chem. Soc.* 143: 3373–3382.

116 Hu, Y., Wong, M.J., and Lipshutz, B.H. (2022). *Angew. Chem, Int. Ed.* 61: e202209784.

117 Ansari, T.N., Sharma, S., Hazra, S. et al. (2021). *JACS Au* 1: 1506–1513.

118 Sharma, S., Parmar, S., Ibrahim, F. et al. (2022). *Adv. Funct. Mater.* 32: 2270202.

119 Kaur, G., Jasinski, J.B., Gallou, F., and Handa, S. (2022). *ACS Appl. Mater. Interfaces* 14: 50947–50955.

120 Rohrbach, S., Smith, A.J., Pang, J.H. et al. (2019). *Angew. Chem. Int. Ed.* 58: 16368–16388.

121 Henderson, A.S., Medina, S., Bower, J.F., and Galan, M.C. (2015). *Org. Lett.* 17: 4846–4849.

122 Baumann, M. and Baxendale, I.R. (2013). *Beilstein J. Org. Chem.* 9: 2265–2319.

123 Wales, S.M., Saunthwal, R.K., and Clayden, J. (2022). *Acc. Chem. Res.* 55: 1731–1747.

124 Jordan, A., Hall, C.G.J., Thorp, L.R., and Sneddon, H.F. (2022). *Chem. Rev.* 122: 6749–6794.

125 Jordan, A., Stoy, P., and Sneddon, H.F. (2021). *Chem. Rev.* 121: 1582–1622.

126 Isley, N.A., Linstadt, R.T.H., Kelly, S.M. et al. (2015). *Org. Lett.* 17: 4734–4737.

127 Roughley, S.D. and Jordan, A.M. (2011). *J. Med. Chem.* 54: 3451–3479.

128 MacMillan, D.S., Murray, J., Sneddon, H.F. et al. (2013). *Green Chem.* 15: 596–600.

129 Hunter, J.H., Anderson, M.J., Castan, I.F.S.F. et al. (2021). *Chem. Sci.* 12: 9475–9484.

130 Winkler, C.K., Schrittwieser, J.H., and Kroutil, W. (2021). *ACS Cent. Sci.* 7: 55–71.

131 Sheldon, R.A., Brady, D., and Bode, M.L. (2020). *Chem. Sci.* 11: 2587–2605.

132 Bell, E.L., Finnigan, W., France, S.P. et al. (2021). *Nat. Rev. Methods Prim.* 1: 46.

133 Gröger, H., Gallou, F., and Lipshutz, B.H. (2023). *Chem. Rev.* 123: 5262–5296.

134 Wu, S., Snajdrova, R., Moore, J.C. et al. (2021). *Angew. Chem. Int. Ed.* 60: 88–119.

135 Sheldon, R.A. and Brady, D. (2021). *ACS Sustain. Chem. Eng.* 9: 8032–8052.

136 Hughes, G. and Lewis, J. (2018). *Chem. Rev.* 118: 1–3.

137 Cortes-Clerget, M., Akporji, N., Zhou, J. et al. (2019). *Nat. Commun.* 10: 1–10.

138 Gallou, F., Gröger, H., and Lipshutz, B.H. (2023). *Green Chem.* 25: 6092–6107.

139 Inukai, T., Kano, T., and Maruoka, K. (2021). *Org. Lett.* 23: 792–796.

140 Shirakawa, S., Yamamoto, K., and Maruoka, K. (2015). *Angew. Chem. Int. Ed.* 54: 838–840.

141 Inukai, T., Kano, T., and Maruoka, K. (2020). *Angew. Chem Int. Ed.* 59: 2211–2214.

142 Han, J., Liu, R., Lin, Z., and Zi, W. (2023). *Angew. Chem. Int. Ed.* 62: e202215714.

143 Kanemitsu, T., Sato, M., Yoshida, M. et al. (2016). *Org. Lett.* 18: 5484–5487.

144 Ahmed, A. (2020). Sustainable organic synthesis in ionic liquids. In: *Applications of Nanotechnology for Green Synthesis* (ed. Inamuddin and A.M. Asiri), 1–18. Cham: Springer International Publishing.

145 Greaves, T.L. and Drummond, C.J. (2008). *Chem. Rev.* 108: 206–237.

146 Wilkes, J.S., Levisky, J.A., Wilson, R.A., and Hussey, C.L. (1982). *Inorg. Chem.* 21: 1263–1264.

147 Cammarata, L., Kazarian, S.G., Salter, P.A., and Welton, T. (2001). *Phys. Chem. Chem. Phys.* 3: 5192–5200.

148 Lourenço, N.M.T. and Afonso, C.A.M. (2003). *Tetrahedron* 59: 789–794.

149 Kim, D.W. and Chi, D.Y. (2004). *Angew. Chem. Int. Ed.* 43: 483–485.

150 Jadhav, V.H., Kim, J.G., Park, S.H., and Kim, D.W. (2017). *Chem. Eng. J.* 308: 664–668.

151 Sepúlveda, F., Carrión, M.C., Phillips, A.D. et al. (2017). *Eur. J. Inorg. Chem.* 2017: 630–638.

152 Li, X., Sun, Y., Wang, S., and Jia, X. (2020). *ACS Appl. Mater. Interfaces* 12: 44094–44102.

4

Physicochemical Aspects of "On-Water" and "In-Water" Reactions

Rachel Napier and Bao N. Nguyen

University of Leeds, School of Chemistry, Woodhouse Lane, Leeds LS2 9JT, UK

4.1 Introduction

The term "on-water reactions" was defined by Sharpless as reactions where insoluble reactant(s) are stirred in aqueous emulsions or suspensions without the addition of any organic cosolvents [1]. The key attribute for these reactions is the lack of solubility of the reactant(s) in water. The term "in-water" was coined by Hayashi to refer to reactions performed under dilute conditions in water or buffer solution in which the reactants are completely soluble. However, "on-water", "in-water", and other terms, e.g. "in the presence of water", have been used interchangeably in the literature to describe widely different reaction conditions [2, 3], which has caused significant confusion in this field of research. These definitions and conditions can have an important impact on the sustainability of the chemical processes they are based on, as demonstrated by Blackmond and Armstrong [4]. More importantly, they can significantly influence and guide the rational discovery of new water-accelerated reactions, compared to the same reactions performed in organic solvents, and green processes through mechanistic understanding. An example is the tandem aldehyde oxidation/Passerini reaction reported by Vigalok et al. (Scheme 4.1) [5]. Water-insoluble aldehyde **1** gave only Passerini product **4**, while partially soluble aldehydes gave a mixture of both Passerini product **4** and the oxidation/Passerini product **3**. The authors concluded that the different pathways in which a reaction can proceed depend on the "in-water" and "on-water" conditions. Crucially, the rate of oxidation of aldehyde **1** by air is clearly dependent on the conditions. The solubility of oxygen in water is well known to be very low (2.56×10^{-4} M in air at 25 °C) [6], and a multiphase system with a larger surface area may be beneficial in promoting mass transfer of oxygen and subsequent oxidation of the aldehyde.

Despite the binary definitions for "in-water" and "on-water" above, real reactions can be much more complex. Many water-accelerated reactions (a blanket term referring to all reactions, which proceed faster when performed in water compared to when performed in an organic solvent) require and/or produce a base or an inorganic salt. This modifies the ionic strength of the aqueous solution, which, in turn,

Organic Transformations in Water: Principles and Applications, First Edition.
Edited by Gopinathan Anilkumar, Nissy Ann Harry, and Sankuviruthiyil M. Ujwaldev.

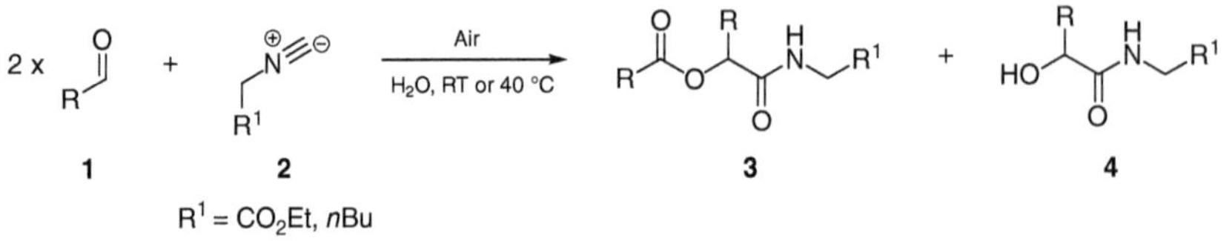

Scheme 4.1 Tandem aldehyde oxidation/Passerini reaction "on-water". Source: Adapted from Shapiro and Vigalok [5].

influences the solubility of the reactants and products and their phase behaviors. As the reaction progresses, the increasing amount of product may also introduce a new phase into the reaction mixture, changing what started as an "in-water" reaction into an "on-water" reaction.

In this chapter, we will discuss the physicochemical aspects of these intriguing and important reactions, particularly focusing on their phase behaviors and how this influences reaction kinetics, mechanisms, and ultimately, process green metrics. Nomenclature-wise, the reactions will be referred to as "water-accelerated reactions" for simplicity. Surfactant-enabled reactions, a uniquely exciting area of research, will be the topic of another chapter. Reactions that employ a mixture of organic solvent(s) and water will also not be covered due to their extremely broad nature.

4.2 Phase Behaviors of Realistic Reactions

4.2.1 Complex and Substrate-Dependent Phase Behaviors

The first example of a water-accelerated reaction, reported by Breslow et al. was a Diels–Alder reaction between cyclopentadiene and methyl vinyl ketone, the latter of which has good solubility in water [7]. Thus, one can reasonably expect the reaction mixture to consist of an emulsion of cyclopentadiene as the organic phase in water, and methyl vinyl ketone partitioning between the two phases. Directly evaluating whether the reaction happens inside the emulsion droplets or at the organic–water interface is nontrivial when at least one reactant can partition across the different phases. Nguyen and coworkers summarized the different partitioning of reactive species in a non-catalytic reaction mixture into four general scenarios (Figure 4.1) [8].

Scenario (a) is a strictly "on-water" reaction, although very few reported reactions fully qualify for this, as many of the reagents in organic synthesis are at least partially soluble in water. Scenario (b) is much more common, as in the Diels–Alder reaction reported by Breslow. In this case, there can be competition between the reaction within the organic emulsion, which can have a high effective concentration of the reactants due to the lack of solvent, and the reaction that happens at the organic–water interface, which can be accelerated through stabilization of the transition state via hydrogen bonding with water molecules. Breslow estimated a maximum fourfold increase in concentration inside the organic droplets compared

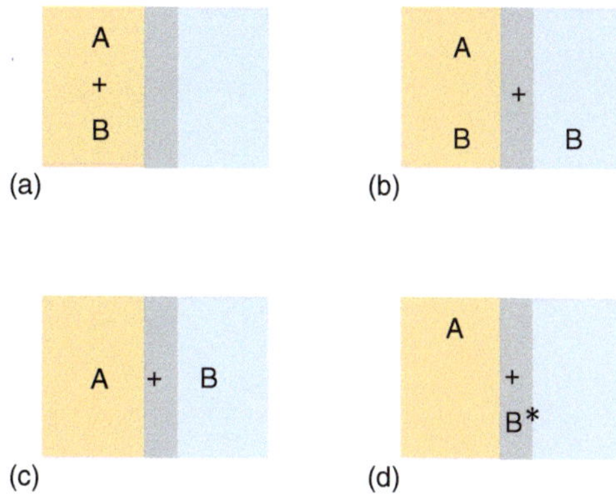

Figure 4.1 Simplified phase complexity scenarios of water-accelerated reactions: (a) all reactants are insoluble in water; (b) one reactant is partially soluble in water; (c) reactants are phase separated; and (d) one reactant is activated at the water–organic interface. Source: Reproduced from Maltby et al. [8]/American Chemical Society/CC BY 4.0.

to organic solvents, which cannot explain the several-hundredfold increase in reaction rate observed in his reactions [7]. In scenarios (c) and (d), the reaction between reactants A and B must unambiguously happen at the water–organic interface, and the reaction rate may be mass-transfer-limited. An example of (d) is the Henry reaction, first reported by Rao and later extensively studied by Nguyen (Scheme 4.2) [8, 9].

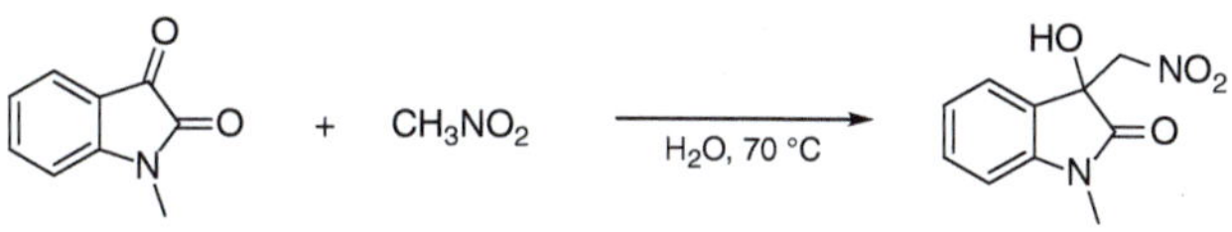

Scheme 4.2 A water-accelerated Henry reaction.

In this reaction, the enolization of nitromethane **6** is catalyzed by water and takes place in the aqueous phase due to the high solubility of nitromethane in water. The enolate must then meet *N*-methylisatin **5** at the water–organic interface for the nucleophilic attack and subsequent steps to form the final product.

When a catalyst is included, the system becomes predictably even more complex. Kobayashi and coworkers outlined seven different scenarios based on the distribution of the catalyst and substrates in the organic and aqueous phases (Figure 4.2) [10].

Finally, the phase behavior of the reaction mixture may change as the reaction progresses, depending on the solubility of the reactants and products in water and on the way the reaction is performed. An example of this is the base-mediated cyclization reaction of propargyl pyridinium salt **8** into imidazo[1,2-*a*]pyridine **9**, a pharmacophore, reported by Willans and Nguyen (Scheme 4.3) [11]. On a 10 g scale, the reaction was performed by slowly adding the starting material **8**, a partially water-soluble powder, into a vigorously stirred aqueous solution of NaOH. As the reaction progressed and more of the product was formed, a buildup of organic phases was observed, altering the number of phases and the "in-water"/"on-water" nature of the reaction (Figure 4.3). The product **9** solidified upon removal of the aqueous layer, giving an analytically pure product without further purification.

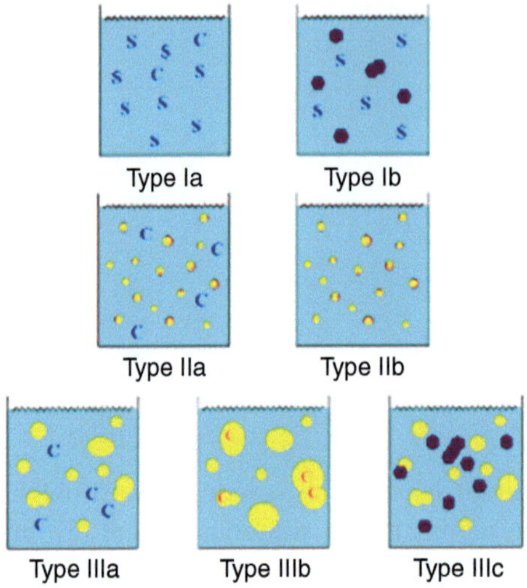

Type	Surfactant is added	All substrates are soluble in water	Catalyst is soluble in water	Interface prescribed for the reaction progress
Ia	No	Yes	Yes	–
Ib	No	Yes	No	catalyst-water
IIa	Yes	–	Yes	micelle surface
IIb	Yes	–	No	micelle surface
IIIa	No	No	Yes	substrate(s)-water
IIIb	No	No	No	substrate(s)-water
IIIc	No	No	No	catalyst-substrate(s)-water

Figure 4.2 Classification of catalytic reactions performed in water: when all substrates are soluble in water (top; Ia and Ib), when any of the substrates form emulsions (middle; IIa and IIb), and when any of the substrates form an aqueous suspension (bottom; IIIa, IIIb, and IIIc). Source: Reproduced from Kitanosono and Kobayashi [10]/John Wiley & Sons.

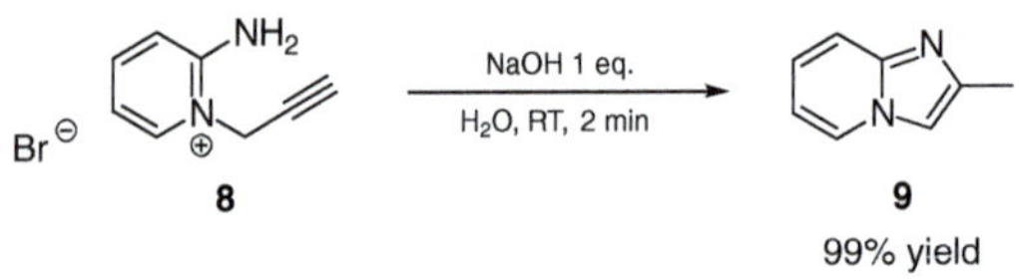

Scheme 4.3 Rapid, metal-free, and aqueous water-accelerated synthesis of imidazo[1,2-*a*] pyridines. Source: Chapman et al. [11]/Royal Society of Chemistry/CC BY 4.0.

4.2.2 Effect of Salts

One of the key early observations that Breslow reported was the impact of salt additives on the reaction rate in water-accelerated reactions [7]. Using LiCl (4.9 M) instead of pure water as the reaction medium led to a 2.5 times increase in reaction rate, while a guanidinium chloride (4.9 M) solution gave no rate increase when

Figure 4.3 A scaled-up setup for reaction from **8** to **9**. (a) Before reaction; (b) during reaction; and (c) phase separation at the end of the reaction. Source: Reproduced with permission from Chapman et al. [11]/Royal Society of Chemistry/CC BY 3.0.

Table 4.1 Rate constants for Diels–Alder reaction between cyclopentadiene and butanone.

Solvent	Additive	$k_2 \times 10^5$ (M^{-1} s^{-1})[a)]
Isooctane	None	5.94 ± 0.3
MeOH	None	75.5
H_2O	None	4400 ± 70
H_2O	LiCl (4.86 M)	10 800
H_2O	Guanidinium chloride (4.86 M)	4300
H_2O	β-Cyclodextrin (10 mM)	10 900
H_2O	α-Cyclodextrin (10 mM)	2610

a) Second-order rate constants.
Source: Rideout and Breslow [7]/American Chemical Society.

compared to water. These observations were explained by the "salting out" effect of LiCl, i.e. its ability to push out nonpolar materials from water, and the "salting in" effect of guanidinium chloride (Table 4.1), i.e. its ability to change the structure of water so that it decreases hydrophobic interactions and increases the solubility of nonpolar materials [12].

This observation has often been tested and exploited in the literature of water-accelerated reactions [13]. The rationale was that the higher ionic strength of the aqueous phase reduces the solubility of the reactants and pushes them together into the organic emulsion droplets, leading to higher effective concentration and faster reaction. The logical problem here is that: (i) in many reactions, at least one reactant is significantly water soluble; and (ii) there is a limit on the effective concentration of reactants inside the organic droplets, which will be quickly reached. Breslow estimated the concentration under neat droplet conditions to be four times that of organic solvent, which can account for a 16 times increase of reaction rate but not the >700 times increase observed under water-accelerated reaction conditions [7]. Lastly, the literature includes cases wherein the chemical

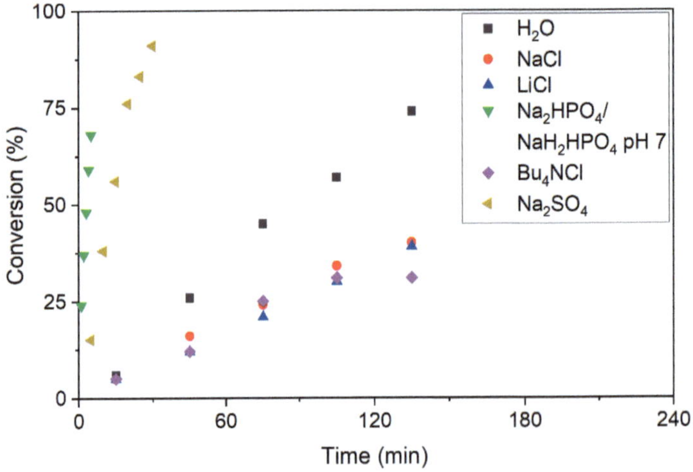

Figure 4.4 Effects of salts on the kinetics of a Henry reaction. Source: Reproduced from Maltby et al. [8]/American Chemical Society/CC BY 4.0.

nature of the salt, rather than its impact on the ionic strength of the aqueous phase, can have a significant effect on the reaction rate. For example, the reaction in Scheme 4.1 was found to proceed significantly faster in a solution of Na_2SO_4 instead of a solution of NaOTs [5].

The complex salt effects can, at least partially, be attributed to the partition of the starting materials/reagents and the direct involvement of the salt in the reaction mechanism, e.g. a proton transfer step. In the Henry reaction in Scheme 4.2, the effect of salts on the reaction kinetics was studied in detail (Figure 4.4).

In this reaction, all chloride salts (NaCl, LiCl, and Bu_4NCl) resulted in a similar decrease in reaction rate due to the increased ionic strength of the aqueous phase and the decreased aqueous solubility of nitromethane. This led to a slower enolization of nitromethane **6**, which was the rate-determining step (RDS) (*vide infra*). On the other hand, using an aqueous phosphate buffer at pH 7 instead of water led to a very rapid reaction, thanks to the well-documented ability of a phosphate anion to act as a proton transfer reagent/catalyst in the enolization of nitromethane [14, 15]. An unexpected result was obtained with sodium sulfate as an additive. Sodium sulfate is better than NaCl at salting out nitromethane from the aqueous phase, and its addition should have resulted in a slower reaction [16], but an acceleration in reaction rate was observed instead. A DFT study by Nguyen explained this by the catalytic activity of HSO_4^-, which is present in minute concentrations at pH 7, in the enolization of nitromethane (*vide infra*).

4.3 Mechanistic Considerations

The discussion above on the effects of salt brings our attention to the mechanistic aspects of water-accelerated reactions. Despite the obvious sustainability benefits of these reactions, mechanistic understanding based on experimental evidence is scarce. Neither of the two most common mechanistic tools for organic chemistry, i.e. spectroscopic observation of intermediates and kinetic modeling of the reaction, are readily applicable to these complex systems, which are subjected to a mixture of physical and chemical controls. NMR characterization of intermediates is difficult due to the multiphasic reaction mixture and the inaccessible lower temperature in D_2O (melting point 3.8 °C). Kinetic profiling of these reactions is nontrivial, again due to their multiphasic nature and the difficulties in reproducing the mixing efficiency and temperature control using standard equipment in organic chemistry laboratories. Finally, reported water-accelerated reactions include many types of organic reactions, e.g. pericyclic [17, 18], multicomponent [13], nucleophilic ring-opening [19–21], Mannich [22], aldol [23], Henry reactions [24], Lewis acid-catalyzed reactions [24], and organocatalytic reactions [18, 25], which follow very different mechanisms, with very different partitioning of reactants across the water–organic interface and phase behaviors. Thus, a deterministic view of a single model to explain the acceleration effect of water is unlikely to explain these reactions.

4.3.1 Rate Determining Steps

The clearly complex nature of water-accelerated reactions means that the RDS can be different between different reactions or even the same reaction under different conditions. Based on the current reported examples where water acceleration can be experimentally confirmed, the RDS can be one of the following:

(i) Mass transfer of the reactants to the water–organic interface
(ii) Intrinsic chemical RDS without the involvement of water molecules
(iii) Intrinsic chemical RDS with the involvement of water molecules

These different types of water-accelerated processes will be discussed in detail in the sections below.

4.3.1.1 Mass Transfer of the Reactants to the Water–Organic Interface as the Rate Determining Step

These are reactions where rate acceleration happens through interaction of the transition state with water molecules near the water–organic interface (scenario (b), (c), and (d) in Figure 4.1). Thus, the reaction rate depends on the availability of the reactants near the interface, and the reaction kinetics are mixing dependent. These reactions should show different rate increases with different mixing powers and a zero-order reaction at low mixing power, i.e. the reaction rate is limited by the surface area. In cases where the reactants are partitioned into different phases, a phase transfer catalyst will effectively accelerate the reaction.

Huck provided important experimental evidence of this type of process with the innovative use of plug flow reactors [26]. By controlling the ratio of the organic and aqueous stream flow rates at the mixing point (a polychlorotrifluoroethylene junction), the size of the organic droplets/plugs in the flow tubing was precisely controlled. There was also no cross-mixing between the droplets, making the design ideal to study the effect of mass transfer to the water–organic interface in

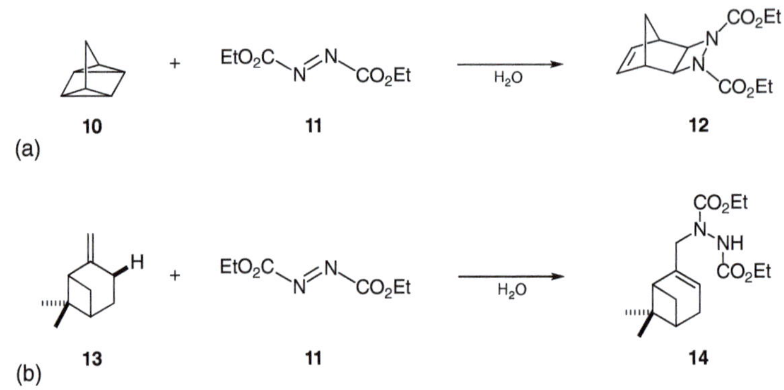

Scheme 4.4 (a) Cycloaddition of quadricyclane with DEAD. (b) The ene reaction between β-pinene and DEAD. Source: Mellouli et al. [26]/John Wiley & Sons.

water-accelerated reactions. Both the reactions between diethyl azodicarboxylate (DEAD) **11** and quadricyclane **10** and between DEAD and β-pinene **13** (Scheme 4.4) showed strong dependence on the surface-to-volume ratio of the droplet, highlighting the mass-transfer RDS in these reactions (Figure 4.5). It is worth noting that due to the biphasic nature of these water-accelerated reactions, the concentration terms used for rate equations can be misleading, and percentage conversion or effective concentration/organic loading have frequently been used instead.

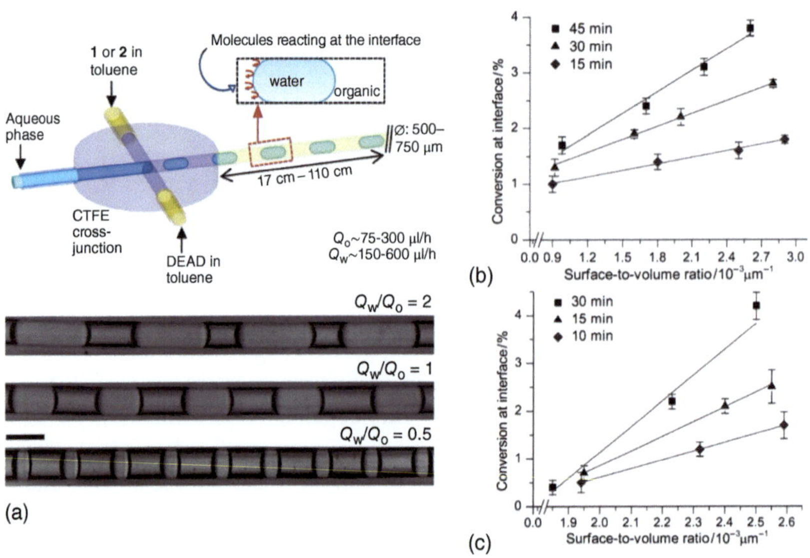

Figure 4.5 (a) Experimental set up: plugs form at the junction between a toluene solution containing the reagents ("organic phase") and an aqueous phase and travel down PTFE tubing where the reaction occurs; plug size and water surface area can be controlled by varying the relative flow rates of the two phases; (b) influence of the surface-to-volume ratio on the percent conversion of DEAD **11** and quadricyclane **10**; (c) influence of the surface-to-volume ratio on the percent conversion of DEAD **11** and β-pinene **13**. Source: Reproduced with permission from Mellouli et al. [26]/John Wiley & Sons.

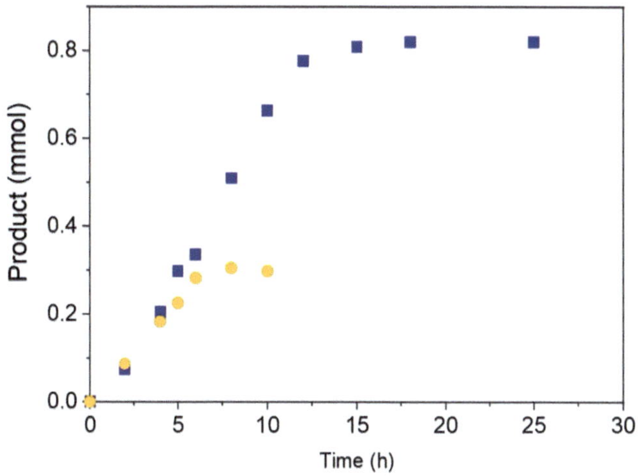

Figure 4.6 Kinetic profile of Henry reaction in batch at a 1 : 1 phase ratio using 25 equivalents (•) or 9 equivalents (■) of nitromethane **6**. Reaction conditions: 0.42 ml solution of *N*-methylisatin in nitromethane with 5 mol% trimethoxybenzene as an internal standard [(•) 0.121 g of **5**/1 ml; (■) 0.331 g of **5**/1 ml], 0.42 ml of 0.1 M phosphate buffer, 40 °C, 850 rpm, 0–25 minutes. Source: Reproduced from Maltby et al. [8]/American Chemical Society/CC BY 4.0.

An example is the Henry reaction in Scheme 4.2 [8]. The kinetics of the reaction are pseudo-zero order, in the presence of excess nitromethane. When the effective loading (concentration is an imprecise term when the organic reagents are not all soluble in water) of all reagents was increased two times in a phosphate buffer at pH 7, the reaction rate did not increase (Figure 4.6). At this high loading, similar to the effective concentration used in industrial processes, the mass transfer of the enolized nitromethane in the aqueous phase to the emulsion interface is the RDS and is only dependent on temperature and mixing power.

4.3.1.2 Intrinsic Chemical Rate Determining Step Without Involvement of Water Molecules

This class of water-accelerated reactions refers to those that are accelerated entirely based on the concentration effect of having all the organic materials inside emulsion droplets (scenario (a) in Figure 4.1). Predictably, reliably proven examples are very few. Nevertheless, Butler and Coyne reported an intriguing reaction, wherein a 1,3-dipole undergoes a [3+2] reaction with an enone (Figure 4.7) [27]. Depending on the solubility of the enone in water, the reaction was observed to be "in-water" or "on-water"; however, this was an unaided visual observation instead of a microscopic observation. In either case, the reaction yield did not differ significantly between reactions performed neat or in MeCN, water, or D_2O (Table 4.2). The authors also found that adding $HClO_4$ or $LiClO_4$ at different concentrations had little effect on the reaction.

The lack of an apparent interaction between the reaction and the aqueous phase was probed with kinetic isotope effect experiments for the reactions between the

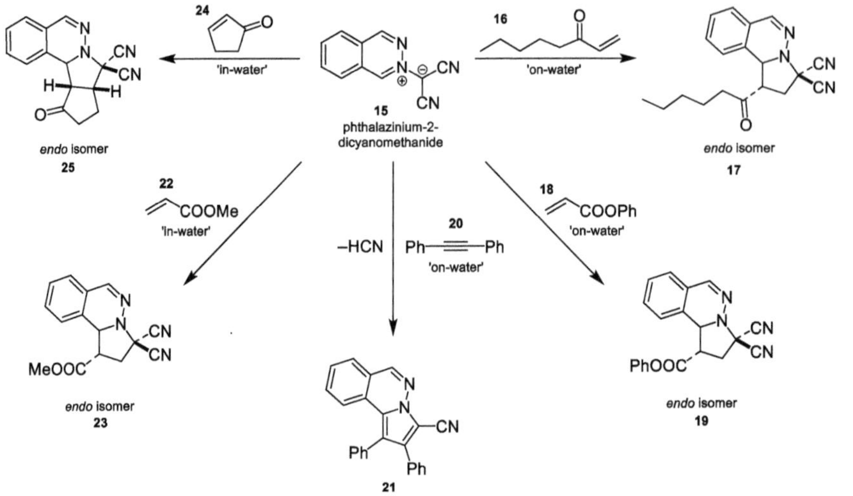

Figure 4.7 1,3-Dipolar cycloaddition reactions of pthalazinium-2-dicyanomethanide (**15**) with a range of dipolarophiles; 2-cyclopenten-1-one (**24**), 1-octen-3-one (**16**), methyl acrylate (**22**), diphenylacetylene (**20**), and phenyl acrylate (**18**), proceeding as either an on-water or an in-water reaction-dependent on the solubility of the dipolarophile. Source: Butler and Coyne [27]/American Chemical Society.

Table 4.2 1,3-Dipolar cycloaddition reactions of phthalazinium-2-dicyanomethanide (**15**) ($s \leq 5 \times 10^{-6}$ mol L^{-1}) at 25 °C with water soluble partners: 2-cyclopenten-1-one (**24**) and methyl acrylate (**22**); and water-insoluble partners: 1-octen-3-one (**16**), phenyl acrylate (**18**), and diphenylacetylene (**20**).

C=C partner	24		22		16		20		18	
Solvent	Time (h)	Yield (%)	Time (h)	Yield (%)	Time (h)	Yield (%)	Time (h)	Yield (%)	Time (h)	Yield (%)
Neat	12	84	1.25	79	1.5	74	24	92	0.5	87
MeCN	45	96	12	65	1.25	99	24	6	1.0	88
H_2O	36	95	12	91	1	95	24	91	0.5	90
D_2O	36	99	12	85	1	85	24	69	0.5	93

Source: Adapted from Butler and Coyne [27].

1,3-dipole and methyl vinyl ketone (in-water) and 1-octen-3-one **16** (on-water), respectively, in H_2O and D_2O. In either case, a very small observable kinetic isotope effect was detected, confirming that the reaction did not involve water molecules or interact with the aqueous phase (Figure 4.8).

4.3.1.3 Intrinsic Chemical Rate Determining Step with Involvement of Water Molecules

A more common scenario is where water is involved in the chemical RDS of the reaction. This provides a way of explaining the water-acceleration effect through hydrogen bonding of the water molecules to the transition state and/or water acting

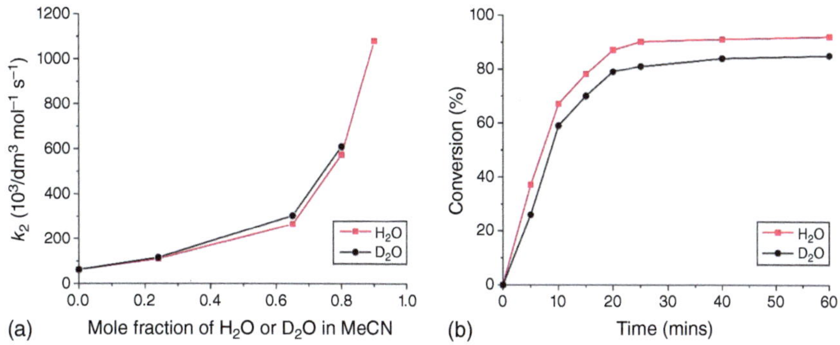

Figure 4.8 (a) Reaction rate constants for reaction between phthalazinium-2-dicyano-methanide and methyl vinyl ketone vs. mole fraction of H_2O (pink) or D_2O (black) in acetonitrile at 37 °C; (b) reaction profiles of reaction between phthalazinium-2-dicyano-methanide (**15**) and 1-octen-3-one **16** in H_2O (pink) and D_2O (black) at 20 °C. Source: Adapted from Butler and Coyne [27].

as a proton transfer reagent in the transition state. It is important to note that this explanation is not excluded from reactions that have mass transfer as the RDS, as explained in Figure 4.12 (*vide infra*). The clearest indications of this scenario are:

(i) A higher acceleration of reaction rate than that which can be accounted for by the concentration effect in neat droplets.
(ii) A non-zero-order kinetics.
(iii) A primary kinetic isotope effect between water and D_2O.

One example is the 1,4-addition of a malonate ester to *trans*-β-nitrostyrene, catalyzed by cinchona-based squaramide catalysts reported by Song (Figure 4.9) [28]. The authors found that using brine solution instead of dichloromethane led to a significant increase in the reaction rate of a non-zero-order reaction. The reaction time under water-accelerated reaction conditions is still 16 hours, ensuring that mass transfer is not the slowest step. Importantly, the two catalysts, QN-SQA and HQN-SQA, perform similarly in dichloromethane, but HQN-SQA gives a much faster reaction than using QN-SQA in brine. The more hydrophobic nature of HQS-SQA may have an impact on the interaction between the transition state of the RDS with water molecules, leading to the observed decrease in activity. The same group also observed improved stereoselectivity in asymmetric additions of thiols to phenylglyoxal, wherein the selectivity was attributed to the attenuation of the water content at the water–organic interface [29].

Another example is the Cu-catalyzed C–N coupling reaction between 5-nitroindole and iodobenzene, reported by Patil and coworkers (Figure 4.10) [30]. The reaction was found to proceed more rapidly in a mixture of water and dimethoxy ethylene (15% DME/H_2O, $k = 3 \times 10^{-5}$ s^{-1}) than in DME ($k = 3 \times 10^{-6}$ s^{-1}) or DMSO, using KOH as the base. Non-zero-order kinetics was observed, consistent with the long reaction time (20 hours), which rules out the mass transfer limit.

It is worth noting that reactions that have chemical RDS, which are relatively slow in both examples above, can switch to mass transfer RDS if the chemical

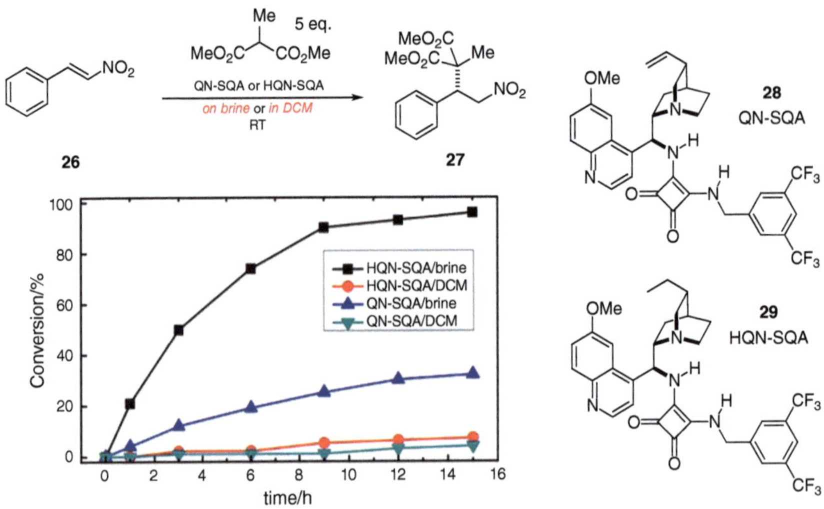

Figure 4.9 Reaction progress of the Michael addition of dimethyl 2-methylmalonate to **26** catalyzed by squaramide catalysts. Source: Adapted from Bae and Song [28].

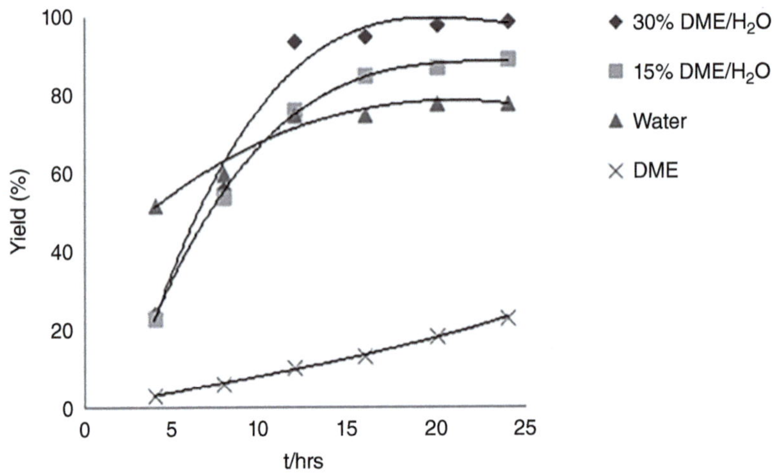

Figure 4.10 Plot of the % yield of product from the reaction of 5-nitroindole and iodobenzene vs. time [h]. Source: Malavade et al. [30]/John Wiley & Sons.

RDS is sufficiently accelerated through increased effective concentration, higher temperature, or a more active catalyst, or if the mass transfer rate is decreased due to slower mixing or increased viscosity. Thus, kinetic findings must be treated with care when significant changes to the reaction conditions or equipment are applied. The Henry reaction in Scheme 4.2 was found to have a kinetic isotope effect of $k_{H_2O}/k_{D_2O} = 2.41$ indicating a chemical RDS [8]. However, when the conditions were changed to a phosphate buffer (significantly increasing the reaction rate), using a lower excess of nitromethane, and an increased organic loading, the reaction was found to be mass transfer limited (Figure 4.6).

Finally, the interaction between the water molecules and the transition state can reasonably be expected to lead to a change in the chemo-, regio-, and stereoselectivity of the reaction compared to those obtained when performing in a non-interacting organic solvent. Examples have been reported for Passerini reactions [5, 31], Michael reactions [32], epoxide ring opening [33], and condensation [34].

4.3.2 Computational Studies of Water-Accelerated Reactions

Given the complexity of water-accelerated reactions and the difficulties in delineating the various effects of concentration, proton transfer, stabilization of transition states, and mass transfer, computational chemistry is an important tool in obtaining a fundamental understanding of these reactions. The widespread use of DFT modeling and increasing computational power allow researchers to directly probe the effect of water molecules on the transition state of the RDS in various reactions. Water involvement as a proton transfer reagent has been demonstrated with many reactions, under water-accelerated conditions or otherwise [35–37]. In this section, we can only highlight a few selected examples to illustrate important points on the methodology, their role in explaining experimental observations, and their limitations.

In 2012, Adamo and coworkers carried out a benchmark study of DFT methods on the structure and kinetics of a number of simple proton transfer reactions [38]. A range of DFT functionals were compared to the reference CCSD/cc-pVTZ method, using the basis set 6-311+G(2d,p) for optimizations and 6-311+G(3df,3pd) for single point calculations. The results indicated that ωB97X, B1LYP, BMK, and PBE0-DH were the best performing functionals, and a compromise needed to be made between accurate structure and accurate energy. One of the most significant improvements in DFT methods since 2012, which is relevant to hydrogen bonding, is the implementation of improved dispersion correction DFT-D3 in most common molecular modeling software [39]. However, due to the low symmetry and multi-molecule nature of the water-stabilized transition states, high-level DFT calculations are quite prone to failure. Thus, early literature on computational studies of water-accelerated reactions must be viewed in this context.

Most DFT studies of water-accelerated reactions focused on calculating the change in the activation energy barrier ($\Delta\Delta G^{\ddagger}$), going from organic solvents into water/organic conditions. These water-accelerated conditions are represented by explicit water molecules, which can directly interact with the transition state, often through hydrogen bonding, and stabilize the developing charges or charge separation. Jorgensen and coworkers carried out the first *ab initio* calculations (MP2/6-31G*) for the aqueous Diels–Alder reaction between cyclopentadiene **30** and butenone **31** by using one explicit water molecule in 1994 (Scheme 4.5, reaction (1)) [40]. They found that the transition state with hydrogen bonding between water and butenone

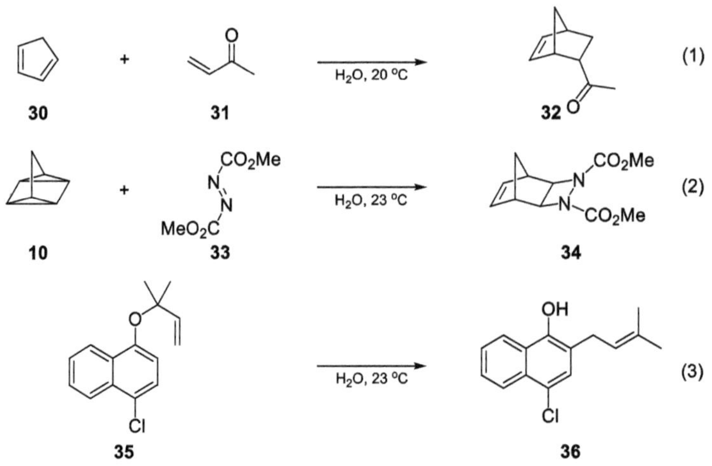

Scheme 4.5 Water-accelerated reactions reported by Sharpless and coworkers. Source: Adapted from Narayan et al. [1].

is 1.5–2.0 kcal mol^{-1} more favorable than the one without. Jung and Marcus reported their DFT studies in 2007, using a relatively low-level functional and basis set, i.e. B3LYP/6-31+G(d), by current standards [41]. For the reaction between DMAD **33** and quadricyclane **10** (Scheme 4.5, reaction (2)), a $\Delta G^{\ddagger} = -7.5$ kcal mol^{-1} was obtained going from neat conditions to aqueous surface conditions by including three water molecules in the transition state (Figure 4.11). A recent DFT study by Nguyen showed that M06-2x/def2-SVP method can reasonably reproduce the observed rate acceleration in reaction for both of these reactions and a Claisen rearrangement (Scheme 4.5, reaction (3)), which were published together in the seminal publication by Sharpless [1, 8]. Other than the more accurate DFT method, four water molecules were included in the transition states for reactions (1) and (2) instead of one and three, respectively. This highlights a key challenge in DFT studies of water-accelerated reactions: there is no specific guideline on the right number of water molecules, and choosing involves trial-and-error experimentation with the transition state. A comprehensive comparison between different numbers of water molecules is still potentially prohibitive in the near future. The other practical challenge in these studies, which is perhaps less discussed, is the selection of the solvent (water vs. organic solvent) and solvent models in modeling software packages. Again, there is little consensus and the solubility of each reactant may significantly influence this decision.

Despite these challenges, DFT studies can provide valuable insights into water-accelerated reactions. Nguyen and coworkers used the DFT calculated $\Delta G^{\ddagger}$ to explain the unusual acceleration effect of Na_2SO_4, despite its well-known salting out effect, in their Henry reaction. Even at pH 7, a small concentration of $[HSO_4^-] = 10^{-5}$ M is expected in the aqueous phase, and DFT studies showed a 7.1 kcal mol^{-1} decrease in $\Delta G^{\ddagger}$ for the reaction catalyzed by water molecules vs. the reaction catalyzed by HSO_4^- anion (Figure 4.12).

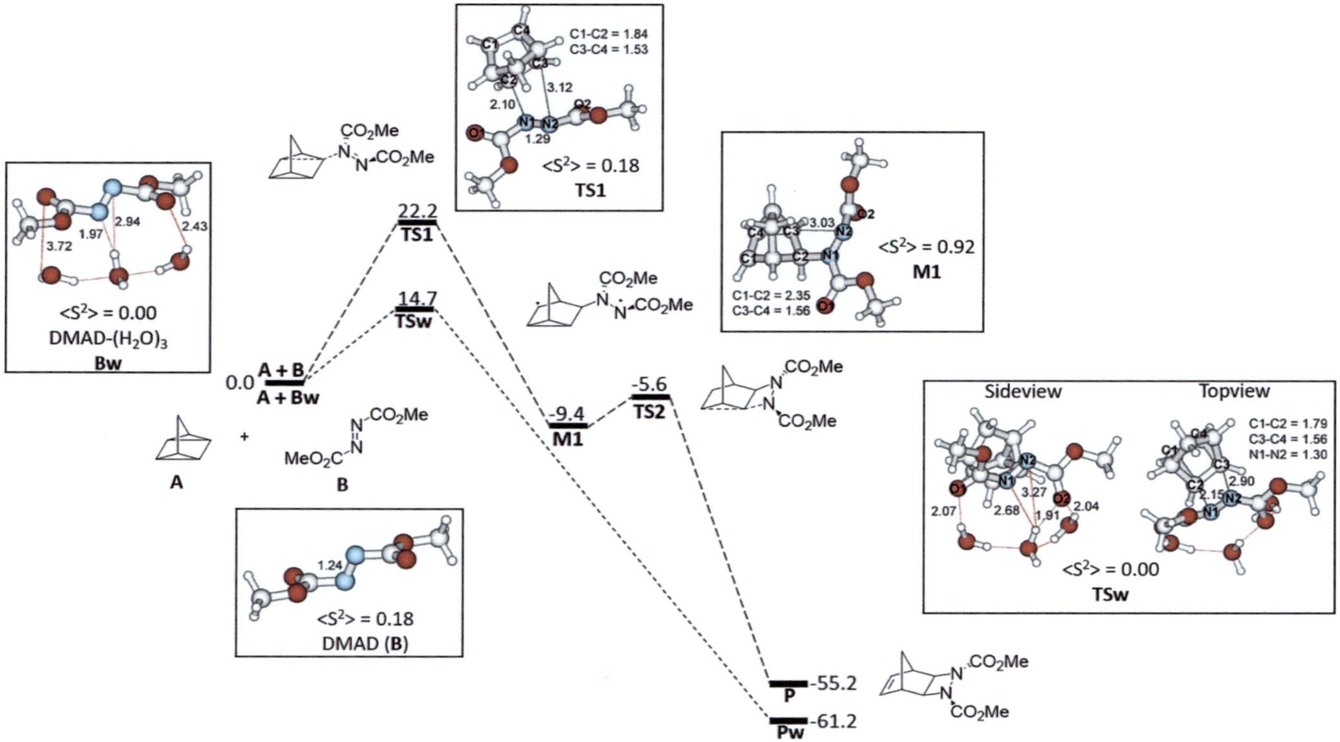

Figure 4.11 Energy diagram (in kcal/mol) of the cycloaddition of DMAD and quadricyclane under neat and aqueous surface conditions and structures for the reactants, transition states, and intermediates with and without water molecules. Source: Adapted from Jung and Marcus [41].

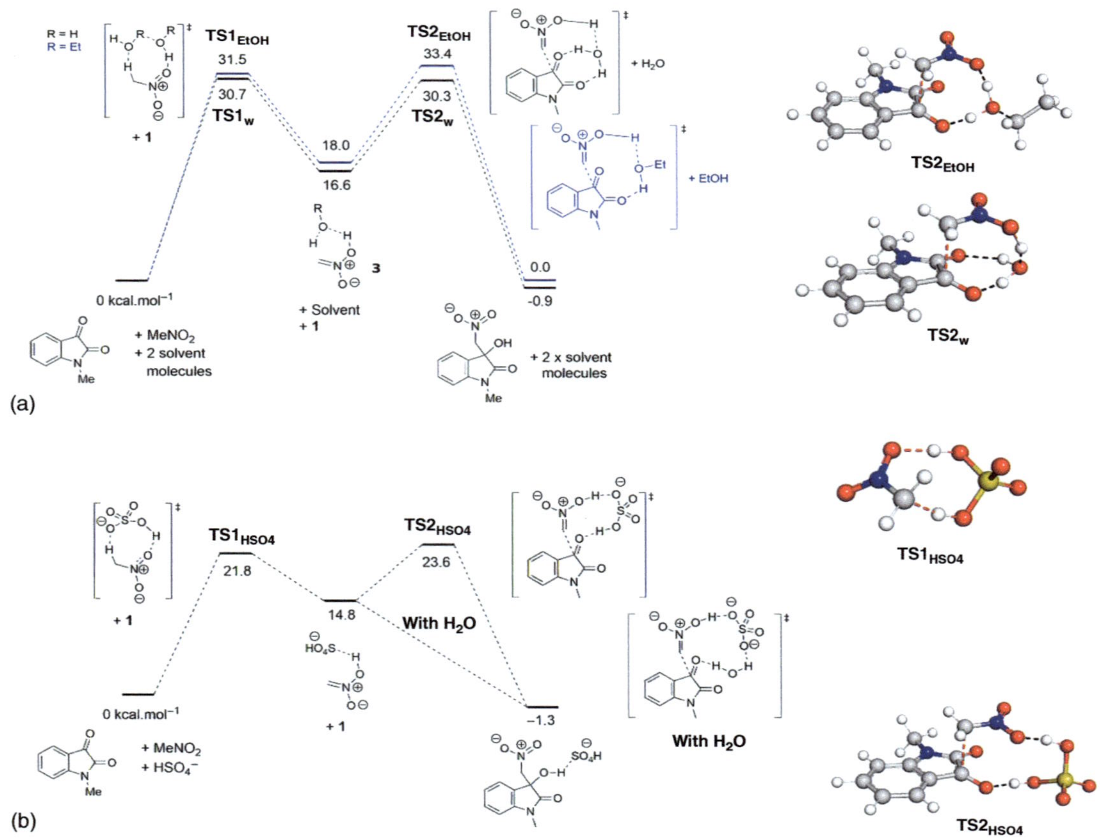

Figure 4.12 Energy profile of Henry reaction under (a) water-accelerated conditions, or in ethanol, and (b) in the presence of HSO_4^-. Source: Adapted from Maltby et al. [8].

4.4 Green Processes Based on Water-Accelerated Reactions

While the full discussion of the green chemistry aspects of water-accelerated reactions is covered elsewhere in this book, it is impossible to discuss their physicochemical details without discussing their impact on the sustainability of the process. Performing a reaction using water as the reaction medium does not automatically make the chemical process based on that reaction, which includes workup, purification of the product, and waste disposal, a green process. The sustainability of water-accelerated processes needs to be made in a holistic manner for the whole process, using currently accepted green metrics, if not Life Cycle Assessment. Water-accelerated reactions do inherently have the benefit of improved space-time yield (STY) due to the higher reaction rate, particularly if higher loadings of organic materials can be used without compromising the purity of the product. The use of water as a reaction medium is also highly beneficial if it can replace hazardous solvents such as carbon tetrachloride, *N*-methyl-2-pyrrolidone, or dimethylacetamide. Nevertheless, the workup and purification of water-accelerated reactions must be carefully assessed when considering their green credentials.

Unless the organic product can be readily separated from the aqueous phase in high purity, as shown in Figures 4.3 and 4.13 [11, 28], an organic extraction with additional organic solvent is likely required at the workup stage. Thus, the partitioning of the starting materials and the product in the extraction, not necessarily during the reaction, can have a significant impact on process mass intensity (PMI) [42]. If an excess of a starting material is used or the reaction yield is not near quantitative, the product will require further purification, and the aqueous phase may be contaminated with organic compounds. The latter can have a significant impact on the possibility of water recycling, which is highly desirable for green synthetic processes.

Figure 4.13 Pictures of the catalyst recovery (HQN-SQA) and product separation procedure from the crude mixture of the Michael addition in Figure 9. (a) Before starting the reaction; (b) after completion of the reaction; (c) after the addition of methylcyclohexane; and (d) after filtration, biphasic filtrate (product (org.)/brine (aq)) and recovered catalyst. Source: Reproduced with permission from Bae and Song [28]/American Chemical Society.

However, when these complications are overcome, truly sustainable processes can be developed, as demonstrated in the continuous flow version of the reaction in Figure 4.2, giving PMI = 4 and STY = 0.64 kg L^{-1} h^{-1} [10].

Lastly, a new class of "on-water" reactions has emerged in recent years based on photochemical reactions [43–46]. These exploit the near transparent nature of water in the near UV and visible regions, rather than just the water-acceleration effect. They are beyond the scope of this current chapter but will be a worthy topic for review in the future.

4.5 Summary and Conclusions

Water-accelerated reactions, which include a diverse list of organic transformations, have great potential in green chemical manufacturing processes. Perhaps because of that, their physicochemical aspects can be very complex and not necessarily static for the duration of the reactions. The general consensus is that, in most cases, interaction of the transition state with water molecules leads to lower activation energy barriers and the rate acceleration effect, which are well described by modern molecular modeling techniques. Examples of the purely concentration effect inside the organic emulsions are known, but few. The partition of reactants and products in the organic and aqueous phases can lead to different physicochemical scenarios, which may lead to mass transfer to the water–organic interface being the RDS. This partition and phase behavior are also dependent on the reaction conditions, particularly the amount of organic materials being put into water. The same partitioning and solubility of reactants and products are critical in designing effective workup/purification protocols to achieve truly sustainable chemical processes.

References

1 Narayan, S., Muldoon, J., Finn, M.G. et al. (2005). *Angew. Chem. Int. Ed.* 44: 3275–3279.
2 Brogan, A.P., Dickerson, T.J., and Janda, K.D. (2006). *Angew. Chem. Int. Ed.* 45: 8100–8102.
3 Hayashi, Y. (2006). *Angew. Chem. Int. Ed.* 45: 8103–8104.
4 Blackmond, D.G., Armstrong, A., Coombe, V., and Wells, A. (2007). *Angew. Chem. Int. Ed.* 46: 3798–3800.
5 Shapiro, N. and Vigalok, A. (2008). *Angew. Chem. Int. Ed.* 47: 2849–2852.
6 Xing, W., Yin, M., Lv, Q. et al. (2014). *Oxygen Solubility, Diffusion Coefficient, and Solution Viscosity* (ed. W. Xing, G. Yin, and J.B.T. Zhang), 1–31. Amsterdam: Elsevier.
7 Rideout, D.C. and Breslow, R. (1980). *J. Am. Chem. Soc.* 102: 7816–7817.
8 Maltby, K.A., Sharma, K., Short, M.A.S. et al. (2023). *ACS Sustain. Chem. Eng.* 11: 8675–8684.

9 SaiPrathima, P., Srinivas, K., and Mohan Rao, M. (2015). *Green Chem.* 17: 2339–2343.

10 Kitanosono, T. and Kobayashi, S. (2020). *Chem. Eur. J.* 26: 9408–9429.

11 Chapman, M.R., Kwan, M.H.T., King, G.E. et al. (2016). *Green Chem.* 18: 4623–4627.

12 Von Hippel, P.H. and Schleich, T. (1969). *Acc. Chem. Res.* 2: 257–265.

13 Pirrung, M.C. and Das Sarma, K. (2005). *Tetrahedron* 61: 11456–11472.

14 Zhang, L., Yuan, P., Chen, J., and Huang, Y. (2018). *Chem. Commun.* 54: 1473–1476.

15 Vilčiauskas, L., Tuckerman, M.E., Bester, G. et al. (2012). *Nat. Chem.* 4: 461–466.

16 Hyde, A.M., Zultanski, S.L., Waldman, J.H. et al. (2017). *Org. Process Res. Dev.* 21: 1355–1370.

17 Brandes, E., Grieco, P.A., and Gajewski, J.J. (1989). *J. Org. Chem.* 54: 515–516.

18 González-Cruz, D., Tejedor, D., de Armas, P. et al. (2006). *Chem. Commun.* 26: 2798–2800.

19 Vilotijevic, I. and Jamison, T.F. (2007). *Science (80-.)* 317: 1189–1192.

20 Azizi, N. and Saidi, M.R. (2005). *Org. Lett.* 7: 3649–3651.

21 Hajra, S., Singha Roy, S., Aziz, S.M., and Das, D. (2017). *Org. Lett.* 19: 4082–4085.

22 Song, C.E., Park, S.J., Hwang, I.-S. et al. (2019). *Nat. Commun.* 10: 851.

23 Kitanosono, T. and Kobayashi, S. (2015). *Chem. Asian J.* 10: 133–138.

24 Du, Z.-H., Yuan, M., Tao, B.-X. et al. (2022). *Synthesis (Stuttg).* 54: 5461–5470.

25 Mase, N., Nakai, Y., Ohara, N. et al. (2006). *J. Am. Chem. Soc.* 128: 734–735.

26 Mellouli, S., Bousekkine, L., Theberge, A.B., and Huck, W.T.S. (2012). *Angew. Chem. Int. Ed.* 51: 7981–7984.

27 Butler, R.N. and Coyne, A.G. (2015). *J. Org. Chem.* 80: 1809–1817.

28 Bae, H.Y. and Song, C.E. (2015). *ACS Catal.* 5: 3613–3619.

29 Park, S.J., Hwang, I.-S., Chang, Y.J., and Song, C.E. (2021). *J. Am. Chem. Soc.* 143: 2552–2557.

30 Malavade, V., Patil, M., and Patil, M. (2020). *Eur. J. Org. Chem.* 2020: 561–569.

31 Sela, T. and Vigalok, A. (2012). *Adv. Synth. Catal.* 354: 2407–2411.

32 Sim, J.H. and Song, C.E. (2017). *Angew. Chem. Int. Ed.* 56: 1835–1839.

33 Gentile, D., Floresta, G., Patamia, V. et al. (2020). *Org. Biomol. Chem.* 18: 1194–1203.

34 Quílez del Moral, J.F., Ruiz Martínez, C., Pérez del Pulgar, H. et al. (2021). *J. Org. Chem.* 86: 3344–3355.

35 Yamabe, S., Zeng, G., Guan, W., and Sakaki, S. (2014). *Beilstein J. Org. Chem.* 10: 1765–1774.

36 Liu, P., Li, C., Wang, S., and Wang, D. (2018). *J. Phys. Chem. B* 122: 3124–3132.

37 Yamabe, S. and Yamazaki, S. (2018). *Int. J. Quantum Chem.* 118: e25510.

38 Mangiatordi, G.F., Brémond, E., and Adamo, C. (2012). *J. Chem. Theory Comput.* 8: 3082–3088.

39 Grimme, S., Antony, J., Ehrlich, S., and Krieg, H. (2010). *J. Chem. Phys.* 132: 154104.

40 Blake, J.F., Lim, D., and Jorgensen, W.L. (1994). *J. Org. Chem.* 59: 803–805.

41 Jung, Y. and Marcus, R.A. (2007). *J. Am. Chem. Soc.* 129: 5492–5502.

42 Jimenez-Gonzalez, C., Ponder, C.S., Broxterman, Q.B., and Manley, J.B. (2011). *Org. Process Res. Dev.* 15: 912–917.
43 Tian, Y.-M., Silva, W., Gschwind, R.M., and König, B. (2024). *Science (80-.)* 383: 750–756.
44 Schulte, R., Löcker, M., Ihmels, H. et al. (2023). *Chem. Eur. J.* 29: e202203203.
45 Ishiyama, T., Tahara, T., and Morita, A. (2022). *J. Am. Chem. Soc.* 144: 6321–6325.
46 Dou, Q. and Zeng, H. (2023). *Curr. Opin. Green Sustain. Chem.* 40: 100766.

5

C–H Activation Reactions in Aqueous Medium

Feng Zhou[1] *and Chao-Jun Li*[2]

[1] *UCB Biosciences Inc., Department of Chemistry, 87 CambridgePark Drive, Cambridge, MA 02140, USA*
[2] *McGill University, Department of Chemistry, 801 Sherbrooke Street West, Montreal, Quebec H3A0B8, Canada*

5.1 Introduction

Carbon–hydrogen (C–H) bond activation has emerged as a transformative strategy in organic synthesis, revolutionizing the way complex molecules are assembled [1]. Historically, organic reactions have been relied on the pre-functionalization of substrates, leading to inefficiencies in both synthetic pathways and waste generation. However, the advent of C–H activation methodologies has provided chemists with a powerful tool to directly access and functionalize inert C–H bonds, thereby streamlining synthetic routes and reducing environmental footprint. To this end, chemists have designed and implemented strategies utilizing transition metal catalysis to achieve the C–H bond activation, thus propelling the advancement of various valuable chemical transformations. This research field is undoubtedly one of the fastest-growing areas, for addressing both synthetic efficiency and the enduring sustainability of chemical processes [2].

Traditionally, organic chemists have depended on organic solvents derived from fossil sources to develop organic chemical reactions [3]. The research on C–H bond activation has also naturally been conducted in organic solvents. As the concerns mount over the exhaustion of nonrenewable fossil resources and the environmental challenges linked to pollutions, the necessity for sustainable development has garnered increasing consensus within the chemistry community [4]. Shifting toward product development from renewable resources and enhancing utilization efficiency emerge as crucial strategies for advancing chemical sustainability for the future [5]. In response to this trend, recent progress has extended chemical transformations to include aqueous settings, marking significant shift toward greener and more sustainable synthetic methods [6].

Water, abundant and environmentally benign, offers a unique advantage as a solvent. However, using water as a solvent also brings significant challenges for organic transformations, especially for those transition-metal-catalyzed reactions [7]. Many organometallic species are susceptible to hydrolysis in water, leading to

Organic Transformations in Water: Principles and Applications, First Edition.
Edited by Gopinathan Anilkumar, Nissy Ann Harry, and Sankuviruthiyil M. Ujwaldev.

the breakdown of the metal-carbon bonds. Balancing the benefits and challenges requires thoughtful design with the consideration of the specific transition metals used, reaction conditions, and the desired outcome [8]. Nevertheless, researchers have unlocked novel strategies for catalyzing C–H bond activation with transition metals under aqueous conditions [9]. This chapter aims to provide a comprehensive overview of the recent developments in this rapidly evolving field, emphasizing the transformative impact of aqueous-phase C–H activation on the future of sustainable chemical synthesis, especially in terms of atom economy [10], step economy [11], and the E-factor [12].

5.2 C–H Bond Activation Reactions in Aqueous Medium

5.2.1 Activation of sp C–H Bonds

Over the past few decades, significant progress has been made in the field of transition-metal-catalyzed aqueous C–H bond activation. Due to their higher acidity, C(sp)–H bonds are particularly amenable to activation in aqueous environments. For instance, the well-known Sonogashira reaction was successfully conducted in aqueous media in the early 1990s [13]. Initially developed for small-molecule organic synthesis, the Sonogashira coupling also serves as a valuable tool for polymer synthesis, enabling the design and creation of tailored polymer architectures with desired properties. In 1998, Li's group reported the synthesis of arylethynylene polymer at high molecular weight in the aqueous medium with reversible hydrogel properties (Scheme 5.1) [14]. Subsequently, the investigation into the activation of C(sp)–H bonds in aqueous environments experienced rapid expansion, encompassing a wide range of transition metals that can activate terminal alkynes, leading to the generation of either π-complexes or metal acetylides (Scheme 5.2). These complexes are capable of undergoing reactions with other functional groups to facilitate the formation of new C–C bonds [15].

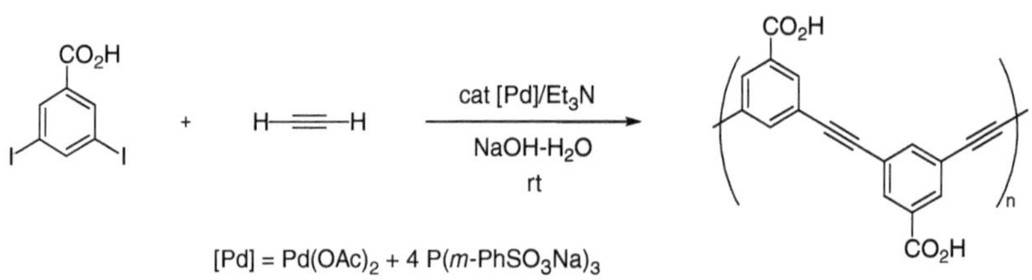

Scheme 5.1 Sonogashira coupling in aqueous medium for the synthesis of arylethynylene polymer.

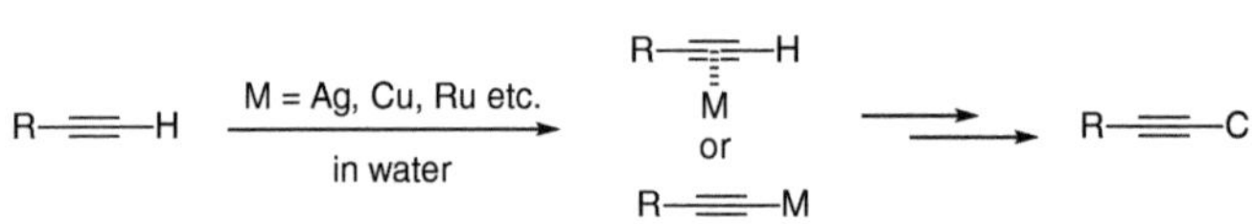

Scheme 5.2 C(sp)–H bond activation by transition metal catalysts.

In 2002, the aldehyde alkynylation in water catalyzed by a bimetallic Ru–In catalytic system was successfully realized by Li and coworkers (Scheme 5.3) [16].

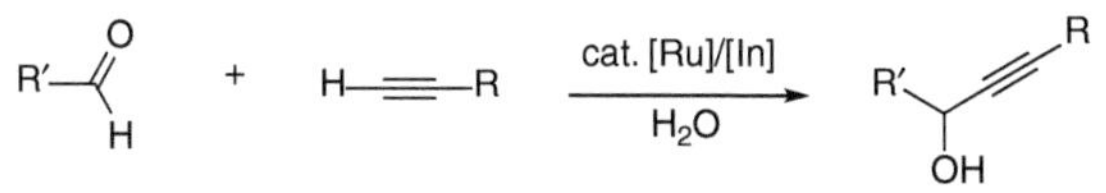

Scheme 5.3 The alkynylation of aldehydes catalyzed by bimetallic Ru-In catalysts in water.

The same group also achieved the convenient synthesis of propargylamines via aldehyde–alkyne–amine coupling (A^3 coupling), conducted either in water or under neat conditions (Scheme 5.4). This method successfully converted a diverse array of substituted aromatic and aliphatic imines into their respective propargylamines [17].

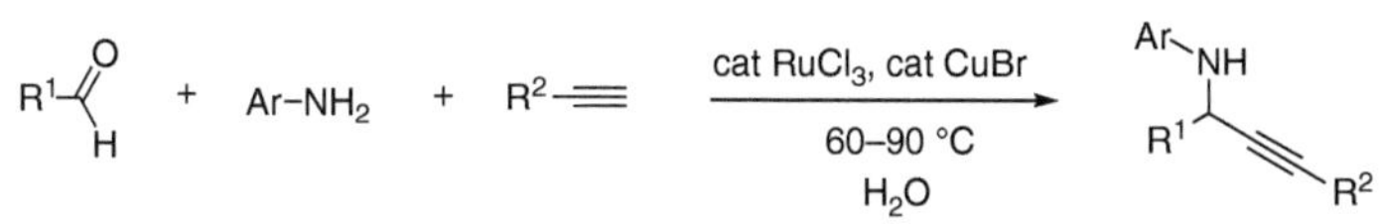

Scheme 5.4 Ru-Cu-catalyzed A^3 coupling for the synthesis of propargylamines in water.

Subsequently, Li and coworkers disclosed a highly efficient gold-catalyzed three-component coupling of aldehyde, alkyne, and amine via C–H activation in water [18]. This operationally simple protocol provided a variety of propargylamines in excellent yields. Both aromatic and aliphatic aldehydes and amines are viable substrates with water as the sole byproduct (Scheme 5.5). The selection of solvent significantly impacts the reaction outcomes. Water presented itself as a superior solvent, enabling remarkably clean reaction pathways with nearly quantitative yields. In contrast, the utilization of organic solvents tetrahydrofuran (THF), toluene, and dimethylformamide (DMF) led to reduced conversions and increased formation of byproducts. After the successful utilization of copper and gold as catalysts, it was discovered that the silver salts, such as AgCl, AgBr, and AgI exhibited promising catalytic activity for the three-component coupling reactions in aqueous environments [19].

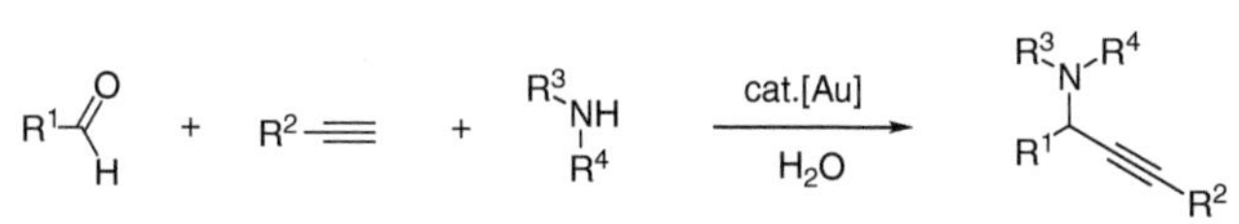

Scheme 5.5 Gold-catalyzed A^3 coupling via C–H activation in water.

Li's group also achieved the asymmetric addition of arylacetylenes to imines by employing a Cu/pyridine-oxazoline catalyst, yielding propargylamine with remarkable enantioselectivity (up to 84% ee in water) in water (Scheme 5.6) [20]. The imines were generated spontaneously within the reaction mixture in most cases, and the protocol was straightforward: combining an aldehyde, an aniline, and an alkyne with

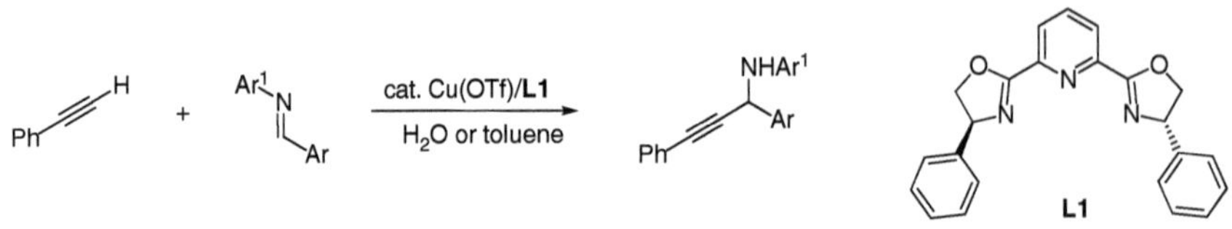

Scheme 5.6 Cu-catalyzed asymmetric addition of phenylacetylene to imines in water.

the catalyst in a single reaction vessel. Later, the excellent diastereoselectivity was achieved (up to >99 : 1), catalyzed by gold (III) salen complex in water by Che's group when the chiral prolinol derivatives were used [21].

Later, Li and coworkers described a method for alkynylation of imines and carbonyls in water with silver as a catalyst and phosphine as a ligand (Scheme 5.7). Without adding phosphine ligand, the exclusive formation of the aldehyde–alkyne–amine coupling product was achieved through C—H bond activation of terminal alkyne; conversely, the presence of a phosphine ligand resulted in the formation of aldehyde–alkyne coupling product in water [22].

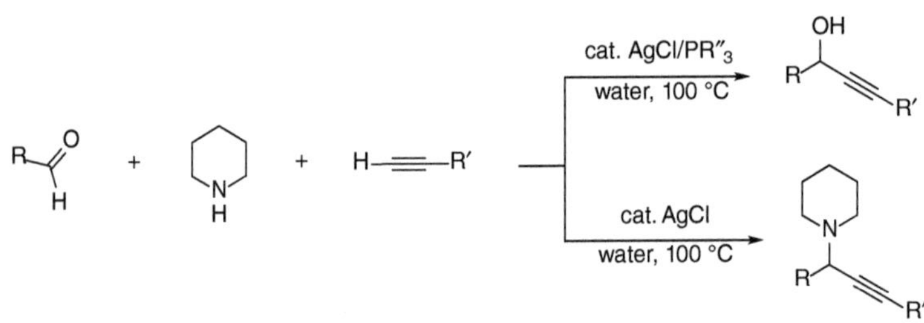

Scheme 5.7 Ag-catalyzed A^3 coupling in water and phosphorus ligand effect on the reactivity.

In addition, the A^3 coupling (aldehyde–alkyne–amine) can be further extended to the manipulation of amino acids and peptides under ambient aqueous conditions by Li and coworkers (Scheme 5.8) [23].

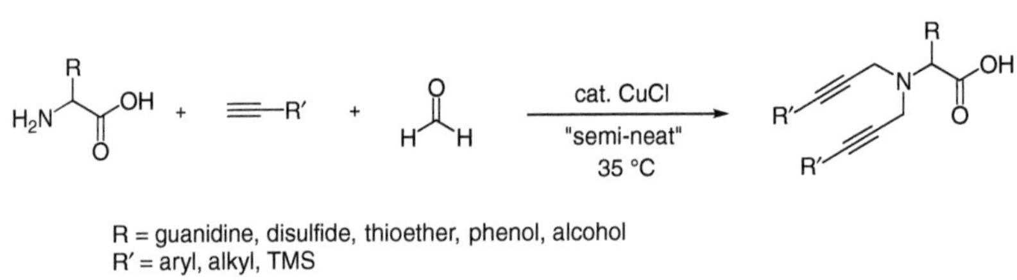

Scheme 5.8 Cu-catalyzed functionalization of amino acids and peptides via A^3 coupling.

Following that, the same group reported a highly efficient synthesis of 1-alkynyl-1H-isochromenes from the alkynylation–cyclization of terminal alkynes with *ortho*-alkynylaryl aldehydes catalyzed by a gold–phosphine complex via C–H bond activation in water (Scheme 5.9) [24].

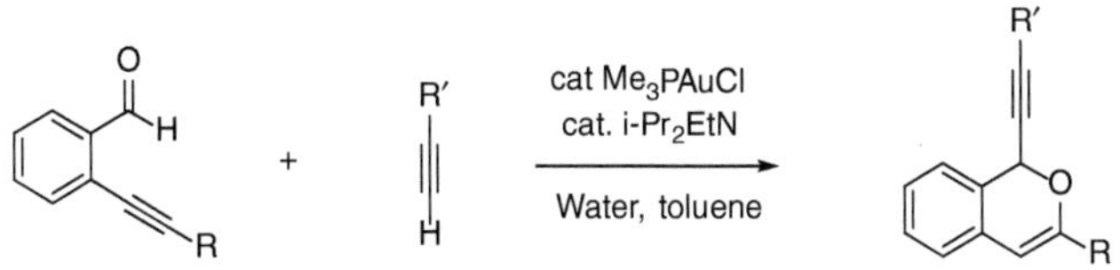

Scheme 5.9 Au-catalyzed cascade reaction of terminal alkyne with *ortho*-alkynylaryl aldehyde.

In 2002, Li's group achieved a CuBr-mediated coupling of alkynes with N-acylimines and N-acyliminium ions in water at slightly elevated temperature (Scheme 5.10) [25]. Two years later, Li and coworkers successfully developed a highly effective coupling of acid chlorides with terminal alkynes in water with sodium lauryl sulfate as surfactant (Scheme 5.11) [26].

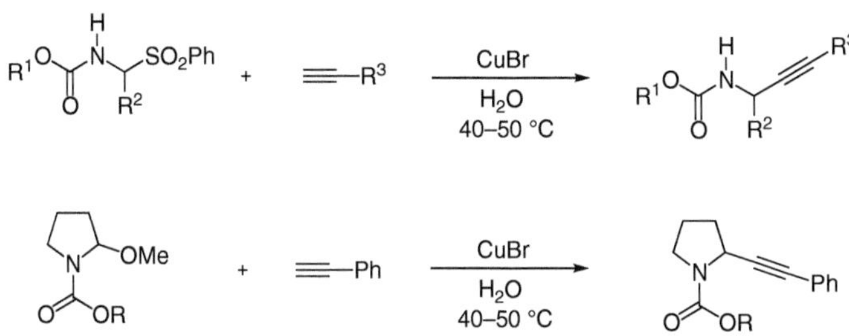

Scheme 5.10 Cu-mediated coupling of alkynes with N-acylimines and N-acyliminium ions.

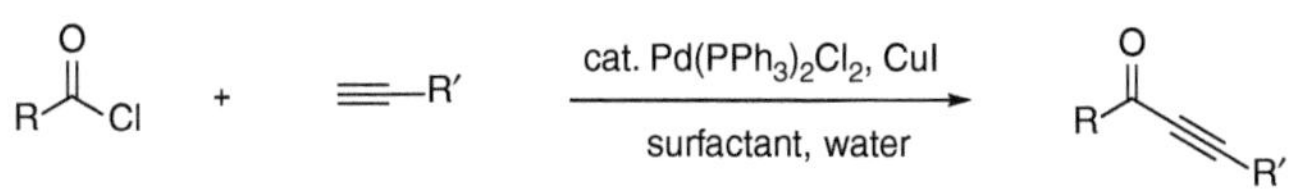

Scheme 5.11 Pd-catalyzed coupling reactions of acid chlorides with terminal alkynes in water.

In 2008, they presented a direct alkynylation reaction of trifluoromethyl ketones catalyzed by AgF/PCy_3 in water (Scheme 5.12). It was observed that the addition of ligand PCy_3 increased the catalyst activity significantly [27].

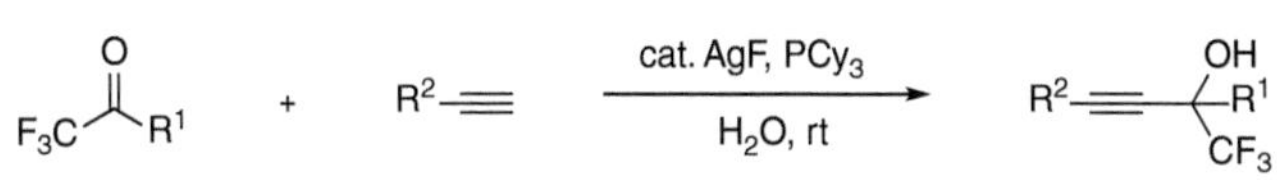

Scheme 5.12 Ag-catalyzed alkynylation of ketones in water.

In 2009, Li's group reported a Cy_3P–silver-complex-catalyzed alkynylation–cyclization of salicylaldehydes with terminal alkynes, yielding the substituted 2,3-dihydrobenzofuran-3-ol derivatives (Scheme 5.13) [28]. Interestingly, the Z(trans)/E(cis) of exocyclic double bond (*Z*/*E*) stereoselectivity control was

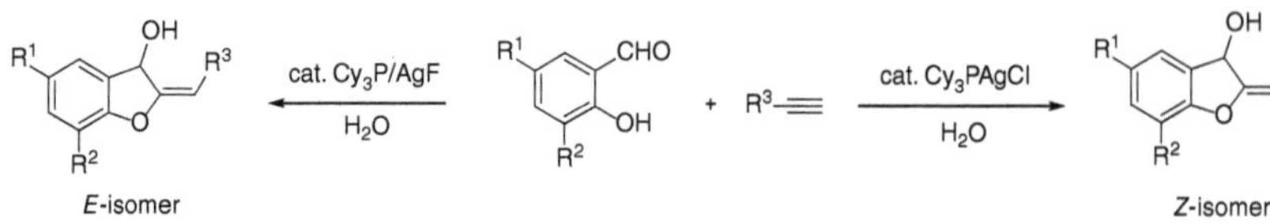

Scheme 5.13 Ag-catalyzed synthesis of dihydrobenzofuran derivatives from salicylaldehydes with terminal alkynes.

primarily influenced by the counter anion in the silver catalyst. Furthermore, a direct addition of terminal alkynes to simple cyclic ketones catalyzed by silver and Ruphos in water was achieved by Li and coworkers in 2012, giving rise to a series of valuable propargylic alcohols (Scheme 5.14) [29].

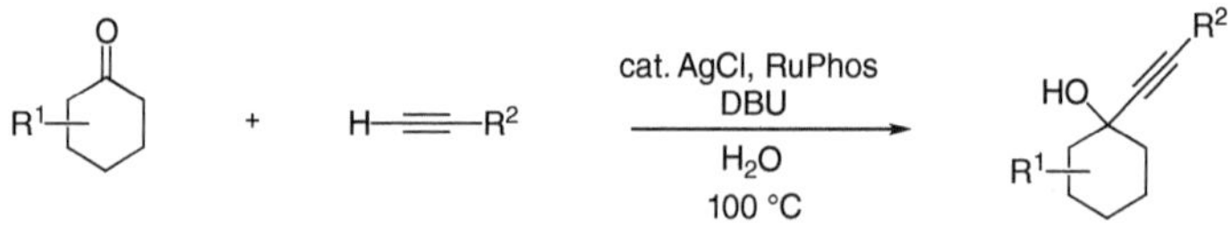

Scheme 5.14 The alkynylation of cyclic ketones catalyzed by silver-Ruphos in water.

Meanwhile, the addition of terminal alkynes to isatins via on-water strategy C(sp)–H activation catalyzed by Ag–NHC complex was developed, providing the efficient synthesis of 3-hydroxy-3-ethynylindolin-2-ones under an air atmosphere (Scheme 5.15) [30]. A significant improvement in reaction rate was observed in the aqueous heterogeneous catalytic system.

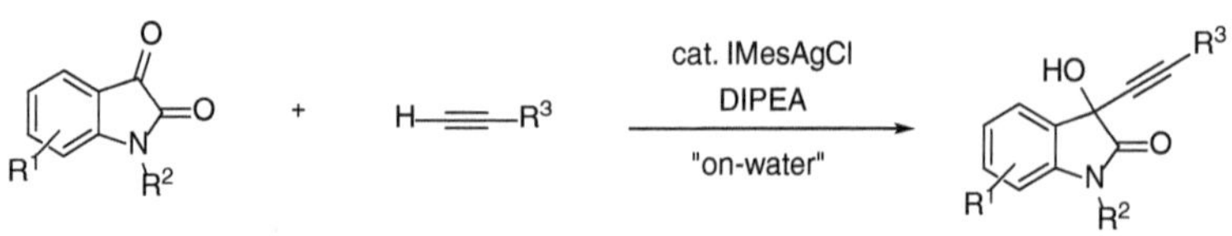

Scheme 5.15 Ag-NHC catalyzed alkynylation of isatins via on-water approach.

In 2015, Li's group disclosed a Pd-N-heterocyclic carbene (NHC)-catalyzed alkynylation of Morita–Baylis–Hillman carbonates with (triisopropylsilyl)acetylene in aqueous medium (Scheme 5.16) [31]. This method provided an efficient synthesis of 1,4-enynes.

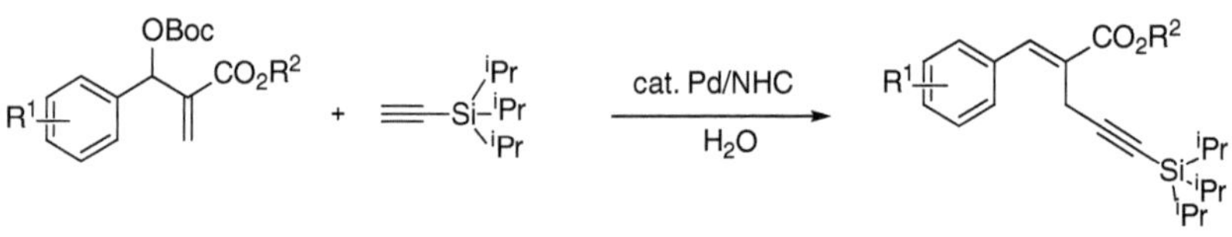

Scheme 5.16 Pd-NHC catalyzed alkynylation of MBH carbonates in water.

For the conjugate addition reactions involving terminal alkynes and unsaturated carbonyl compounds, Carreira and coworkers documented a method in which alkynyl copper reagents, formed through the reaction of terminal alkynes with catalytic $Cu(OAc)_2$ in the presence of sodium ascorbate, reacted with Meldrum's acid-derived Michael acceptors at ambient temperature in aqueous environments, yielding the respective products in high yields (Scheme 5.17) [32]. The enantioselective version in water was achieved by the groups of Carreira and Aponick using atropoisomeric *P*, *N*-ligands and the *ee* can reach up to 98% [33].

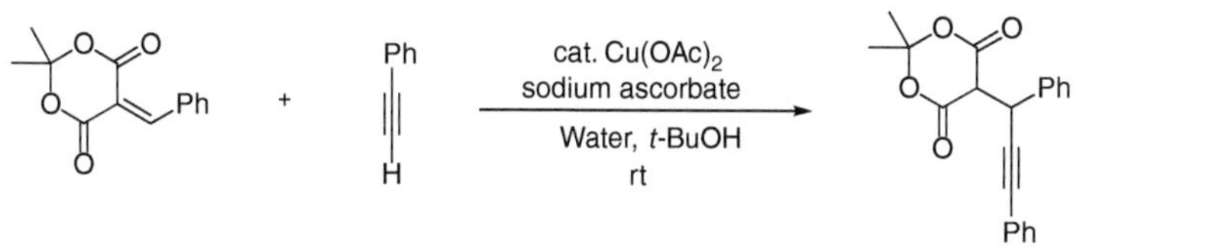

Scheme 5.17 Cu-catalyzed conjugate addition of terminal alkyne and alkylidene Meldrum's acid.

Li and coworkers disclosed a Pd-catalyzed conjugate addition of terminal alkynes to enones in water, yielding the corresponding γ,δ-alkynyl ketones in high yields (Scheme 5.18) [34]. Later, Li's group achieved the 1,4-addition of terminal alkynes to acrolein catalyzed by $Pd(OAc)_2/PMe_3$ on water (Scheme 5.19) [35]. The experimental conditions were fine-tuned to promote 1,4-addition over the acrolein polymerization. A variety of 4-alkynals were synthesized in moderate-to-good yields.

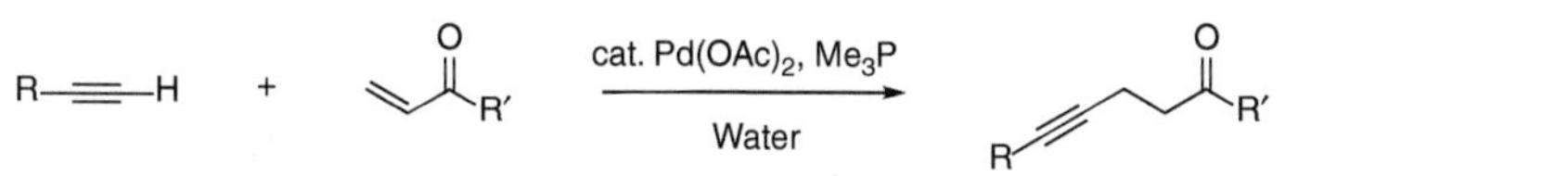

Scheme 5.18 Pd-catalyzed conjugate addition of terminal alkynes with enones in water.

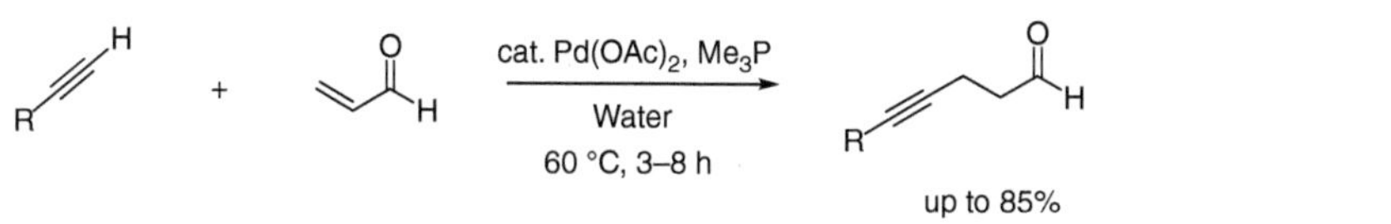

Scheme 5.19 Pd-catalyzed conjugate addition of terminal alkynes to acrolein in water.

Li's group published a bimetallic Cu–Pd-catalyzed addition of terminal alkynes to activate alkynes in water, affording the synthetically highly valuable conjugated enyne products (Scheme 5.20). Notably, this chemical transformation was devoid of any interference from the homocoupling of the terminal alkynes [36].

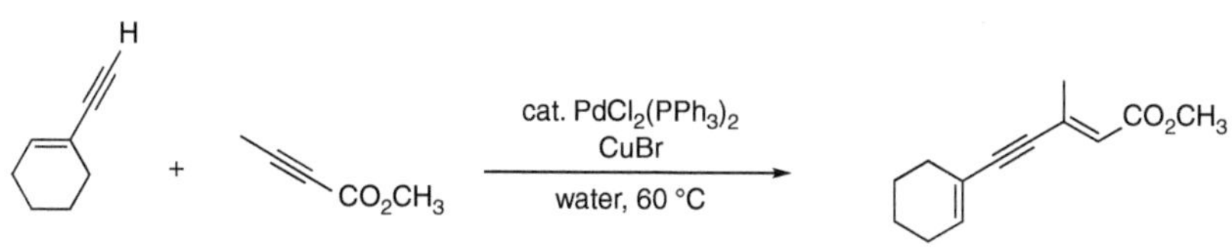

Scheme 5.20 Cu-Pd-catalyzed nucleophilic addition of terminal alkyne to activated alkyne.

In 2015, Li's group disclosed a greener approach for the coupling of terminal alkynes and alkyl iodides, in which the transformation proceeded smoothly under ultraviolet irradiation in basic aqueous media without using any transition metals (Scheme 5.21) [37].

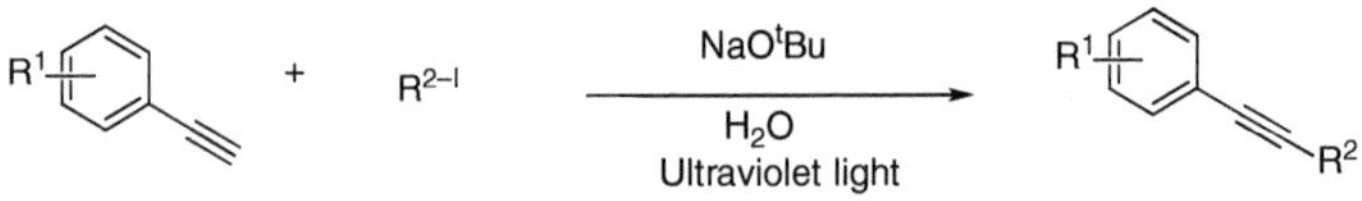

Scheme 5.21 Transition-metal-free coupling of alkynes and alkyl iodides under UV irradiation.

5.2.2 Activation of sp^2 C–H Bonds

For less acidic $C(sp^2)$–H activation, the strategies for its activation encompass a diverse array of approaches aimed at selectively activating and functionalizing sp^2 hybridized carbon-hydrogen bonds. These strategies often involve the use of transition metals, directing groups (DGs), and reaction conditions tailored to achieve site-selective C–H bond functionalization. In 2006, Li and coworkers reported the $In(OTf)_3$-catalyzed addition of aromatic *C*-nucleophiles to 1,4-quinone derivatives in water for the synthesis of mono- and diaryl-substituted 1,4-quinones, and water was found to be beneficial for the transformation (Scheme 5.22) [38].

In 2010, Li and coworkers disclosed a highly efficient coupling of indole compounds with 1,4-benzoquinones using on-water approach without using any catalyst or additive (Scheme 5.23). The optimal yields of the products were

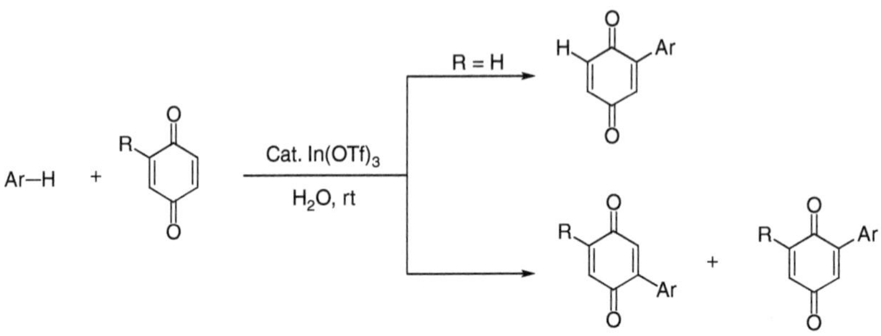

Scheme 5.22 The synthesis of aryl-substituted 1,4-benzoquinone via $In(OTf)_3$-catalyzed addition.

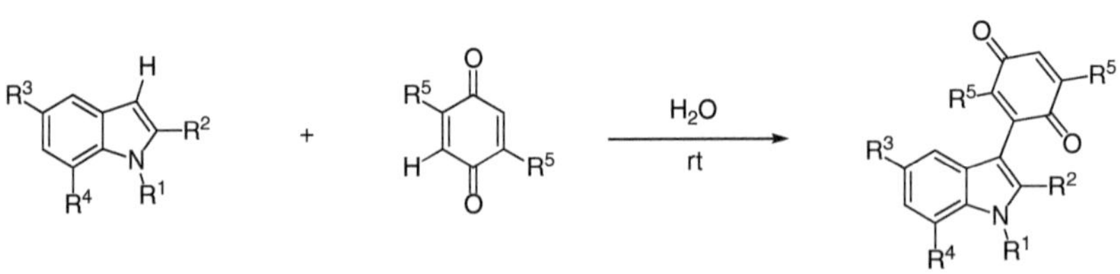

Scheme 5.23 The direct coupling of indoles with 1,4-benzoquinones using "on-water" approach.

obtained with the "on-water" conditions [39]. In the same year, Li's group reported a self-catalytic and on-water method for the synthesis of 1-naphthoyl tetrahydroisoquinoline via aza-Friedel Crafts reaction from 3,4-dihydroisoquinoline and 1- or 2-naphthols (Scheme 5.24) [40]. The use of water as a solvent led to higher yields.

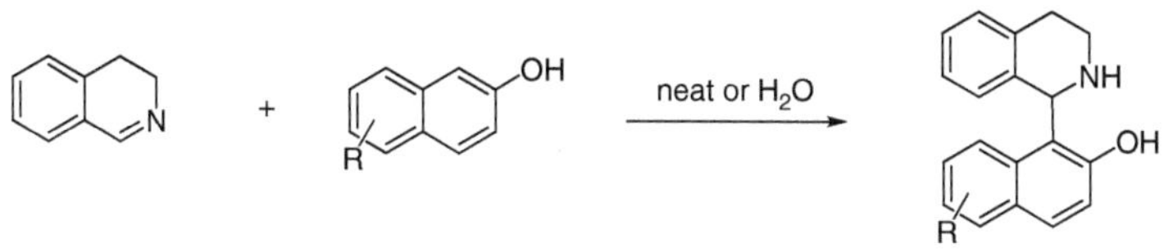

Scheme 5.24 The addition of naphthols to 3,4-dihydroisoquinoline via aza-Friedel-Crafts reaction.

In 2012, Li's group successfully developed a novel Rh-catalyzed cyclization reaction between benzoic acid derivatives and aldehydes, providing 3-substituted phthalides in moderate-to-good yields (Scheme 5.25) [41]. It is worth mentioning that the reaction can be conducted in water, and only water was generated as a byproduct.

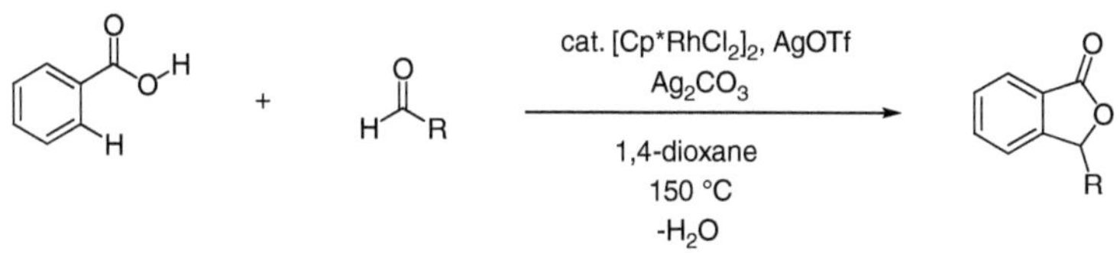

Scheme 5.25 Rh-catalyzed synthesis of 3-substituted phthalides via a cascade cyclization.

Dixneuf and coworkers presented a Ru-catalyzed and pyridine-directed C(sp^2)–H bond activation in water, enabling efficient *ortho*-phenylation (Scheme 5.26) [42]. It was observed that the selectivity is superior in water than in an organic solvent environment for monophenylation and bisphenylations.

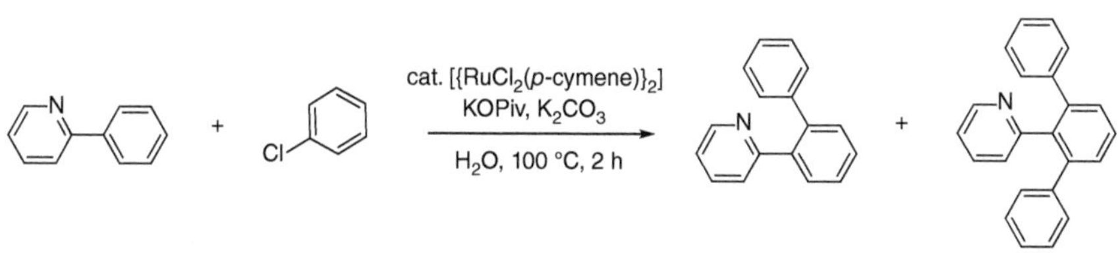

Scheme 5.26 The *ortho*-phenylation catalyzed by Ru via C(sp^2)–H bond activation in water.

In 2012, Ackermann's group described a tandem cyclization reaction utilizing Ru-catalysis, involving aniline derivatives and alkynes to afford the efficient indole synthesis through C(sp^2)–H bond activation in water (Scheme 5.27) [43]. The same group later reported a Ru-catalyzed arylation reaction in water using pyrazole as

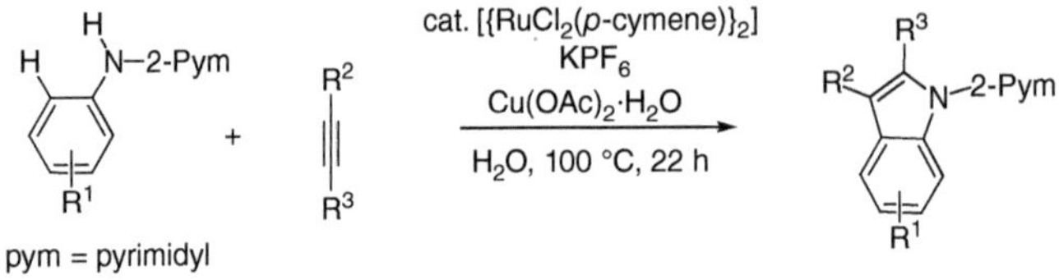

Scheme 5.27 Ru-catalyzed tandem cyclization between aniline derivative and alkyne.

the DG and phenol as the aryl source (Scheme 5.28) [44]. It is worth mentioning that the introduction of a carboxylic acid ligand into the ruthenium catalyst enhanced its aqueous solubility and improved its catalytic activity. Subsequently, his group successfully developed the Ru-catalyzed $C(sp^2)$–H bond activation for the isoquinolone synthesis and oxidative alkenylation reactions in water [45].

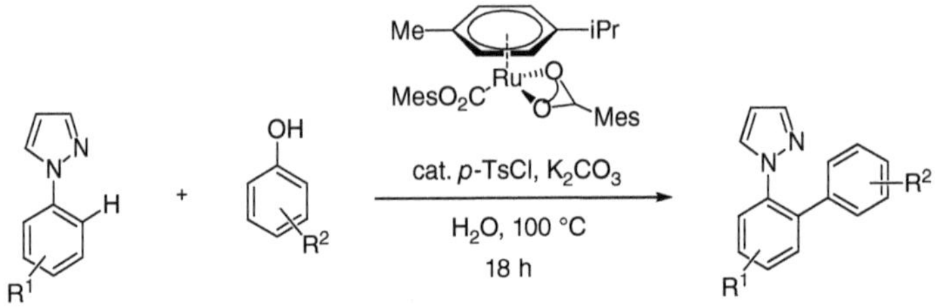

Scheme 5.28 Ru-catalyzed direct arylation in water with phenol as the aryl source.

Later, the Ru-catalyzed C–H bond activation reaction was further developed by Mehta and coworkers and a new water-friendly catalyst sodium η-6-*p*-cymene dichloro diruthenium triformato complex (MCAT-53) ($[Ru_2Cl_2(HCOO)_3$(p-cymene)])Na was demonstrated as an efficient catalyst for the synthesis of heterobiaryl compounds (Scheme 5.29).

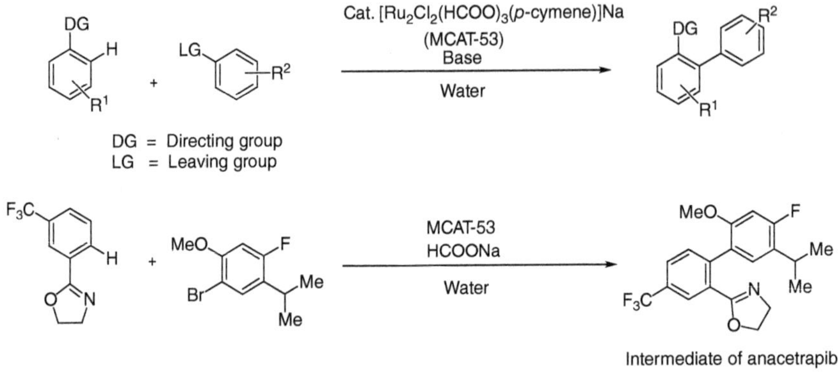

Scheme 5.29 Ru-based MCAT-53 catalyst for the synthesis of anacetrapib intermediate.

An advanced intermediate of cholesteryl ester transfer protein (CETP) inhibitor, anacetrapib, was synthesized as a single regioisomer in water catalyzed by this catalyst [46].

In 2013, Zhou's group reported a highly *para*-selective palladium-catalyzed arylation of unprotected phenols with aryl iodides via C–H bond activation in water under mild conditions. When *para*-methylphenol was used, the C–H bond activation occurs at the *ortho*-position of the substituted phenol, yielding the cyclic lactone products (Scheme 5.30) [47].

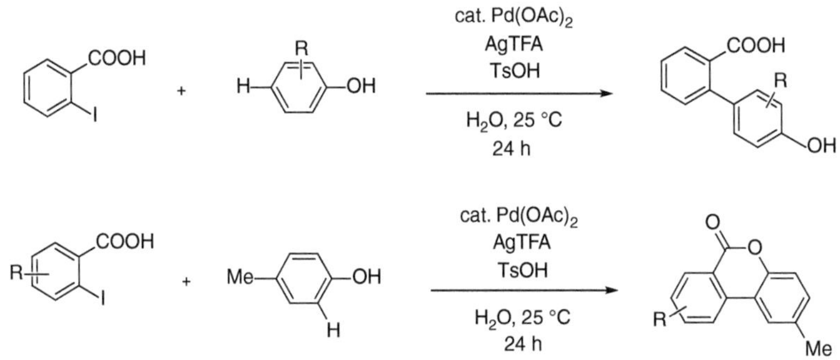

Scheme 5.30 Pd-catalyzed *para*-selective arylation of phenols with aryl iodides in water and Pd-catalyzed *ortho*-arylation of *para*-substituted phenols in water.

The same group later disclosed a Pd-catalyzed *ortho*-arylation reaction of benzoic acid via C–H bond activation in water with water-soluble diaryliodonium salts as the aryl source (Scheme 5.31) [48]. The notable feature of this reaction is the utilization of water-soluble aryl sources, which demonstrate excellent performance with high-valent iodine salts of various anions. The reaction does not require additional oxidants and thus avoids the use of expensive silver salts. Later, a Pd-catalyzed dehydrogenative coupling reaction between acetanilides and benzylic alcohols under aqueous conditions was reported by his group [49].

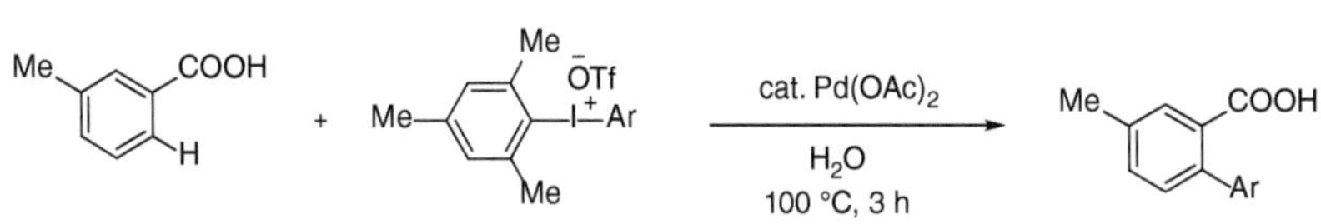

Scheme 5.31 Pd-catalyzed *ortho*-arylation of benzoic acid in water.

Differing from the Ru- and Pd-catalyzed C(sp^2)–H bond activation in water, Loh and coworkers disclosed a Rh-catalyzed and pyrimidine-directed 2-phenylation of indole derivatives via C(sp^2)–H bond activation in water in 2014. Trimethoxylphenylsilane was utilized as the phenylation reagent in this process (Scheme 5.32) [50].

In 2015, the Rh-catalyzed homo-coupling of aryl carboxylic acids in water was accomplished by Li's group. This process involved dual C(sp^2)–H bond activation with MnO_2 serving as the terminal oxidant (Scheme 5.33) [51]. Changing the oxidant from MnO_2 to $NaClO_2$ with acetic acid as an additive, the coupling proceeded smoothly, providing the corresponding products in good-to-excellent yields with broad functional group tolerance (Scheme 5.34) [52].

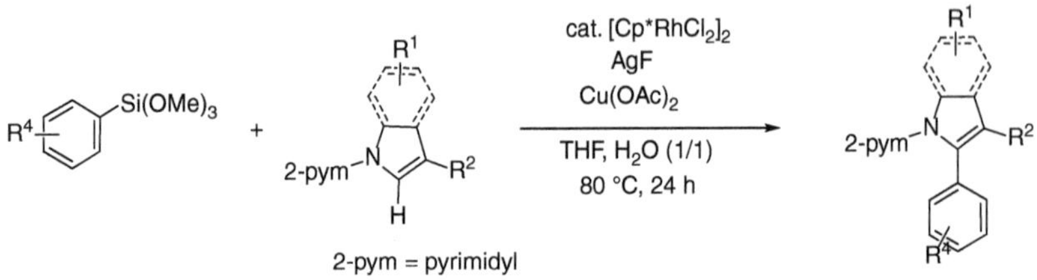

Scheme 5.32 Rh-catalyzed 2-phenylation of indole derivatives in water.

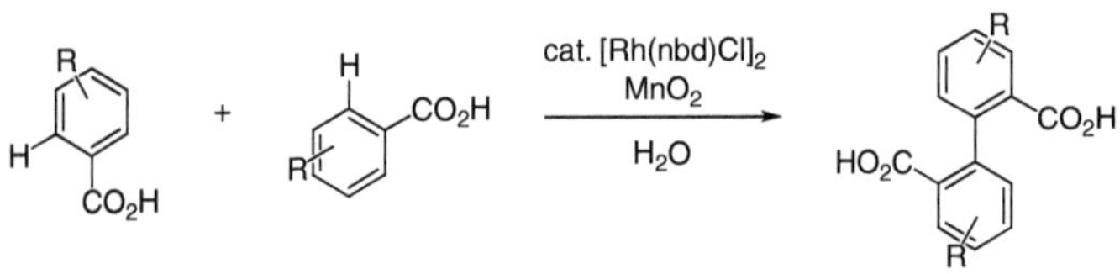

Scheme 5.33 The homo-coupling of aryl carboxylic acid catalyzed by Rh in water.

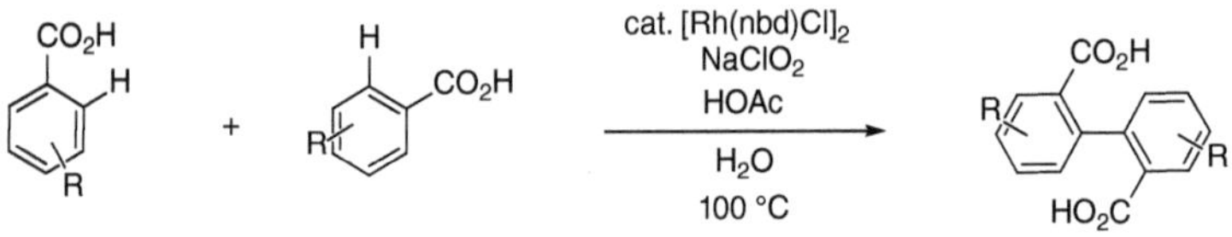

Scheme 5.34 The regioselective synthesis of biaryl acids catalyzed by Rh in water.

Subsequently, Li and coworkers accomplished the direct cross-dehydrogenative coupling (CDC) of two different aromatic acids for the synthesis of non-symmetrical diaryl acids in water (Scheme 5.35) [53].

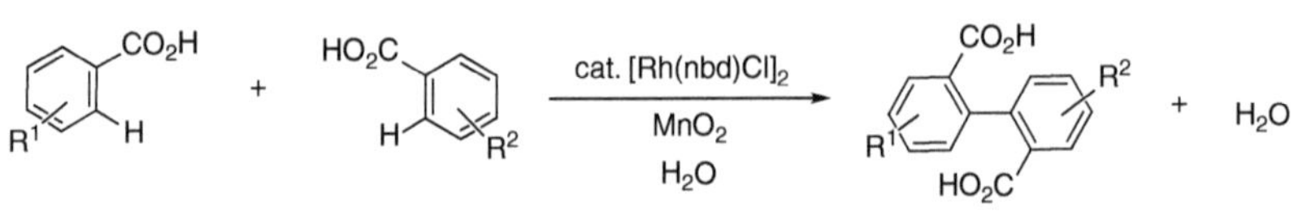

Scheme 5.35 The regiospecific CDC of aryl carboxylic acids catalyzed by Rh in water.

Later, a metal-free intramolecular annulation reaction for the synthesis of xanthones with tetrabutylammonium bromide as a promoter in water was discovered by Li and coworkers (Scheme 5.36) [54]. The reaction features direct oxidative coupling of an aldehyde C–H bond and aromatic C–H bond.

In 2015, a palladium-catalyzed *ortho*-acylation reaction with aldehydes in the presence of *tert*-butyl hydroperoxide as the oxidant under aqueous conditions was demonstrated by Deng's group. Excellent regioselectivity and wide functional group tolerance were observed for those transformations (Scheme 5.37) [55].

In 2016, Nallasamy and coworkers described a Pd–pincer-complex-catalyzed C–H bond activation under aerobic conditions for the synthesis of functionalized carbazoles in neat water (Scheme 5.38). It is worth noting that the catalytic system can be used for up to six consecutive runs with only 0.01 mol% catalyst loading, and an excellent yield was achieved without the column chromatography [56].

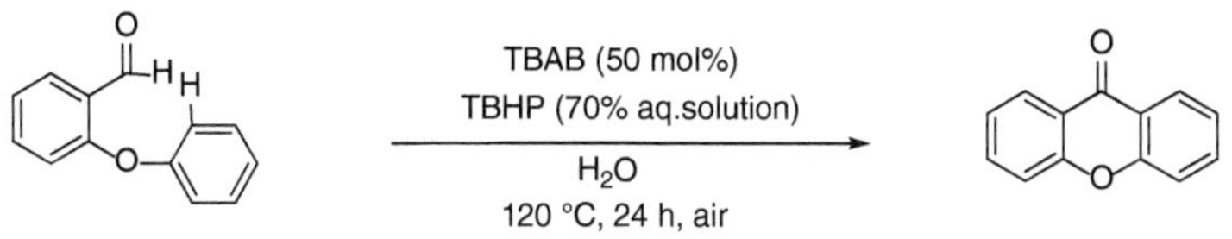

Scheme 5.36 The metal-free oxidative coupling for the synthesis of xanthone in aqueous medium.

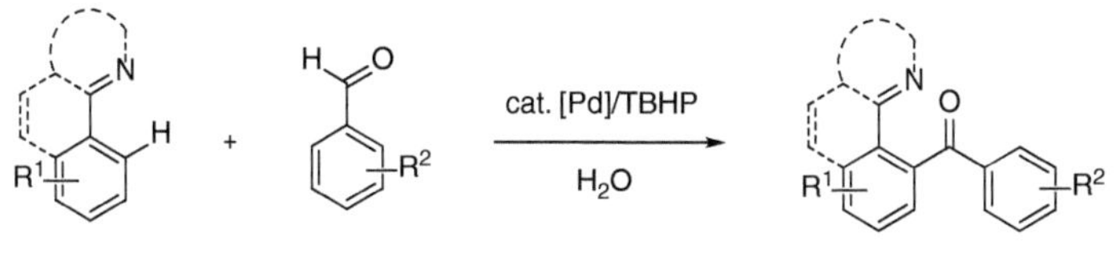

Scheme 5.37 Palladium-catalyzed *ortho*-acylation reactions in water.

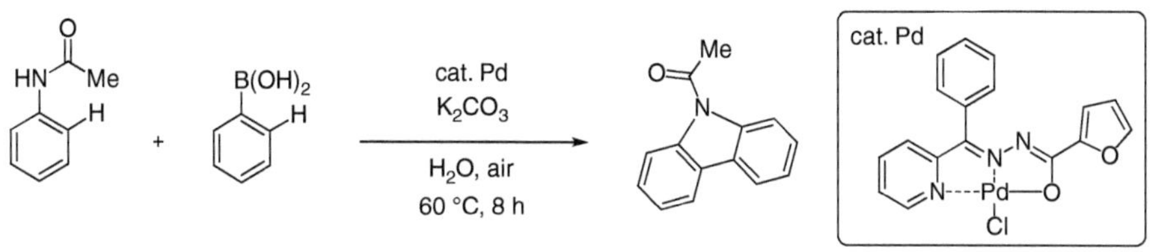

Scheme 5.38 Palladium-catalyzed C(sp^2)–H bond activation in water for carbazole synthesis.

In 2017, Cheng's group presented a rhodium-catalyzed benzamides' C(sp^2)–H bond activation in water for the efficient synthesis of isoquinoline derivatives in high yields with air as the sole oxidizing reagent (Scheme 5.39) [57]. Notably, the products precipitated out upon completion of the transformation, and the use of water as the solvent resulted in different regioselectivities when using various alkynes.

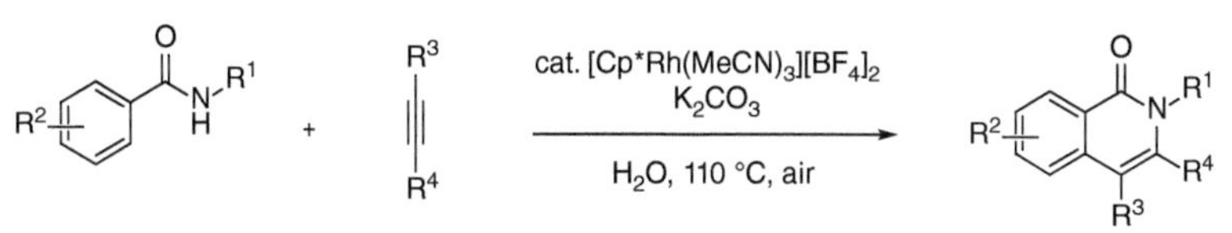

Scheme 5.39 Rh-catalyzed synthesis of isoquinoline derivatives in water.

In 2019, Li and coworkers reported a Pd-catalyzed C–H bond functionalization of tryptamine with a variety of ketones in water using two-in-one strategy, in which the *in situ*-generated imine functions as the DG for C–H activation, followed by cyclization in water to form various tetrahydro-β-carbolines (Scheme 5.40) [58]. Notably, the natural products komavine **1** and spiroindolone **2** were highly efficiently synthesized by this methodology. Additionally, this strategy can be extended to electron-deficient aromatic systems, which would not be viable under standard Pictet–Spengler conditions.

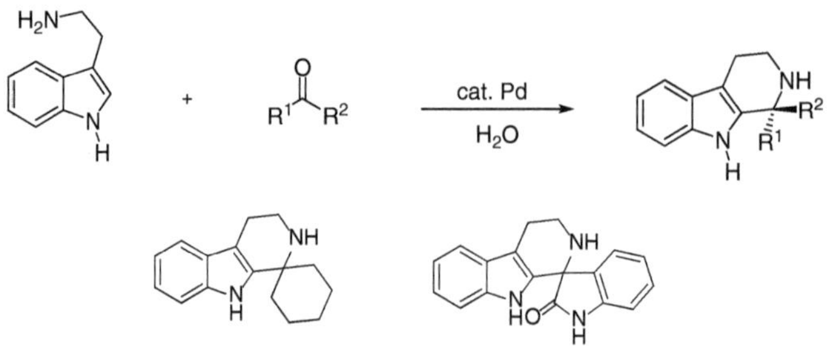

Scheme 5.40 Pd-catalyzed synthesis of tetrahydro-β-carbolines in water.

Shi and coworkers investigated the aqueous-environment-enabled alkylation of aromatic acids with maleimides, rendering them convertible. The reaction displayed diverse chemoselectivity with substituted benzoic acids (Scheme 5.41). In the proposed protocol, the carboxyl group acted as a crucial DG or a traceless DG, producing two distinct products: 2-alkyl-substituted benzoic acids or alkyl-substituted benzenes [59].

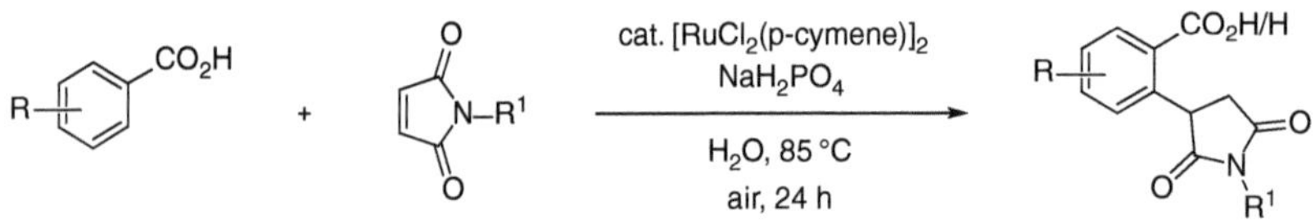

Scheme 5.41 Ru-catalyzed C–H alkylation of aromatic acids with maleimides in water.

In 2020, Roy and coworkers reported their successful discovery of Pd(II)-catalyzed C(sp^2)–H arylation and methylation of the benzamide and phenyl acetamide derivatives with 3-amino-1-methyl-1*H*-pyridin-2-one as a bidentate N, O-DG in water to synthesize pyridone derivatives (Scheme 5.42) [60]. The method was applied to a variety of substrates, providing the desired products in moderate-to-good yields.

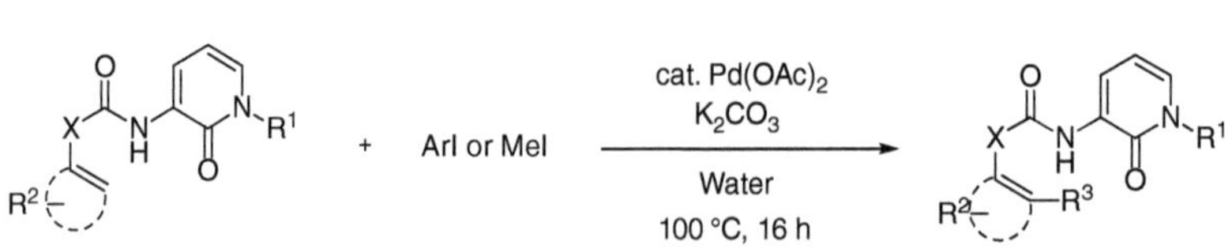

R^1 = Me, benzyl
R^2 containing ring = (Un)substituted aryl, heteroaryl
X = $(CH_2)_n$, n = 0, 1

Scheme 5.42 Pd-catalyzed chelation-assisted arylation and methylation reactions in water.

Recently, Chandra Nandi and coworkers achieved a visible-light-induced rhodium-catalyzed oxidative [4+2] annulation for the synthesis of 1,2-benzothiazines via C–H bond activation in water in good-to-high yields (Scheme 5.43) [61].

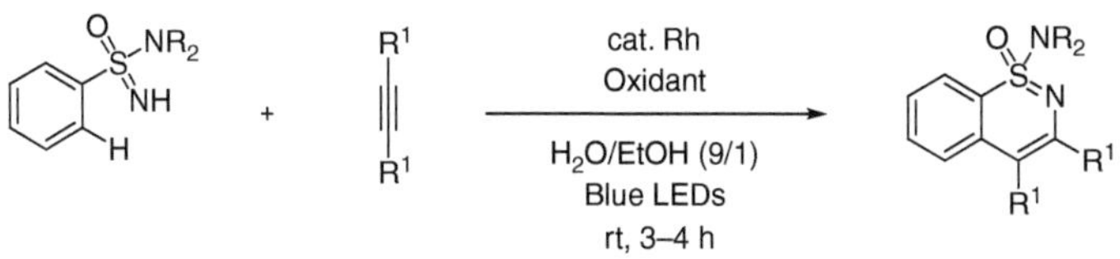

Scheme 5.43 Rh-catalyzed and a visible-light-induced C(sp^2)–H bond activation for the synthesis of 1, 2-benzothiazines in water.

Singh's group conducted research on ligand effect on the Ru-catalyzed C–H bond activation for arylation reactions in the aqueous medium, in which the 2-hydroxypyridine-based ligand was observed to greatly accelerate the C–H bond activation rate (Scheme 5.44). Mechanistic studies by nuclear magnetic resonance (NMR), mass spectral analysis, and discrete Fourier transform (DFT) revealed that the ligand coordinated to Ru(II) assisted a concerted metalation-deprotonation process [62].

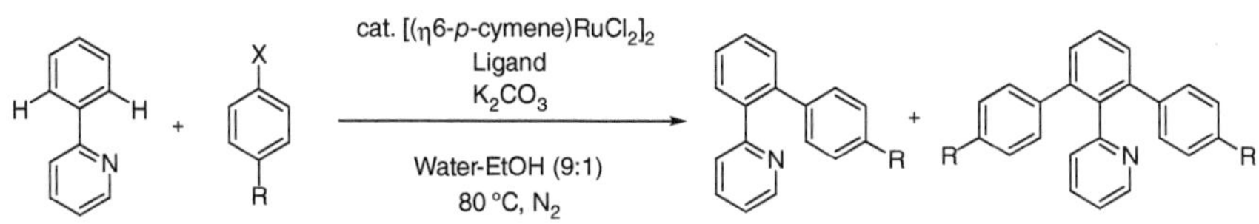

Scheme 5.44 Ligand effect on the Ru-catalyzed arylation reactions in water.

5.2.3 Activation of sp^3 C–H Bonds

The challenges in sp^3 C–H bond activation stem from the relatively low reactivity of aliphatic C–H bonds compared to the sp^2 or sp C–H bonds, as well as the potential for undesired side reactions and lack of selectivity. Additionally, the inherent complexity and diversity of sp^3-rich substrates present significant challenges in achieving site-selective C–H activation. One strategy is to tailor transition metal catalysts to promote sp^3 C–H activation. The catalysts with appropriate ligands and coordination environments can enhance reactivity and selectivity toward sp^3 C–H bonds. In 2002, Li's group successfully developed a Ru-catalyzed addition reaction of aldehyde with allyl alcohol via allylic C–H activation in water under air atmosphere (Scheme 5.45) [63]. This method offered a convenient synthesis of β-hydroxyl ketone compounds under mild conditions.

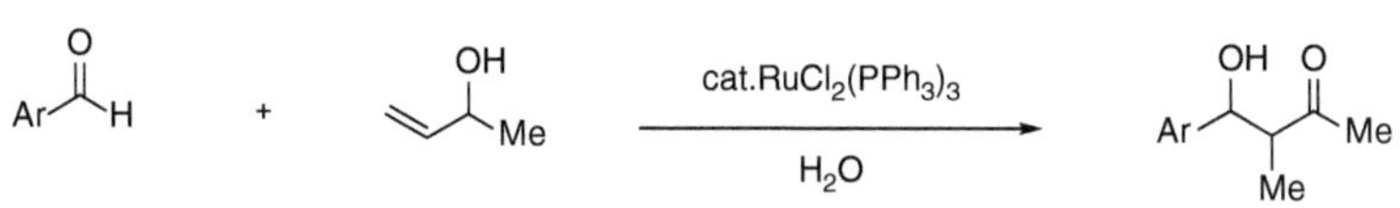

Scheme 5.45 Ru-catalyzed *in situ* olefin migration reaction in water.

Another approach that was successfully developed was the sp^3 C–H activation adjacent to heteroatoms, such as nitrogen, oxygen, etc. In 2007, Li's group accomplished the first oxidative CDC of tetrahydroisoquinoline with nitroalkanes and malonates in water, in which CuBr and molecular oxygen were used as the catalyst and oxidant, respectively, affording the corresponding coupling products in high yields (Scheme 5.46) [64]. It was proposed that the reaction entails copper-catalyzed oxidation of amines to imine-type intermediates, succeeded by a copper-catalyzed Henry-type reaction when nitroalkanes were employed.

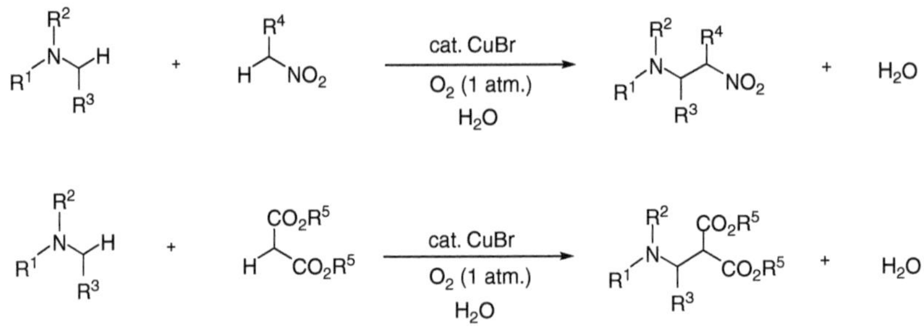

Scheme 5.46 Cu-catalyzed CDC reactions of tertiary amines with nitroalkanes and malonates.

Later, Zhang et al. reported the CDC reaction between N-aryl tetrahydroisoquinolines and 3,5-dialkyl-4-nitroisoxazoles in water catalyzed by copper (Scheme 5.47) [65].

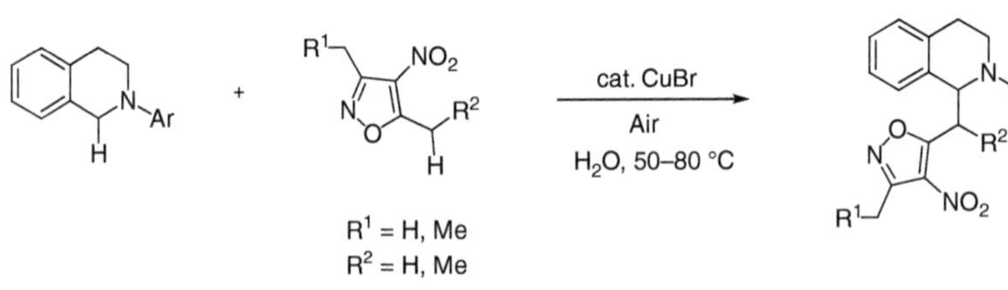

Scheme 5.47 Cu-catalyzed CDC reaction between N-aryl tetrahydroisoquinolines and 3,5-dialkyl-4-nitroisoxazoles in water.

This approach utilizes air as the oxidizing reagent, presenting an effective method for the synthesis of important isoxazoles derivatives. In 2018, Zhang's group presented the CDC reaction of N-aryl tetrahydroisoquinolines and nitroalkanes under visible light irradiation catalyzed by composite photocatalyst (Fe_3O_4-RB/LDH) (Scheme 5.48) [66]. Notably, the coupling reaction conducted in water provided a superior yield than in organic solvents. Then, a series of sp^3 C–H coupling reactions involving tetrahydroisoquinoline derivatives and various coupling partners were documented [67].

Introducing DG through the chelation strategy represents another effective approach for sp^3 C–H activation. In 2015, Rao and coworkers successfully developed a Pd-catalyzed hydroxylation of β-C(sp^3)–H bonds in amides, in which 8-aminoquinoline was utilized as a DG in water (Scheme 5.49) [68]. It is noteworthy that water serves as the source of oxygen, as evidenced by the isotope labeling experiment.

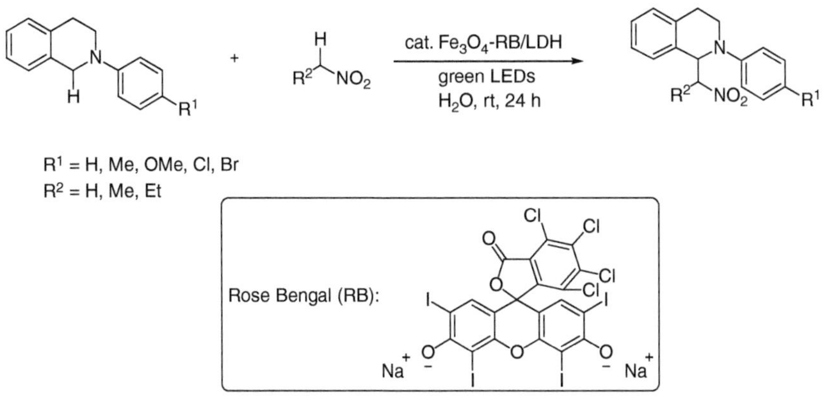

Scheme 5.48 The photocatalyst (Fe_3O_4-RB/LDH) catalyzed CDC reaction under visible light.

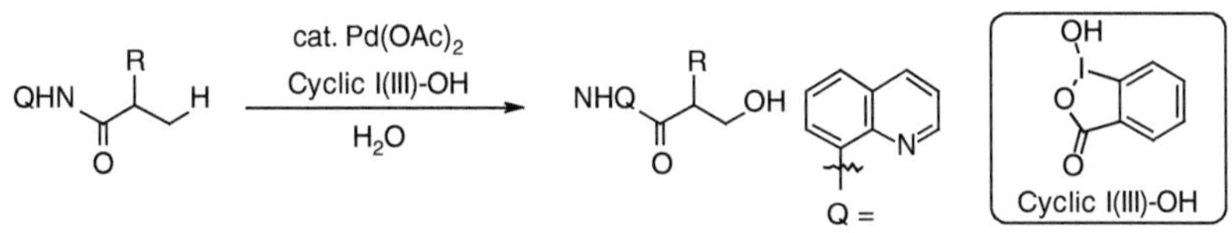

Scheme 5.49 The β-C(sp^3)–H bond hydroxylation of amides catalyzed by Pd in water.

Importantly, a method featuring mild reaction conditions for the synthesis of both natural and unnatural aromatic α-amino acids was reported by Chen and coworkers, in which the β-arylation of alanine was achieved by Pd-catalyzed C(sp^3)–H bond activation at room temperature with N-quinolylcarboxamide as DG employing water as a co-solvent [69], as outlined in Scheme 5.50.

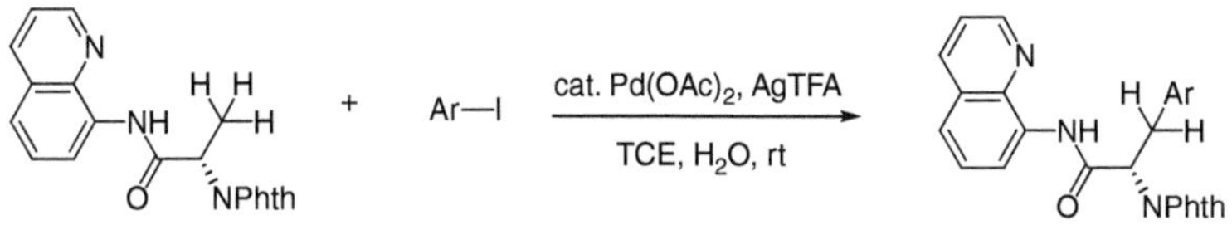

Scheme 5.50 Pd-catalyzed β-arylation of alanine with water as a co-solvent.

In addition to the sp^3 C–H bond activation reactions catalyzed by transition metals, an alternative method for activating C(sp^3)–H bonds involves metal-free catalysis. In 2008, Li and coworkers introduced an innovative pathway for the direct addition of cycloalkanes to imines mediated by peroxide. This method demonstrated tolerance toward a variety of functionalities and aqueous conditions [70], as depicted in Scheme 5.51.

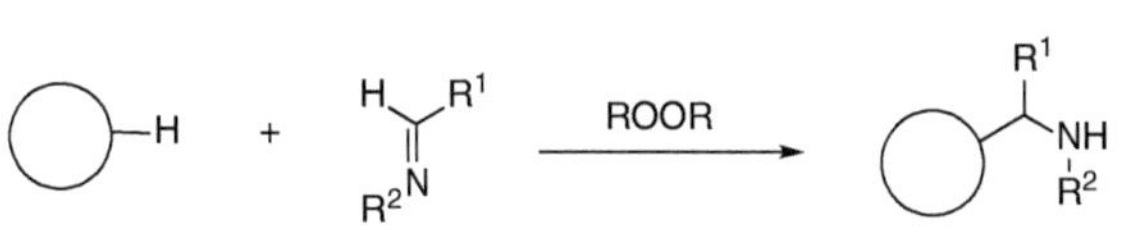

Scheme 5.51 The direct addition of cycloalkanes to imines is mediated by the peroxide.

In 2013, Itoh and coworkers reported a metal-free catalytic oxidation of CDC reactions involving two sp^3 C–H bonds, utilizing catalytic amount of I_2 and hydrogen peroxide as the oxidant (Scheme 5.52) [71].

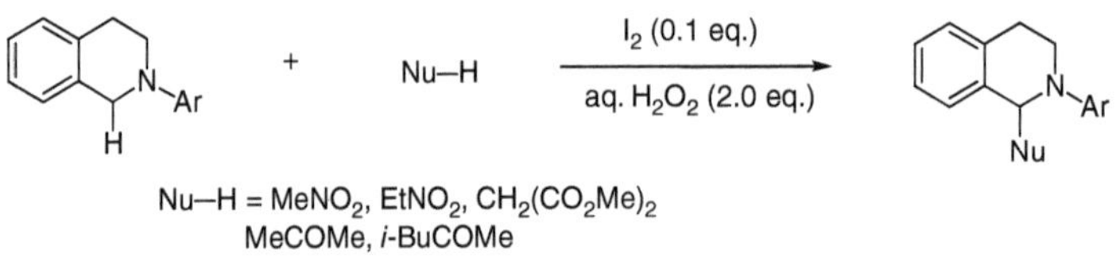

Scheme 5.52 I_2-catalyzed oxidative CDC reactions of N-aryl tetrahydroisoquinolines and nitroalkanes, malonates, and ketones with H_2O_2 as the oxidant.

Another oxidative CDC reaction of N-aryl tetrahydroisoquinolines and 2-methylazaarenes on-water under metal-free condition was realized by Xie and coworkers in 2016 (Scheme 5.53) [72]. It is worth mentioning that the mild oxidant 4-acetylamino-2,2,6,6-tetramethylpiperidine-1-oxoammonium tetrafluoroborate can be recovered and reused directly. The photocatalytic and metal-free CDC reaction of tertiary amines and ketones was successfully developed in water by Zhang and coworkers [73], and the initial studies also showcased that the bifunctional organic sponge photocatalyst exhibited the capability to catalyze the asymmetric transformation with favorable enantioselectivities.

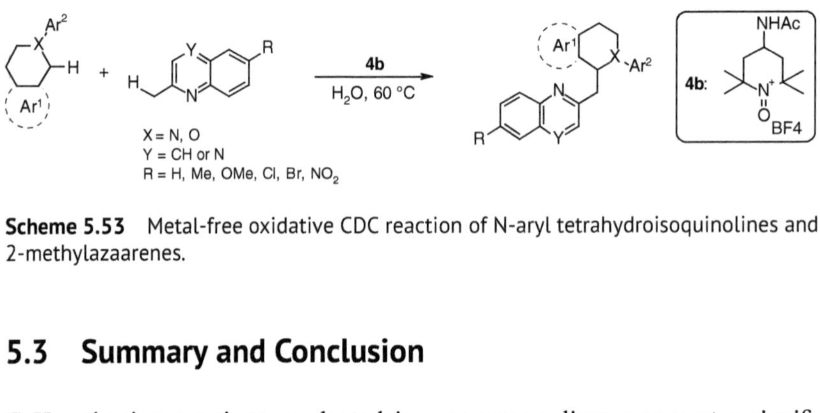

Scheme 5.53 Metal-free oxidative CDC reaction of N-aryl tetrahydroisoquinolines and 2-methylazaarenes.

5.3 Summary and Conclusion

C–H activation reactions conducted in aqueous medium represent a significant breakthrough in synthetic organic chemistry. By overcoming the challenges posed by the polar and protic nature of water, researchers have demonstrated the feasibility and potential advantages of performing these reactions in aqueous environments. By leveraging strategies, such as using appropriate transition metal catalysts, DGs with chelation, and employing water-soluble substrates and catalysts, researchers have expanded the scope of C–H activation reactions, paving the way for greener and more sustainable synthetic methodologies with potential applications in pharmaceuticals, materials science, and other fields. This chapter offers a concise overview of the rapid progress in C–H bond activation reactions, encompassing C(sp)–H, C(sp^2)–H, and C(sp^3)–H. The research on C(sp^3)–H activation is comparatively less extensive than that on C(sp)–H and C(sp^2)–H primarily due to the inherent challenges associated with C(sp^3)–H activation. The authors of this chapter foresee the discovery of more useful and practical C–H bond activation reactions in aqueous environments in the near future, especially in the field of C(sp^3)–H activation. Further research in this fascinating area holds promise to unlock new opportunities for more efficient and environmentally friendly synthesis.

Acknowledgments

We are grateful to the Canada Research Chair Foundation (to C.–J. L.), the Canadian Foundation for Innovation, FRQNT Centre in Green Chemistry and Catalysis, and the Natural Science and Engineering Research Council of Canada for support of our research.

References

1 (a) Yu, J.-Q., Ackermann, L., and Shi, Z. (2010). C—H activation. In: *Topics in Current Chemistry*. Berlin: Springer. (b) Dixneuf, P.H. and Doucet, H. (2016). C—H bond activation and catalytic functionalization II. In: *1a,b,c*. Cham: Springer. (c) Dixneuf, P.H. and Doucet, H. (2015). C—H bond activation and catalytic functionalization I. In: *Topics in Organometallic Chemistry*. Cham: Springer.

2 Sheldon, R.A. (2012). *Chem. Soc. Rev.* 41: 1437–1451.

3 Tobiszewski, M., Tsakovski, S., Simeonov, V. et al. (2015). *Green Chem.* 17: 4773–4785.

4 (a) Henderson, R.K., Jiménez-González, C., Constable, D.J.C. et al. (2011). *Green Chem.* 13: 854–862. (b) Prat, D., Pardigon, O., Flemming, H.-W. et al. (2013). *Org. Process Rev. Dev.* 17: 1517–1525.

5 (a) Byrne, F.P., Jin, S., Paggiola, G. et al. (2016). *Sustainable Chem. Processes* 4: 7. (b) Gandeepan, P., Kaplaneris, N., Santoro, S. et al. (2019). *ACS Sustain. Chem. Eng.* 7: 8023–8040.

6 (a) Jung, Y. and Marcus, R.A. (2007). *J. Am. Chem. Soc.* 129: 5492–5550. (b) Li, C.-J. and Chan, T.-H. (1997). *Organic Reactions in Aqueous Media*. New York: John Wiely & Sons. (c) Li, C.-J. and Chan, T.-H. (2007). *Comprehensive Organic Reactions in Aqueous Media*. Hoboken, NJ: John Wiely & Sons, Inc. (d) Li, C.J. (1993). *Chem. Rev.* 93: 2023–2035. (e) Li, C.-J. (2005). *Chem. Rev.* 105: 3095–3166. (f) Simon, M.-O. and Li, C.-J. (2012). *Chem. Soc. Rev.* 41: 1415–1427. (g) Cicco, L., Sblendorio, S., Mansueto, R. et al. (2016). *Chem. Sci.* 7: 1192–1199. (h) Chanda, A. and Fokin, V.V. (2009). *Chem. Rev.* 109: 725–748. (i) Lipshutz, B.H., Ghorai, S., and Cortes-Clerget, M. (2018). *Chem. Eur. J.* 24: 6672–6695.

7 García-Álvarez, J., Hevia, E., and Capriati, V. (2018). *Chem. Eur. J.* 24: 14854–14863.

8 Zhou, F. and Li, C.-J. (2019). *Chem. Sci.* 10: 34–46.

9 Herrerías, C.I., Yao, X., Li, Z., and Li, C.-J. (2007). *Chem. Rev.* 107: 2546–2562.

10 Trost, B.M. (1991). *Science* 254: 1471–1477.

11 Wender, P.A., Verma, V.A., Paxton, T.J., and Pillow, T.H. (2008). *Acc. Chem. Res.* 41: 40–49.

12 Sheldon, R.A. (1994). *Chem. Tech.* 24: 38–47.

13 Casalnuovo, A.L. and Calabrese, J.C. (1990). *J. Am. Chem. Soc.* 112: 4324–4330.

14 Li, C.-J., Slaven, W.T. IV, Chen, Y.P. et al. (1998). *Chem. Commun.* (13): 1351–1352.

15 Li, C.-J. (2010). *Acc. Chem. Res.* 43: 581–590.

16 Wei, C. and Li, C.-J. (2002). *Green. Chem.* 4: 39–41.

17 Li, C.-J. and Wei, C. (2002). *Chem. Commun.* 268–269.

18 (a) Wei, C. and Li, C.-J. (2003). *J .Am. Chem. Soc.* 125: 9584–9585. (b) Huang, B., Yao, X., and Li, C.-J. (2006). *Adv. Synth. Catal.* 348: 1528–1532.

19 Wei, C., Li, Z., and Li, C.-J. (2003). *Org. Lett.* 5: 4473–4475.

20 (a) Wei, C. and Li, C.-J. (2002). *J. Am. Chem. Soc.* 124: 5638–5639. (b) Wei, C., Mague, J.T., and Li, C.-J. (2004). *Proc. Natl. Acad. Sci.* 101: 5749–5754.

21 Lo, V.K.Y., Liu, Y.G., Wong, M.-K., and Che, C.-M. (2006). *Org. Lett.* 8: 1529–1532.

22 Yao, X. and Li, C.-J. (2005). *Org. Lett.* 7: 4395–4398.

23 Uhlig, N. and Li, C.-J. (2012). *Org. Lett.* 14: 3000–3003.

24 Yao, X. and Li, C.-J. (2006). *Org. Lett.* 8: 1953–1955.

25 Zhang, J., Wei, C., and Li, C.-J. (2002). *Tetrahedron Lett.* 43: 5731–5734.

26 Chen, L. and Li, C.-J. (2004). *Org. Lett.* 6: 3151–3153.

27 Deng, G. and Li, C.-J. (2008). *Synlett* (10): 1571–1573.

28 Yu, M., Skouta, R., Zhou, L. et al. (2009). *J. Org. Chem.* 74: 3378–3333.

29 Jia, Z., Li, X., Chan, A.S.C., and Li, C.-J. (2012). *Synlett* 23: 2758–2762.

30 Fu, X.-P., Liu, L., Wang, D. et al. (2011). *Green Chem.* 13: 549–553.

31 Li, Y., Liu, L., Kong, D. et al. (2015). *J. Org. Chem.* 80: 6283–6290.

32 Knoepfel, T.F. and Carreira, E.M. (2003). *J. Am. Chem. Soc.* 125: 6054–6055.

33 (a) Knoepfel, T.F., Zarotti, P., Ichikawa, T., and Carreira, E.M. (2005). *J. Am. Chem. Soc.* 127: 9682–9683. (b) Mishra, S., Liu, J., and Aponick, A. (2017). *J. Am. Chem. Soc.* 139: 3352–3355.

34 Chen, L. and Li, C.-J. (2004). *Chem. Commun.* 2362–2364.

35 Wang, H., Hearne, Z., Knauber, T. et al. (2015). *Tetrahedron* 71: 5866–5870.

36 Chen, L. and Li, C.-J. (2004). *Tetrahedron Lett.* 45: 2771–2774.

37 Liu, W., Li, L., and Li, C.-J. (2015). *Nat. Commun.* 6: 6526–6531.

38 Zhang, H.B., Liu, L., Chen, Y.J. et al. (2006). *Adv. Catal. Synth.* 348: 229–235.

39 Zhang, H.B., Liu, L., Chen, Y.J. et al. (2006). *Eur. J. Org. Chem.* 869–873.

40 Macleod, P.D., Li, Z., and Li, C.-J. (2010). *Tetrahedron* 66: 1045–1050.

41 Shi, X.-Y. and Li, C.-J. (2012). *Adv. Synth. Catal.* 354: 2933–2938.

42 Arockiam, P.B., Fischmeister, C., Bruneau, C., and Dixneuf, P.H. (2010). *Angew. Chem. Int. Ed.* 49: 6629–6632.

43 Ackermann, L. and Lygin, A.V. (2012). *Org. Lett.* 14: 764–767.

44 Ackermann, L., Pospech, J., and Potukuchi, H.K. (2012). *Org. Lett.* 14: 2146–2149.

45 (a) Ackermann, L. and Fenner, S. (2011). *Org. Lett.* 13: 6548–6551. (b) Ackermann, L. and Pospech, J. (2011). *Org. Lett.* 13: 4153–4155.
46 Mehta, A., Saha, B., Koohang, A.A., and Chorghade, M.S. (2018). *Org. Process Rev. Dev.* 22: 1119–1130.
47 Wu, Z., Luo, F., Chen, S. et al. (2013). *Chem. Commun.* 49: 7653–7655.
48 Wu, Z., Chen, S., Hu, C. et al. (2013). *ChemCatChem* 5: 2839–2842.
49 Luo, F., Yang, J., Li, Z. et al. (2015). *Eur. J. Org. Chem.* 2015: 2463–2469.
50 Lu, M.-Z., Lu, P., Xu, Y.-H., and Loh, T.-P. (2014). *Org. Lett.* 16: 2614–2617.
51 Gong, H., Zeng, H., Zhou, F., and Li, C.-J. (2015). *Angew. Chem. Int. Ed.* 54: 5718–5721.
52 Liu, Y., Zhu, Y.-Q., and Li, C.-J. (2018). *Front. Chem. Sci. Eng.* 12: 3–8.
53 Zhang, H., Deng, G.-J., Li, S. et al. (2016). *RSC Adv.* 6: 91617–91620.
54 Rao, H.-H., Ma, X.-Y., Liu, Q.-Z. et al. (2013). *Adv. Synth. Catal.* 355: 2191–2196.
55 Xiao, F., Chen, S., Huang, H., and Deng, G. (2015). *Eur. J. Org. Chem.* 2015: 7919–7925.
56 Arumugam, V., Kaminsky, W., and Nallasamy, D. (2016). *Green Chem.* 18: 3295–3301.
57 Upadhyay, N.S., Thorat, V.H., Sato, R. et al. (2017). *Green Chem.* 19: 3219–3224.
58 Zeng, H., Wang, Z., and Li, C.-J. (2019). *Angew. Chem. Int. Ed.* 58: 2859–2863.
59 Pu, F., Liu, Z., Zhang, L. et al. (2019). *ChemCatChem* 11: 4116–4122.
60 Mitra, T., Kundu, M., and Roy, B. (2020). *J. Org. Chem.* 85: 345–359.
61 Natarajan, K., Ravindra, S., Padma Priya, V.R. et al. (2024). *Eur. J. Org. Chem.* 27: e202301217.
62 Binnani, C., Arora, S., Priya, B. et al. (2023). *Chem. Asian J.* 18: e202300569.
63 Wang, M. and Li, C.-J. (2002). *Tetrahedron Lett.* 43: 3589–3591.
64 Basle, O. and Li, C.-J. (2007). *Green Chem.* 9: 1047–1050.
65 Zhang, Y., Wei, B.-W., Wang, W.-X. et al. (2017). *RSC Adv.* 7: 1229–1232.
66 Huang, Y., Xin, Z., Yao, W. et al. (2018). *Chem. Commun.* 54: 13587–13590.
67 (a) Hirao, T., Amaya, T., and Ito, T. (2012). *Heterocycles* 86: 927–932. (b) Liang, W., Zhang, T., Liu, Y. et al. (2018). *ChemSusChem* 11: 3586–3590. (c) Meng, Q.-Y., Liu, Q., Zhong, J.-J. et al. (2012). *Org. Lett.* 14: 5992–5995. (d) Peng, K. and Dong, Z.-B. (2021). *Adv. Synth. Catal.* 363: 1185–1201.
68 Hu, J., Lan, T., Sun, Y. et al. (2015). *Chem. Commun.* 51: 14929–14932.
69 Wang, B., Nack, W.A., He, G. et al. (2014). *Chem. Sci.* 5: 3952–3957.
70 Deng, G. and Li, C.-J. (2008). *Tetrahedron Lett.* 49: 5601–5604.
71 Nobuta, T., Tada, N., Fujiya, A. et al. (2013). *Org. Lett.* 15: 574–577.
72 Fang, L., Li, Z., Jiang, Z. et al. (2016). *Eur. J. Org. Chem.* 2016: 3559–3567.
73 Zhang, T., Liang, W., Huang, Y. et al. (2017). *Chem. Commun.* 53: 12536–12539.

6

Recent Developments in Multicomponent Reactions in Water

Bibi E. M. Nanlohy and Floris P. J. T. Rutjes

Radboud University, Institute for Molecules and Materials, Heyendaalseweg 135, 6525 AJ Nijmegen, The Netherlands

6.1 Introduction

MCRs represent chemical processes in which three or more starting materials participate in a one-pot reaction, resulting in a single product which contains all atoms of the starting materials in its final structure. The starting materials react in a sequence of steps rather than simultaneously in one step [1]. In the context of MCRs, it is desirable that all reactions within the MCR mixture exhibit reversibility, while the terminal step responsible for product formation remains irreversible (Figure 6.1). This strategic arrangement provides a driving force that transfers all intermediates and starting materials toward the synthesis of a single final compound [2]. In contrast to conventional multistep synthesis, MCRs provide a synthetic method with a reduction in the number of synthetic steps, including their workup and purification. While in multistep synthesis, the formation of bonds proceeds stepwise, in MCRs, multiple bonds form in one operation without isolation of intermediates or addition of further reactants [3]. As opposed to the constraints of multistep synthesis, which primarily relies on two starting materials, MCRs provide a more expansive way for synthesizing products with a wide range in complexity and variety in functional groups. This versatility can be accomplished by varying the starting materials in MCRs [4]. Altogether, MCRs hold a distinct advantage attributable to their lower costs, shorter reaction time, higher degree of atom economy, environmental friendliness, reduction of solvent, and reduction in time for workup and purification processes [5]. For this reason, and the urge for greener methods, the application of MCRs in industry has become increasingly relevant [2, 6].

The use of solvents in chemical synthesis contributes to 80–90% of the overall reaction mass and is an important contributor to the waste generated in synthesis procedures [7]. Moreover, the utilization of numerous organic solvents often involves toxicity concerns, increasing the adverse environmental impact of such practices. Therefore, the urge to find sustainable reaction media is necessary with

Organic Transformations in Water: Principles and Applications, First Edition.
Edited by Gopinathan Anilkumar, Nissy Ann Harry, and Sankuviruthiyil M. Ujwaldev.

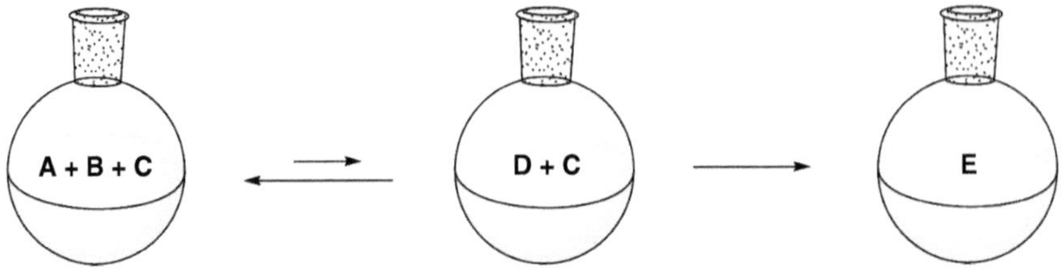

Figure 6.1 Schematic overview of a three-component reaction (3-CR). For the sake of convenience, regular arrows have been used for all steps throughout this chapter.

water attracting considerable attention over the years [8]. From both an economic and environmental point of view, water stands out as an advantageous solvent. It is cheap, nontoxic, nonflammable, and readily available, rendering it an attractive candidate for green chemistry initiatives [9]. Nevertheless, water is known for notoriously poorly dissolving organic compounds, causing challenges in synthetic procedures. Conversely, the physiochemical properties of water including its polarity, hydrogen bonding, hydrophobic effect, and trans-phase interactions may influence the reaction course [10].

The initial recognition of the advantages associated with organic transformations conducted in aqueous environments dates back to 1980 [11]. Breslow explored a rate enhancement when Diels–Alder reactions were performed in aqueous media instead of organic solvents, which he attributed to the hydrophobic effect, leading small organic compounds to form aggregates to reduce their contact surface area with water. Water molecules will then surround these aggregates forming cavities in the cluster structure of water. Importantly, apart from hydroxyl groups which participate in intermolecular hydrogen bonding, approximately 25% of hydroxyl groups remain uninvolved in such associations [12]. This remaining fraction can interact with organic reactants in their transition state, thereby maximizing the packing density of organic molecules. Consequently, this results in a reduction of activation energies and an acceleration of reaction rates. Additionally, the high cohesive energy density of water (550 $Cal\,cm^{-3}$) and enforced hydrophobic interactions serve to further amplify reaction rates [13]. The cohesive energy density is the energy required to eliminate an organic molecule from the bulk, thereby creating a cavity based on the volume of the eliminated molecule. The relatively high value for water can be attributed to the small molecular size [14]. Furthermore, organic transformations characterized by negative reaction volumes will proceed faster in aqueous media [15]. Notably, aside from cycloaddition reactions, MCRs feature negative reaction volumes attributable to the condensation of multiple molecules into a single product. Consequently, MCRs are anticipated to proceed faster in water [16].

The utilization of multiwalled carbon nanotubes magnetic nanocomposites (MWCNTs MNCs) has raised interest within the field of MCRs over the past two years. This approach has been employed in various syntheses recently [17]. MNCs are recognized for their broad catalyst support applications due to their capacity for reusability and recoverability [18]. Unlike the conventional homogeneous catalysts necessitating extensive workup and filtration procedures, MNCs can be easily recovered using an external magnet. Due to the paramagnetic character of the support of MNCs, the catalyst can easily be recovered and reused without the need for a filtration step [19]. Additionally, advantages of MNCs are their high surface area, good dispersion, and stability [18]. However, MNCs tend to easily form large aggregates and thus lose their nanoparticle (NP) properties. To avert aggregate formation of the MNCs, it becomes imperative to stabilize the NPs, which can be achieved through either organic or inorganic coatings. Alternatively, surface modification of MNCs onto MWCNTs can be pursued [17a]. MWCNTs are known for their unique optical, electrical, and mechanical properties [20]. Their surface area spans from 150 to 1500 $m^2 g^{-1}$ rendering them as good sorbents [21]. Furthermore, MWCNTs can be easily functionalized via covalent or non-covalent strategies. Initially discovered in 1991 by Iijima, MWCNTs can be considered as cylinder-shaped graphene sheets capped by fullerene-like structures [22]. The MWCNTs MNCs in MCRs serve as catalyst, predominantly acting as a Lewis acid or base and are primarily employed for the synthesis of heterocycles.

The functionalization process of MWCNTs proceeds in three steps (Figure 6.2). The first step involves the activation of MWCNTs through treatment with sulfuric acid and nitric acid, leading to the formation of hydroxyl and carboxylic acid functional groups on the surface of the nanotubes. The second step entails the bond formation between, for example, Cu and the introduced hydroxyl and carboxylic acid groups on the surface of the MWCNTs, while formaldehyde is used to reduce Cu(II) to Cu(0). Subsequently, the chosen metals are fixed on the Cu@MWCNTs [23].

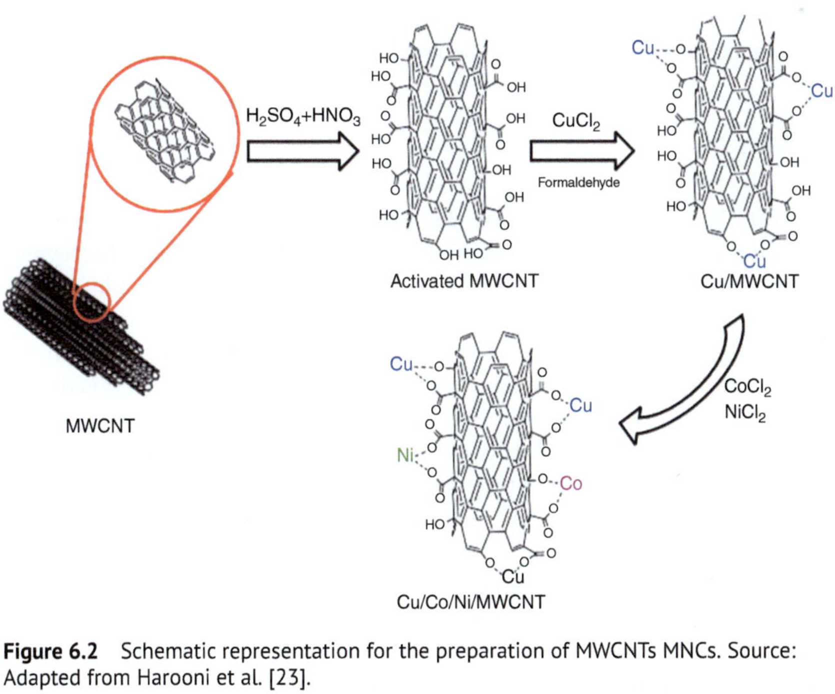

Figure 6.2 Schematic representation for the preparation of MWCNTs MNCs. Source: Adapted from Harooni et al. [23].

MCRs involving the use of MWCNTs MNCs are typically performed in water, ethanol [24], or are conducted under solvent-free conditions [25].

The first MCRs conducted in water were the Ugi and Passerini reactions performed by Pirrung and Sarma [16]. An increase in rate acceleration was observed when performing the MCRs in aqueous media in lieu of traditional organic solvents. This phenomenon can be attributed to the hydrophobic effect and cohesive energy density of water. Shapiro and Vigalok reported the Passerini reaction using "in-water" and "on-water" conditions by varying the reactants' solubilities [26]. "In-water" reactions are specified as a homogeneous mixture in water, while "on-water" reactions are specified as a heterogeneous mixture in water [27]. Over the past two decades, the investigation of MCRs in water has been widely explored. The first comprehensive review pertaining to MCRs in water was published in 2009 by Kumaravel and Vasuki [28]. Subsequently, two additional reviews appeared focusing on the preparation of heterocycles in water via isocyanide-based MCRs (I-MCRs), covering the developments in the last two decades [8, 14]. In 2022, Nazeri et al. published a review focusing on the preparation of pseudopeptides in water via I-MCRs [29]. Javanbakht et al. presented a comprehensive review elucidating the functionalization and modification of materials in water via MCRs [30]. In parallel, Pagadala et al. reviewed ultrasound-assisted MCRs in water for the preparation of nitrogen, sulfur, and oxygen heterocycles [31]. Lastly, in 2023, Russo et al. provided a comprehensive review of visible light-triggered I-MCR in water [32].

This review describes the most recent advancements pertaining to I-MCRs and Knoevenagel-initiated MCRs in water, encompassing the timeframe from 2020 to 2023. Knoevenagel-initiated MCRs utilizing isocyanides as starting material will be covered in the I-MCR part. Furthermore, the most pivotal MCRs concerning heterocyclic compounds from prior literature are described as well.

6.2 I-MCRs

The largest class of MCRs concerns the I-MCRs of which the most common members include the Strecker, Passerini, Ugi, and Groebke–Blackburn–Bienayme (GBB) reactions (Scheme 6.1). I-MCRs rely on an isocyanide functionality in the starting material, a highly reactive functional group, exhibiting a unique reactivity profile, causing them to undergo reactions with both nucleophilic and electrophilic species at the same atom, distinguishing them from conventional functional groups [3].

Over the past few decades, I-MCRs have garnered substantial scientific interest due to their capacity to generate heterocyclic bioactive scaffolds [33]. Their relevance is particularly notable in the pharmaceutical industry, in which they have been extensively explored in both the production of natural products and the rational design of drug-like compounds [14]. Consequently, an array of comprehensive reviews about I-MCRs have been published over the years [3, 14, 29, 32, 34, 35]. Nasiriani et al. reviewed the synthesis of various types of heterocycles via I-MCRs in aqueous medium (Scheme 6.2), some of which will be described in the sequel.

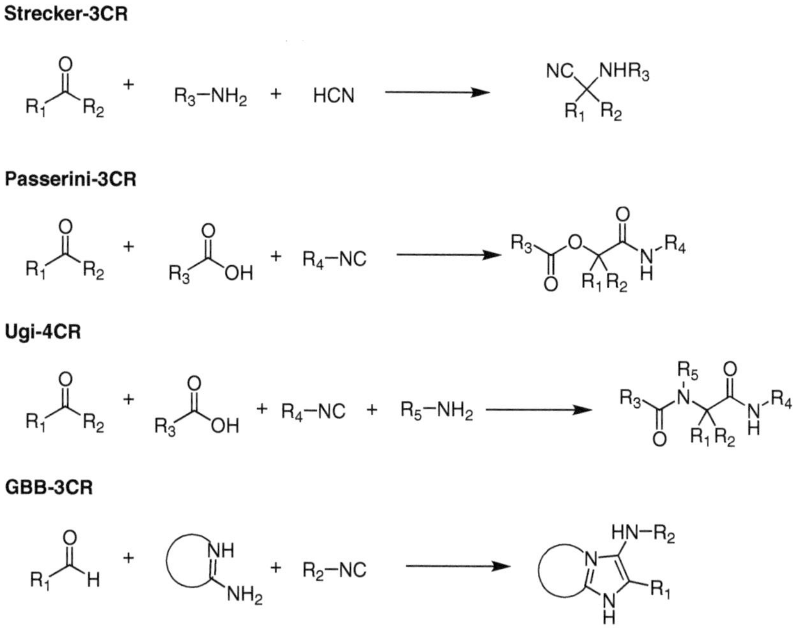

Scheme 6.1 The most common I-MCRs.

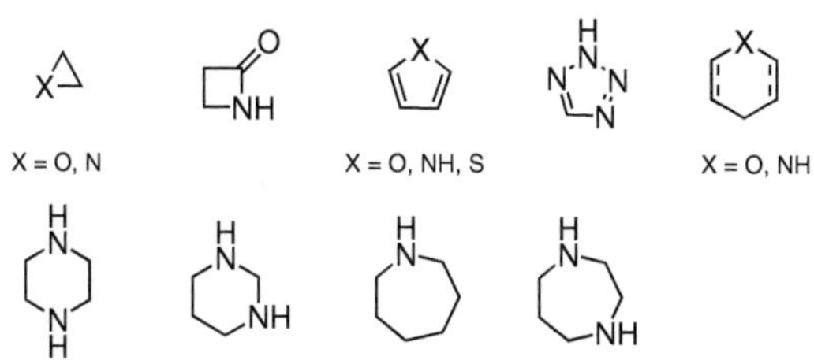

Scheme 6.2 Overview of heterocyclic scaffolds described in the review of Nasiriani et al. Source: Nasiriani et al. [14]/Springer Nature.

6.2.1 Four-Membered Heterocycles

β-Lactam structures are used in antibiotic, antidepressant, anticancer, and antitubercular agents [36]. The first synthesis of a β-lactam via an I-MCR in water was reported by Pirrung et al., who concurrently reported the first MCR performed in water at 50 °C. Their synthesis follows an Ugi four-center three-component reaction (Ugi-4C-3CR), as shown in Scheme 6.3 [16]. First, the condensation between amine **2** and aldehyde **1** yielded iminium ion **3**. This reacted with isocyanide **4** to form nitrilium ion **5**, after which the carboxylate was attacked to form intermediate **6** that underwent an irreversible Mumm rearrangement to yield β-lactam **7**.

Kanizsai et al. reported the synthesis of β-lactams **10a** and **10b** from oxanorbornene-based amino acid **8** in both methanol and water at room temperature, providing similar diastereoselectivity and yields in water compared to methanol [37]. One year later, the same research group reported the synthesis of bi- and tricyclic β-lactams **11**, by varying β-amino acid **9** [38] (Scheme 6.4).

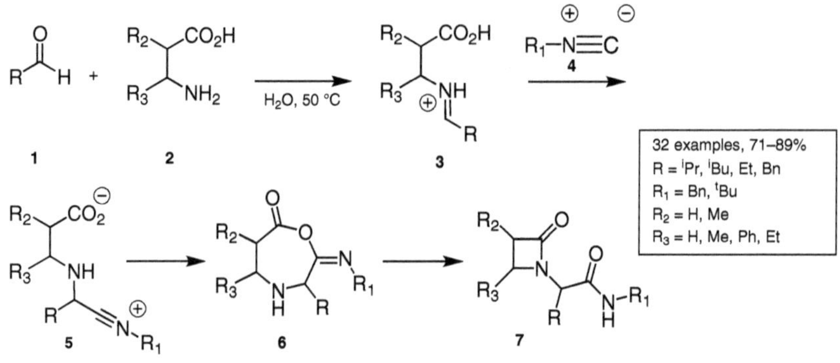

Scheme 6.3 Preparation of β-lactam derivatives via an Ugi-4C-3CR in water.

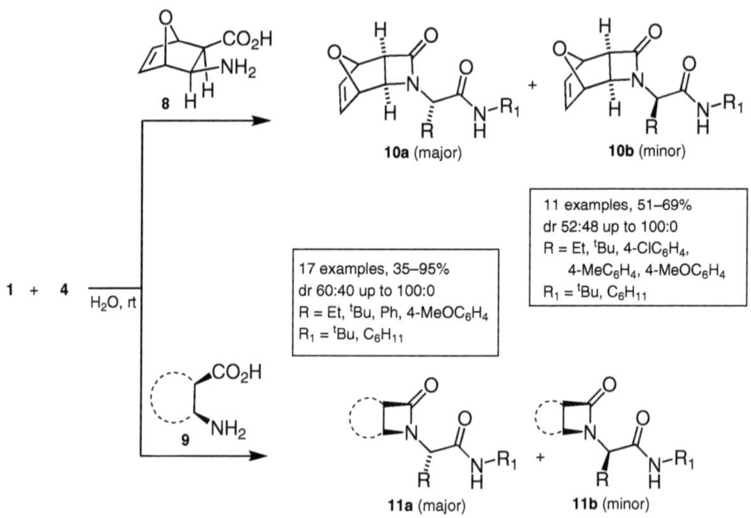

Scheme 6.4 Preparation of various β-lactam derivatives in water.

6.2.2 Five-Membered Heterocycles

6.2.2.1 Oxygen Heterocycles

Furan derivatives are of great importance within the field of medicinal chemistry owing to their noteworthy biological properties, such as antibiotic, anti-infective, antitumor, antimicrobial, and antispasmodic activity [39]. Their synthesis was achieved via a 3-CR involving a Knoevenagel condensation, and a [1+4] cycloaddition or a Michael addition in water at 75 °C (Scheme 6.5). The first step involved a Knoevenagel condensation with cyclic 1,3-diketone **12** and aldehyde **1**, yielding Knoevenagel intermediate **13**. Subsequently, intermediate **13** and isocyanide **4** reacted in a [1+4] cycloaddition or Michael addition to give intermediate **14**, which tautomerizes to tetrahydrobenzofuran **15** [40].

By varying the cyclic 1,3-diketone **12** in the 3-CR shown in Scheme 6.5, a variety of furan derivatives was obtained (Scheme 6.6). Shaabani et al. documented an

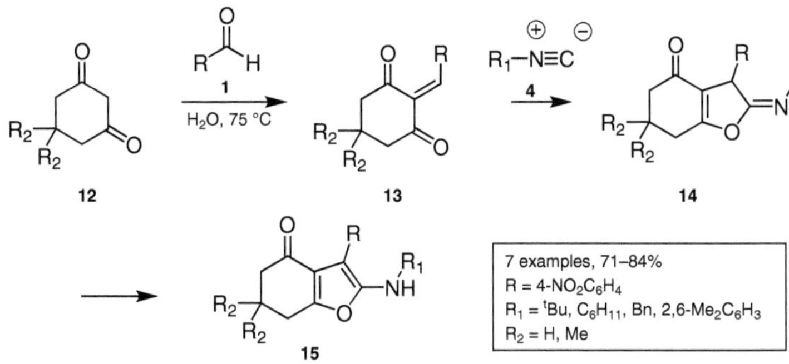

Scheme 6.5 Preparation of furan derivatives via a Knoevenagel/Michael addition/cyclization reaction in water.

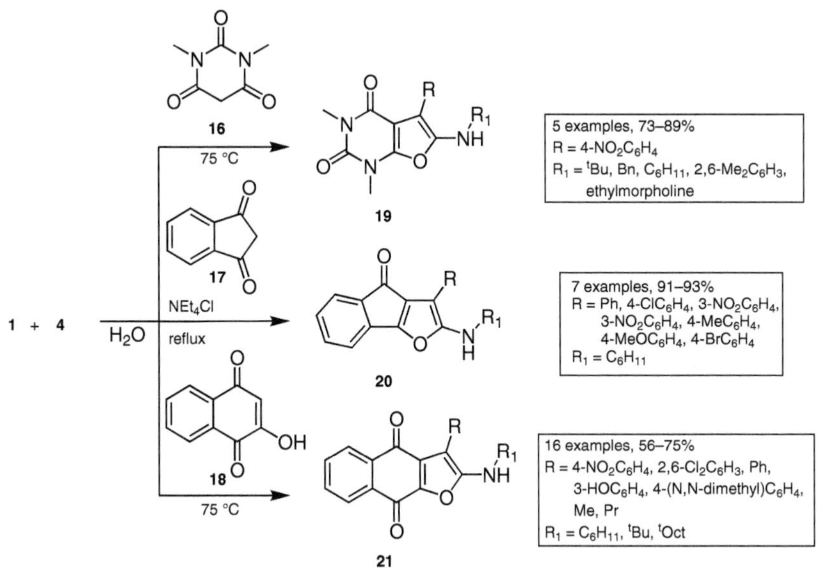

Scheme 6.6 Preparation of various furan derivatives using different cyclic diketones in water.

analogous synthetic approach, employing *N*,*N*-dimethylbarbituric acid (**16**) to yield the formation of furan derivative **19** in water at 75 °C [41]. Furthermore, the utilization of 1,3-indandione (**17**) in the presence of tetraethyl ammonium chloride as catalyst under reflux conditions yielded furan derivative **20** [42]. Teimouri and Bazhrang reported the same synthesis using 2-hydroxy-1,4-naphthoquinone (**18**) in water at 75 °C to yield furanonaphthoquinone **21** [43].

Mousavi et al. reported the preparation of novel furo-[2,3-*d*]pyrimidines with potential applicability as COVID-19 drug candidates. This achievement was realized by a green synthetic route performed in aqueous media (Scheme 6.7). Their protocol made use of the low-cost catalyst zirconium(IV)oxychloride octahydrate ($ZrOCl_2 \cdot 8\ H_2O$) [44]. The use of this catalyst has garnered substantial attention within

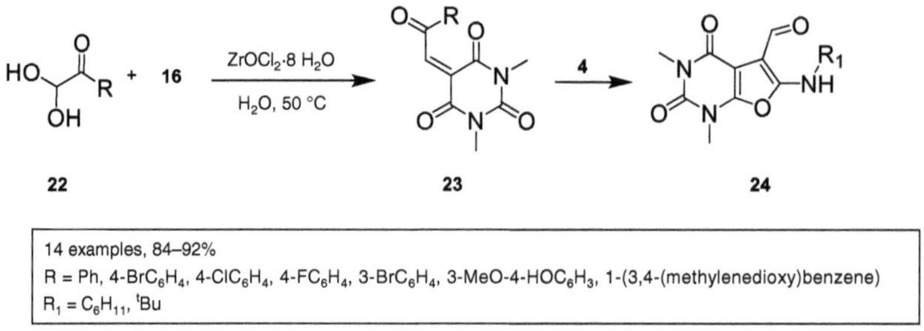

Scheme 6.7 Preparation of furo[2,3-*d*]pyrimidines via a 3-CR in the presence of $ZrOCl_2 \cdot 8\ H_2O$ in water.

the field of organic chemistry due to its convenient handling, high coordination stability, and moisture stability [45]. It was used to activate glyoxal monohydrate **22** to its anhydrate form, in which it underwent a Knoevenagel condensation with 1,3-dimethylbarbituric acid (**16**) to form Knoevenagel intermediate **23**. The final product **24** was attained via a sequence involving a Michael addition and heteroannulation, facilitated by the utilization of isocyanide **4** and Knoevenagel product **23** in the presence of the catalyst.

6.2.2.2 Nitrogen Heterocycles

The synthesis of pyrroles through an I-MCR conducted in an aqueous milieu at 80 °C has been documented by Pagadala et al. (Scheme 6.8) [46]. The Michael addition with Knoevenagel intermediate **26** and isocyanide **4** yielded intermediate **27**, which upon reaction with primary amine **28** gave amidine **29**. Intramolecular attack of **29** yielded the functionalized pyrrole **30**.

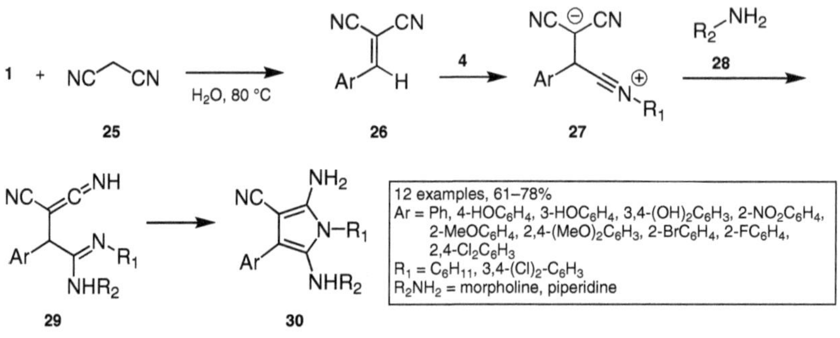

Scheme 6.8 Preparation of pyrrole derivatives via a 4-C Knoevenagel/Michael addition/cyclization reaction in water.

Zhu et al. developed a 3-CR to synthesize thioimidazolidinones in the presence of pyridine as a catalyst (Scheme 6.9) [47]. The 3-CRs were performed in a mixture of MeCN and water (3 : 1) at room temperature. The first step involved a Knoevenagel condensation to give intermediate **27**, which by attack of thiol **31** gave rise to intermediate **32**. Subsequent Ugi–Smiles rearrangement gave intermediate **33**, after which nucleophilic addition of the amine resulted in the final compound **34**.

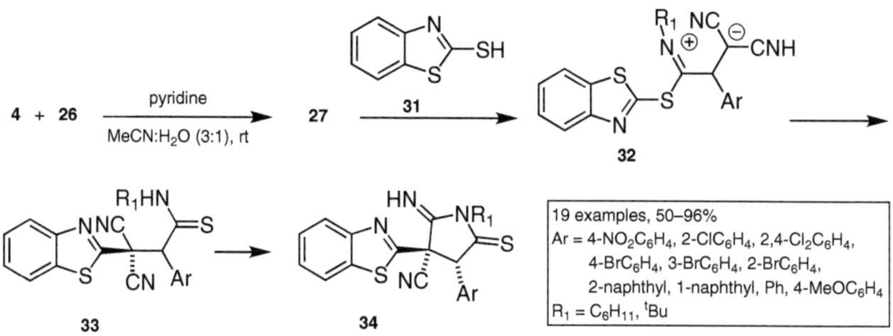

Scheme 6.9 Preparation of thioimidazolidinonesvia a 3-CR in water.

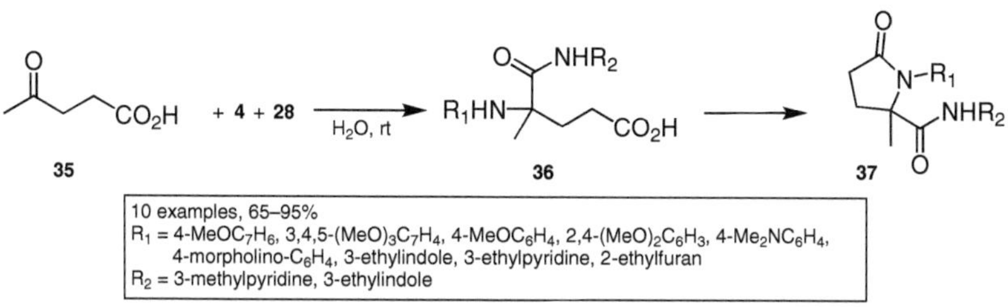

Scheme 6.10 Preparation of five-membered lactams via an Ugi reaction in water.

Mironov et al. reported aqueous Ugi reactions to form five-membered lactams (Scheme 6.10) [48]. The first step covered the Ugi reaction with levulinic acid **35**, primary amine **28**, and isocyanide **4** to give intermediate **36**. Nucleophilic attack of the amine group to the carboxylic acid group yielded final product **37** by elimination of water.

Kordmahalleh et al. described an aqueous 5-CR catalyzed by TiO_2/Fe_3O_4@ MWCNTs to form spiropyrrolopyrrolizine derivatives **45** (Scheme 6.11) [49]. In this catalytic system, Ti and Fe act as Lewis acids, while the oxygen atoms act as Lewis base within the context of MWCNTs MNCs. The MWCNTs MNCs caused activation of isatoic anhydride (**38**), which then reacted with *tert*-butyl isocyanide (**4**) to produce intermediate **39** via a cyclization. Subsequently, the primary amine **40** reacted with the carbonyl in compound **39** to create the spirocyclic intermediate. Activation by MWCNTs MNCs of acetylenic compound **42** led to the attack of the previously formed intermediate to create adduct **43**, which in turn reacted with alkyl bromide **44** to create the final compound **45** upon elimination of HBr and *tert*-butyl amine, followed by cyclization.

The synthesis of novel selenazolines via I-MCRs in water at 40 °C was described by Liu et al. [50]. The synthesis commenced with a reaction between selenium powder (**46**) and isocyanide **4**, yielding isoselenocyanate **47**, which was *in situ* reacted with 3-aminooxetane **48** to form the nucleophilic selenide **49**. The desired product **50** was then formed via an intramolecular ring-opening reaction of the oxetane. Scheme 6.12 shows the reaction with various isocyanide moieties **4**, while Scheme 6.13 shows the products using various 3-aminooxetane moieties **52**.

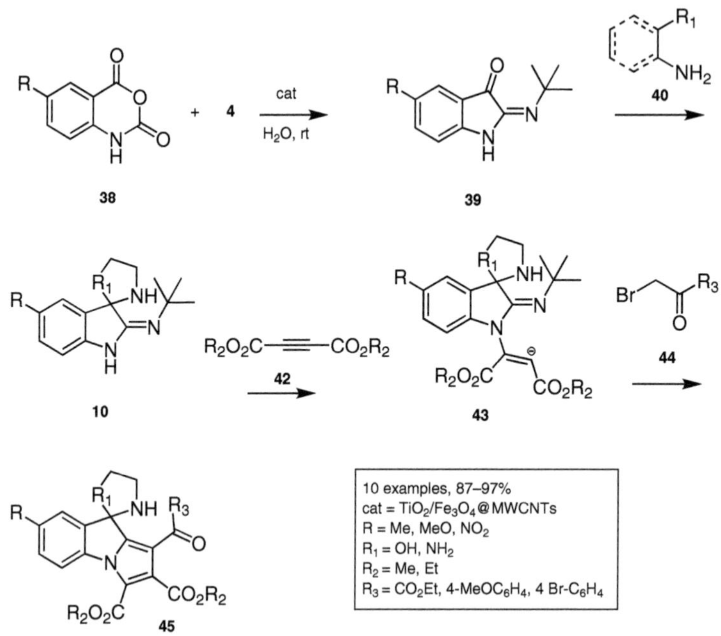

Scheme 6.11 Spiropyrrolopyrrolizines via a 5-CR catalyzed by TiO_2/Fe_3O_4/MWCNTs in water.

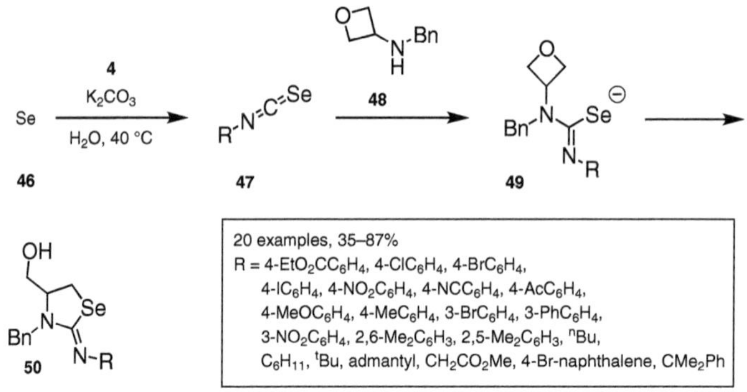

Scheme 6.12 Preparation of selenazolines via an I-MCR in water.

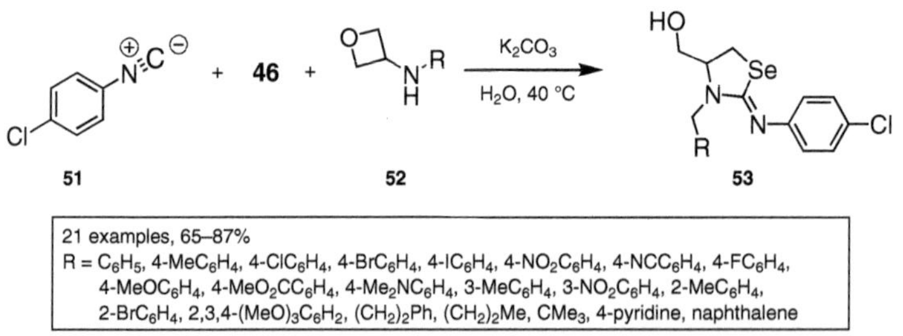

Scheme 6.13 Preparation of selenazolines using various 3-aminooxetanes via an I-MCR in water.

6.2.2.3 Sulfur Heterocycles

The synthesis of thiophenes via an I-MCR is presented in Scheme 6.14. The first step involved reaction of compound **54** and isocyanide **4** to provide imine intermediate **55**. S-Alkylation with intermediate **55** and α-haloketone **56** followed by intramolecular attack gave dihydrothiophene **58**. Finally, elimination of the amine produced the functionalized thiophene **59**. Moghaddam et al. reported the synthesis of thiophene derivatives using potassium carbonate as catalyst at 60 °C [51], while four years later, the same research group reported a similar synthesis at room temperature [52].

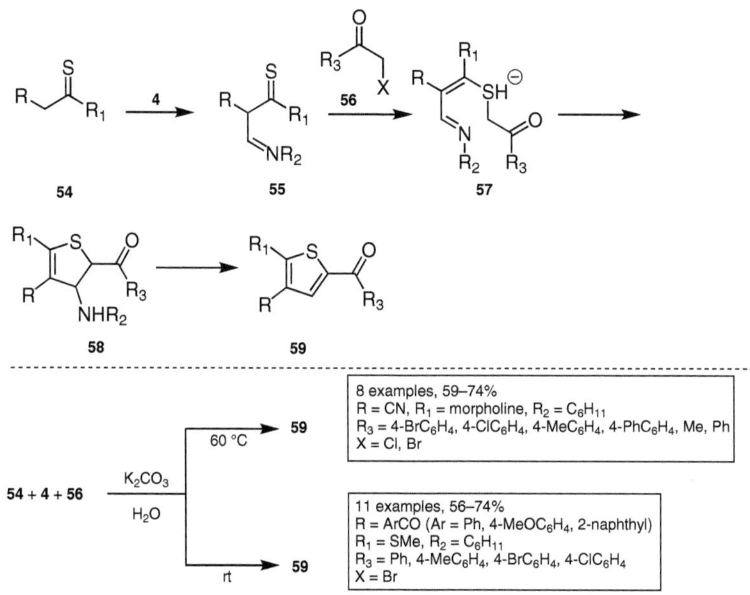

Scheme 6.14 Preparation of thiophene derivatives via a 3-CR in water.

6.2.3 Six-Membered Heterocycles

6.2.3.1 Oxygen Heterocycles

The synthesis of pyranderivatives via an aqueous I-MCR in the presence of $BF_3 \cdot SiO_2$ was reported by Akbari et al. (Scheme 6.15) [53]. First, zwitterionic intermediate **60** was formed by reacting dialkyl acetylenedicarboxylate **42** with isocyanide **4**. Subsequently, the nucleophilic addition of zwitterionic intermediate **60** with dialkyl acetone-1,3-dicarboxylate (**61**) resulted in the formation of intermediate **62**, which cyclized to give the functionalized pyran **63**.

The same methodology has also been applied in the synthesis of pyranocoumarin derivatives (Scheme 6.16). The use of 4-hydroxycoumarin (**64**) instead of dialkyl acetone-1,3-dicarboxylate (**61**) resulted in the formation of pyranocoumarin **65**. Sarma et al. made use of a phase-transfer catalyst (PTC) in the 3-CR to form pyranocoumarins **65** in water at 80 °C [54]. The use of the PTC was to address the issue of the poor solubility of the starting materials in water [55].

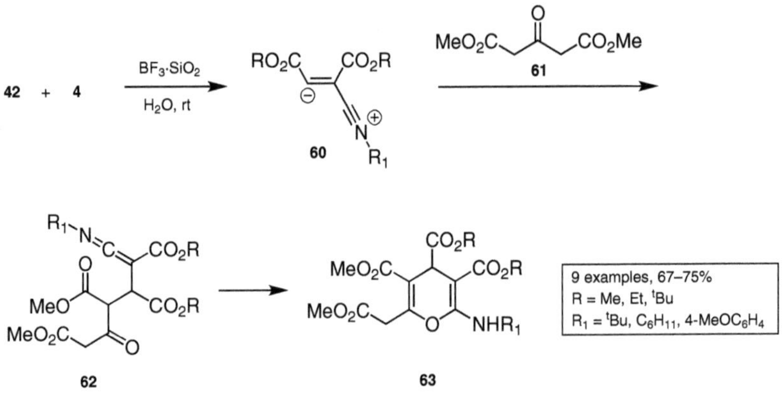

Scheme 6.15 Preparation of pyran derivatives using a 3-CR catalyzed by $BF_3 \cdot SiO_2$ in water.

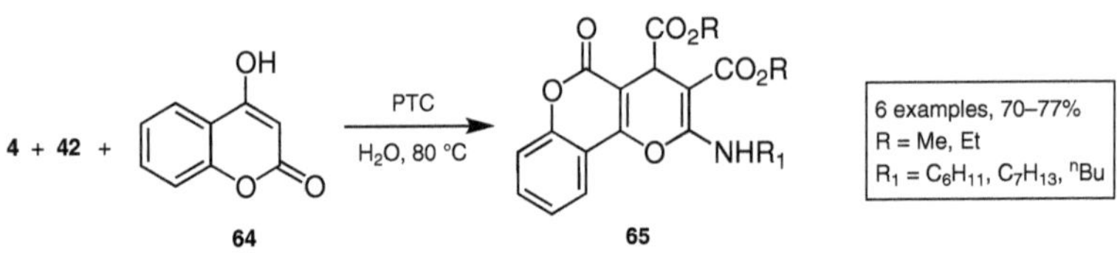

Scheme 6.16 Preparation of pyrancoumarin derivatives using a 3-CR in the presence of a PTC in water.

6.2.3.2 Nitrogen Heterocycles

Shaabani et al. reported the preparation of pyrazine derivatives through a 3-CR in water at room temperature (Scheme 6.17) [56]. The first step involved the condensation of diaminomaleonitrile (**66**) with 3-oxopentanedioic acid (**67**) to give imine intermediate **68**. Subsequent decarboxylation yielded the second imine intermediate **69**, which upon reaction with isocyanide **4** gave intermediate **70**, that in turn cyclized to final compound **71**.

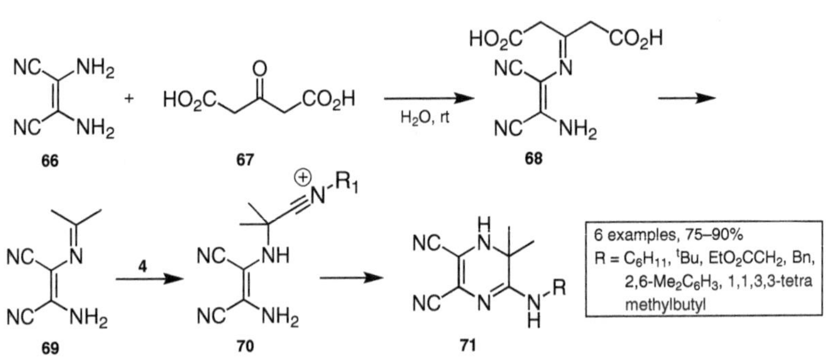

Scheme 6.17 Preparation of pyrazine derivatives via a 3-CR in water.

The same pyrazine derivatives **71** were prepared via a 3-CR in water at room temperature using Meldrum's acid (**72**), which upon hydrolysis provided acetone (**73**) and malonic acid (**74**) (Scheme 6.18). The acetone condensed with

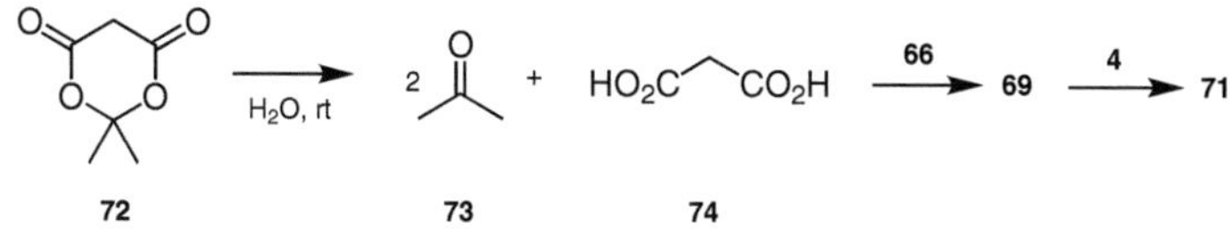

Scheme 6.18 Preparation of pyrazine derivatives via a 3-CR using Meldrum's acid in water.

diaminomaleonitrile (**66**) to yield intermediate **69**, which subsequently reacted with isocyanide **4** to yield pyrazine **71**. Notably, malonic acid (**74**) served as a catalyst in this 3-CR [56].

The methodology in Scheme 6.17 was also used in the synthesis of dihydroquinoxalin-2-amine derivatives **77** (Scheme 6.19). Kolla and Lee reported a 3-CR using *o*-phenylenediamine (**75**), ketone **76**, and isocyanide **4** in the presence of ethylenediaminetetraaceticacid (EDTA) as a catalyst in water at 80 °C [57].

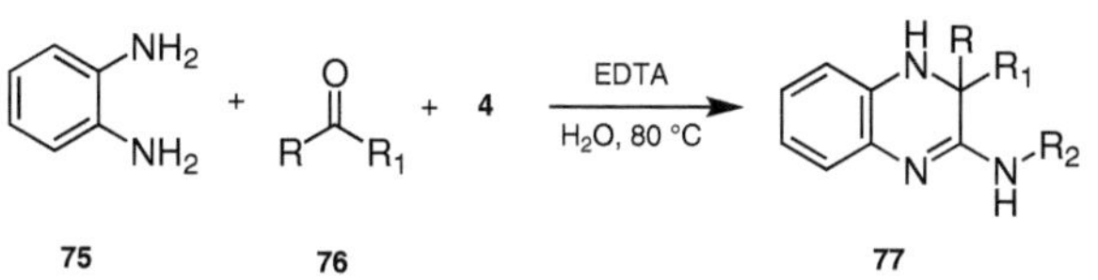

18 examples, 20–95%
R = Me, H
R_1 = Et, Me, Ph
R, R_1 = C_6H_{12}, C_5H_{10}, C_4H_8
R_2 = ^tBu, C_6H_{11}, 1,1,3,3-tetramethylbutyl

Scheme 6.19 Preparation of dihydroquinoxalin-2-amine derivatives via a 3-CR in the presence of EDTA in water.

A method for the preparation of pyrimidoisoquinolines via a 4-CR supported by MWCNTs MNCs in water at room temperature was reported by Amiri et al. (Scheme 6.20). This synthetic protocol employs potassium fluoride supported on clinoptilolite (CP), an affordable zeolite. Zeolitic supports have recently gained interest in catalytic applications due to their high thermal and hydrothermal stability [58]. Furthermore, zeolites have found adaptation and utility within the field of MCRs [59]. The first step covered the nucleophilic attack of ammonium

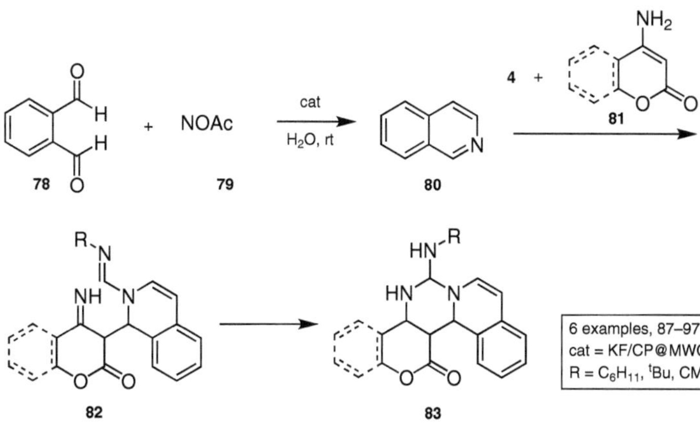

6 examples, 87–97%
cat = KF/CP@MWCNTs
R = C_6H_{11}, ^tBu, $CMe_2CH_2CMe_3$

Scheme 6.20 Preparation of pyrimidoisoquinoline via a 4-CR supported with KF/CP@ MWCNTs in water.

acetate (**79**) with phthaladehyde (**78**), through activation by KF/CP@MWCNTs, yielding isoquinoline (**80**). This isoquinoline reacted with a mixture of isocyanide **4** and 4-aminocoumarin (**81**) leading to intermediate **82**, which in turn cyclized in the presence of KF/CP@MWCNTs to give pyrimidoisoquinoline **83**.

6.2.4 Seven-Membered Heterocycles

The synthesis of diazepines was reported by Shaabani et al., who made use of a 5-CR catalyzed by *p*-toluenesulfonic acid in water at room temperature (Scheme 6.21) [60]. The first step, a double condensation with diaminomaleonitrile (**66**) and ketone **84**, resulted in diimine intermediate **85**. Subsequently, an imine-enamine cyclization of this intermediate gave rise to intermediate **86**, which by nucleophilic attack of isocyanide **4** provided the nitrilium intermediate **87**. Nucleophilic attack of water finally resulted in the formation of diazepine **88**.

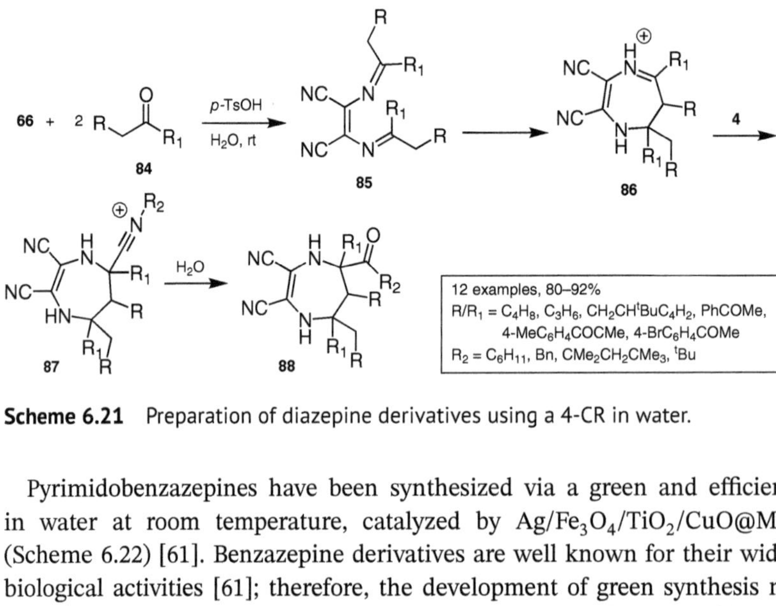

12 examples, 80–92%
R/R$_1$ = C_4H_8, C_3H_6, $CH_2CH^tBuC_4H_2$, PhCOMe, 4-MeC$_6$H$_4$COMe, 4-BrC$_6$H$_4$COMe
R$_2$ = C_6H_{11}, Bn, $CMe_2CH_2CMe_3$, ^{t}Bu

Scheme 6.21 Preparation of diazepine derivatives using a 4-CR in water.

Pyrimidobenzazepines have been synthesized via a green and efficient 7-CR in water at room temperature, catalyzed by Ag/Fe_3O_4/TiO_2/CuO@MWCNTs (Scheme 6.22) [61]. Benzazepine derivatives are well known for their widespread biological activities [61]; therefore, the development of green synthesis methods is of substantial value. The synthesis commenced with alkylation of isoquinoline (**80**) with alkyl bromide **44** to form iminium ion **89**. Subsequently, 2-amino benzoic acid **90** reacted with diethyl carbonate (**91**), both of which were activated by the catalyst, creating isatoic anhydride **92**. The iminium ion **85** and isatoic anhydride **92** then reacted to form intermediate **93**, which in the presence of acetylenic compound **94** and ammonium acetate (**79**) led to intermediate **95**. In the final step, intermediate **95** and *tert*-butyl isocyanide (**4**) further cyclized to form the pyrimidobenzazepines **96**.

The synthesis of oxazolopyrimidoazepines in the presence of Ag/Fe_3O_4/TiO_2/CuO@MWCNTs in water at room temperature was described by Avarsaji et al. (Scheme 6.23). It is worth noting the role of Ag in the synthesis acting as a strong Lewis acid [62]. The first steps in the 6-CR were similar to those depicted in

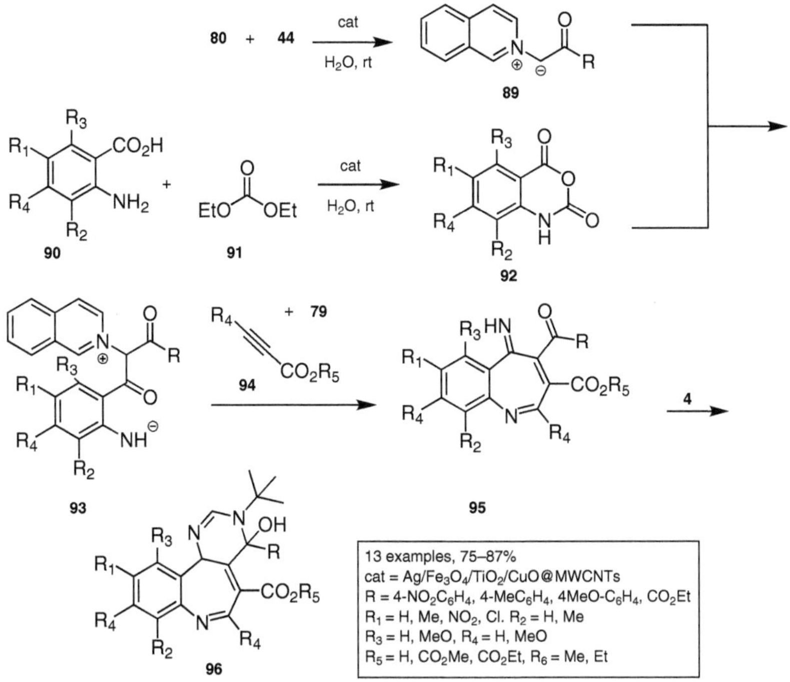

Scheme 6.22 Preparation of pyrimidobenzazepine via a 7-CR catalyzed by $Ag/Fe_3O_4/TiO_2/CuO@MWCNTs$ in water.

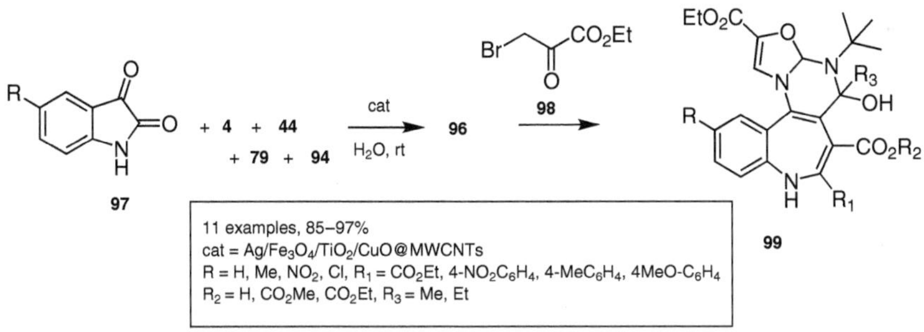

Scheme 6.23 Preparation of oxazolopyrimidoazepines via a 6-CR in the presence of $Ag/Fe_3O_4/TiO_2/CuO@MWCNTs$ in water.

Scheme 6.22, leading to intermediate **96**. The last step required the nucleophilic attack of **96** with ethyl bromopyruvate (**98**), which cyclized to the functionalized oxazolopyrimidoazepines **99**.

6.3 Knoevenagel-Initiated MCRs

Knoevenagel condensation-based MCRs represent a class of domino reactions wherein the Knoevenagel product serves an intermediate, typically followed by

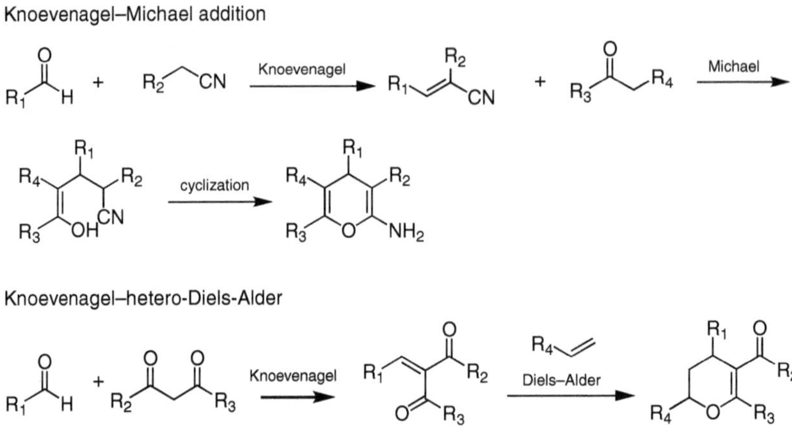

Scheme 6.24 Tandem Knoevenagel–Michael addition/cyclization and Knoevenagel–hetero-Diels–Alder MCRs.

a Michael addition/cyclization or a cycloaddition, leading to complexity and heterocycle formation (Scheme 6.24). On the one hand, the Knoevenagel product is an electron-deficient alkene which is a good Michael acceptor [63], but, on the other hand, when the Knoevenagel condensation involves a 1,3-dicarbonyl compound, a 1,3-heterodiene species is formed, poised for participation in a Diels–Alder reaction. Owing to the diversity in aldehydes, the resulting Knoevenagel product can exhibit a high degree of chemical functionality. Knoevenagel-initiated MCRs are well conducted in aqueous media and therefore widely used in MCRs in water [64]. Consequently, various reviews dedicated to Knoevenagel-initiated MCRs have emerged in recent years [63, 65], including a chapter about Knoevenagel-initiated MCRs in water by Gu [33].

6.3.1 Five-Membered Heterocycles

6.3.1.1 Oxygen Heterocycles

The preparation of *trans*-2,3-dihydrofurans via a Knoevenagel-initiated MCR in water at room temperature was described by Bantitaba (Scheme 6.25). Their protocol made use of ultrasound irradiation and SiO_2 NPs as a catalyst to shorten the reaction time and enhance the yield of the final product [66]. Their synthesis started with the formation of pyridinium ylide **100** through a substitution reaction employing pyridine and alkyl bromide **44**. Knoevenagel intermediate **101** reacted with 4-hydroxycoumarin (**64**), forming a zwitterionic intermediate, which via annulation and by loss of pyridine yielded product **102**.

6.3.1.2 Nitrogen Heterocycles

The synthesis of pyrazole derivatives via a Knoevenagel-initiated MCR was described by Sapkal and Kamble (Scheme 6.26). Their 3-CR was conducted in an aqueous medium at room temperature using the hydrotrope sodium toluene-4-sulfonate (NaPTS). Hydrotropes generally increase the solubility of

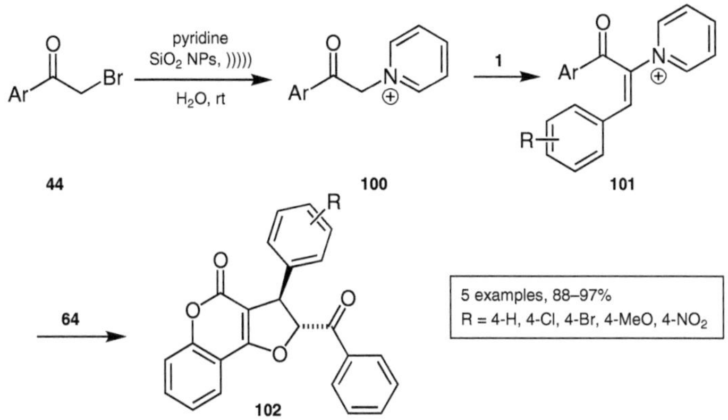

Scheme 6.25 Preparation of *trans*-2,3-substituted dihydrofurans via a 3-CR catalyzed by SiO_2 NPs in water.

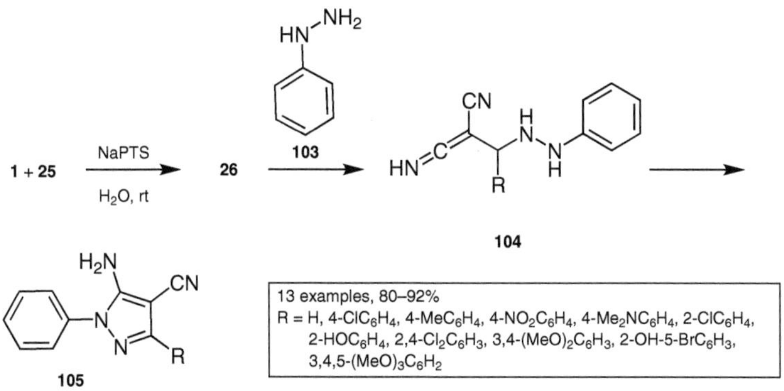

Scheme 6.26 Preparation of 5-aminopyrazole-4-carbonitrile via a 3-CR in the presence of NaPTS in water.

poorly soluble organic compounds [67]. The synthesis commenced with a reaction between Knoevenagel intermediate **26** and phenylhydrazine (**103**) in the presence of NaPTS to form intermediate **104**, which in turn cyclized into the final compound **105**.

6.3.1.3 Nitrogen and Oxygen Heterocycles

Novel isoxazole derivatives were synthesized via a 3-CR in water at room temperature using ultrasound irradiation (Scheme 6.27), as described by Daroughehzadeh and Kiyani. The nucleophilic addition of hydroxylamine **106** to the carbonyl of ketoester **107** resulted in oxime intermediate **108**. Subsequently, triphenylphosphine-activated aldehyde **109** and oxime **108** reacted in a Knoevenagel condensation to yield intermediate **110**. Finally, product **111** was generated via cyclization and elimination of ethanol [68].

A mechanistically identical approach was adapted in Scheme 6.28. Zhang et al. reported a comparable synthetic pathway (Scheme 6.28) using vitamin B1 as a catalyst [69]. Reddy and Reddy reported a synthetic protocol using ultrasonic

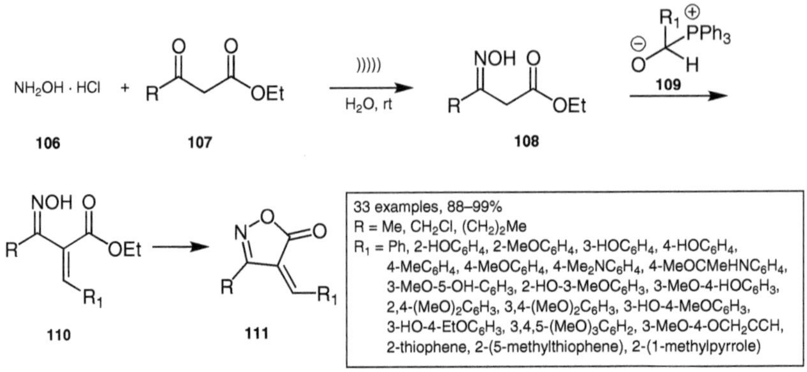

Scheme 6.27 Preparation of isoxazoles via a 3-CR in water.

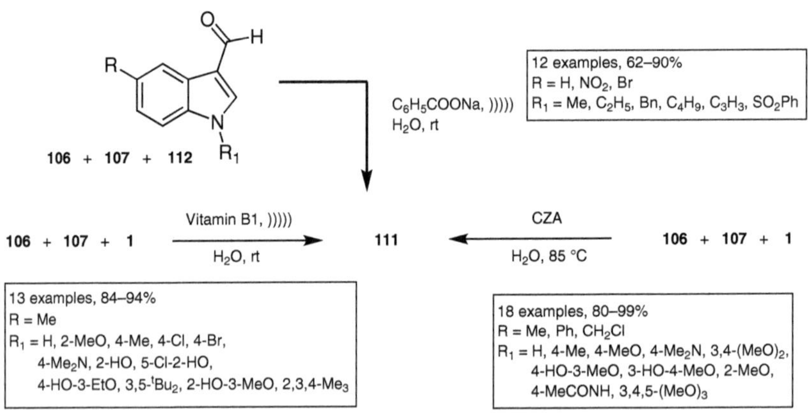

Scheme 6.28 Overview of isoxazoles prepared via a 3-CR using different catalysts in water.

irradiation and sodium benzoate as catalyst in water at room temperature to produce indolylidine isoxazole derivatives **111** [70]. Likewise, Ostadzadeh and Kiyani reported the synthesis of isoxazole-5(4*H*)-one derivatives **111** in an aqueous medium at 85 °C using citrazinic acid (CZA) as a catalyst [71].

6.3.2 Six-Membered Heterocycles

6.3.2.1 Oxygen Heterocycles

Aldehyde **1** and malonontrile (**25**) represent the primary precursors for the synthesis of fused pyranopyrimidines, tetrahydrobenzopyrans, and chromenes. These compounds were prepared via a tandem Knoevenagel/Michael addition/cyclization one-pot reaction, with water-compatible Michael donors, such as **16**, **64**, **12**, **116**, **118**, or **120** resulting in libraries of different heterocycles upon cyclization (Scheme 6.29).

The synthesis of pyranopyrimidines **113** was described by Kondabanthini et al. (Scheme 6.30). Their protocol made use of ultrasonic irradiation in water at 30 °C together with sulfonic acid-functionalized Wang resin, which served as a polymeric and recoverable catalyst [72]. Alternatively, novel chitosan-ZnO (CS-ZnO) was used

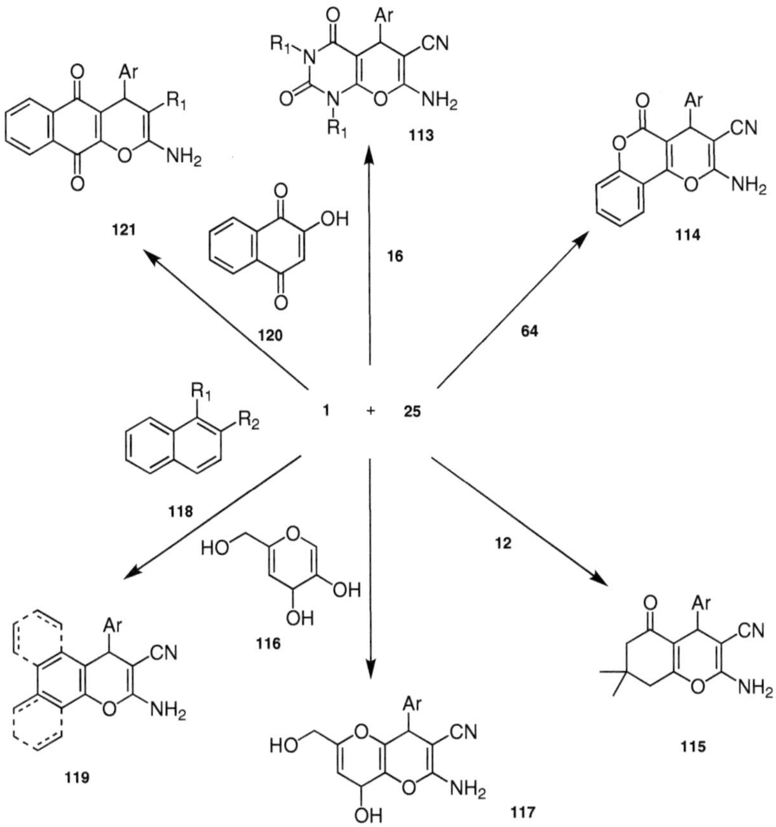

Scheme 6.29 Overview of fused pyranopyrimidines, tetrahydrobenzopyrans, and chromenes derived from aldehydes and malononitrile.

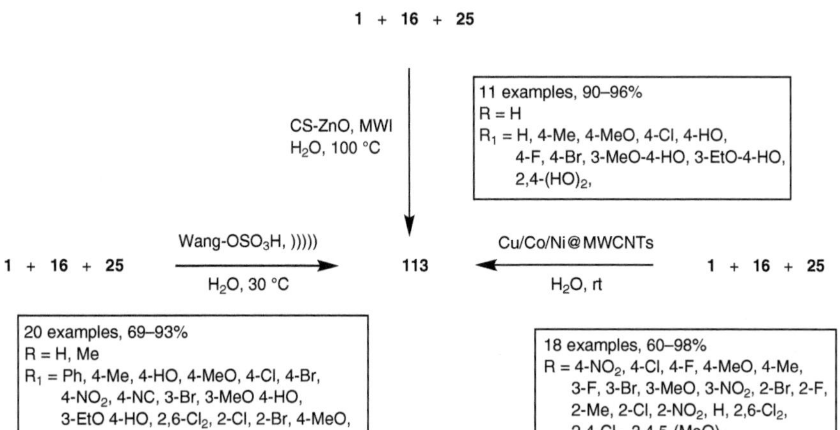

Scheme 6.30 Synthesis of 7-amino-6-cyano-1,5-dihydro-*1*H-pyrano[2,3-*d*]pyrimidine-2,4(3*H*)-diones through a 3-CR in water.

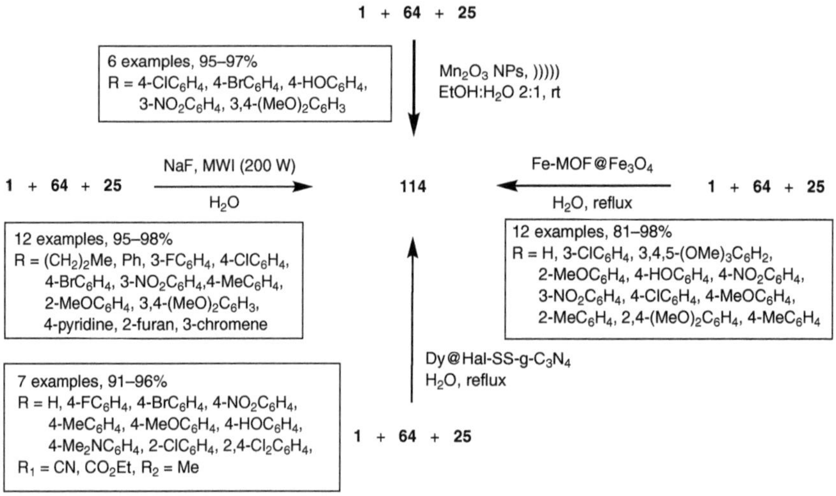

Scheme 6.31 Preparation of pyrano[3,2-*c*]chromene derivatives via a 3-CR in water.

as a catalyst in water at 100 °C under microwave irradiation (MWI) resulting in shorter reaction times and improved yields [73]. Harooni et al. reported a one-pot synthesis for the preparation of pyranopyrimidines catalyzed by MWCNTs MNCs in aqueous media. In this case, Cu/Co/Ni@MWCNTs served as a catalyst for the activation of the carbonyl group in **1** facilitating the nucleophilic attack of malononitrile (**25**) on the activated carbonyl center. Michael addition with 1,3-diethyl barbituric acid (**16**) onto the Knoevenagel adduct resulted in the formation of a new intermediate, which underwent Thorpe–Ziegler cyclization to give the final pyranopyrimidines **113**.

Shaikh et al. reported an ultrasound-assisted synthetic protocol for the preparation of pyrano[3,2-*c*]chromene derivatives in the presence of Mn_2O_3 MNPs as a catalyst in a mixture of ethanol and water (2 : 1) (Scheme 6.31). The authors aimed to develop a green method to derive 2-amino-4*H*-chromene as conventional synthetic procedures made use of harmful organic solvents, long reaction times, and low yields [74]. Furthermore, Sheikhhosseini and Yahyazadehfar reported the synthesis of chromene compounds **114** employing a Fe-hydroxyquinoline@Fe_3O_4 catalyst under reflux conditions. Hereby, the hydroxyquinoline was used to prepare a metal–organic framework (MOF) together with Fe to modify and protect the NPs of Fe_3O_4 [75]. Rahimzadeh et al. synthesized similar compounds in the presence of NPs composed of dysprosium and balsalazide trapped between functionalized graphitic carbon nitrides (g-C_3N_4) and halloysite (Dy@Hal-SS-g-C_3N_4) as a heterogeneous catalyst under reflux conditions [76]. The catalyst was used for activation of the carbonyl group in aromatic aldehyde **1** and the nitrile in **25**. Analogously, Fabitha et al. described an approach using a base-catalyzed Knoevenagel/Michael addition/cyclization mechanism, employing sodium fluoride as the catalyst. The application of MWI shortened the reaction time and led to higher yields [77].

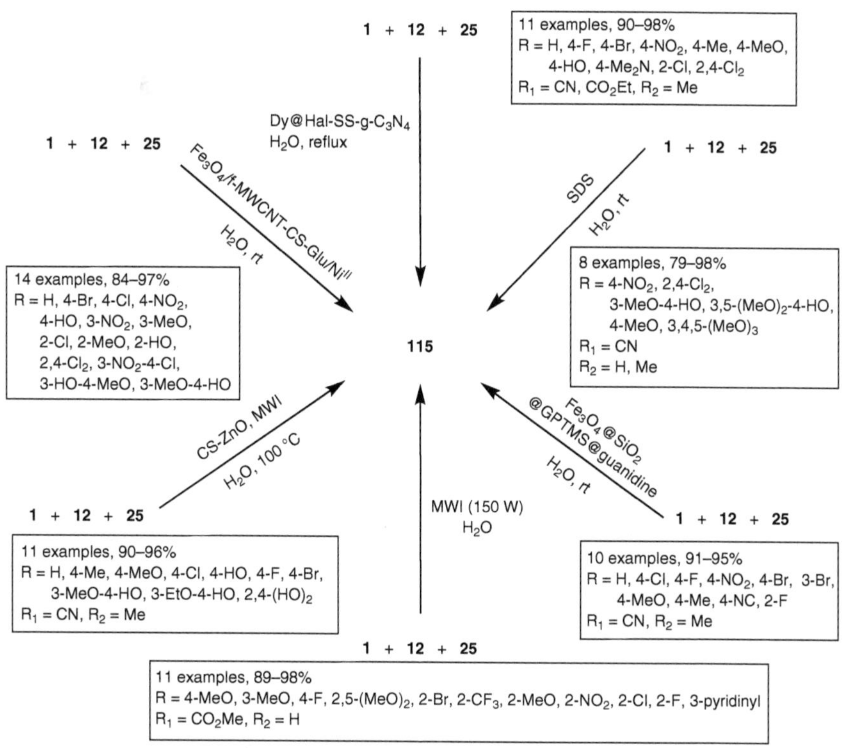

Scheme 6.32 Preparation of tetrahydrobenzo[*b*]pyran derivatives via a 3-CR in water.

Rahimzadeh et al. made use of a 3-CR to generate tetrahydrobenzo[*b*]pyran derivatives **115** in the presence of Dy@Hal-SS-g-C_3N_4 as heterogeneous catalyst under reflux conditions (Scheme 6.32) [76]. The utilization of sodium dodecyl sulfate (SDS) as catalyst in the synthesis to form compound **115** was described by Banerjee et al. SDS served as a metal-free surfactant-type catalyst for the Knoevenagel/Michael addition/cyclization mechanism [78]. Furthermore, preparation of 2-amino-4*H*-benzo[*b*]pyran **115** was also described in the presence of 3-glycidoxypropyltrimethoxysilane (GPTMS)/guanidine-protected MNPs as catalyst. The heterogeneous nanocatalyst was prepared by conjugating GPTMS-guanidine to MNPs, thereby facilitating the activation of the carbonyl in **1** and nitrile groups in **25** [79]. Khumalo et al. described an aqueous microwave-assisted synthesis to form tetrahydrobenzo[*b*]pyran derivatives **115** [80]. The previously described CS-ZnO catalyst in combination with MWI at 100 °C was also used in the synthesis of compounds **115**, again resulting in shorter reaction times and improved yields [72]. The novel green synthesis of coumarin derivatives using a matrix composed of L-glutamic acid cross-linked-chitosan supported on MWCNTs with Ni(II) was reported by Galehban et al. This synthetic procedure follows the Knoevenagel/Michael addition/Thorpe–Ziegler cyclization sequence [25b].

A similar procedure of Shaikh et al. using Mn_2O_3 MNPs as a catalyst (Scheme 6.31) was used to prepare pyrano[3,2-*b*]pyran derivatives **117** in a 2 : 1 mixture of

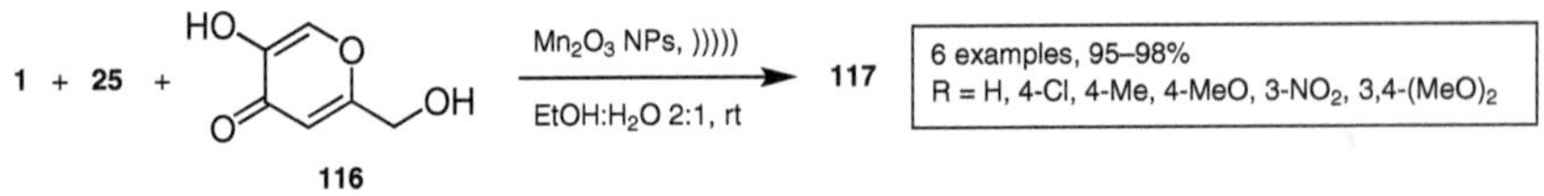

Scheme 6.33 Preparation of pyrano[3,2-*b*]pyran derivatives via ultrasound-assisted one-pot synthesis in ethanol and water.

ethanol and water (Scheme 6.33). The introduction of kojic acid (**116**) instead of 4-hydroxycoumarin (**64**) resulted in the formation of pyrano[3,2-*b*]pyran derivative **117** [74].

The sodium-fluoride-catalyzed microwave-assisted protocol in water of Fabitha et al. was extended to the synthesis of 2-amino-4*H*-benzo[*h*]chromene-3-carbonitriles **119** by substituting the phenolic reactant with either α- or β-naphthol **118** (Scheme 6.34) [77].

$$\mathbf{1} + \mathbf{118} + \mathbf{25} \xrightarrow[\text{H}_2\text{O}]{\text{NaF, MWI (200 W)}} \mathbf{119}$$

22 examples, 86–96%
R = Ph, 2-FC_6H_4, 3-FC_6H_4, 4-FC_6H_4, 2-ClC_6H_4, 3-ClC_6H_4, 4-ClC_6H_4, 2-BrC_6H_4, 3-BrC_6H_4, 4-BrC_6H_4, 2-$NO_2C_6H_4$, 3-$NO_2C_6H_4$, 4-$NO_2C_6H_4$, 2-MeC_6H_4, 3-MeC_6H_4, 4-MeC_6H_4, 2-$MeOC_6H_4$, 3-$MeOC_6H_4$, 4-$MeOC_6H_4$, 4-pyridine, 3-chromene, 2-furan
R_1 = R_2 = H, OH

Scheme 6.34 Preparation of 2-amino-4*H*-benzo[*h*]chromene-3-carbonitrile derivatives via a microwave-assisted 3-CR in the presence of sodium fluoride in water.

The synthesis of chromene derivatives **121** was performed in the research of Rahimzadeh et al. (Scheme 6.35). Their protocol made use of a 3-CR catalyzed by Dy@Hal-SS-g-C_3N_4 as heterogeneous catalyst under reflux conditions [76].

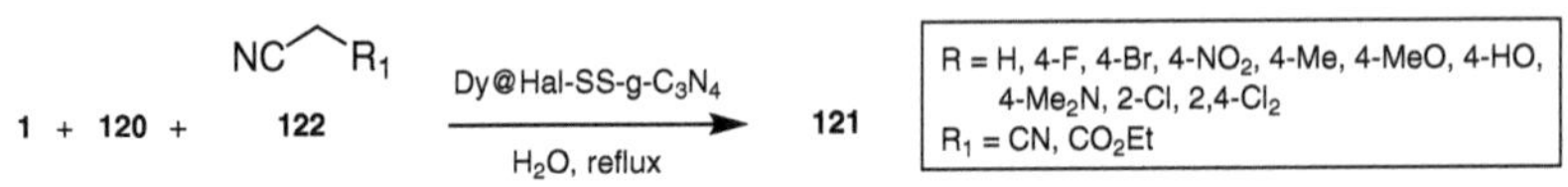

Scheme 6.35 Preparation of chromene derivatives via a 3-CR in the presence of a heterogenous catalyst in water.

Pyranopyridocarbazole derivatives **124** were synthesized through an ultrasound-assisted 3-CR in the presence of Fe_3O_4 MNPs as catalyst in water at room temperature (Scheme 6.36). The reaction involved a Michael addition with the Knoevenagel intermediate and 4-hydroxypyridocarbazolone (**123**), after which the ensuing intermediate cyclized to yield the final product **124**. Parmar et al. compared the ultrasound-assisted method with conventional reflux conditions in water. The use of ultrasound irradiation provided reaction conditions with shorter reaction times and higher yields for compounds **124** [81].

The synthesis of spirooxindoles via a tandem Knoevenagel/Michael addition/cyclization reaction in water at 90 °C or room temperature was described by

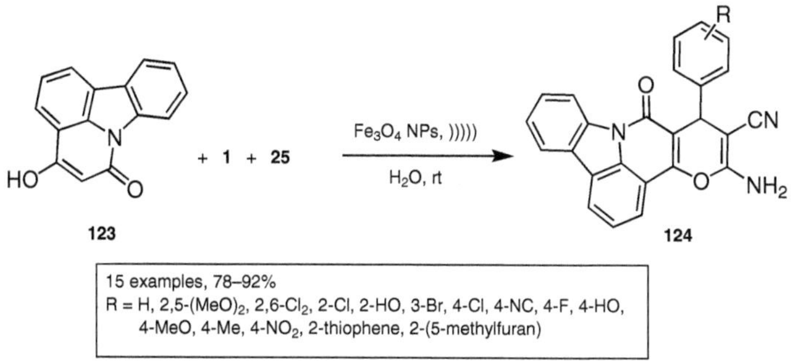

Scheme 6.36 Preparation of pyranopyridocarbazole derivatives via an ultrasound-assisted 3-CR in water.

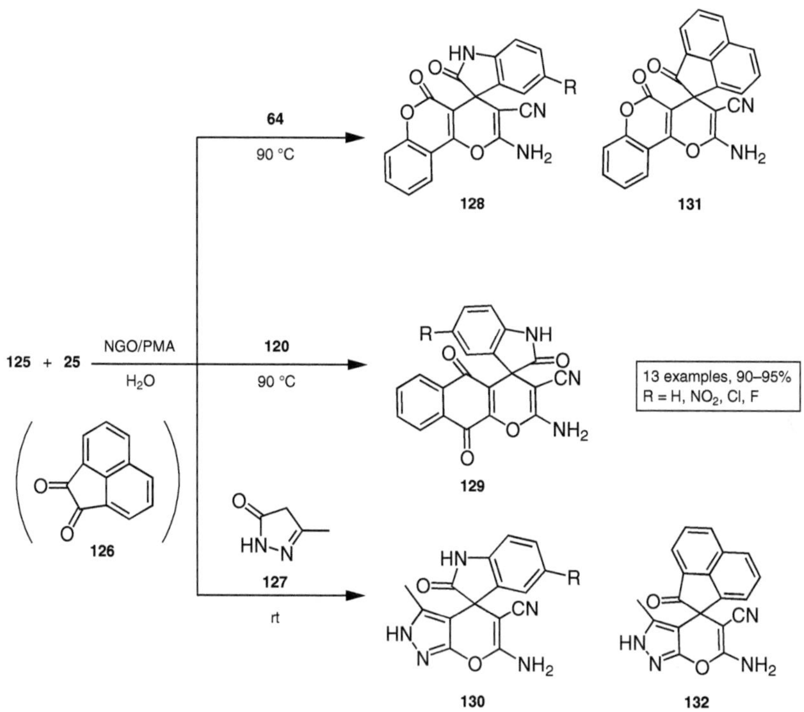

Scheme 6.37 Preparation of spirooxindole derivatives via N-doped graphene oxide immobilized phosphomolybdic acid (NGO/PMA)-assisted one-pot synthesis in water.

Palizi and Moradi. Their 3-CR made use of immobilized phosphomolybdic acid on N-doped graphene oxide (NGO/PMA) as a catalyst to prepare five different spirooxindole derivatives (Scheme 6.37) [82]. Notably, the catalyst was employed for the activation of the carbonyl groups in the Knoevenagel/Michael addition/cyclization mechanism. The use of isatin (**125**) as starting material resulted in the generation of products **128–130**, while the use of acenaphthoquinone (**126**) as starting material led to the formation of products **131** and **132**.

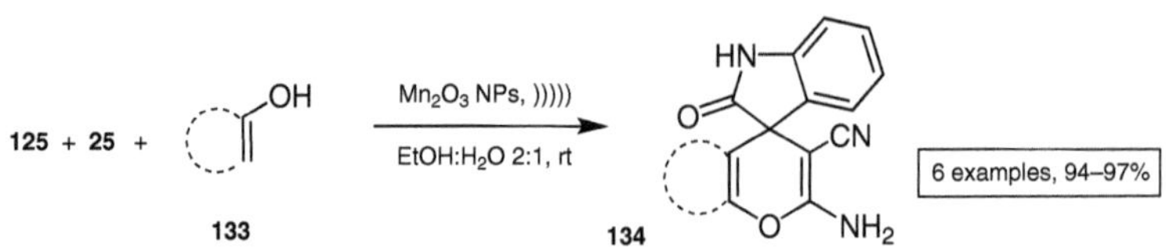

Scheme 6.38 Preparation of spirooxindoles via an ultrasound-assisted 3-CR in ethanol and water.

Furthermore, Shaikh et al. reported the ultrasound-assisted synthesis of spirooxindole derivatives, employing Mn_2O_3 MNPs as a catalyst in a mixture of ethanol and water (2 : 1) (Scheme 6.38) [74]. This procedure followed the previously described Knoevenagel condensation/Michael addition/Thorpe–Zeigler cyclization pathway. In this context, malononitrile (**25**) served as the primary starting material, in combination with an enol **133** and isatin (**125**) to yield spirooxindole **134**.

Tabarsa et al. described a synthetic approach for the preparation of spiropyrrolooxindoles through a 5-CR utilizing Ag/KF/CP@MWCNTs as a catalyst in water at room temperature (Scheme 6.39) [83]. First, isatin (**125**) and malononitrile (**25**) reacted in the presence of Ag/KF/CP@MWCNTs yielding intermediate **135**. Simultaneously, ethyl acetoacetate (**136**) and primary amine **28** reacted in the presence of Ag/KF/CP@MWCNTs resulting in intermediate **137**. Subsequently, both intermediates reacted further and cyclized to intermediate **138**, which reacted

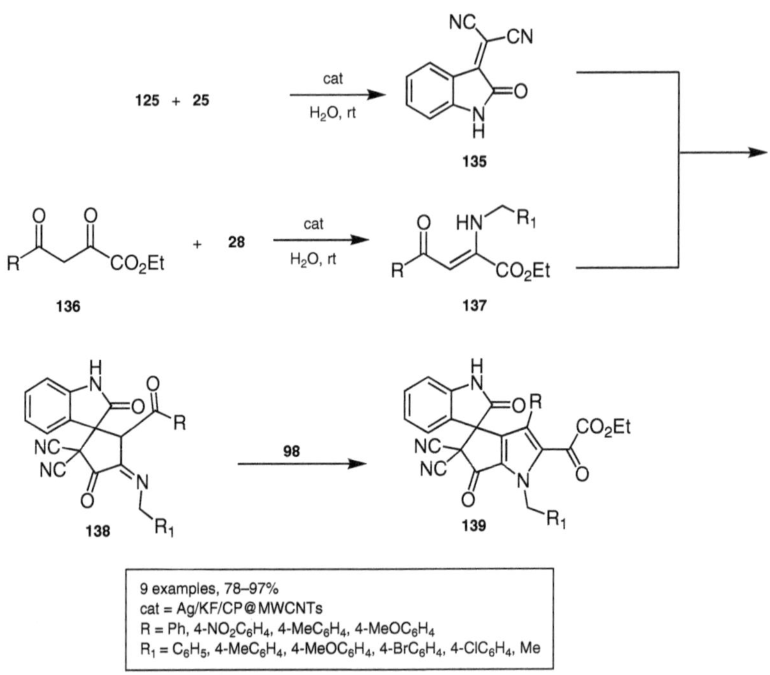

Scheme 6.39 Preparation of spiropyrrolooxindole via a 5-C Knoevenagel/cyclization reaction in the presence of Ag/KF/CP@MWCNTs in water.

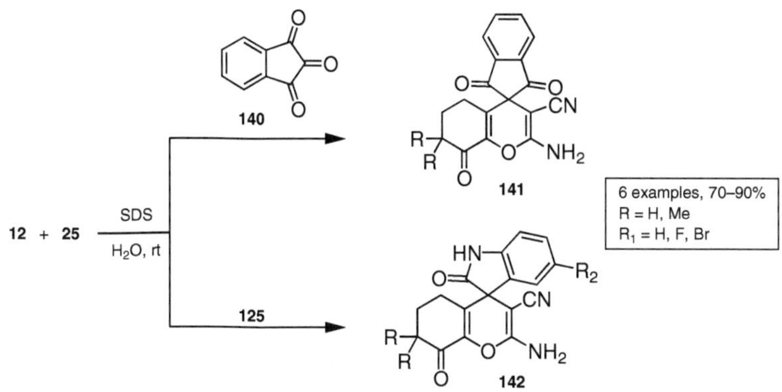

Scheme 6.40 Preparation of spiropyran derivatives via an SDS-catalyzed 3-CR in water.

with ethyl bromopyruvate (**98**) leading after cyclization and water elimination to the formation of spiropyrrolooxindole **139**.

The synthesis of spiropyrans through a 3-CR in the presence of SDS in water at room temperature was described by Banerjee et al. (Scheme 6.40) [78]. The use of either ninhydrin (**140**) or isatin (**125**) as starting material resulted in the synthesis of spiropyrans **141** and **142**.

Pyrazolopyranopyrimidine derivatives **147** were obtained through a 4-CR catalyzed by starch@Fe_3O_4 and the utilization of ultrasonic irradiation in water at room temperature (Scheme 6.41) [84]. Starch, being a natural polysaccharide of D-glucose, possesses hydroxyl functional groups that enhance the surface activity and stability of NPs [85]. The first step was the condensation between ethyl acetoacetate (**143**) and diamine **144** to form imine intermediate **145**. Subsequently, enolate **146** was generated through elimination of ethanol. The following steps involved the tandem Knoevenagel/Michael addition/cyclization mechanism using barbituric acid (**16**) and aromatic aldehyde **1** to yield final product **147**.

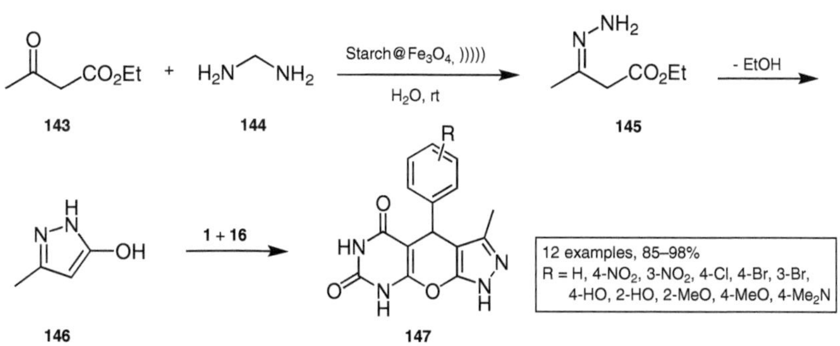

Scheme 6.41 Preparation of pyrazolopyranopyrimidines using ultrasound conditions and starch@Fe_3O_4 in water.

Patel et al. reported an ultrasound-assisted 4-CR for the preparation of 1,4-dihydropyrano[2,3-*c*]pyrazole derivatives **149** in water at room temperature (Scheme 6.42).

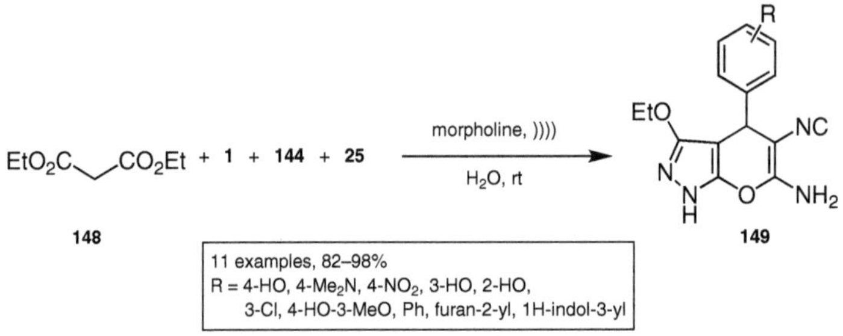

Scheme 6.42 Preparation of 1,4-dihydropyrano[2,3-*c*]pyrazole derivatives via an ultrasound-assisted one-pot synthesis in morpholine and water.

Morpholine was used as catalyst and the reaction was conducted in an aqueous medium, involving the Knoevenagel/Michael addition/cyclization reaction sequence [86].

6.3.2.2 Nitrogen Heterocycles

Kumar et al. reported a 3-CR employing MWI for the synthesis of aminocyanopyridines **151** in water at 50 °C (Scheme 6.43). The synthesis involved a reaction between the chromene moiety of the Knoevenagel intermediate and *in situ*-generated ammonia, to form an intermediate which upon cyclization formed product **151**. Comparison was made between the conventional reflux-based synthesis procedure and the microwave-assisted protocol. Notably, the microwave-assisted synthesis demonstrated advantages by affording higher yields of the final products while achieving remarkably short reaction times of only a few minutes [87].

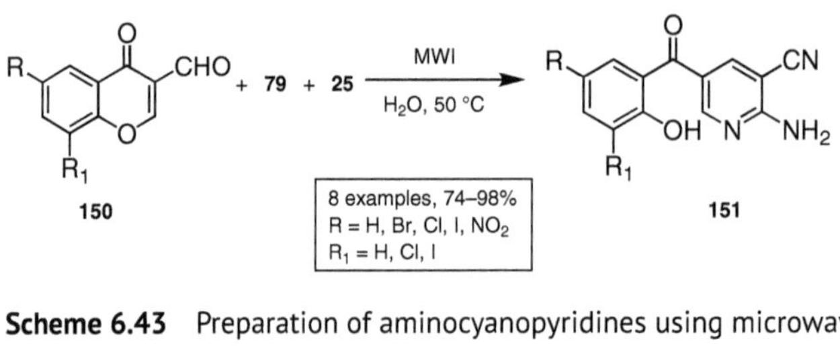

Scheme 6.43 Preparation of aminocyanopyridines using microwave conditions in water.

The synthesis of pyrimidine derivatives via a 3-CR in water at 70 °C was documented by Kaurav et al. (Scheme 6.44). This synthetic route followed the same Knoevenagel/Michael addition/cyclization approach, this time, in the presence of the amino acid L-proline as the catalyst [88].

The synthesis of pyrido[2,3-*d*]pyrimidines was performed using ultrasonication conditions in water at room temperature (Scheme 6.45). This process was facilitated by silver (Ag) immobilized on the surface of titanium dioxide (TiO_2) as a catalyst for the Knoevenagel/Michael addition/cyclization mechanism [89]. Presence of the amine group in 6-amino-1,3-dimethyluracil (**154**) resulted in the formation of pyrimidine **156**, while the use of barbituric acid (**16**) led to the formation of the previously described pyran derivatives. Derivatives of 1,6-naphthyridine-2,5-dione **157** were prepared via a 3-C Knoevenagel/Michael addition/cyclization reaction,

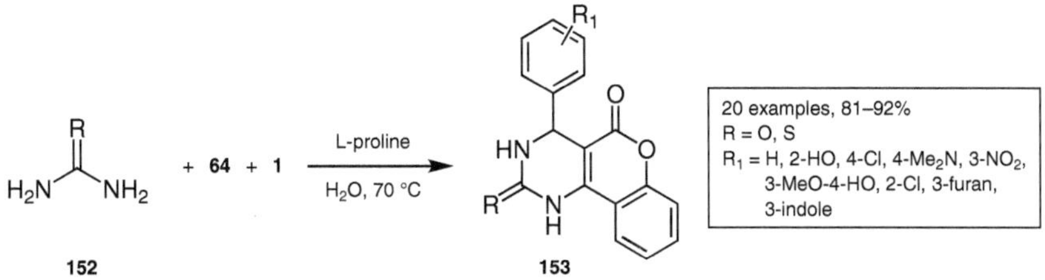

Scheme 6.44 Preparation of fused pyrimidines in the presence of *L*-proline in water.

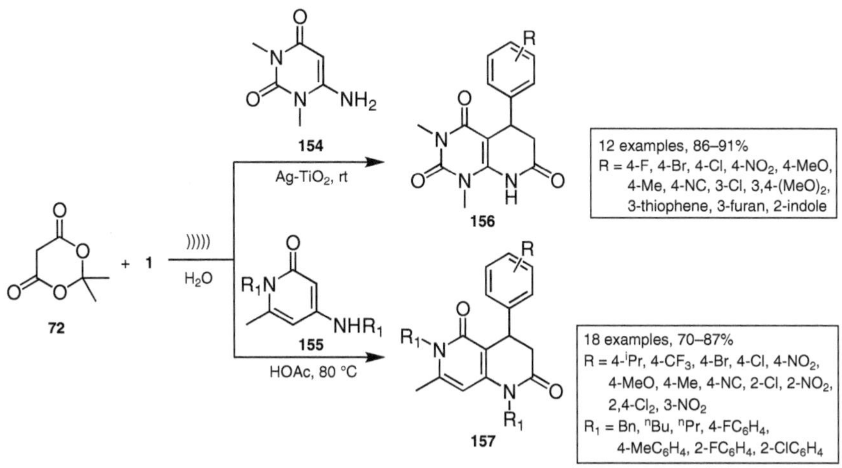

Scheme 6.45 Preparation of pyrido[2,3-*d*]pyrimidines and 1,6-naphthyridine-2,5-diones via ultrasound-assisted synthesis in water.

catalyzed by acetic acid in water at 80 °C. This protocol harnessed the benefits of ultrasonic irradiation, resulting in shorter reaction times and higher yields [90].

The microwave-assisted synthesis of thiadiazoloquinazolinone-coumarin hybrids **160** in water at 120 °C was reported by Wadhwa et al. (Scheme 6.46). This synthetic approach followed the same Knoevenagel/Michael addition/cyclization mechanism in the presence of $Bi(NO_3)_3 \cdot 5\ H_2O$, which under the reaction conditions decomposed into nitric acid and $Bi(OH)_3$ [91]. Both resulting species were thought to actively facilitate the mechanistic progression.

The preparation of 1,4-dihydropyridines **163** followed the same Knoevenagel/Michael addition/cyclization strategy as described before. Khumalo et al. reported a synthetic protocol using MWI and triethylamine (NEt_3) as a catalyst in an aqueous medium (Scheme 6.47) [92]. Intermediate **162** was generated via a Michael addition, while subsequent condensation of the alcohol with ammonium acetate (**79**) yielded an amine, which in turn cyclized into compound **163** through the elimination of water.

Functionalized 1,4-dihydropyridines **167** were synthesized through a catalyst-free and microwave-assisted 4-CR in water at room temperature (Scheme 6.48) [93]. Following the tandem Knoevenagel/Michael addition/cyclization reaction, 1,4-dihydropyridine **167** was obtained in excellent yields.

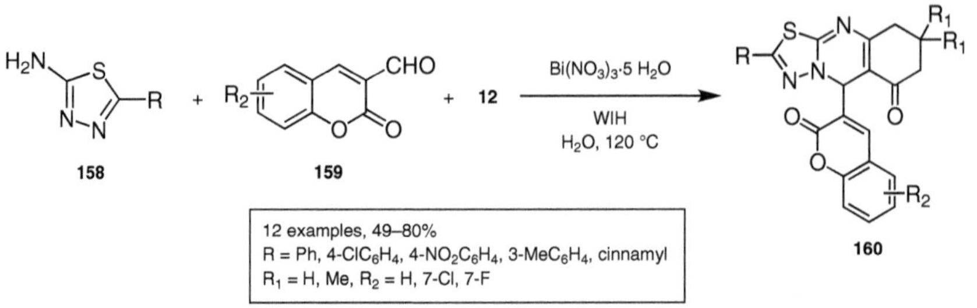

Scheme 6.46 Preparation of thiadiazoloquinazolinone-coumarin hybrids using microwave conditions in water.

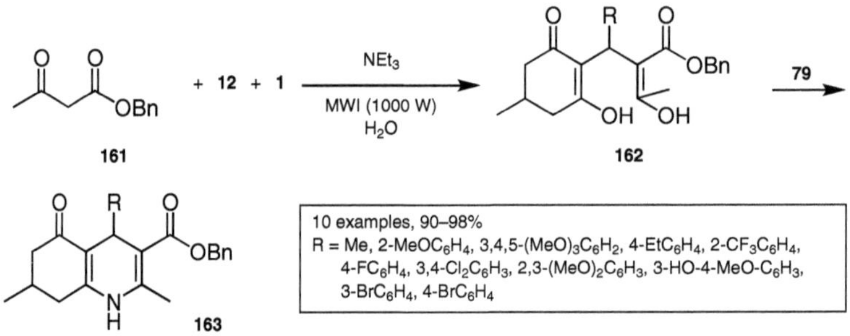

Scheme 6.47 Preparation of 1,4-dihydropyridines using microwave conditions in water.

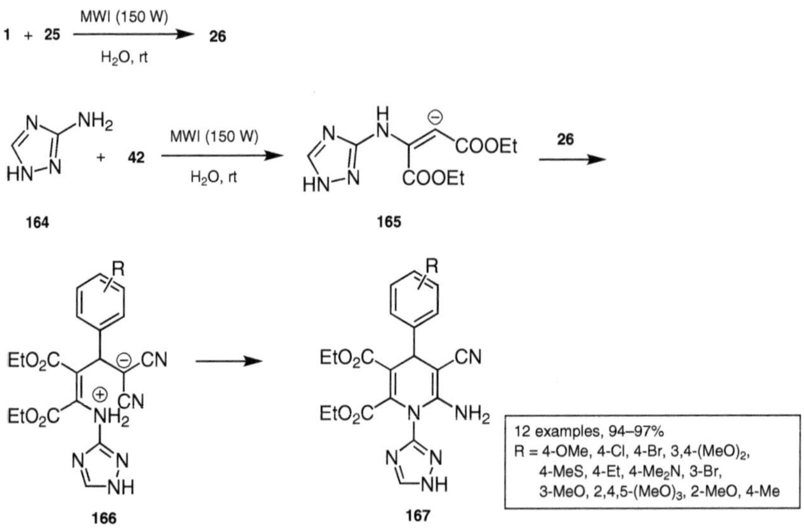

Scheme 6.48 Preparation of 1,4-dihydropyridines using microwave conditions in water.

The preparation of tetrahydrodipyrazolopyridines via a 6-CR in water at room temperature was described by Keihanfar and Mirjalili (Scheme 6.49) [94] First, intermediate **145** was formed twice via the same method as depicted in Scheme 6.41. Intermediate **146**, aromatic aldehyde **1**, and ammonium acetate (**79**) underwent Michael addition and cyclization, forming the targeted products **168**.

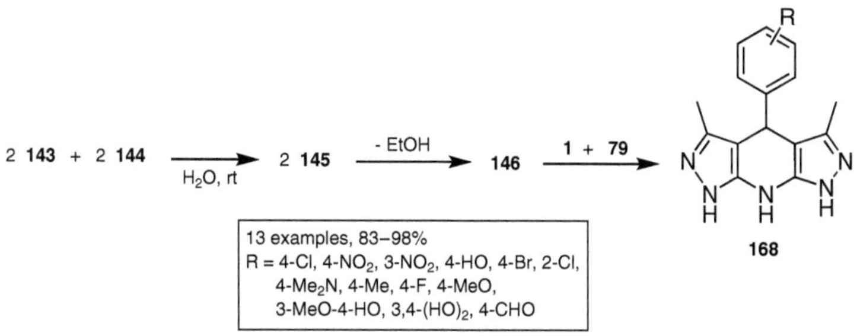

Scheme 6.49 Preparation of tetrahydrodipyrazolopyridines via a 6-CR in water.

Hossaini et al. reported the synthesis of spiropyridoindolepyrrolidines through a green 6-CR catalyzed by $Ag/Fe_3O_4/CdO/MWCNTs$ in water at room temperature (Scheme 6.50) [95]. The synthesis commenced with a substitution between isatin (**125**) and acetyl chloride (**169**), giving intermediate **170** which upon reaction with secondary amine **171** formed iminium ion **172**. Intermediate **174** was formed after cyclization of iminium ion **172** and vinylidene Meldrum's acid (**173**), accompanied

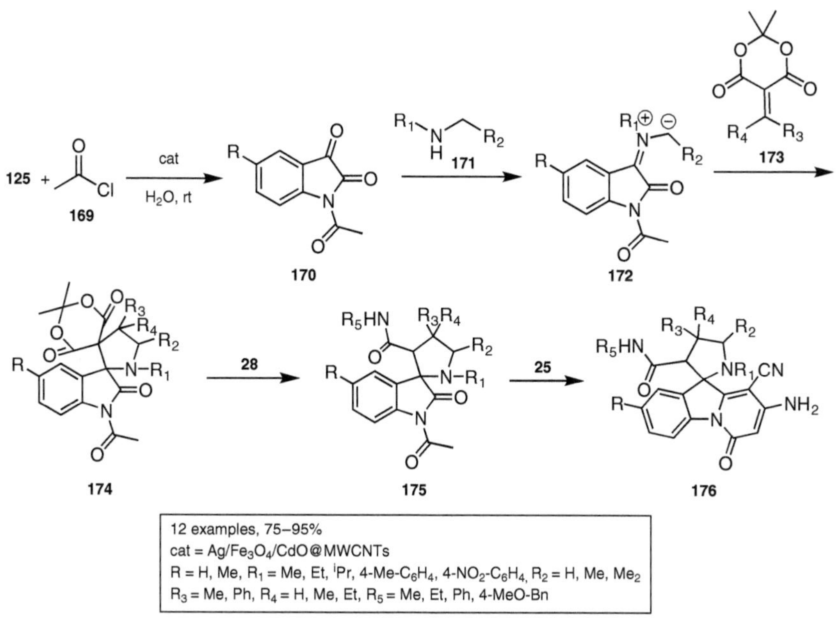

Scheme 6.50 Preparation of spiropyridoindolepyrrolidines via a 6-CR in the presence of $Ag/Fe_3O_4/CdO/MWCNTs$ in water.

by elimination of acetone and carbon dioxide. Primary amine **28** and intermediate **174** reacted to form amide **175**. The final step involved a Knoevenagel condensation and cyclization of malononitrile (**25**) and amide **175** in the presence of the catalyst, affording spiropyridoindolepyrrolidines **176** [95].

The synthesis of spiropyridoindoles using MWCNTs in water at room temperature was investigated by Tabarsaei et al. (Scheme 6.51). Ag exhibited the most potent catalytic effect as a Lewis acid compared to Cd and Fe [96]. The first steps led to

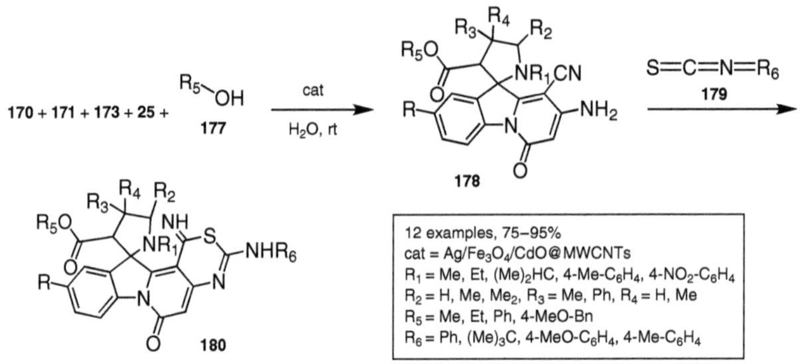

12 examples, 75–95%
cat = $Ag/Fe_3O_4/CdO@MWCNTs$
R_1 = Me, Et, $(Me)_2HC$, 4-Me-C_6H_4, 4-NO_2-C_6H_4
R_2 = H, Me, Me_2, R_3 = Me, Ph, R_4 = H, Me
R_5 = Me, Et, Ph, 4-MeO-Bn
R_6 = Ph, $(Me)_3C$, 4-MeO-C_6H_4, 4-Me-C_6H_4

Scheme 6.51 Preparation of spiropyridoindoles via a 6-CR in the presence of $Ag/Fe_3O_4/CdO@MWCNTs$ in water.

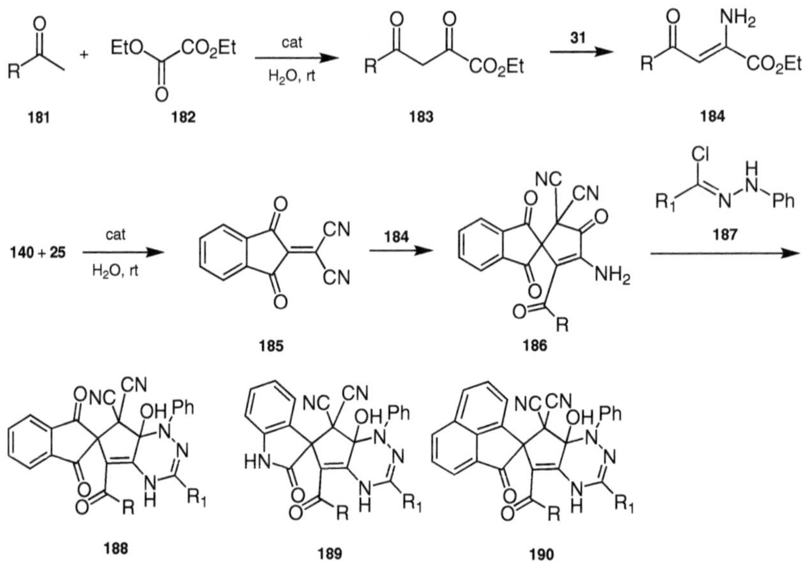

14 examples, 80–95%
cat = $Ag/Fe_3O_4/CdO@MWCNTs$, NEt_3
R = 4-MeO-C_6H_4, 4-Me-C_6H_4, 4-Et-C_6H_4, 4-NO_2-C_6H_4
R_1 = Ph, 4-MeO-C_6H_4, 4-Me-C_6H_4

Scheme 6.52 Preparation of spiro-1,2,4-triazine derivativesvia a 6-CR in the presence of $Ag/Fe_3O_4/CdO@MWCNTs$ in water.

the formation of intermediate **178** following a similar procedure as depicted in Scheme 6.50. Subsequently, intermediate **178** reacted with isothiocyanate **179** and cyclized in the presence of the catalyst yielding spiropyridoindole **180**.

Spiro-1,2,4-triazine derivatives have been synthesized via a 6-CR in the presence of NEt_3 and $Ag/Fe_3O_4/CdO@MWCNTs$ as catalyst in water at room temperature (Scheme 6.52) [97]. Acetophenone **181** and diethyl oxalate (**182**) reacted in the presence of $Ag/Fe_3O_4/CdO@MWCNTs$ to give 1,3-dicarbonyl compounds **183**. Subsequently, ammonium acetate (**79**) reacted with intermediate **183** to form enamine **184**. Knoevenagel intermediate **185** underwent cyclization with this enamine to yield spirocycle intermediate **186**, which reacted further with hydrazonoyl chloride **187** and cyclized to form spiro-1,2,4-triazine **188**. Alternative spiro-1,2,4-triazines **189** and **190** were synthesized following the same strategy, but using other starting materials, replacing compound **140** with either isatin (125) or acenaphthylene-1,2-dione [97].

6.4 Summary and Conclusions

In the past decades, the incorporation of MCRs in the synthesis of heterocyclic scaffolds has attracted increasing attention. MCRs, characterized by their ability to combine three or more available starting materials in one-pot, have, in many instances, proven to be a more efficient approach for the synthesis of particular heterocycles than their conventional multistep synthetic alternatives. The use of MCRs in the field of organic synthesis has contributed to reduced time, manpower, and the use of chemicals, leading to a reduction in the E-factor and mass intensity, key considerations within the principles of green chemistry. An additional important aspect is the use of green solvents in organic synthesis which has also contributed significantly to the use of water in MCRs. Beyond adhering to the aims of green chemistry, performing MCRs in water has the benefit that it facilitates accelerating the reaction rate and reduces workup and purification steps as in many instances, a simple filtration step was sufficient to obtain the final product. With the use of water also, a strongly increased application of MWCNTs MNCs is witnessed in I-MCRs and Knoevenagel-initiated MCRs. Considering the need for developing greener synthesis methods and the continuing relevance of heterocycles in the search for new drugs, it is highly anticipated that new synthetic methods employing MCRs in water will continue to appear from future research.

References

1 Zhu, J. and Bienaymé, H. (2005). *Multicomponent Reactions: Front Matter*, vol. 2 (ed. J. Zhu and H. Bienaymé), i–xvi. Weinheim, Germany: Wiley-VCH.
2 Weber, L. (2002). *Drug Discov. Today* 7: 143–147.
3 Dömling, A. (2006). *Chem. Rev.* 106: 17–89.
4 Dömling, A. and Ugi, I. (2000). *Angew. Chem. Int. Ed.* 39: 3168–3210.

5 Fan, W., Verrier, C., Wang, L. et al. (2020). *Recent Trends in Carbohydrate Chemistry*, vol. 1 (ed. A.P. Rauter, B.E. Christensen, L. Somsák, et al.), 73–100. Elsevier.

6 Cioc, R.C., Ruijter, E., and Orru, R.V.A. (2014). *Green Chem.* 16: 2958–2975.

7 Constable, D.J.C., Jimenez-Gonzalez, C., and Henderson, R.K. (2007). *Org. Process Res. Dev.* 11: 133–137.

8 Rahmati, A. and Pashmforoush, N. (2015). *J. Iran. Chem. Soc.* 12: 993–1036.

9 Simon, M. and Li, C. (2012). *Chem. Soc. Rev.* 41: 1415–1427.

10 Cortes-Clerget, M., Yu, J., Kincaid, J.R.A. et al. (2021). *Chem. Sci.* 12: 4237–4266.

11 Rideout, D.C. and Breslow, R. (1980). *J. Am. Chem. Soc.* 102: 7816–7817.

12 Jung, Y. and Marcus, R.A. (2007). *J. Am. Chem. Soc.* 129: 5492–5502.

13 Lubineau, A., Augé, J., and Queneau, Y. (1994). *Synthesis* 1994: 741–760.

14 Nasiriani, T., Javanbakht, S., Nazeri, M.T. et al. (2022). *Top. Curr. Chem.* 380: 50.

15 Lubineau, A. (1986). *J. Org. Chem.* 51: 2142.

16 Pirrung, M.C. and Sarma, K.D. (2004). *J. Am. Chem. Soc.* 126: 444–445.

17 (a) Abdolmohammadi, S., Hossaini, Z., and Heravi, M.R.P. (2022). *Mol. Divers.* 26: 1983–1993. (b) Shylesh, S., Schünemann, V., and Thiel, W.R. (2010). *Angew. Chem. Int. Ed.* 49: 3428–3459.

18 Aghahosseini, H. and Ramazani, A. (2020). *Eurasian Chem. Commun.* 2: 410–419.

19 Polshettiwar, V., Luque, R., Fihri, A. et al. (2011). *Chem. Rev.* 111: 3036–3075.

20 Morales-Cid, G., Fekete, A., Simonet, B.M. et al. (2010). *Anal. Chem.* 82: 2743–2752.

21 Hussain, C.M. and Mitra, S. (2011). *Anal. Bioanal. Chem.* 399: 75–89.

22 (a) Zhang, B., Zheng, X., Li, H., and Lin, J. (2013). *Anal. Chim Acta* 784: 1–17. (b) Iijima, S. (1991). *Nature* 354: 56–58.

23 Harooni, N.S., Ghasemi, A.H., and Naeimi, H. (2023). *J. Clust. Sci.* 34: 2189–2203.

24 (a) Nikoofar, K., Mehrikaram, F., Lashanizadegan, M., and Aghaei, A. (2021). *Polycycl. Aromat. Compd.* 41: 663–675. (b) Saeidiroshan, H. and Moradi, L. (2019). *J. Organomet. Chem.* 893: 1–10. (c) Alshabanah, L.A., Gomha, S.M., Al-Mutabagani, L.A. et al. (2021). *Polymers* 13: 1728.

25 (a) Hezarcheshmeh, N.K. and Azizian, J. (2022). *Mol. Divers.* 26: 2011–2024. (b) Galehban, M.H., Zeynizadeh, B., and Mousavi, H. (2022). *RSC Adv.* 12: 16454–16478.

26 Shapiro, N. and Vigalok, A. (2008). *Angew. Chem. Int. Ed.* 47: 2849–2852.

27 Butler, R.N. and Coyne, A.G. (2010). *Chem. Rev.* 110: 6302–6337.

28 Kumaravel, K. and Vasuki, G. (2009). *Curr. Org. Chem.* 13: 1820–1841.

29 Nazeri, M.T., Nasiriani, T., Farhid, H. et al. (2022). *ACS Sustain. Chem. Eng.* 10: 8115–8134.

30 Javanbakht, S., Nasiriani, T., Farhid, H. et al. (2022). *Front. Chem. Sci. Eng.* 16: 1318–1344.

31 Pagadala, R., Kasi, V., Shabalala, N.G., and Jonnalagadda, S.B. (2022). *Arab. J. Chem.* 15: 103544.

32 Russo, C., Brunelli, F., Tron, G.C., and Giustiniano, M. (2023). *Chem. Eur. J.* 29: e202203150.

33 Gu, Y. (2012). *Green Chem.* 14: 2091–2128.

34 (a) Hulme, C.C., Hulme, C., and Gore, V. (2003). *Curr. Med. Chem.* 10: 51–80. (b) Sadjadi, S., Heravi, M.M., and Nazari, N. (2016). *RSC Adv.* 6: 53203–53272.

35 van Berkel, S.S., Bögels, B.G.M., Wijdeven, M.A. et al. (2012). *Eur. J. Org. Chem.* 19: 3543–3559.

36 Banik, B.K., Becker, F.F., and Banik, I. (2004). *Bioorg. Med. Chem.* 12: 2523–2528.

37 Kanizsai, I., Szakonyi, Z., Sillanpää, R., and Fülöp, F. (2006). *Tetrahedron Lett.* 47: 9113–9116.

38 Kanizsai, I., Gyónfalvi, S., Szakonyi, Z. et al. (2007). *Green Chem.* 9: 357–360.

39 Banerjee, R., Kumar, H.K.S., and Banerjee, M. (2015). *Int. J. Res. Phytochem. Pharmacol.* 5: 48–57.

40 Shaabani, A., Teimouri, M.B., and Bijanzadeh, H.R. (2004). *Monatsh. Chem.* 135: 441–446.

41 Shaabani, A., Teimouri, M.B., and Bijanzadeh, H.R. (2002). *Tetrahedron Lett.* 43: 9151–9154.

42 Heravi, M.M., Baghernejad, B., and Oskooie, H.A. (2009). *Mol. Divers.* 13: 385–387.

43 Teimouri, M.B. and Bazhrang, R. (2008). *Monatsh. Chem.* 139: 957–961.

44 Mousavi, H., Zeynizadeh, B., and Rimaz, M. (2023). *Bioorg. Chem.* 135: 106390.

45 Nikoofar, K. and Khademi, Z. (2016). *Res. Chem. Intermed.* 42: 3929–3977.

46 Pagadala, R., Kommidi, D.R., Kankala, S. et al. (2015). *Org. Biomol. Chem.* 13: 1800–1806.

47 Zhu, X., Xu, X., Sun, C. et al. (2010). *J. Comb. Chem.* 12: 822–828.

48 Mironov, M.A., Ivantsova, M.N., Tokareva, M.I., and Mokrushin, V.S. (2005). *Tetrahedron Lett.* 46: 3957–3960.

49 Kordmahalleh, M.Y., Moradi, A.V., Hossaini, Z., and Golsefidi, M.A. (2022). *Appl. Organomet. Chem.* 36: e6457.

50 Liu, H., Ye, Z., Cai, Z., and Ji, S. (2023). *Green Chem.* 25: 4239–4243.

51 Moghaddam, F.M., Bardajee, G.R., and Dolabi, M. (2010). *J. Sulfur Chem.* 31: 387–393.

52 Moghaddam, F.M., Khodabakhshi, M.R., and Latifkar, A. (2014). *Tetrahedron Lett.* 55: 1251–1254.

53 Akbari, A., Azami-Sardooei, Z., and Hosseini-Nia, A. (2013). *J. Korean Chem. Soc.* 57: 455–460.

54 Sarma, R., Sarmah, M.M., Lekhok, K.C., and Prajapati, D. (2010). *Synlett* 2010: 2847–2852.

55 Koldobskii, G., Ostrovskii, V., and Osipova, T. (1983). *Chem. Heterocycl. Compd.* 19: 1141–1154.

56 Shaabani, A., Hajishaabanha, F., Mofakham, H. et al. (2012). *Helv. Chim. Acta.* 95: 246–254.

57 Kolla, S.R. and Lee, Y.R. (2010). *Tetrahedron* 66: 8938–8944.

58 Amiri, S.S., Ghazvini, M., Khandan, S., and Afrashteh, S. (2022). *Polycycl. Aromat. Compd.* 42: 4736–4751.

59 Marqués-López, E., Herrera, R.P., and Rios, R. (2015). *Multicomponent Reactions* (ed. R.P. Herrera and E. Marqués-López), 358–381. Weinheim, Germany: Wiley-VCH.

60 Shaabani, A., Maleki, A., Mofakham, H., and Moghimi-Rad, J. (2008). *J. Org. Chem.* 73: 3925–3927.

61 Kohestani, T., Sayyed-Alangi, S.Z., Hossaini, Z., and Baei, M.T. (2023). *Polycycl. Aromat. Compd.* 43: 3392–3415.

62 Avarsaji, M., Hossaini, Z., Moradi, A.V. et al. (2023). *Polycycl. Aromat. Compd.* 43: 5996–6017.

63 Han, Y.F. and Min, X. (2010). *Curr. Org. Chem.* 14: 379–413.

64 Bigi, F. (2012). *Curr. Org. Synth.* 9: 31–39.

65 (a) Voskressensky, L.G., Festa, A.A., and Varlamov, A.V. (2014). *Tetrahedron* 70: 551–572. (b) Majumdar, K.C., Taher, A., and Nandi, R.K. (2012). *Tetrahedron* 68: 5693–5718. (c) Tokala, R., Bora, D., and Shankaraiah, N. (2022). *ChemMedChem* 17: e202100736.

66 Banitaba, S.H. (2023). *Polycycl. Aromat. Compd.* 43: 2321–2334.

67 Sapkal, A. and Kamble, S. (2020). *J. Heterocycl. Chem.* 57: 3597–3604.

68 Daroughehzadeh, Z. and Kiyani, H. (2023). *Polycycl. Aromat. Compd.* 1–22.

69 Zhang, D., Liu, C., Ren, L. et al. (2023). *ChemistrySelect* 8: e202204658.

70 Reddy, C.V.R. and Reddy, G.G. (2020). *Chem. Afr.* 3: 61–74.

71 Ostadzadeh, H. and Kiyani, H. (2023). *Org. Prep. Proced. Int.* 1–11.

72 Kondabanthini, S., Akshinthala, P., Katari, N.K. et al. (2023). *J. Mol. Struct.* 1276: 134753.

73 Nesaragi, A.R., Gasti, T., Metre, T.V. et al. (2022). *ChemistrySelect* 7: e202200604.

74 Shaikh, S., Yellapurkar, I., and Ramana, M.M.V. (2021). *Chem. Afr.* 4: 821–834.

75 Sheikhhosseini, E. and Yahyazadehfar, M. (2023). *Front. Chem.* 10.

76 Rahimzadeh, G., Tajbakhsh, M., Daraie, M., and Mohammadi, M. (2022). *Appl. Organomet. Chem.* 36: e6829.

77 Fabitha, K., Arya, C.G., Chandrakanth, M., and Banothu, J. (2023). *Res. Chem. Intermed.* 49: 997–1014.

78 Banerjee, B., Priya, A., Kaur, M. et al. (2023). *Cat. Lett.* 12: 3547–3560.

79 Taherkhani, H., Ramazani, A., Sajjadifar, S. et al. (2021). *ChemistrySelect* 6: 11362–11374.

80 Khumalo, M.R., Maddila, S.N., Maddila, S., and Jonnalagadda, S.B. (2019). *BMC Chem.* 13: 132.

81 Parmar, S., Sharma, V., Jain, S., and Verma, D. (2020). *ChemistrySelect* 5: 11621–11629.

82 Palizi, N. and Moradi, L. (2023). *Appl. Organomet. Chem.* 37: e7196.

83 Tabarsa, N., Mehrabian, R.Z., Alangi, S.Z.S., and Hossaini, Z. (2023). *Polycycl. Aromat. Compd.* 1–22.

84 Dharmendra, D., Chundawat, P., Vyas, Y., and Ameta, C. (2022). *J. Chem. Sci.* 134: 47.

85 Verma, P., Pal, S., Chauhan, S. et al. (2020). *J. Mol. Struct.* 1203: 127410.

86 Patel, S., Patel, S., Vaghani, H., and Kumbhani, J. (2023). *Russ. J. Org. Chem.* 59: 712–718.

87 Kumar, L., Verma, N., Kumar, N. et al. (2022). *New J. Chem.* 46: 20580–20591.

88 Kaurav, M.S., Sahu, P.K., Sahu, P.K. et al. (2019). *RSC Adv.* 9: 3755–3763.

89 Bhardwaj, D., Dhawan, K., and Singh, R. (2022). *J. Iran. Chem. Soc.* 19: 1003–1013.

90 Li, C., Qi, C., and Zhang, F. (2020). *Synlett* 31: 1313–1317.

91 Wadhwa, P., Kharbanda, A., Bagchi, S., and Sharma, A. (2018). *ChemistrySelect* 3: 2837–2841.

92 Khumalo, M.R., Maddila, S.N., Maddila, S., and Jonnalagadda, S.B. (2019). *ChemistrySelect* 4: 12503–12506.

93 Kerru, N., Maddila, S., and Jonnalagadda, S.B. (2021). *Front. Chem.* 9: 1–7.

94 Keihanfar, M. and Mirjalili, B.B.F. (2022). *BMC Chem.* 16: 1–6.

95 Hossaini, Z., Mohammadi, M., and Sheikholeslami-Farahani, F. (2022). *Front. Chem.* 10: 1–23.

96 Tabarsaei, N., Moradi, A.V., Alangi, S.Z.S. et al. (2022). *Polycycl. Aromat. Compd.* 43: 8110–8133.

97 Ezzatzadeh, E., Soleimani-Amiri, S., Hossaini, Z., and Barani, K.K. (2022). *Front. Chem.* 10: 1–14.

7

Pericyclic Reactions in Aqueous Medium

Surabhi Mishra, Sumran Raikwar, and Beeraiah Baire

Indian Institute of Technology Madras, Department of Chemistry, Adyar, Chennai 600036, India

7.1 Introduction

7.1.1 Pericyclic Reactions

Pericyclic reactions are a special class of organic reactions that involve a concerted, cyclic transition state. In other words, the bond breaking and making happen simultaneously in a cyclic manner (but not necessarily synchronously) and do not involve any distinct intermediates [1]. The term "pericyclic", introduced by Woodward and Hoffmann in 1969, originated from the word "peri", meaning around and indicating the cyclic movement of the electrons (i.e. bonds) during these processes [2]. Pericyclic reactions can be either unimolecular or bimolecular in nature and are categorized into four major classes [3], such as cycloaddition reactions, electrocyclic reactions, sigmatropic rearrangements, and group transfer reactions [4].

7.1.1.1 Cycloaddition Reactions

Cycloaddition reaction is an addition reaction between two π-reactants resulting in the formation of a cyclic adduct (i.e. addition product) *via* the construction of two new σ-bonds at the expense of two π-bonds (Scheme 7.1a). These are bimolecular (intermolecular) reactions in general, for example, the Diels–Alder (D–A) reaction, but many unimolecular (intramolecular) reactions are also well known. Cycloaddition reactions can also show reversibility, depending on the nature of the substrates and reaction conditions, and these are called *retro*-cycloaddition reactions. The *retro*-cycloaddition is a unimolecular process and typically involves the loss of two σ-bonds and the creation of two new π-bonds.

7.1.1.2 Electrocyclic Reactions

Electrocyclic reactions are unimolecular in nature and include both ring-forming and ring-opening processes. An electrocyclic ring closure reaction represents a ring-forming process where a new σ-bond is established between the two ends of a conjugated π-system, resulting in a cyclic system with shorter π-bonds. Conversely,

Organic Transformations in Water: Principles and Applications, First Edition.
Edited by Gopinathan Anilkumar, Nissy Ann Harry, and Sankuviruthiyil M. Ujwaldev.

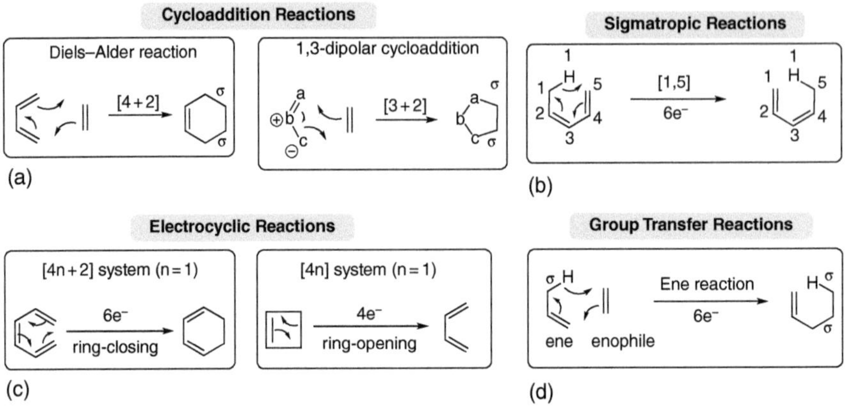

Scheme 7.1 General representation of various pericyclic reactions.

in the ring opening, a σ-bond of a cyclic system is broken, leading to a longer conjugated π-system (Scheme 7.1b).

7.1.1.3 Sigmatropic Rearrangements

Sigmatropic rearrangement involves an intramolecular migration of a σ-bond (located adjacent to one or two π-components), leading to the simultaneous reorganization of the π-system(s) (Scheme 7.1c). This process proceeds through the breaking of an existing σ-bond and the formation of a new one. Given its reversible nature, the preferred direction of the rearrangement is dictated by thermodynamics. These sigmatropic rearrangements are inherently unimolecular.

7.1.1.4 Group Transfer Reactions

These are a class of pericyclic reactions in which one or two groups are transferred from one molecule to another (Scheme 7.1d). This process involves the breaking and making of σ-bonds, accompanied by modifications in the associated π-systems. These reactions are bimolecular. Only a few chemical reactions are classified under this category, such as the ene reaction and diimide reduction.

Pericyclic reactions are very important in organic synthesis and have been widely used to construct complex molecular frameworks in a stereochemically controlled manner. The concept of molecular orbital (MO) theory is very important to understand pericyclic reactions. Understanding the energetics and symmetries of the MOs of reacting units is crucial for predicting the outcome of any of the pericyclic reactions. There are three well-known mechanistic approaches for analyzing pericyclic reactions:

(i) Orbital symmetry correlation approach, developed by Woodward and Hoffmann [5], as well as by Longuet-Higgins and Abrahamson [6].
(ii) Transition state aromaticity approach, developed independently by Zimmerman [7] and Dewar [8].
(iii) Frontier molecular orbital (FMO) approach, developed first by Woodward and Hoffmann [5b] and then by Fukui [9].

7.1.2 Water in Organic Synthesis

Water has been called the universal solvent. However, the utilization of water as a reaction medium for organic reactions has only gained significant attention in recent few decades within the field of organic synthesis. This delayed interest can be attributed to the prevalent belief that substances do not interact unless dissolved, and like needs (dissolves). In this context, several articles, books, and reviews have already been published on the importance of water in organic reactions [10–12]. In recent times, the usage of water as a reaction medium for organic synthesis has become a vital point due to several compelling reasons. Water stands out as the most abundant liquid on the planet and possesses characteristics such as being cost-effective, readily available, nontoxic, and nonflammable. In nearly all investigated cases, a marked enhancement in reaction rates has been observed. Over the past few decades, pericyclic reactions have been extensively investigated under various conditions and solvents, including water. In this chapter, we have limited our discussion to different types of pericyclic reactions, which employed water as the reaction medium and described the effect of water on the reaction rates, regio- and stereochemical outcomes of this unique class of organic reactions. Since there is a lot of development on the chemistry of cycloaddition reactions "in-water" and "on-water", we have presented our discussion in the order of cycloaddition reactions, electrocyclic reactions, sigmatropic rearrangements, and group transfer reactions.

7.2 Cycloaddition Reactions in Aqueous Medium

In the literature, various types of cycloaddition reactions such as [4+2], [3+2] [4+3], [2+2], [2+2+2], and [5+2] are precedent. These reactions differ in the nature of the reacting units involved, particularly the number of atoms involved in a cycloaddition reaction from each of the reactants. Most of them have been studied in the aqueous medium.

7.2.1 The In-Water [4+2] Cycloaddition Reactions (or) Diels–Alder Reactions

The [$4\pi+2\pi$] cycloaddition reaction, also known as the classical D–A reaction, is one of the most extensively studied organic reactions. This involves the interaction of a diene (possessing 4π electrons over 4-atoms) with a dienophile (possessing 2π electrons over 2-atoms), leading to the formation of a six-membered ring through the simultaneous construction of two new carbon–carbon σ-bonds. It was discovered in 1928 by Otto Diels and Kurt Alder, when they successfully demonstrated the reaction between cyclopentadiene and maleic anhydride, resulting in the formation of the bicyclo[2.2.1]heptene derivative [13]. Their pioneering work was recognized with the Nobel Prize in 1950 (Scheme 7.2a). The D–A reaction holds a significant synthetic importance as it allows the controlled synthesis of six-membered rings with precise regio- and stereocontrol [14].

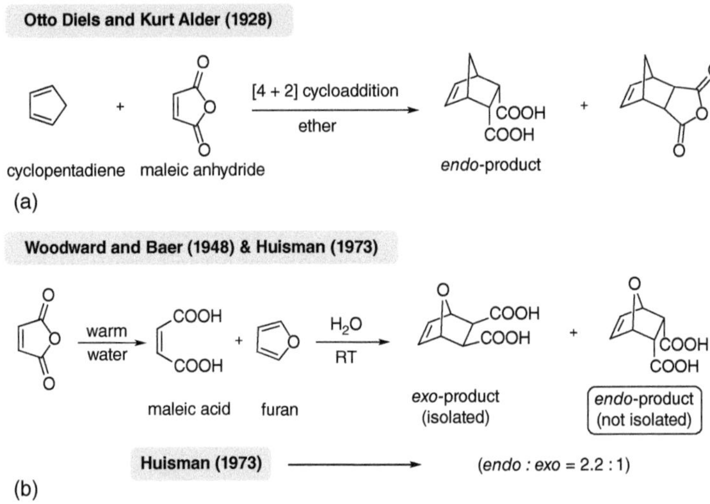

Scheme 7.2 (a) The first ever [4+2] cycloaddition by Diels and Alder; (b) early examples of in-water [4+2] cycloaddition reactions by Woodward and Baer, and Huisman.

The D–A reaction was also one of the first organic reactions to be performed in water by Otto Diels and Kurt Alder themselves in 1931 [15]. This accidental discovery was observed when they refluxed a mixture of furan and maleic acid in water. However, the significance of the reaction was unrecognized at the time since water was not a commonly used solvent. Later in 1948, Woodward and Baer [16], observed a substantial change in the selectivity of *endo-/exo*-products when the reaction was employed in water, as compared to traditional organic solvents (Scheme 7.2b). In 1973, Huisman and coworkers [17] performed the same D–A reaction between furan and maleic acid in various solvents, including H_2O, D_2O (55%), CD_3OD (33%), and DMSO-d_6 (10%), at room temperature for five days and observed similar *endo-/exo*-product ratios (2.2 : 1) in all the solvents. Surprisingly, until the rediscovery by Breslow in 1980, no efforts were made on this reaction to understand the effect of water.

Breslow and coworkers performed a comprehensive study on the D–A reactions to understand the impact of water on both reaction rates and *endo-/exo*-product selectivity [18]. Varying reaction rates, yields, and the degree of selectivity were evidently observed for D–A reactions conducted in solvents of different natures. The factors influencing the magnitude of reaction rate and selectivity in these reactions were attributed to solvophobic effects, interactions involving hydrogen bond donors (HBD), and dipolarity–polarizability. Accordingly, the enhanced reaction rates observed in aqueous reaction media were explained by the high solvophobic power (Sp) of aqueous media. Conversely, in fluorinated alcohols, the increase in reaction rate was attributed to the HBD ability. Solvent solvophobicity has been invoked in some media to elucidate the heightened *endo/exo* selectivity, while HBD power appeared to be the sole factor influencing regioselectivity in certain cases.

During their investigations in 1980 [18] of the cycloaddition reaction between cyclopentadiene and methyl vinyl ketone (MVK) in three different solvents, Rideout

Table 7.1 Solvent-dependent rate constants for the Diels–Alder reaction of cyclopentadiene with dienophiles.

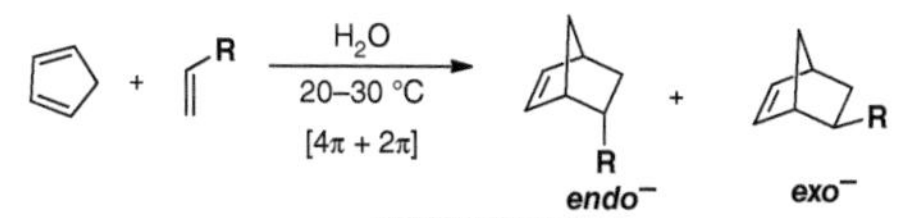

entry	solvent	temp (°C)	$K_2 \times 10^5$ ($M^{-1}s^{-1}$)[a]	endo-	exo-
(a) R = COMe					
1	Isooctane	20	5.94	minor	major
2	MeOH	20	75.5	minor	major
3	H_2O	20	4400	minor	major
4[b]	H_2O	20	10800	minor	major
5[c]	H_2O	20	4300	minor	major
(b) R = CN					
6	Isooctane	30	1.9	minor	major
7	MeOH	30	4.0	minor	major
8	H_2O	30	59.3	minor	major

[a]second order rate constants

[b]additive added LiCl (4.66 M), [c]additive added $C(NH_2)_3^+$ Cl^-

(c) Between anthracene-9-carbinol and *N*-ethylmaleimide

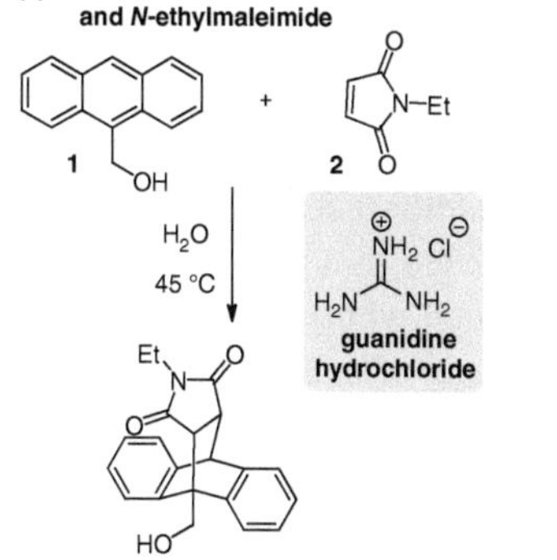

entry	solvent	temp (°C)	$K_2 \times 10^5$ ($M^{-1}s^{-1}$)
1	Isooctane	45	796
2	MeOH	45	344
3	H_2O	45	22600

and Breslow noted a gradual increase in the reaction rate from isooctane ($k = 6 \times 10^5$ M^{-1} s^{-1}) to methanol ($k = 76 \times 10^5$ M^{-1} s^{-1}). However, in the case of water ($k = 4400 \times 10^5$ M^{-1} s^{-1}), there was a remarkable jump in the rate of the cycloaddition reaction (entries 1–3, Table 7.1a). Similarly, an enhanced reaction rate was also observed during the D–A reaction between cyclopentadiene and acrylonitrile in water, i.e. 31 times higher than that in isooctane (entry 6 vs. entry 8, Table 7.1b). This work was marked as a significant milestone in the history of organic chemistry. They speculated that the observed rate enhancement may not only be solely attributable to solvent polarity but also involve another factor. This is evident as the reaction rate trend did not align with the polarity trend (MeOH > water > isooctane; polarity). Subsequent experiments (entries 1–3, Table 7.1c) involving the cycloaddition reaction between 9-hydroxymethylanthracene **1** and *N*-ethylmaleimide **2** provided strong support for the proposed hidden factor, as a drop-in reaction rate was observed from isooctane to methanol, while a significant enhancement was observed in water. This observation led Breslow to propose the hydrophobic effect of water as the unrevealed rate-enhancing factor.

To support this phenomenon, Breslow conducted additional control experiments by introducing *pro*-hydrophobic and *anti*-hydrophobic agents in 1991 [19]. The incorporation of LiCl (*pro*-hydrophobic agent, 4.86 M aqueous solution) into the reaction medium involving cyclopentadiene and MVK led to a 2.5-fold increase in the reaction rate (entry 4, Table 7.1a) compared to the normal water condition (entry 3, Table 7.1a). This effect resembles a "salts-out" phenomenon, where external salts enhance the hydrophobicity of the reaction medium. Conversely, a threefold decrease in reaction rate was observed with the addition of a 4.86 M aqueous solution of guanidinium chloride (an *anti*-hydrophobic agent) (entry 5,

Table 7.1a). This reduction in rate is associated with a decrease in the hydrophobicity of the medium due to the large guanidinium ion, resembling a "salts-in" effect. Subsequently, numerous research groups have independently determined the second-order rate constants for D–A reactions involving various dienes and dienophiles [20]. In most cases, they observed significant rate accelerations in aqueous media compared to the routinely used organic solvents.

Following the study of Rideout and Breslow on the rate enhancement of cycloaddition reactions in water, it became evident that rate accelerations were achievable in a wide range of bimolecular and intramolecular organic reactions (in-water, where the reactants are soluble in the aqueous medium). However, the challenge of substrate solubilities in water persisted, posing difficulties for practical synthetic applications of this observation. Various strategies were explored to address this issue, including the utilization of binary aqueous solvent mixtures. Additionally, innovative approaches were also explored, such as the incorporation of a pro-hydrophobic functional group into either of the substrates, i.e. diene or dienophile. For example, in 1983, Grieco et al. [21] reported an intermolecular D–A reaction between sodium (*E*)-3,5-hexadienoate **3** and methyl (acetoxymethyl) acrylate **4** in water. After 24 hours, an *endo*-selective (*endo:exo*; 2 : 1) product mixture **5a/b** was obtained with a 67% overall yield at 60 °C (entry 1, Scheme 7.3). In contrast, the corresponding reaction (without structural modification) with ester **6** took 96 hours in refluxing toluene (entry 2), resulting in an *exo*-selective (*endo:exo*; 1 : 2) product mixture **5a/b**.

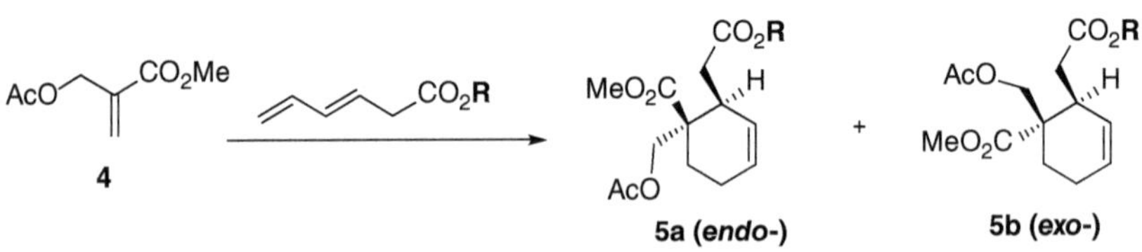

entry	R	condition	*endo* : exo ratio	yield
1	Na (**3**)	H_2O, 60 °C, 24 h	2:1	67%
2	Me (**6**)	$PhCH_3$, reflux, 96 h	1:2	84%

Scheme 7.3 Enhanced reaction rates in water for an intermolecular Diels–Alder reaction.

Another such example was reported in 2002 by Yoshida and coworkers [22], where they performed a D–A reaction between 2-PyMe$_2$Si-substituted 1,3-diene **7** (2-PyMe$_2$Si = 2-pyridyldimethylsilyl; the pyridyl group initially forms a pyridinium salt with the added HCl to increase the solubility of the substrate **7**, in water) with *p*-benzoquinone **8**, in water at room temperature (entries 1 and 2, Scheme 7.4). The same reaction in organic solvents (Et_2O/toluene) found to be very slow and less efficient, indicating the dramatic rate acceleration in water (Scheme 7.4). Later, many reports have appeared in the literature based on this approach [23]. Additional notable characteristics observed in aqueous D–A reactions include the beneficial impact of water-tolerant Lewis acids when employed as catalysts [24] and the utilization of surfactants to create micellar environments [25].

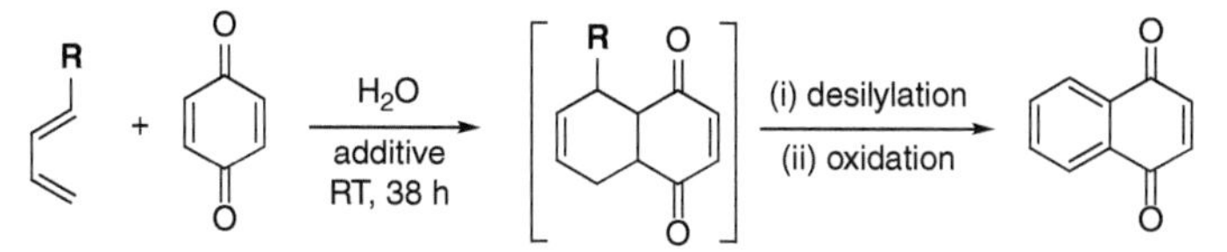

7 8

entry	solvent	additive (equiv.)	time (h)	yield (%)
1	water	HCl (1.0)	38	quant.
2	water	HCl (0.2)	47	2
3	water	–	89	4
4	water	H_2SO_4 (1.0)	38	73
5	Et_2O/toluene	HCl (1.0)	38	22
6	Et_2O/toluene	–	38	6

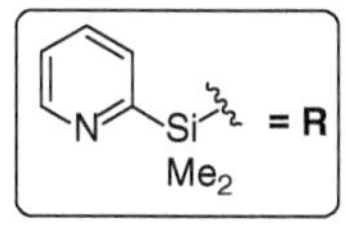

Scheme 7.4 D–A reaction of 2-PyMe_2Si-substituted 1,3-dienes with *p*-benzoquinone in water.

In addition to Breslow's hydrophobic effect, several research groups have suggested a new factor, i.e. hydrogen bonding by the solvent, for the water-mediated rate enhancement in D–A reactions [26]. Engbert's group in 1992 studied the Gibbs free energy of activation for the D–A reaction between cyclopentadiene and MVK in water and several other organic solvents. They observed that the reaction rates were higher in water compared to other solvents. They attributed these observations to the combined influence of the hydrophobic effect of water on reactants and the hydrogen bonding effect of water with the transition state [27].

Subsequently, the same group explored the role of hydrogen bonding in D–A reactions in-water by employing two different dienophiles [28], namely, MVK and methyl vinyl sulfone (MVS), which differ in their hydrogen bonding abilities. Generally, sulfones are weaker hydrogen-bond acceptors and relatively poor carbanion stabilizers on the α-carbon than their ketone counterparts, and the reaction rates are expected to align with this trend (rate CO > SO_2). As expected, the [4+2] cycloaddition reaction rate of MVS in water was four times slower than that of MVK (compare entry 1 of Table 7.2ab). Furthermore, they conducted the reactions in a series of HBD organic solvents, such as 1,1,1,3,3,3-hexafluoroisopropanol (HFIP), 2,2,2-trifluoroethanol (TFE), 1-propanol, ethanol, and acetonitrile for both substrates and compared their reaction rates with those of in-water reactions (Table 7.2a,b, entries 1–6). The increase in Gibbs energy of activation and rate retardation was observed from water to HFIP to TFE to 1-propanol to ethanol to acetonitrile. These observations strongly support the role of hydrogen-bond interactions with the transition states in accelerating the reaction rate during the D–A reactions in-water. Although HFIP and TFE are stronger HBDs than water, the observed reaction rates are lower than in water. This discrepancy is explained by the fact that the Gibbs energies calculated for the reactions in-water account for both hydrophobic and hydrogen-bonding effects, whereas for HFIP and TFE, it only considers the hydrogen-bonding effect without the hydrophobic effect.

Table 7.2 Rate constants and Gibbs energies of activation for the D–A reactions in different solvents with distinct hydrogen bonding abilities.

(a)

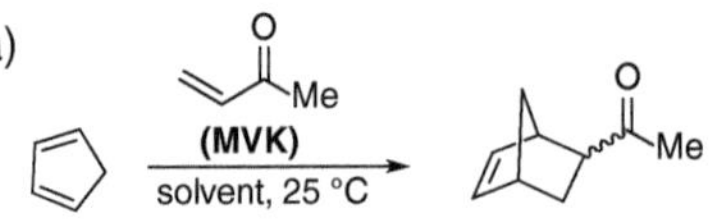

entry	solvent	rate constant ($M^{-1}s^{-1}$)	Gibbs energy ($kJmol^{-1}$)
1	water	5.08×10^{-2}	80.3
2	HFIP	1.75×10^{-2}	83.0
3	TFE	8.72×10^{-3}	84.7
4	1-propanol	9.12×10^{-4}	90.3
5	ethanol	8.39×10^{-4}	90.5
6	acetonitrile	1.75×10^{-4}	94.4

(b)

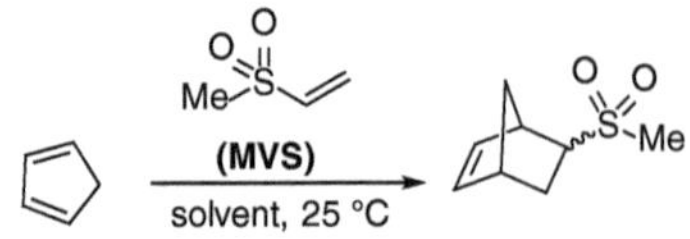

entry	solvent	rate constant ($M^{-1}s^{-1}$)	Gibbs energy ($kJmol^{-1}$)
1	water	6.29×10^{-4}	91.2
2	HFIP	2.00×10^{-4}	94.0
3	TFE	9.50×10^{-5}	95.9
4	1-propanol	3.04×10^{-5}	98.7
5	ethanol	2.20×10^{-5}	99.5
6	acetonitrile	8.85×10^{-6}	101.7

Similarly, several research groups have also performed systematic studies on various D–A reactions (using different dienes and dienophiles) to support the concept of hydrogen bonding-enhanced reaction rates [29, 30].

7.2.2 Effect of Water on *Endo-/Exo*-Selectivity in [4+2] Cycloaddition Reactions

One of the very crucial factors in [4+2] cycloaddition reactions is the *endo-/exo*-stereoselectivity, which can arise when substituted dienes and dienophiles are employed. Under ordinary circumstances, the kinetic product, specifically the *endo*-isomer, tends to be favored over its thermodynamic counterpart, the *exo*-isomer. However, in a few instances, the *exo*-isomer has been identified as the predominant product. This distribution of products can be effectively elucidated by examining the energy levels of their respective transition states. It is proposed by Woodward and Katz that, the involvement of secondary orbital interactions, particularly in dienophiles with an additional π-electron density in conjugation, contributes to the understanding of this selectivity [31]. In addition, several research groups have proven (experimentally as well as theoretically) that the polarities of the solvents play an important role on the *endo-* vs. *exo*-selectivity in D–A reactions. They observed an increased preference for the *endo*-product, which was attributed to the higher polarity of its transition state compared to the *exo*-transition state [32–36]. However, when focusing exclusively on aqueous solvent mixtures for D–A reactions, the presence of water (acting as a polar solvent) in D–A reactions, favors *endo*-product over *exo*-product.

Breslow's research group [37] has also observed a similar trend of *endo-/exo*-product distribution when they performed the D–A reaction between cyclopentadiene and MVK in diverse solvents, including water, which differ in their polarity. They observed an increment in *endo-/exo*-ratio as well as the rate of reactions in aqueous medium as compared to organic solvents (Scheme 7.5). It was observed that when water was employed as a solvent, the *endo-/exo*-ratio went up to (21.4, 1) in comparison to organic solvents. Later, they also observed further enhancement of the *endo*-product selectivity in the presence of the pro-hydrophobic additive LiCl (salt-in effect) and a decrease in *endo*-selectivity when the *anti*-hydrophobic agent guanidinium chloride (GnCl) (salt-out effect) was added to the reaction medium (Scheme 7.5).

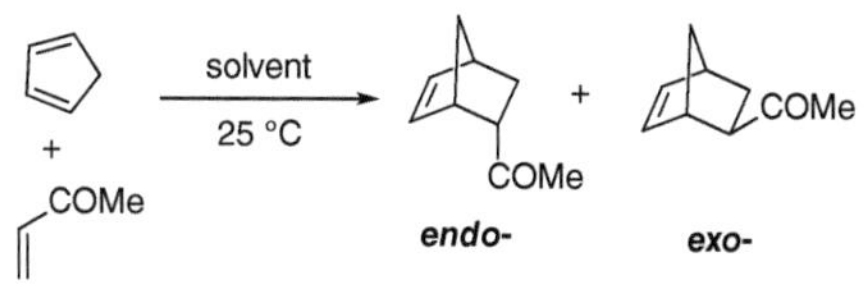

entry	solvent	*endo*-	*exo*-
1	EtOH (0.15 M)	8.5	1
2	H_2O (0.15 M)	21.4	1
3	H_2O + LiCl	28.0	1
4	H_2O + GuCl	22.0	1

Scheme 7.5 Comparison of *endo-/exo*-product ratios during the Diels–Alder reactions of cyclopentadiene in organic solvents and water.

7.2.3 Intramolecular Diels–Alder Reactions In-Water

In comparison to the bimolecular [4+2] cycloaddition reactions, the intramolecular D–A reactions are less explored in water. In 1995, the research group of Engberts [38] reported probably the first intramolecular D–A reactions in-water by employing the *N*-furfuryl-*N*-alkylacrylamides **9** and **10** (Scheme 7.6). Here, water was identified as a promoter to enhance the rate of the reactions. At 25 °C, the rate constants for compound **9** ($R^1 = CO_2H$) were measured as $k_{DCM} = 2.36 \times 10^{-5}\ s^{-1}$ and $k_{water} = 2.50 \times 10^{-2}\ s^{-1}$, reflecting a remarkable enhancement of 1064 times in water (entries 1–4). In contrast, for the compound **10** ($R^1 = H$), the reaction in EtOH was unsuccessful even after three weeks at 60 °C (entry 5). However, the reaction was successful in water at 60 °C, exhibiting a rate constant of $k_{water} = 2.05 \times 10^{-5}\ s^{-1}$ (entry 6).

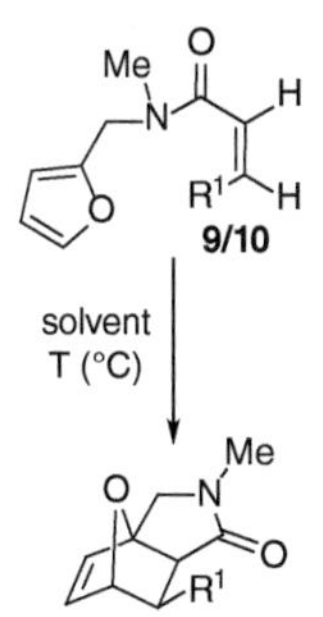

entry	R^1 (substrate)	solvent	T (°C)	rate constant (s^{-1})
1	CO_2H (**9**)	CH_2Cl_2	25	2.36×10^{-5}
2	CO_2H (**9**)	MeCN	25	9.27×10^{-5}
3	CO_2H (**9**)	EtOH	25	3.09×10^{-4}
4	CO_2H (**9**)	H_2O	25	2.50×10^{-2}
5[a]	H (**10**)	EtOH	60	–
6	H (**10**)	H_2O	60	2.05×10^{-5}

[a]No observable conversion even after 3 weeks time

Scheme 7.6 First example of intramolecular Diels–Alder reactions in water.

Later, several reports have appeared in the literature based on the intramolecular D–A reactions in water medium [39].

In addition to the bimolecular as well as intramolecular [4+2] cycloaddition reactions, several other variants [40], like inverse electron demand D–A reactions [41], asymmetric D–A reactions [42], *retro*-D–A reactions [43], were also studied in water medium by various research groups.

7.2.4 On-Water [4+2] Cycloaddition Reactions

In 2005, Sharpless introduced the term "on-water" reactions to describe heterogeneous reactions where one or both reactants are not soluble in water [44]. The "on-water" method involves simply stirring the insoluble reactant(s) with water to create an aqueous suspension. Sharpless conducted pericyclic reactions in aqueous suspension (on-water) and observed significantly enhanced kinetics and yields in most cases when compared to those in organic solvents. For example, the cycloaddition reaction between *trans*, *trans*-2,4-hexadienyl acetate **11** and *N*-propylmaleimide **12** gives the cycloadduct **13** (Scheme 7.7a). Notably, the protic organic solvent MeOH outperformed aprotic solvents like toluene, CH_3CN, as indicated in table below.

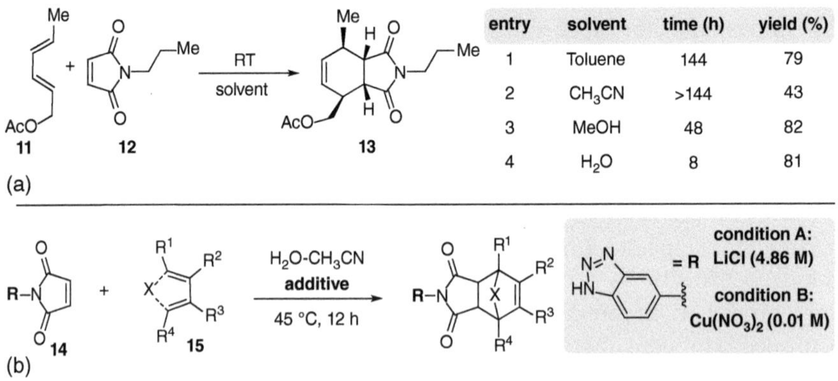

entry	solvent	time (h)	yield (%)
1	Toluene	144	79
2	CH_3CN	>144	43
3	MeOH	48	82
4	H_2O	8	81

Scheme 7.7 "On-water" Diels–Alder reaction.

Another such "on-water" D–A reaction was reported in 2001 (Scheme 7.7b), by the research group of Graham, where the reaction between the benzotriazole maleimide **14** (BTM) against seven diene partners **15** was performed in aqueous acetonitrile [45].

7.3 Miscellaneous Cycloaddition Reactions In-Water

In addition to the extensive studies on the [4+2] cycloaddition reactions in-water, a few examples of other types of cycloaddition reactions such as [3+2], [4+3], [2+2], [2+2+2], and [5+2] have also been reported in the aqueous medium.

7.3.1 The [3+2] or 1,3-Dipolar Cycloaddition Reactions

This category of pericyclic reactions holds significant synthetic value in organic synthesis for the construction of five-membered heterocycles like 1,2,3-triazoles

[46], tetrazoles [47], pyrazoles [48], pyrazolines [49], isoxazoles [50], isoxazolidines [51], pyrroles, and pyrrolidines [52]. Over the past few decades, in-water 1,3-dipolar cycloaddition reactions have gained recognition as a valuable and unconventional synthetic method.

The role of water in [3+2] cycloaddition reactions was investigated by Engberts and coworkers in 1995 [46d]. The kinetic study of the reaction between norbornene and phenyl azide giving triazoles **16** was performed in organic solvents as well as aqueous medium. A rate enhancement from 7.4 to 37 was observed by changing the solvent from pure ethanol to (75 : 25) H_2O: EtOH mixture (entries 1–3). A substantial acceleration to 83 was observed with the (92 : 8) H_2O: EtOH solvent mixture. Experimentally recorded kinetic data clearly indicated that the 1,3-dipolar cycloaddition reactions experience significant rate acceleration in solutions with higher water content (Scheme 7.8).

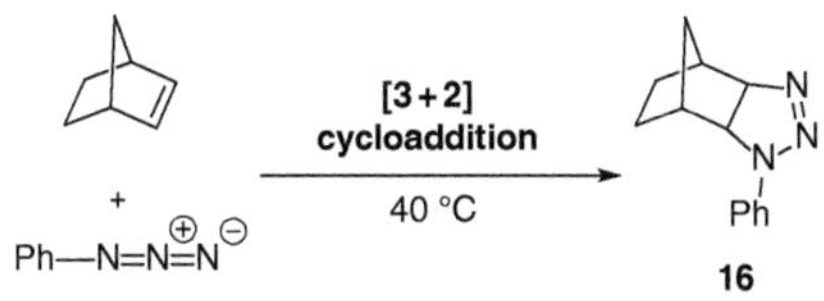

entry	solvent	K_2 (× 10^5 M^{-1} s^{-1})
1	$CHCl_3$	6.8
2	EtOH	7.4
3	H_2O : EtOH (75 : 25)	37
4	H_2O : EtOH (92 : 8)	83

Scheme 7.8 Rate enhancement for the [3+2] cycloaddition of norbornene and phenyl azide in-water solvent.

Ju and coworkers [53] investigated the Huisgen cycloaddition reaction involving 5-azidovalerate **17** and a series of electron-deficient alkynes **18**, in an aqueous environment (Scheme 7.9a) to synthesize the corresponding 1,2,3-triazoles **19** with yields ranging from 67% to 94%.

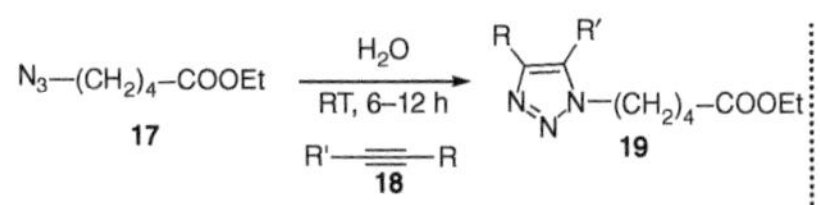

entry	alkyne	yield (%)
1	EtO_2C—≡—CO_2Et	81
2	Me—≡—CO_2Et	94
3	H—≡—CO_2Et	82
4[a]	H—≡—CO_2Et	90
5	H—≡—CH_2SO_2Ph	67
6[a]	H—≡—CH_2SO_2Ph	85

[a]additive CuCl (0.1 equiv.) was added

(a)

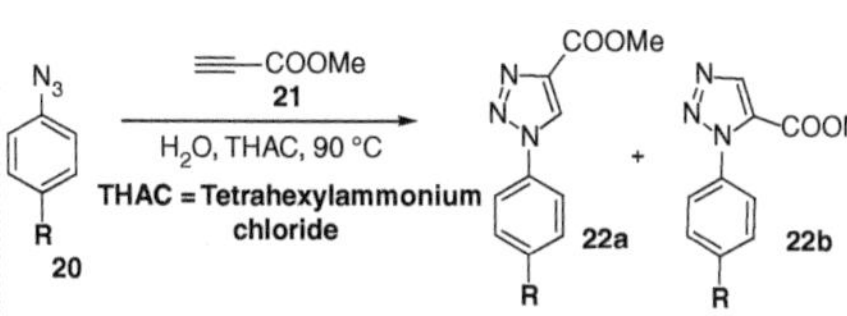

entry	R	time (min)	yield a+b (%)	ratio a : b
1	H	60	92	82 : 18
2	Me	40	>95	75 : 25
3	MeO	30	92	78 : 22
4	F	120	94	82 : 18
5	NO_2	300	>95	60 : 40

(b)

Scheme 7.9 The water-mediated, 1,3-dipolar cycloaddition reactions of (a) 5-azidovalerate with alkynes and (b) phenyl azide derivatives with methyl propiolate.

Molteni and Ponti [54] conducted various 1,3-dipolar cycloadditions in aqueous medium using *p*-substituted phenyl azide derivatives **20** against methyl propiolate **21** (Scheme 7.9b). Tetrahexylammonium chloride (THAC) was served as a phase transfer catalyst, and the reactions were carried out at 90 °C. The reaction times varied with the substituent: 60 minutes (R = H), 40 minutes (R = Me), 30 minutes (R = MeO), 120 minutes (R = F), and 300 minutes (R = NO_2), and they were notably shorter than the uncatalyzed reactions reported by Wang and Qin [46b]. The isolated yields of the triazoles **22** were more than 92% and found to be independent of the nature of the substituent. Later, in the subsequent years, many in-water [3+2] cycloaddition reactions were reported in the literature [55].

7.3.2 [2+2+2] Cycloaddition Reactions

The [2+2+2] cycloaddition represents a sophisticated, atom-efficient, and group-tolerant method for synthesizing aromatics and heteroaromatic systems. Despite the extensive exploration of the [2+2+2] cycloaddition reaction in recent decades, there are relatively few instances where water has been employed as a solvent [56]. In 2000, Eaton and coworker reported a water-soluble cobalt complex **23** catalyzed [2+2+2] cycloaddition reaction [57]. This process involves two alkyne units and a nitrile in a mixture of water and methanol (7 : 3) at 85 °C, to give a chemospecific synthesis of functionalized pyridines in good yields (Scheme 7.10a).

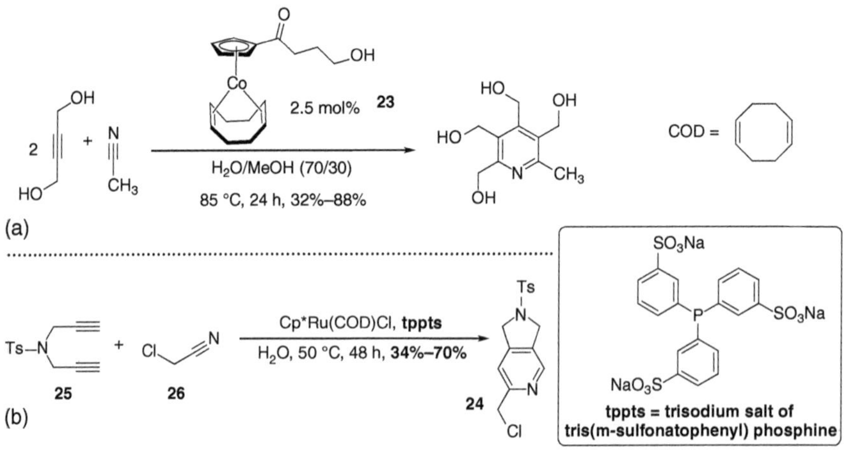

Scheme 7.10 Synthesis of functionalized pyridines using [2+2+2] cycloaddition reactions in water by employing (a) a water-soluble cobalt complex and (b) a Ru-catalyst.

A highly efficient [2+2+2] cycloaddition-based approach for the synthesis of functionalized pyridines **24** from diynes **25** and nitriles **26** in aqueous medium was also demonstrated (Scheme 7.10b) in 2012 by the research group of Wan. This method utilizes Cp*Ru(COD)Cl/tppts as the catalyst [58].

7.3.3 [4+3] Cycloaddition Reactions

The [4+3] cycloaddition reactions, involving allylic (oxyallyl) cations and dienes (like furan or cyclopentadiene), represent an important approach for the direct construction of seven-membered rings, both in inter- and intramolecular manners.

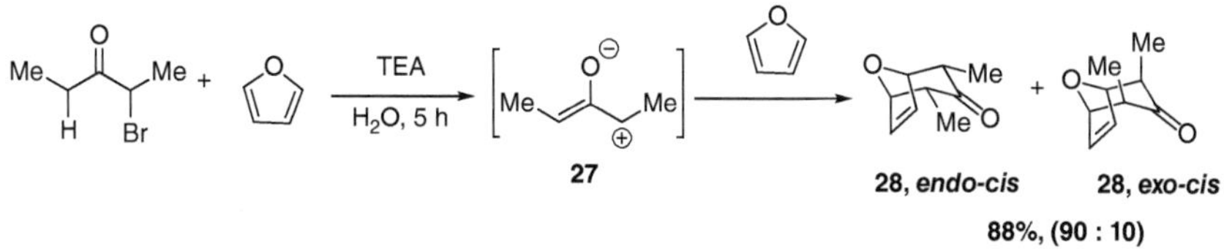

Scheme 7.11 [4+3] Cycloaddition reactions in water.

However, examples of [4+3] cycloaddition reactions in-water are very scarce in the literature (Scheme 7.11). A water-promoted [4+3] cycloaddition reaction involving an *in situ* generated oxyallyl cation **27** and furan, leading to the formation of the *oxa*-bicyclo[3.2.1]octenone skeleton **28**, was disclosed by Lubinea and Bouchain in 1997 [59]. This pioneering work was marked as the first instance of [4+3] cycloaddition utilizing water as the solvent. The process resulted in excellent yields (60–88%) and favorable *endo*-selectivity at room temperature. However, conducting the same reaction in any organic solvent gave very poor yields (<30%) for the products and highlighted the rate-enhancing effect of water as observed in [4+2] cycloaddition reactions.

7.3.4 [5+2] Cycloaddition Reactions

In 2003, the research group of Wender [60] reported a novel water-soluble rhodium catalyst for an intramolecular [5+2] cycloaddition of vinylcyclopropanes and alkynes in-water for the construction of fused seven-membered rings (Scheme 7.12). This method was further extended to intramolecular [4+2] cycloaddition reactions of dienes and alkenes in-water.

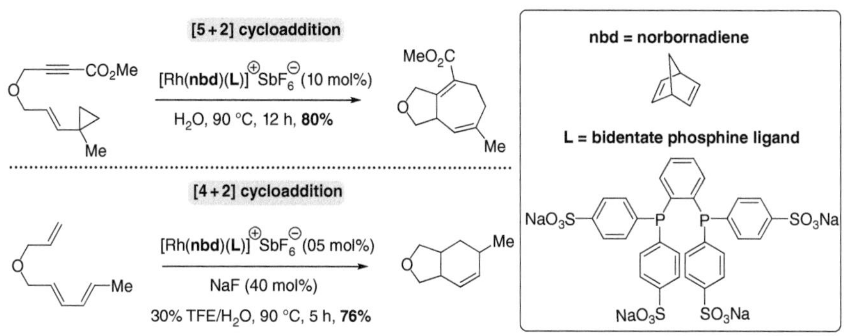

Scheme 7.12 [5+2] Cycloaddition reactions in-water.

7.3.5 [2+2] Cycloaddition Reactions

Photochemical [2+2] cycloadditions represent a highly effective method for synthesizing four-membered rings. Interestingly, the use of water as a reaction medium can significantly enhance the potential of these photochemical cycloadditions. For instance, in 1982, the research group of Ramamurthy reported the

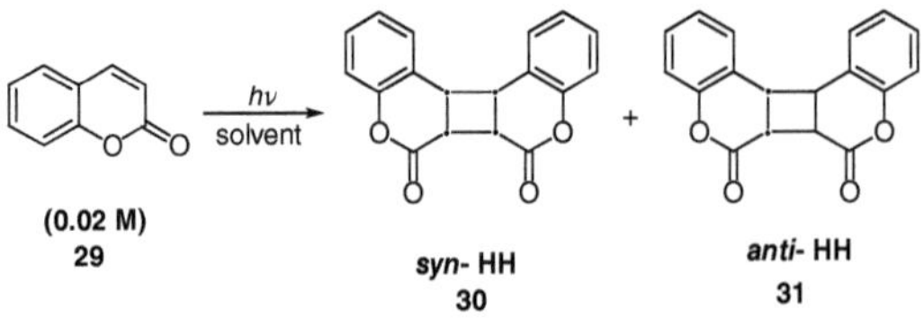

solvent	product	quantum yield
methanol	*anti*- HH	$<10^{-5}$
benzene	*anti*- HH	$<10^{-5}$
water	*syn*- HH	2×10^{-3}
SDS	*syn*- HH	0.3×10^{-3}
CTAB	*syn*- HH	1×10^{-3}

SDS = sodium dodecyl sulfate; CTAB = cetrimonium bromide

Scheme 7.13 [2+2] Cycloaddition of coumarin in-water.

photodimerization of coumarin **29** in pure water at room temperature, resulting in the formation of *syn*-adduct **30** exclusively (Scheme 7.13). In contrast, when the same reaction was performed in organic solvents like methanol and benzene, a strong preference for the *anti*-adduct **31** was observed. Additionally, the quantum yield of these reactions was significantly lower compared to that observed in-water [61].

7.4 Electrocyclic Reactions In-Water (Ring Closure and Ring Opening)

In this section of the chapter, we present the details of electrocyclic reactions performed in an aqueous medium. When compared to the cycloaddition reactions, a very small number of examples of electrocyclization reactions have been studied in the water medium.

7.4.1 4π-Electrocyclization Reactions

The first electrocyclic reaction in an aqueous medium was reported in 1979, by Gaoni's research group (Scheme 7.14a). The 3-alkyl-3-azabicyclo[3.2.0]hept-1(5)-enes **32** were synthesized in good yields through the photo-4π-electrocyclization of cyclic amines **33** in an aqueous acidic solution. These bicyclic amines **32** were found to be unstable but can be stored under cold conditions. At 200 °C, these bicyclic amines **32** went through the thermal electrocyclic ring opening to provide the starting material **33** [62].

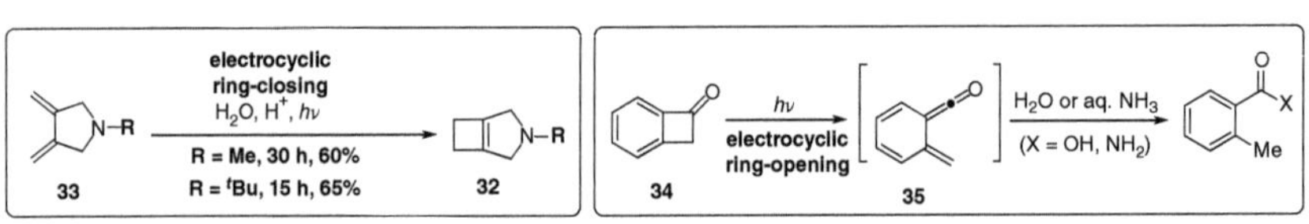

(a) (b)

Scheme 7.14 (a) First 4π-electrocyclic reaction in aqueous medium and (b) flash photolysis of benzocyclobutenone in-water.

Later in 2003, Zhan and coworkers demonstrated the flash-photolytic, 4π-electrocyclic ring opening of benzocyclobutenone **34** in-water for the synthesis of 2-methylbenzoic acid through the *in situ* generation of 6-methylene-2,4-cyclohexadienyli-deneketene **35** [63]. When the reaction was performed in aqueous ammonia, the formation of benzamide was observed in addition to benzoic acid (Scheme 7.14b).

Among the other unique features of water, its ability to enhance the high loss factor (tan $\delta = 0.123$) makes it a valuable solvent for microwave-mediated synthesis. Microwave irradiation rapidly heats water, making it a less polar pseudo-organic solvent and increasing H^+ concentration, suggesting it may function as a weak acid [64]. In 2017, Nardi et al. demonstrated an efficient synthesis of 4,5-diaminocyclopent-2-enones **36** from furfural and different secondary amines (Scheme 7.15). Microwave assistance was pivotal for the initial step, which involved the exclusive formation of 4,5-bisubstituted cyclopentenone **36** (the kinetic product) in just five minutes. In the following step, the aqueous environment transformed the kinetic product **36** into 2,4-bisubstitued cyclopentenone **37** (a thermodynamic product). The reaction gives *trans*-diastereomers through an intermediate enamine **38**, which undergoes a thermal conrotatory 4π-electrocyclic ring closure [65]. In 2018, Afonso also developed the same reaction in water by employing $Cu(OTf)_2$ as a catalyst for the selective synthesis of *trans*-4,5-diaminocyclopent-2-enones **36**.

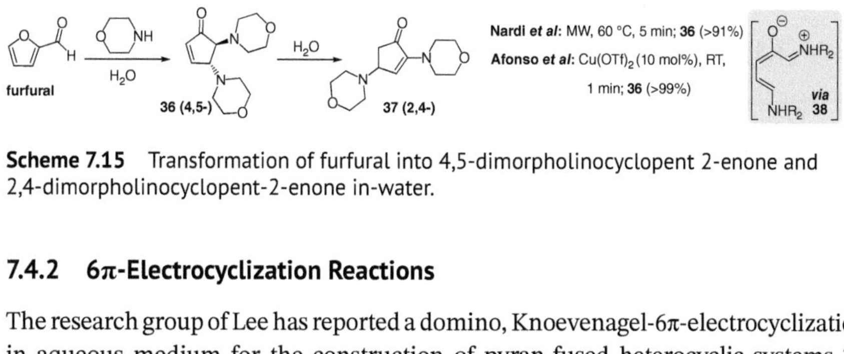

Scheme 7.15 Transformation of furfural into 4,5-dimorpholinocyclopent 2-enone and 2,4-dimorpholinocyclopent-2-enone in-water.

7.4.2 6π-Electrocyclization Reactions

The research group of Lee has reported a domino, Knoevenagel-6π-electrocyclization in aqueous medium for the construction of pyran-fused heterocyclic systems **39** [66]. This reaction was found to be inefficient in various organic solvents, and the best yield (91%) was achieved in water at 80 °C after four hours (Scheme 7.16a). This strategy was further extended for the synthesis of pyran-fused quinoline natural products such as ferprenin **40**, flindersine **41**, haplamine **42**, and zanthosimuline **43** and their derivatives (Scheme 7.16b).

A microwave promoted 6π-electrocyclization reaction in-water was used for the construction of imidazo[1,2-*a*]pyridine-fused quinoline-2-thiones **44** by employing the corresponding *ortho*-heteroaryl-anilines and carbon disulfide (CS_2). The reaction was found to be sluggish and inefficient in various organic solvents in the basic medium (DBU and K_3PO_4) [67]. The reaction may proceed through the isothiocyanate intermediate **45** (Scheme 7.16c).

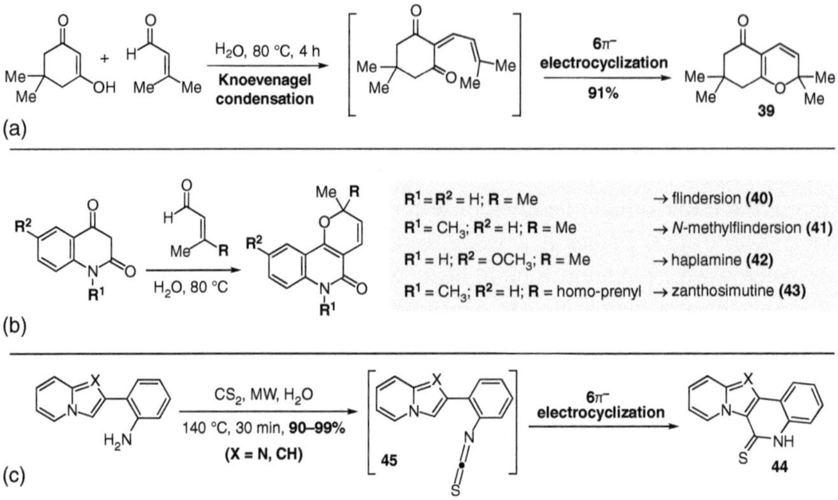

Scheme 7.16 (a) One-pot, domino Knoevenagel-6π-electrocyclization in water; (b) Synthesis of pyran-fused quinoline natural products; (c) Synthesis of quinoline-2-thiones *via* a 6π-electrocyclization in water.

7.5 Sigmatropic Rearrangements In-Water

7.5.1 [3,3]-Sigmatropic Rearrangements

The [3,3]-sigmatropic rearrangements [68], e.g. Claisen rearrangement, Cope rearrangement, etc. are helpful techniques in organic synthesis for the generation of new functional groups and novel cyclic systems. In the case of Cope rearrangement, the reaction involves the conversion of 1,5-dienes (an all-carbon system) [69], on the other hand, Claisen rearrangement [70] involves an allyl vinyl ether for the generation of γ,δ-unsaturated carbonyl compounds.

In 1973, one of the earliest known aqueous Claisen rearrangements was reported by R. Knowles (Scheme 7.17a). This involves the conversion of chorismate **46** to prephenate **47**, an important intermediate in the "Shikimate pathway" for generating aromatic amino acids [71].

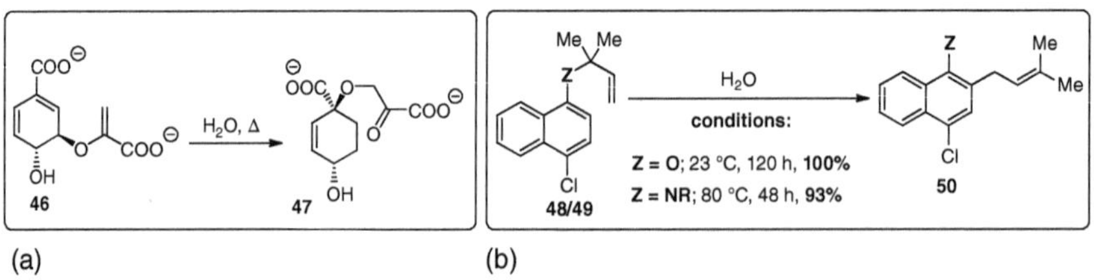

Scheme 7.17 (a) First aqueous Claisen rearrangement of chorismate to prephenate and (b) aza-Claisen rearrangement in-water.

Several research groups have also demonstrated the water influenced rate acceleration during the Claisen rearrangement of several structurally distinct substrates [72].

In 2005, Sharpless and coworkers [44a] found that the chloronaphthylether **48** (Scheme 7.17b, Z = O), exhibited a significant increase in reaction rate when performed on-water Claisen rearrangement compared to when performed in neat as well as in other organic solvents (100% yield on water vs. 73% in neat vs. 56% in MeOH). Later in 2013, McErlean and coworkers conducted an on-water, *aza*-Claisen rearrangement using similar substrates **49** (Scheme 7.17b, Z = NR), and isolated an impressive yield of 93% within 48 hours for **50**, when compared to the reaction in organic solvents, where no reaction was observed [73].

7.5.2 [2,3]-Sigmatropic Rearrangements

In 1994, the research group of Otera discovered a stereoselective [2,3]-sigmatropic rearrangement of γ-hydroxyalkyl γ-alkoxyallylic sulfoxides **51** in-water [74]. The γ-hydroxyl group has a longer-range effect on the sulfonyl group, thereby resulting in the more selective generation of E-isomer (Scheme 7.18a).

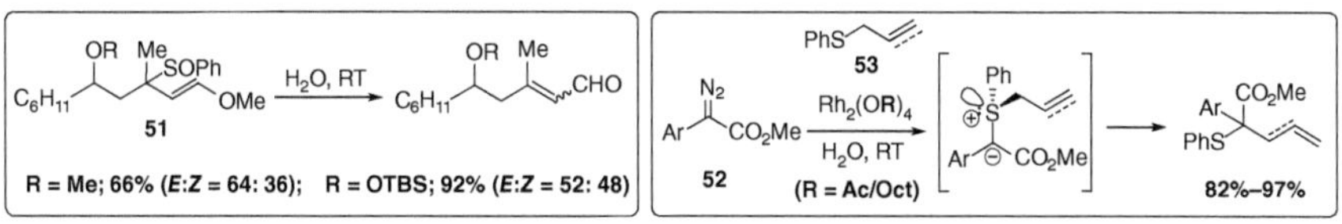

(a) (b)

Scheme 7.18 The [2,3]-sigmatropic rearrangement of (a) allylic sulfoxides (Mislow-Evans rearrangement) and (b) sulfur ylides in water.

In 2007, Wang and coworkers developed a Rh-catalyzed, [2,3]-sigmatropic rearrangement of sulfonium ylides in aqueous medium (Scheme 7.18b). These ylides are generated in situ from the reaction between Rh-carbenoid (formed from the corresponding aryldiazoacetates **52**) and allylic/propargylic sulfides **53** [75].

7.6 Ene Reactions

No examples of in-water ene reactions were reported in the literature, whereas only one example of a corresponding on-water ene reaction was reported. The research group of Sharpless reported an ene reaction of highly water-insoluble azodicarboxylates **54** against cyclohexene [76] and α-pinene [77] by utilizing the "on-water" approach. In the case of cyclohexene, vigorous stirring with the azo-diester **54** in water provided the enhanced rate and yield compared to the same reaction in organic solvents or neat conditions (Scheme 7.19). In benzene, after

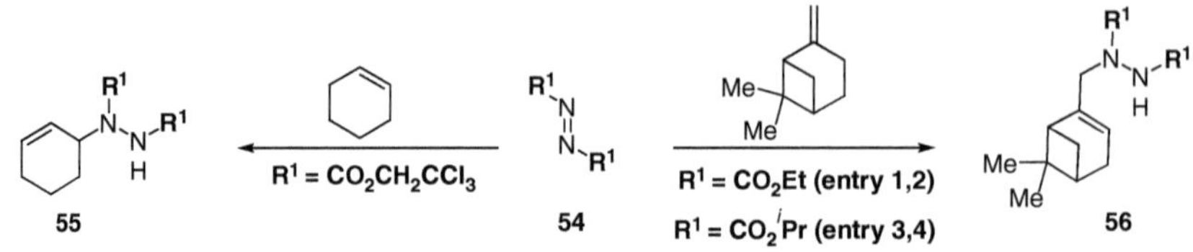

entry	solvent	temp (°C)	time (h)	yield (%)
1	benzene	80	24	70
2	neat	50	36	62
3	"on-water"	50	8	91

entry	solvent	temp (°C)	time (h)	yield (%)
1	neat	23	36	90
2	"on-water"	23	3	82
3	CH_2Cl_2	RT	32	89
4	"on-water"	RT	13	80

Scheme 7.19 On-water ene reaction of azodicarboxylates and olefins.

heating at 80 °C for 24 hours resulted in 70% of the adduct, whereas when the reaction was performed "on-water", 91% of the adduct **55** was obtained within eight hours at 50 °C. However, the reactions of α-pinene gave comparable yields of **56** in organic solvents vs. neat vs. on-water conditions.

7.7 Summary and Conclusions

To summarize, water, the greenest solvent, has been well utilized as the solvent as well as promoter for various types of pericyclic reactions, such as cycloaddition reactions, electrocyclic reactions, sigmatropic rearrangements, and group transfer reactions, in organic synthesis. We have detailed the hydrophobic effect and hydrogen bonding effect of water on the reaction rates, regio- and stereochemical selectivities of this special class of reactions. Due to page constraints, in each case we have presented only limited examples but cited the additional relevant references appropriately for the sake of readers. We hope that this chapter will give an flavor of the most up-to-date developments in the field of pericyclic reactions in aqueous medium.

References

1 (a) Houk, K.N., Gonzalez, J., and Li, Y. (1995). *Acc. Chem. Res.* 28: 81–90. (b) Rzepa, H.S. (2007). *J. Chem. Educ.* 84: 1535–1540. (c) Arrieta, A., de Cozar, A., and Cossio, F.P. (2011). *Curr. Org. Chem.* 15: 3594–3608.

2 Woodward, R.B. and Hoffmann, R. (1969). *Angew. Chem. Int. Ed. Engl.* 8: 781–853.

3 Fleming, I. (2015, pp. 253–362). *Pericyclic Reactions*, vol. 2. Oxford: Oxford University Press.

4 Sankararaman, S. (2005). *Pericyclic Reactions – A Textbook, Reactions, Applications and Theory*, 1–18. Wiley-VCH.

5 (a) Hoffmann, R. and Woodward, R.B. (1968). *Acc. Chem. Res.* 1: 17–22. (b) Hoffmann, R. and Woodward, R.B. (1965). *J. Am. Chem. Soc.* 87: 2046–2048.

6 Longuet-Higgins, H.C. and Abrahamson, E.W. (1965). *J. Am. Chem. Soc.* 87: 2045–2046.

7 Zimmerman, H.E. (1971). *Acc. Chem. Res.* 4: 272–280.

8 Dewar, M.J.S. (1971). *Angew. Chem. Int. Ed. Engl.* 10: 761–776.

9 Fukui, K. (1971). *Acc. Chem. Res.* 4: 57–64.

10 (a) Otto, S. and Engberts, J.B.F.N. (2003). *Org. Biomol. Chem.* 1: 2809–2820. (b) Otto, S. and Engberts, J.B.F.N. (2000). *Pure Appl. Chem.* 72: 1365–1372. (c) Wittkopp, A. and Schreiner, P.R. (2009, pp. 253–362). *Catalysis of Diels–Alder Reactions in Water and in Hydrogen-Bonding Environments*. John Wiley & Sons Ltd. (d) Garner, P.P. (1998). *Organic Synthesis in Water* (ed. P.A. Grieco), 1–46. London: Blackie Academic & Professional. (e) Du, H. and Ding, K. (2013). *Comprehensive Enantioselective Organocatalysis: Catalysts, Reactions, and Applications*, vol. 1 (ed. P.I. Dalko), 1131–1162. Wiley-VCH Verlag GmbH & Co. KGaA. (f) Moulay, S. and Touati, A. (2010). *C. R. Chim.* 13: 1474–1151. (g) Kumar, A. (2001). *Chem. Rev.* 101: 1–19.

11 (a) Hong, S.H. and Grubbs, R.H. (2006). *J. Am. Chem. Soc.* 128: 3508–3509. (b) Tomasek, J. and Schatz, J. (2013). *Green Chem.* 15: 2317–2338. (c) Lipshutz, B.H. and Ghorai, S. (2014). *in Olefin Metathesis in Water and Aqueous Media: Theory and Practice* (ed. K. Grela), 515–521. Hoboken: Wiley. (d) Leadbeater, N.E. and Marco, M. (2003). *J. Org. Chem.* 68: 888–892. (e) Bhattacharya, S. and Sengupta, S. (2004). *Tetrahedron Lett.* 45: 8733–8736. (f) Jagtap, S. and Deshpande, R. (2013). *Tetrahedron Lett.* 54: 2733–2736. (g) Hoffmann, I., Blumenroder, B., Thumann, S.O. et al. (2015). *Green Chem.* 17: 3844–3857. (h) Christoffel, F. and Ward, T.R. (2018). *Catal. Lett.* 148: 489–511. (i) Hooshmand, S.E., Heidari, B., Sedghi, R., and Varma, R.S. (2019). *Green Chem.* 21: 381–405.

12 (a) Chanda, A. and Fokin, V.V. (2009). *Chem. Rev.* 109: 725–748. (b) Butler, R.N. and Coyne, A.G. (2010). *Chem. Rev.* 110: 6302–6337. (c) Butler, R.N. and Coyne, A.G. (2016). *Org. Biomol. Chem.* 14: 9945–9960. (d) Kitanosono, T., Masuda, K., Xu, P., and Kobayashi, S. (2018). *Chem. Rev.* 118: 679–746. (e) Li, C.J. and Chen, L. (2006). *Chem. Rev.* 35: 68–82. (f) Herrerias, C.I., Yao, X., Li, Z., and Li, C.H. (2007). *Chem. Rev.* 107: 2546–2562. (g) Dambacher, J., Zhao, W., El-Batta, A. et al. (2005). *Tetrahedron Lett.* 46: 4473–4477.

13 Diels, O. and Alder, K. (1928). *Ann. Chem.* 460: 98–122.

14 (a) Brieger, G. and Bennett, J.N. (1980). *Chem. Rev.* 80: 63–97. (b) Kloetzel, M.C. (2011). *Organic Reactions*, vol. 4 (ed. R. Adams), 1–59. Hoboken, NJ, USA: John Wiley & Sons, Inc. (c) Tripoli, R., Cayzer, T.N., Willis, A.C. et al. (2007). *Org. Biomol. Chem.* 5: 2606–2616.

15 Diels, O. and Alder, K. (1931). *Ann. Chem.* 490: 243–257.

16 Woodward, R.B. and Baer, H. (1948). *J. Am. Chem. Soc.* 70: 1161–1166.

17 Eggelte, T.A., Koning, H.D., and Huisman, H.O. (1973). *Tetrahedron* 29: 2491–2493.

18 Rideout, D.C. and Breslow, R. (1980). *J. Am. Chem. Soc.* 102: 7816–7817.

19 Breslow, R. and Rizzo, C.J. (1991). *J. Am. Chem. Soc.* 113: 4340–4341.

20 (a) Meijer, A., Otto, S., and Engberts, J.B.F.N. (1998). *J. Org. Chem.* 63: 8989–8994. (b) Sarma, D., Pawar, S.S., Deshpande, S.S., and Kumar, A. (2006). *Tetrahedron Lett.* 47: 3957–3958. (c) Windmon, N. and Dragojlovic, V. (2008). *Green Chem. Lett. Rev.* 1: 155–163.

21 Grieco, P.A., Yoshida, K., and Garner, P. (1983). *J. Org. Chem.* 48: 3137–3139.

22 Itami, K., Nokami, T., and Yoshida, J.I. (2002). *Adv. Synth. Catal.* 344: 441–451.

23 (a) Yoshida, K. and Grieco, P.A. (1985). *Chem. Lett.* 14: 155–158. (b) Grieco, A., Galatsis, P., and Spohn, R.F. (1986). *Tetrahedron* 42: 2847–2853. (c) Keana, J.F.W., Guzikowski, A.P., Morat, C., and Volwerk, J.J. (1983). *J. Org. Chem.* 48: 2661–2666. (d) Yoshida, K. and Grieco, P.A. (1984). *J. Org. Chem.* 49: 5257–5260.

24 Fringuelli, F., Piermatti, O., Pizzo, F., and Vaccaro, L. (2001). *Eur. J. Org. Chem.* 2001: 439–455.

25 Otto, S. and Engberts, J.B.F.N. (2001). *Reactions and Synthesis* (ed. J. Texter), 247–263. New York: Micellar Media, Marcel Dekker.

26 (a) Ruiz-Lopez, M.F., Assfeld, X., Garcia, J.I. et al. (1993). *J. Am. Chem. Soc.* 115: 8780–8787. (b) Furlani, T.R. and Gao, J. (1996). *J. Org. Chem.* 61: 5492–5497. (c) Kong, S. and Evanseck, J.D. (2000). *J. Am. Chem. Soc.* 122: 10418–10427.

27 Blokzijl, W. and Engberts, J.B.F.N. (1992). *J. Am. Chem. Soc.* 114: 5440–5442.

28 Blokzijl, W., Blandamer, M.J., and Engberts, J.B.F.N. (1991). *J. Am. Chem. Soc.* 113: 4241–4246.

29 Cativiela, C., Garcia, J.I., Gil, J. et al. (1997). *J. Chem. Soc. Perkin Trans. 2* 1997: 653–660.

30 Shrinidhi, A. (2016). *Chemistry Select* 1: 3016–3021.

31 (a) Woodward, R.B. and Katz, T.J. (1958). *Tetrahedron* 5: 70–89. (b) Gleiter, R. and Bohm, M.C. (1983). *Pure Appl. Chem.* 55: 237–244.

32 Berson, J.A., Hamlet, Z., and Mueller, W.A. (1962). *J. Am. Chem. Soc.* 84: 297–304.

33 Blankenburg, V.B., Fiedler, H., Hampel, M. et al. (1974). *J. Prakt. Chem.* 316: 804–816.

34 (a) Schneider, H. and Sangwan, N.K. (1986). *J. Chem. Soc. Chem. Commun.* 1986: 1787–1789. (b) Schneider, H. and Sangwan, N.K. (1989). *J. Chem. Soc. Perkin Trans. 2* 1989: 1223–1227.

35 (a) Desimoni, G., Faita, G., Righetti, P. et al. (1989). *J. Chem. Soc. Perkin Trans. 2* 1989: 437–441. (b) Burdisso, M., Desimoni, G., Faita, G. et al. (1989). *J. Chem. Soc. Perkin Trans.* 2: 1989, 845–1989, 850. (c) Desimoni, G., Faita, G., Righetti, P.P., and Toma, L. (1990). *Tetrahedron* 46: 7951–7970. (d) Desimoni, G., Faita, G., Pasini, D., and Righetti, P.P. (1992). *Tetrahedron* 48: 1667–1674. (e) Abraham, M.H., Grellier, P.L., and McGill, R.A. (1988). *J. Chem. Soc. Perkin Trans.* 2: 1988, 339–1988, 345.

36 (a) Cativiela, C., Mayoral, J.A., Avenoza, A., and Peregrina, J.M. (1990). *J. Phys. Org. Chem.* 3: 414–418. (b) Cativiela, C., Garcia, J.I., Mayoral, J.A. et al. (1992). *J. Phys. Org. Chem.* 5: 230–238. (c) Cativiela, C., Garcia, J.I., Mayoral, J.A., and Salvatella, L. (1994). *J. Chem. Soc. Perkin Trans.* 2: 1994, 847–1994, 851.

(d) Cativiela, C., Garcia, J.I., Mayoral, J.A., and Salvatella, L. (1996). *Chem. Soc. Rev.* 25: 209–218.

37 Breslow, R., Maitra, U., and Rideout, D. (1983). *Tetrahedron Lett.* 24: 1901–1904.

38 Engberts, J.B.F.N. (1995). *Pure Appl. Chem.* 67: 823–828.

39 (a) Sternbach, D.D. and Rossana, D.M. (1982). *J. Am. Chem. Soc.* 104: 5853–5854. (b) Wang, W.B. and Roskamp, E.J. (1992). *Tetrahedron Lett.* 33: 7631–7634. (c) Vivar, A.R.D., Guerrero, C., Diaz, E. et al. (1976). *Phytochemistry* 15: 525–527. (d) Grieco, P.A. and Kaufman, M.D. (1999). *J. Org. Chem.* 64: 6041–6048. (e) Iafe, R.G. and Houk, K.N. (2008). *J. Org. Chem.* 73: 2679–2686.

40 (a) Schneider, H.J. and Sangwan, N.K. (1987). *Angew. Chem. Int. Ed. Eng.* 26: 896–897. (b) Cativiela, C., Garcia, J.I., Mayoral, J.A. et al. (1991). *J. Phys. Org. Chem.* 4: 448–452. (c) Griesbeck, A.G. (1988). *Tetrahedron Lett.* 29: 3477–3480. (d) Rraun, R., Schuster, F., and Sauer, J. (1986). *Tetrahedron Lett.* 27: 1285–1288. (e) Rao, K.R., Srinivasan, T.N., and Bhanumathi, N.A. (1990). *Tetrahedron Lett.* 31: 5959–5960. (f) Chrzanowska, M., Grajewska, A., and Rozwadowska, M.D. (2016). *Chem. Rev.* 116: 12369–12465. (g) Mao, B., Mastral, M.F., and Feringa, B.L. (2017). *Chem. Rev.* 117: 10502–10566. (h) Evans, D.A., Chapman, K.T., and Bisaha, J. (1984). *Tetrahedron Lett.* 25: 4071–4074. (i) Paquette, L.A. (1984). *Asymmetric Synthesis*, vol. 3 (ed. J.D. Morrison), 455–502. Orlando: Academic Press. (j) Wu, Y.D. and Yu, Z.X. (2001). *J. Am. Chem. Soc.* 123: 5777–5786. (k) Liao, W. and Yu, Z.X. (2014). *J. Org. Chem.* 79: 11949–11960. (l) Dols, P.P.M.A., Klunder, A.J.H., and Zwanenburg, B. (1994). *Tetrahedron* 28: 8515–8538. (m) Waldmann, H. and Drager, M. (1990). *Liebigs Ann. Chem.* 7: 681–685. (n) Masamune, S., Choy, W., Petersen, J.S., and Sita, L.R. (1985). *Angew. Chern. Int. Ed. Engl.* 24: 1–30. (o) Geary, P.J., Pryce, R.L., Roberts, S.M. et al. (1990). *J. Chem. Soc. Chem. Commun.* 1990: 204–205. (p) Lubineau, A., Auge, J., and Lubin, N. (1990). *J. Chem. Soc. Perkin Trans.* 1: 1990, 3011–1990, 3015.

41 (a) Attanasi, O.A., Crescentinia, L.D., Filipponea, P. et al. (2001). *Helv. Chim. Acta* 84: 513–525. (b) Fringuelli, F., Matteucci, M., Piermatti, O. et al. (2001). *J. Org. Chem.* 66: 4661–4666.

42 (a) Ahrendt, K.A., Borths, C.J., and MacMillan, D.W.C. (2000). *J. Am. Chem. Soc.* 122: 4243–4244. (b) Northrup, A.B. and MacMillan, D.W.C. (2002). *J. Am. Chem. Soc.* 124: 2458–2460. (c) Hayashi, Y., Samanta, S., Gotoh, H., and Ishikawa, H. (2008). *Angew. Chem. Int. Ed.* 47: 6634–6637. (d) Han, B., Li, J.L., Ma, C. et al. (2008). *Angew. Chem. Int. Ed.* 47: 9971–9974.

43 (a) Wijnen, J.W. and Engberts, J.B.F.N. (1997). *J. Org. Chem.* 62: 2039–2044. (b) Wijnen, J.W. and Engberts, J.B.F.N. (1997). *Liebigs Ann. Recueil* 6: 1085–1088.

44 (a) Narayan, S., Muldoon, J., Finn, M.G. et al. (2005). *Angew. Chem. Int. Ed.* 44: 3275–3279. (b) Klijn, J.E. and Engberts, J.B.F.N. (2005). *Nature* 435: 7467–7470.

45 Grondin, A., Robson, D.C., Smith, W.E., and Graham, D. (2001). *J. Chem. Soc. Perkin Trans.* 2: 2001, 2136–2001, 2141.

46 (a) Lia, P. and Wang, L. (2007). *Lett. Org. Chem.* 4: 23–26. (b) Wang, Z.X. and Qin, H.L. (2003). *Chem. Commun.* 2003: 2450–2451. (c) Fuger, B., Sklute, G.,

Marek, I. et al. (2008). *Synlett* 1: 116–118. (d) Wijnen, J.W., Steiner, R.A., and Engberts, J.B.F.N. (1995). *Tetrahedron Lett.* 36: 5389–5392. (e) Youcef, A.R., Dos, M., Roussel, S. et al. (2009). *J. Org. Chem.* 74: 4318–4323.

47 Myznikov, L.V., Roh, J., Artamonova, T.V. et al. (2007). *Russ. J. Org. Chem* 43: 765–767.

48 Yet, L. (2008). *Comprehensive Heterocyclic Chemistry III*, vol. 4, 1–141. Elsevier.

49 (a) Molteni, G., Orlandi, M., and Broggini, V. (2000). *J. Chem. Soc., Perkin Trans. 1* 2000: 3742–3745. (b) Pang, W., Zhu, S.F., Jiang, H.F., and Zhu, S.Z. (2007). *J. Fluorine Chem.* 128: 1379–1384.

50 (a) Bala, K. and Hailes, H.C. (2005). *Synthesis* 2005: 3423–3427. (b) Rispens, T. and Engberts, J.B.F.N. (2005). *J. Phys. Org. Chem.* 18: 908–917. (c) Chatterjee, N., Pandit, P., Halder, S. et al. (2008). *J. Org. Chem.* 73: 7775–7778. (d) Van Mersbergen, D., Wijnen, J.W., and Engberts, J.B.F.N. (1998). *J. Org. Chem.* 63: 8801–8805. (e) Gonza'lez-Cruz, D., Tejedor, D., de Armas, P. et al. (2006). *Chem. Commun.* 2006, 2798–2006, 2800. (f) McKay, C.S., Kennedy, D.C., and Pezacki, J.P. (2009). *Tetrahedron Lett.* 50: 1893–1896. (g) Engberts, J.B.F.N., Fernandez, E., Garcia-Rio, L., and Leis, J.R. (2006). *J. Org. Chem.* 71: 6118–6123.

51 (a) Gholami, M.R. and Yangjeh, A.H. (1999). *J. Chem. Res. (S)* 1999: 226–227. (b) Gholami, M.R. and Yangjeh, A.H. (2000). *Intern. J. Chem. Kin.* 32: 431–434. (c) Pandey, P.S. and Pandey, I.K. (1997). *Tetrahedron Lett.* 38: 7237–7240. (d) Hamza, K., Touati, A., Ait-Yahia, S. et al. (2010). *Asian J. Chem.* 21: 1231–1244. (e) Holmberg, K. (2003). *Curr. Opin. Colloid Interface Sci.* 8: 187–196. (f) Chatterjee, A., Maiti, D.K., and Bhattacharya, P.K. (2003). *Org. Lett.* 5: 3967–3969. (g) Chatterjee, A. and Bhattacharya, P.K. (2006). *J. Org. Chem.* 71: 345–348.

52 (a) Butler, R.N., Coyne, A.G., Cunningham, W.J., and Burke, L.A. (2002). *J. Chem. Soc. Perkin Trans.* 2: 1807–1815. (b) Butler, R.N., Cunningham, W.J., Coyne, A.G., and Burke, L.A. (2004). *J. Am. Chem. Soc.* 126: 11923–11929. (c) Butler, R.N., Coyne, A.G., Cunningham, W.J. et al. (2005). *Helv. Chim. Acta* 88: 1611–1629. (d) Butler, R.N., Coyne, A.G., and Moloney, E.M. (2007). *Tetrahedron Lett.* 48: 3501–3503. (e) Lubineau, A., Bouchain, G., and Queneau, Y. (1995). *J. Chem. Soc., Perkin Trans.* 1: 2433–2437. (f) Grigg, R., Mongkolaussavaratana, T., Pounds, C.A., and Sivagnanam, S. (1990). *Tetrahedron Lett.* 31: 7215–7218.

53 Li, Z., Seoa, T.S., and Ju, J. (2004). *Tetrahedron Lett.* 45: 3143–3146.

54 Molteni, G. and Ponti, A. (2006). *Arkivoc* 16: 49–56.

55 (a) Portmann, R., World Patent WO Patent 9802423, **1998**. (Novartis AG); (b) Kolb, H.C., Finn, M.G., and Sharpless, K.B. (2001). *Angew. Chem. Int. Ed.* 40: 2004–2021. (c) Shin, J.A., Lim, Y.G., and Lee, K.H. (2012). *J. Org. Chem.* 77: 4117–4122. (d) Rostovtsev, V.V., Green, L.G., Fokin, V.V., and Sharpless, K.B. (2002). *Angew. Chem. Int. Ed.* 41: 2708–2711. (e) Sharpless, K. B., Fokin, V., Rostovtsev, V. A., Green, L., Himo, F., US Patent 7,375, 234 B2 May 20, 2008; (f) Appukkuttan, P., Dehaen, W., Fokin, V.V., and Van der Eycken, E. (2004). *Org. Lett.* 6: 4223–4225. (g) Kumar, D., Reddy, V.B., and Varma, R.S. (2009).

Tetrahedron Lett. 50: 2065–2068. (h) Grundmann, C. and Dean, M.J. (1965). *J. Org. Chem.* 30: 2809–2812. (i) Dignam, R.J., Hegarty, A.F., and Quain, P.L. (1978). *J. Org. Chem.* 43: 388–393. (j) Demko, Z.P. and Sharpless, K.B. (2001). *J. Org. Chem.* 66: 7945–7950. (k) Jiang, N. and Li, C.J. (2004). *Chem. Commun.* 2004: 394–395. (l) Molteni, G., Ponti, A., and Orlandi, M. (2002). *New J. Chem.* 26: 1340–1345. (m) Inoue, Y., Araki, K., and Shiraishi, S. (1991). *Bull. Chem. Soc. Jpn.* 64: 3079–3083.

56 (a) Yong, L. and Butenschon, H. (2002). *Chem. Commun.* 2002: 2852–2853. (b) Kinoshita, H., Shinokubo, H., and Oshima, K. (2003). *J. Am. Chem. Soc.* 125: 7784–7785.

57 Fatland, A.W. and Eaton, B.E. (2000). *Org. Lett.* 2: 3131–3133.

58 Xu, F., Wang, C., Li, X., and Wan, B. (2012). *ChemSusChem* 5: 854–857.

59 Lubineau, A. and Bouchain, G. (1997). *Tetrahedron Lett.* 38: 8031–8032.

60 Wender, P.A., Love, J.A., and Williams, T.J. (2003). *Synlett* 2003: 1295–1298.

61 Muthuramu, K. and Murthy, V.R. (1982). *J. Org. Chem.* 47: 3976–3979.

62 Gaoni, Y. and Sadeh, S. (1980). *J. Org. Chem.* 45: 871–880.

63 Chiang, Y., Kresge, A.J., and Zhan, H.Q. (2003). *Can. J. Chem.* 81: 607–611.

64 (a) Li, S.W. and Batey, R.A. (2007). *Chem. Commun.* 2007: 3759–3761. (b) Griffiths, K., Gallop, C.W.D., Abdul-Sada, A. et al. (2015). *Chem. Eur. J.* 21: 6358–6361. (c) Nardi, M., Costanzo, P., De Nino, A. et al. (2017). *Green Chem.* 19: 5403–5411.

65 Gomes, R.F.A., Esteves, N.R., Coelho, J.A.S., and Afonso, C.A.M. (2018). *J. Org. Chem.* 83: 7509–7513.

66 Jung, E.J., Park, B.H., and Lee, Y.R. (2010). *Green Chem.* 12: 2003–2011.

67 Li, X.Y., Liu, Y., Chen, X.L. et al. (2020). *Green Chem.* 22: 4445–4449.

68 (a) Domingo, L.R., Ríos-Gutiérrez, M., Chamorro, E., and Pérez, P. (2016). *Chemistry Select* 1: 6026–6039. (b) Domingo, L.R., Ríos-Gutiérrez, M., Silvi, B., and Pérez, P. (2018). *Eur. J. Org. Chem.* 2018: 1107–1120.

69 Cope, A.C. and Hardy, E.M. (1940). *J. Am. Chem. Soc.* 62: 441–444.

70 Claisen, L. (1912). *Chem. Ber.* 45: 3157–3166.

71 (a) Rhoads, S.J. and Raulins, N.R. (1975). *Organic Reactions*, vol. 22 (ed. W.G. Dauben), 1–253. New York: Wiley. (b) Copley, S.D. and Knowles, J.R. (1987). *J. Am. Chem. Soc.* 109: 5008–5013.

72 (a) Wipf, P. and Ribe, S. (2001). *Org. Lett.* 3: 1503–1503. (b) Yang, X., Stern, C.L., and Marks, T.J. (1994). *J. Am. Chem. Soc.* 116: 1001–1003. (c) Mehta, B.K., Yanagisawa, K., Shiro, M., and Kotsuki, H. (2003). *Org. Lett.* 3: 1605–1608. (d) Nicolaou, K.C., Xu, H., and Wartmann, M. (2005). *Angew. Chem. Int. Ed.* 44: 756–761. (e) Tisdale, E.J., Slobodov, I., and Theodorakis, E.A. (2004). *Proc. Natl. Acad. Sci. USA* 101: 12030–12035. (f) Tisdale, E.J., Chowdhury, C., Vong, B.G. et al. (2002). *Org. Lett.* 4: 909–912. (g) Tisdale, E.J., Vong, B.G., Li, H. et al. (2003). *Tetrahedron* 59: 6873–6887. (h) Tisdale, E.J., Slobodov, I., and Theodorakis, E.A. (2003). *Org. Biomol. Chem.* 1: 4418–4422. (i) Severance, D.L. and Jorgensen, W.L. (1992). *J. Am. Chem. Soc.* 114: 10966–10968. (j) Gajewski, J.J., Jurayj, J., Kimbrough, D.R. et al. (1987). *J. Am. Chem. Soc.* 109: 1170–1186. (k) White, W.N. and Wolfarth, E.F. (1970). *J. Org. Chem.* 35: 2196–2199.

(l) Brandes, E., Grieco, P.A., and Gajewski, J.J. (1989). *J. Org. Chem.* 54: 515–516. (m) Gajewski, J.J. (1992). *J. Org. Chem.* 57: 5500–5506. (n) Pirrung, M.C. (2006). *Chem. Eur. J.* 12: 1312–1317. (o) Cramer, C.J. and Truhlar, D.G. (1992). *J. Am. Chem. Soc.* 114: 8794–8799. (p) Grieco, P.A., Brandes, E.B., McCann, S., and Clark, J.D. (1989). *J. Org. Chem.* 54: 5849–5851. (q) McMurry, J.E., Andrus, A., Ksander, G.M. et al. (1981). *Tetrahedron* 37: 319–327. (r) Narwid, T.A., Cooney, K.E., and Uskokovic, M.R. (1974). *Helv. Chim. Acta* 57: 764–771. (s) Wipf, P., Waller, D.L., and Jonathan, T.R. (2005). *J. Org. Chem.* 70: 8096–8102. (t) Eronen, A., Nieger, M., Kajander, T.A., and Repo, T. (2023). *J. Org. Chem.* 88: 12914–12923.

73 Beare, D.K. and McErlean, C.S.P. (2013). *Org. Biomol. Chem.* 11: 2452–2459.

74 Sato, T. and Otera, J. (1994). *Tetrahedron Lett.* 35: 6701–6704.

75 Liao, M. and Wang, J. (2007). *Green Chem.* 9: 184–188.

76 Narayan, S., Muldoon, J., Finn, M.G. et al. (2005). *Angew. Chem. Int. Ed.* 44: 3339–3343.

77 (a) Narayan, S., Fokin, V.V., and Sharpless, K.B. (2007). *Organic Reactions in Water: Principles, Strategies and Applications* (ed. U.M. Lindstrom), 350–365. Blackwell, Oxford. (b) Pirrung, M.C., Sarma, K.D., and Wang, J. (2008). *J. Org. Chem.* 73: 8723–8730.

8

Olefin Metathesis In-Water: Recent Progress and Challenges

Bengi Özgün Öztürk

Hacettepe University, Faculty of Science, Chemistry Department, Beytepe-Ankara, 06800, Turkey

8.1 Introduction

Olefin metathesis emerged as an efficient catalytic method to construct advanced organic structures through the formation of carbon–carbon double bonds in an efficient manner [1–4]. The invention of an air-stable and functional group-tolerant ruthenium-based Grubbs type catalyst has opened a new era in metathesis chemistry, expanding both the scope and application areas of Grubbs first (G1), second (G2), Hoveyda–Grubbs first (HG1)-, and second (HG2)-generation catalysts in both academia and industry (Figure 8.1) [5–7].

Olefin metathesis is usually carried out in classical organic solvents such as dichloromethane, chloroform, toluene, and other polar anhydrous solvents [8]. The adaptation of catalytic reactions to aqueous media is of great importance considering the environmentally negative impact of chlorinated organic solvents [9]. However, the hydrophobic nature of both the Grubbs/Hoveyda–Grubbs catalyst and the olefinic substrates tackles the adaptation of olefin metathesis to aqueous media. Therefore, for the last two decades, there have been on-going challenges for the adaptation of metathesis catalytic systems to aqueous media [10].

The first example of aqueous olefin metathesis was introduced by Grubbs and coworker. The integration of ionic groups on the phosphine group has led to the water-soluble Grubbs first-generation analogs (**1, 2**) (Figure 8.2) [11]. Although **1, 2**, and **3** exhibit water solubility to an extent, their stability in an aqueous medium in metathesis reactions such as ring-closing metathesis (RCM), cross-metathesis, and self-metathesis reactions is limited, and only a stable performance was observed in ring-opening metathesis polymerization (ROMP) reactions of hydrophobic and hydrophilic norbornene-based monomers using these first-generation catalysts. Phosphine-free Hoveyda–Grubbs analog **4** bearing a hydrophilic non-ionic group (PEG) on the N-heterocyclic carbene (NHC) moiety exhibits considerable performance in RCM reactions of hydrophilic dienes bearing cationic amine groups (Scheme 8.1).

Organic Transformations in Water: Principles and Applications, First Edition.
Edited by Gopinathan Anilkumar, Nissy Ann Harry, and Sankuviruthiyil M. Ujwaldev.

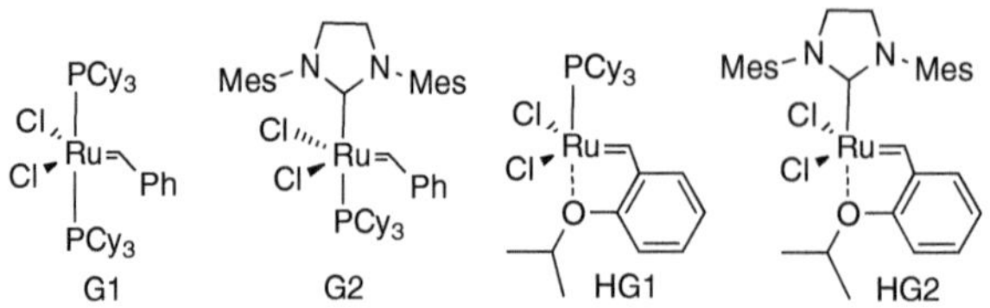

Figure 8.1 Ruthenium-based olefin metathesis catalysts.

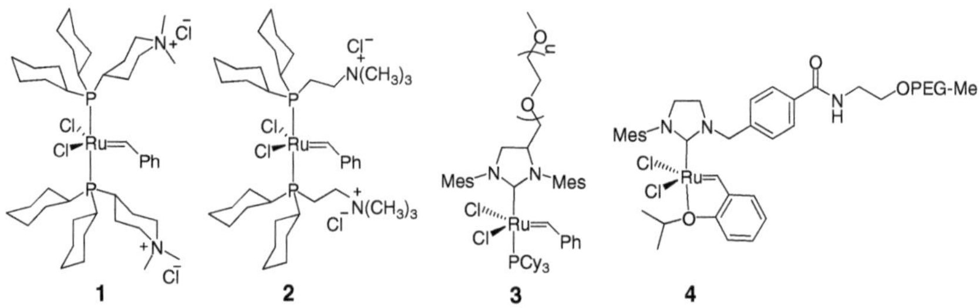

Figure 8.2 Ionic and non-ionic group-tagged ruthenium-based metathesis catalysts.

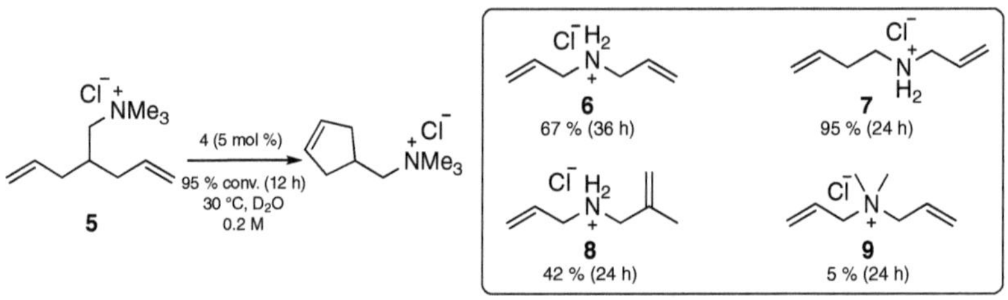

Scheme 8.1 RCM reactions of hydrophilic dienes catalyzed by a PEG-substituted Grubbs catalyst.

8.2 Ionic, Non-Ionic, and Amphiphilic Group-Tagged Olefin Metathesis Catalysts

As an alternative to phosphine ligand modification, benzylidene or alkylidene can be modified with ionic-tagged groups to improve the hydrophilic character of metathesis catalysts. Schanz and his research group synthesized cationic phosphine (**10**) and dimethyl amine-substituted benzylidene analogs of Grubbs catalysts through alkylidene exchange reactions of Grubbs first-generation catalysts (**11–12**) (Figure 8.3) [12]. ROMP reactions of cationic norbornene derivatives were carried out using a 2,2,2-trifluoroethanol (TFE)/dichloromethane (CH_2Cl_2) mixture (v/v: 1/1) using catalysts **10–12**. Catalyst **10** has formed a uniform solution in the reaction mixture due to its ionic nature; however, catalysts **11** and **12** required a protonation step using

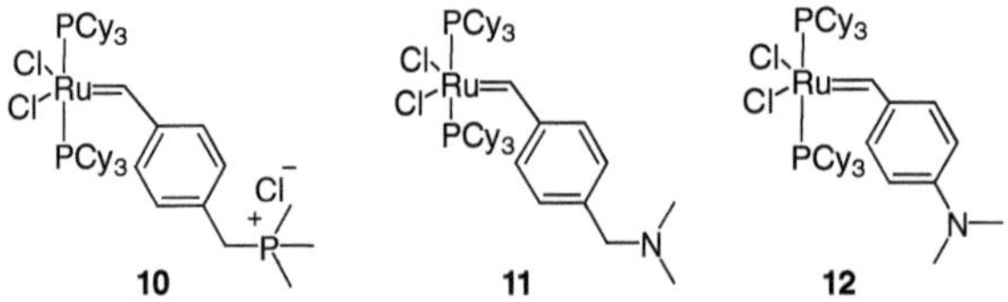

Figure 8.3 Water-soluble first-generation Grubbs catalyst derivatives.

two mol equivalents of HCl. Due to the protonation of amine groups on **11** and **12**, which affects the electronic environment of the catalysts, the performance of **11** and **12** was boosted in ROMP reactions. The second role of the acid addition is to prevent the rebinding of dissociated tricyclohexylphosphine (PCy_3) ligands through the protonation of PCy_3 to $H^+PCy_3^-$. The strategy of integrating protonable pH sensitive dimethylamino groups on metathesis catalysts was expanded to Hoveyda–Grubbs second-generation analogs by Grela and his research team [13].

The modification of the chelating ortho-isopropoxybenzylidene group with the diethylamine group results in pH-sensitive second-generation ruthenium catalyst systems. The main principle in this catalytic system is to transform the non-polar nature of the catalyst to more polar character through protonation. Also following the protonation, the electron-donating nature of $-NEt_2$ has shifted to an electron-withdrawing group effect due to the formation of $-NEt_2H^+$. The protonation has resulted in an activity increase in RCM and enyne-metathesis reactions in organic media. Following these innovative studies, Grela group has prepared several HG2 analogs bearing quaternary ammonium groups on various ligand positions (Figure 8.4). [14, 15]. These catalysts showed superior performance in RCM and enyne metathesis reactions of hydrophobic olefinic substrates in aqueous media and can be recycled easily. Complex **13** showed high silica gel affinity in organic media, allowing the easy separation of the catalyst from the reaction media and obtaining pure metathesis products below 1 ppm Ru contamination. To improve the water solubility of **13**, a second quaternary ammonium group was attached to the chelating ortho-isopropoxy benzylidene moiety. The isopropyl group on ether oxygen was substituted with a cationic 1,1-dimethylpiperidine group, which led to improved water solubility of the complex. In neat water, the RCM reaction of ionic diene molecules reached 96, 88, and 95% conversions. Complexes **13** and **14** can easily be recycled from the reaction media through extraction with diethyl ether. The product remains in the ether phase, while catalysts **13** and **14** remain in the aqueous phase [13a]. Grubbs and coworker utilized quaternary ammonium-substituted NHC and chelating isopropoxybenzyldine groups to catalyze ROMP and RCM reactions in neat water [11a]. The catalyst **15**, however, showed poor performance in cross-metathesis of allyl alcohol derivatives in water, yielding the isomerization product propionaldehyde in most cases. The catalyst **15** also failed to catalyze

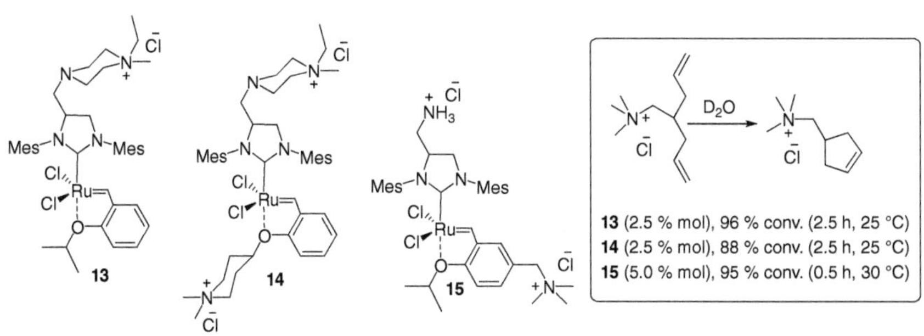

Figure 8.4 Hoveyda–Grubbs catalysts bearing quaternary ammonium groups.

the cross-metathesis of various olefinic reactants bearing different hydrophilic substituents such as amino acids, carbohydrates, and ammonium salts.

Schanz and his team have shown that 1-methylimidazole can be used as an inhibitor for Grubbs catalysts through the formation of an 18 electron complex with the substitution of the PCy_3 group with two moles of 1-methylimidazole groups (Scheme 8.2) [16]. The so-called "latent" catalysts can be activated through the addition of acids (HCl or H_3PO_4) to protonate imidazole groups, yielding 14 electron-active species in the reaction medium. The catalytic performance of **16** was tested in RCM reactions of diethyl diallyl malonate in d_6-benzene in the presence of 12 mol equivalent of H_3PO_4, reaching an 82.8 % conversion after one hour. However, the catalyst is not recyclable and still requires relatively higher catalytic loadings (5% mol) to reach quantitative conversion values in RCM reactions. To address this issue, poly(vinylimidazole)-supported **G1** catalyst was developed to obtain a water-compatible latent metathesis catalyst (**17**) for RCM and ROMP applications of hydrophobic substrates [17]. Catalyst **17** showed latent behavior and can be activated through the introduction of excess HCl in water in the RCM of diethyl diallylmalonate, reaching a conversion of 99% after 24 hours with 5% Ru loading. The catalyst can be recycled up to three run in RCM. For ROMP reactions of hydrophobic norbornene derivatives, the addition of a non-ionic surfactant (Tween20) is required to prevent the coagulation of the growing ROMP polymer in an aqueous medium during metathesis reactions. Following this study, ruthenium indenylidene catalysts bearing imidazole or morpholine-substituted bidentate (N,O) Schiff base ligands (**18, 19**) (Scheme 8.6) were reported for aqueous olefin metathesis applications [18, 19]. The activity of the catalysts **17–19** can be tuned by varying the HCl/Ru (mol/mol) ratio in aqueous ROMP reactions of hydrophobic monomers, allowing a high control over the molecular weight of the resulting polymers.

Another strategy in aqueous olefin metathesis is to use surfactant-like non-ionic polymers as dispersive agents for both hydrophobic catalysts and olefinic reagents [20, 21]. The non-ionic groups can be tagged directly on various ligand positions, including benzylidene and NHC moieties, on metathesis catalysts or can be used as an additive to stabilize the non-modified metathesis catalysts within hydrophobic compartments of micelle structures. Hydrophilic polymer substitution on Grubbs or Hoveyda–Grubbs catalysts alters the polarity of the complex, which allows the recycling of the catalysts with a suitable polar organic solvent after aqueous metathesis reactions. The integration of hydrophilic PEG groups on the chelating isopropoxybenzylidene moiety of HG2 analogs was reported by several research groups [22–26]. Owing to the hydrophilic nature of the PEG group, the catalysts can be dispersed in an aqueous medium, including acetone/water and methanol/water mixtures for CM and RCM applications (Figure 8.5). In contrast to PCy_3 groups on G1 and G2, NHC ligands do not dissociate from the ruthenium complex during metathesis reactions and stabilize 14 electron intermediates through electron and steric effects exerted by the NHC structure. The substituents on the N-atoms of the NHC group play an important role in the stability of the complex in metathesis reactions. 2,4,6-Mesitylene or 2,6-disopropylphenyl groups are usually preferred in

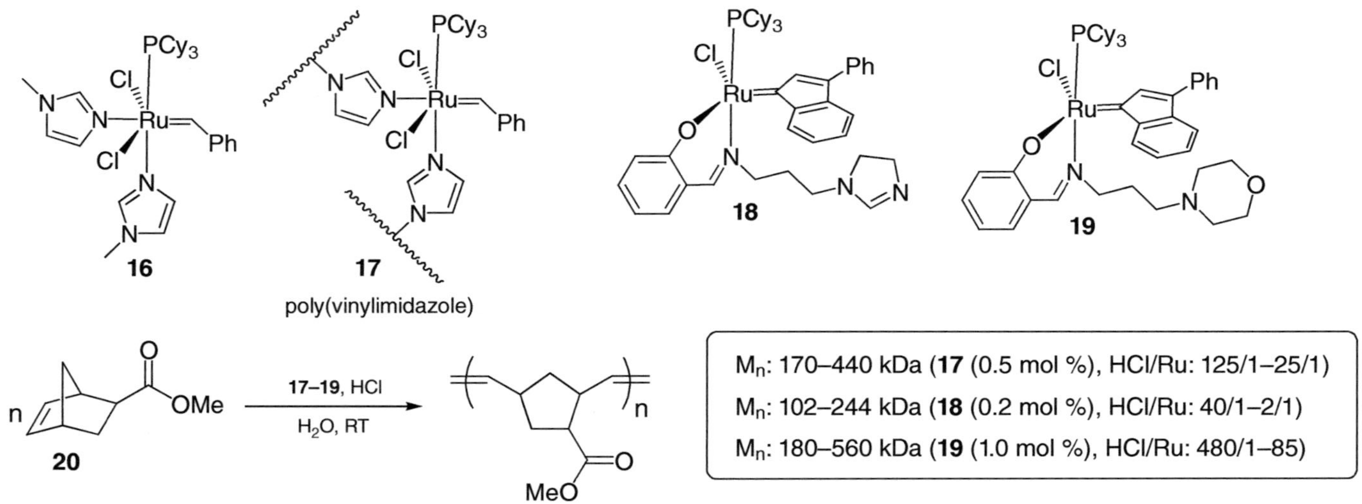

Scheme 8.2 Latent ruthenium metathesis catalysts for aqueous ROMP and RCM reactions.

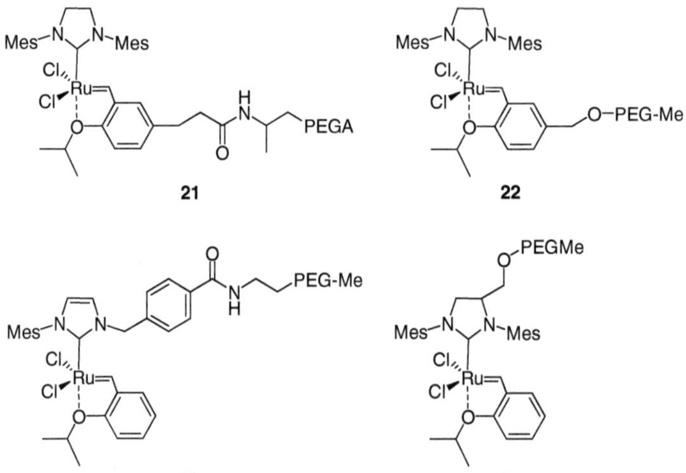

21 22 23 24

Figure 8.5 PEG-substituted Hoveyda–Grubbs second-generation catalysts.

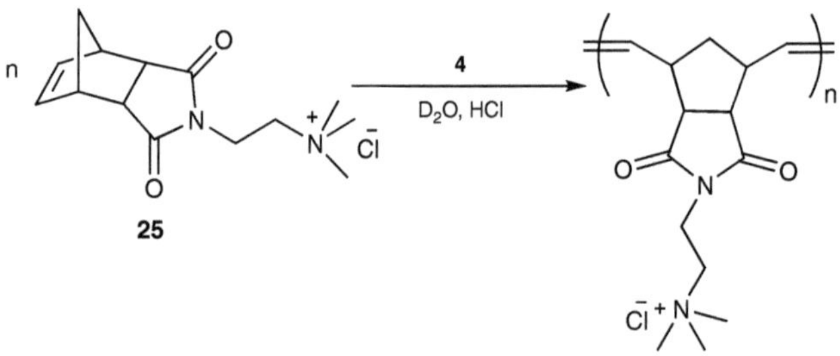

25

Scheme 8.3 ROMP of hydrophilic norbornene derivative in water using PEG-substituted Grubbs catalyst (4).

commercial Grubbs and Hoveyda–Grubbs-type metathesis catalysts. The modification of the NHC group can be carried out through different positions of the NHC group [24–26]. Grubbs and his research group substituted one of the mesitylene groups with the benzyl end-functionalized PEG group to obtain a water-soluble metathesis catalyst [25]. Although the catalyst performed well in ROMP of both *exo* and *endo*-norbornene derivative **25** in HCl/H_2O mixture, it showed poor performance in RCM and CM reactions (Scheme 8.3). The main problem of **23** in metathesis reactions is the steric environment of the NHC ligand. When compared to SIMes ligands, the NHC group on **23** lacks one mesitylene substituent on one of the nitrogen groups, which dramatically decreases the stability of the activated complex in aqueous media, yielding lower turnover numbers in the catalytic cycle in RCM and CM reactions. In addition, the double bond in the backbone of the NHC ligand drastically decreases the σ-donating ability of the ligand when compared to saturated analog SIMes. To address the current challenge, the position of the PEG group was altered and substituted on the NHC backbone (**24**), which did not significantly affect the electron-donating ability of the NHC ligand but improved the polarity of the complex, leading to a hydrophilic metathesis catalyst [26].

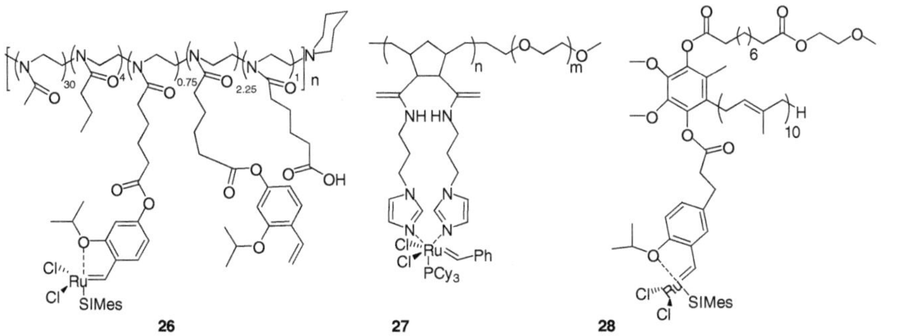

26 27 28

Figure 8.6 Grubbs and Hoveyda–Grubbs catalysts immobilized on amphiphilic polymers.

As an alternative to hydrophilic polymers, amphiphilic block polymers are also attached to the Grubbs and Hoveyda–Grubbs catalysts to obtain a water-compatible catalytic system [27]. The Grubbs catalyst is mainly attached to the hydrophobic block through covalent bonding, thus stabilizing the Grubbs catalyst in the hydrophobic domains of micelle structures in-water. Additionally, the hydrophobic compartments in aqueous micelle structures provide hydrophobic confined spaces for hydrophobic olefinic substrates, improving the reaction rates to some extent. Isopropoxy styrene modified block poly(2-oxazoline) derivatives are used to immobilize Hoveyda–Grubbs second-generation catalyst in hydrophobic domains of the block polymer (Catalyst **26**) (Figure 8.6) [28].

The presence of two chelating isopropoxy stryrene-substituted blocks on the hydrophobic chain increases the robustness of the catalyst in the aqueous medium. A latent ruthenium catalyst was reported in 2018 using imidazole-substituted amphiphilic block ROMP polymers [29]. Diester-substituted ROMP polymers were end-functionalized using allyl-end-capped PEG polymers as chain-transfer agents. The post-polymerization modification of ester groups with a 3-aminopropylimidazole group, followed by *in situ* hydrogenation of the ROMP polymer, yielded imidazole-substituted amphiphilic polymers. The integration of Grubb's first-generation catalyst has led to the formation of a latent catalytic system. The catalyst **27** has formed spherical micelle structures with an average size of 89.69 (±34) nm, as confirmed by transmission electron microscopy (TEM) analysis. The addition of HCl (aq.) to the reaction media activates the catalyst for RCM reactions in the aqueous medium. The integration of lipophilic and hydrophilic chains and isopropoxy styrene groups through an aromatic binder molecule has led to a unique Hoveyda–Grubbs-type catalyst (**28**) that showed excellent reusability up to ten runs in RCM of hydrophobic dienes in water [30] (Figure 8.6). The RCM performance of the amphiphilic catalysts; **26**, **27**, and **28** was tested in the RCM of diethyl diallyl malonate (**29**) in neat water, and all three catalysts reached higher conversion values even under low catalytic loadings (Figure 8.6). In an alternative to NHC and alkylidene ligand modification, Buchmeister and his research team utilized fluoro group functionalized poly(2-oxazoline) derivatives for the immobilization of Hoveyda–Grubbs second-generation catalyst (**29**) through anionic ligand exchange reactions (Scheme 8.4). The catalyst **29** was used for the

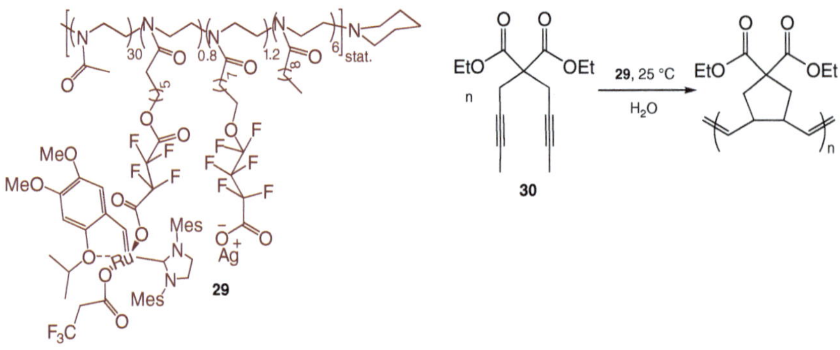

Scheme 8.4 Hoveyda–Grubbs second-generation catalyst supported on poly(4-oxaline)-based amphiphilic polymers.

enyne metathesis polymerization of diynes in aqueous media, yielding nano-sized micellar macromolecular particles [31].

The addition of supramolecular additives such as surfactants and amphiphiles is another strategy to conduct olefin metathesis in water. Nonionic or ionic surfactants can be employed to obtain a stable emulsion in water for hydrophobic Grubbs and Hoveyda–Grubbs catalysts and olefinic substrates [32]. Ionic surfactants; sodium dodecyl sulfate (SDS), sodium dodecyl sulfonate ($SDSO_3Na$) and non-ionic surfactants; polyoxyethylenesorbitanemonopalmitate (Tween-40) and decaoxyethyelene dodecyl ether (Brij30 or 35) were employed on the RCM of diethyldiallyl malonate in water using a substrate/surfactant ratio between 2.5–10 (mol mol^{-1}) in the presence of G1 at room temperature (Figure 8.7). The control experiments showed that the RCM reaction reached 51% conversion within one hour in the absence of any surfactants. No improvement in yield was observed when non-ionic surfactants (Tween-40, Brij-35) and anionic surfactants; $SDSSO_3Na$ were used in water media (Table 8.1). However, the performance of the system was drastically increased when anionic surfactant SDS was employed under identical reaction conditions, reaching 96 % conversion within 30 minutes of reaction time under 5% mol G1 loading in water media [33]. Following this study, Lipshutz utilized a vitamin E-based surfactant; polyoxyethanyl α-tocopheryl sebacate (PTS), in an aqueous RCM of diethyl diallylmalonate derivatives and cross-metathesis of allylbenzene and *tert*-butylacrylate using G1 and G2. The CM reaction proceeded with high *trans* (*E*) selectivity, and among the other tested surfactants, only the PTS-based micellar system yielded the corresponding product in 97% yield (Table 8.2) [34]. This strategy was also extended to the RCM of different diene molecules with varying chain lengths to form six- or seven-membered ring structures in water [35]. The same research group also showed the effect of pH changes in aqueous CM reactions using 0.02 M $KHSO_4$ and a PTS/water mixture at 22 C in the presence of G2 (2 mol %). Recently, the effect of pH and salt concentration on the metathesis activity of Grubbs-type catalysts in water was investigated by Foster et al. [36]. The addition of KCl, NaCl, or tetrabutylammonium chloride (TBAC) improved the stability and performance of the catalyst in aqueous ROMP reactions. The main role of the chloride ion is to prevent

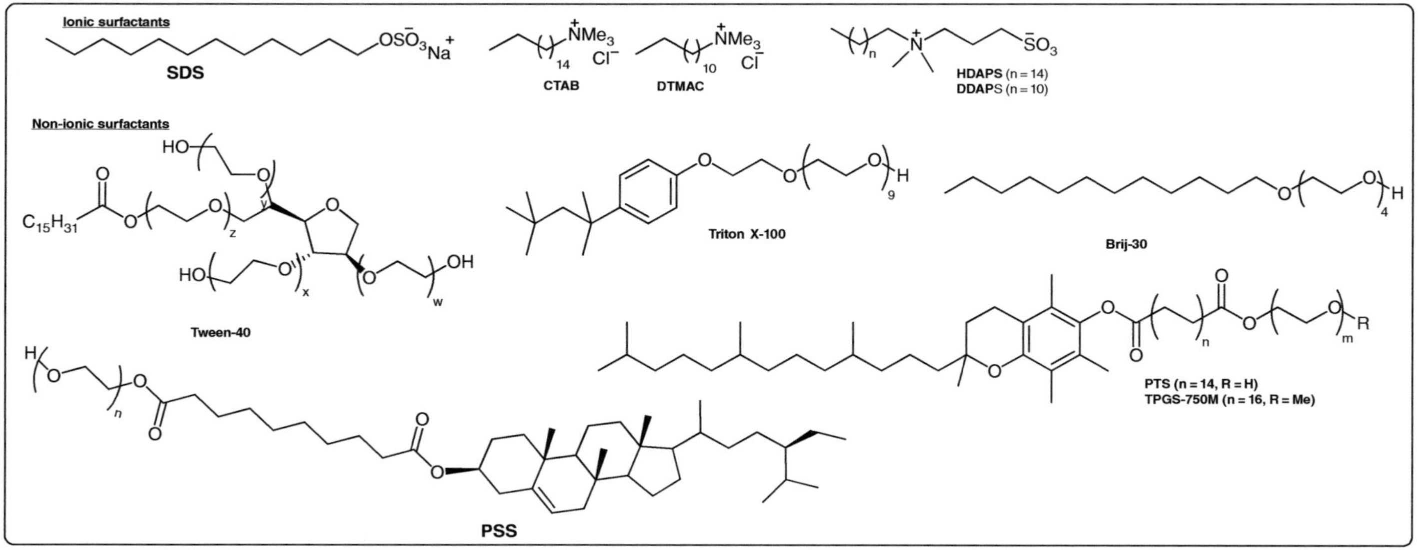

Figure 8.7 Various ionic and non-ionic surfactants for aqueous olefin metathesis reactions.

Table 8.1 G1-catalyzed RCM of diethyl diallylmalonate using various surfactants in water.

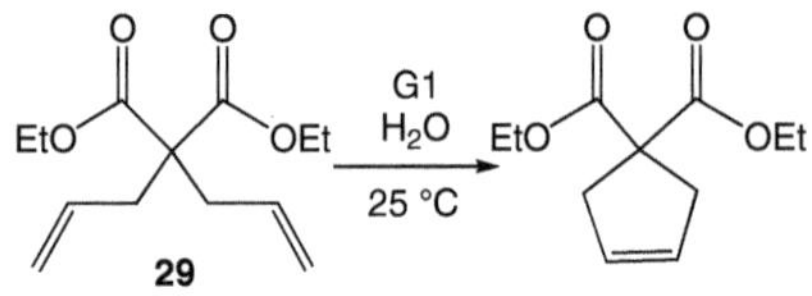

Surfactant	Catalyst/ surfactant	Catalyst % (mol %)	Time (min.)	Conv. %
-	-	5	60	51
Tween-40	5	5	30	59
Brij-35	5	5	30	43
SDS	5	5	30	96
$SDSO_3Na$	5	5	30	46

Table 8.2 G2-catalyzed cross-metathesis of allyl benzene using various non-ionic surfactants in water.

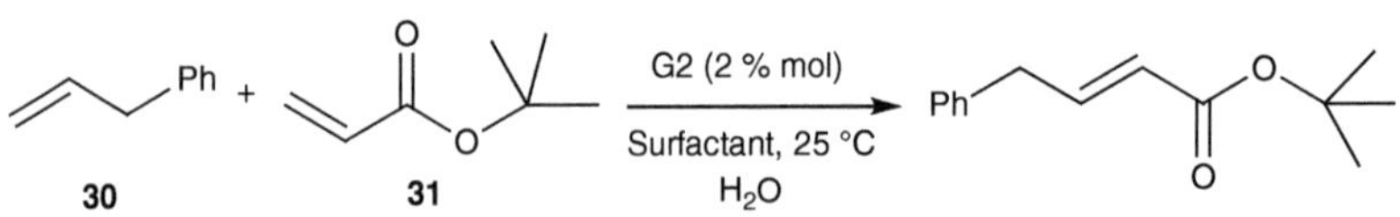

Entry	Surfactant	Time (h)	Yield %
1	SDS	12	68
2	Brij-30	12	63
3	PSS	12	78
4	PTS	12	97
5	TPGS	12	67

the anionic ligand displacement with hydroxide or water molecules in aqueous media, which significantly decreases the performance of the catalyst through slower turnover numbers. At pH = 2 (HCl) with a salt concentration of 100 mM TBAC, complete conversion of the hydrophilic monomer was obtained in ROMP reactions in aqueous medium.

In 2018, the ruthenium indenylidene-based metathesis catalyst, commercially known as UmicoreM1, was used along with Tween-20 and DTMAC to catalyze sequential cross-metathesis/hydrogenation of 10-undecenal with methyl acrylate [37] (Scheme 8.5). The combination of non-ionic/cationic surfactants with a rate of 5% Tween-20 and 4% DTMAC yielded the corresponding CM/TH product in 72% yield with a catalytic loading of 5% mol UmicoreM1.

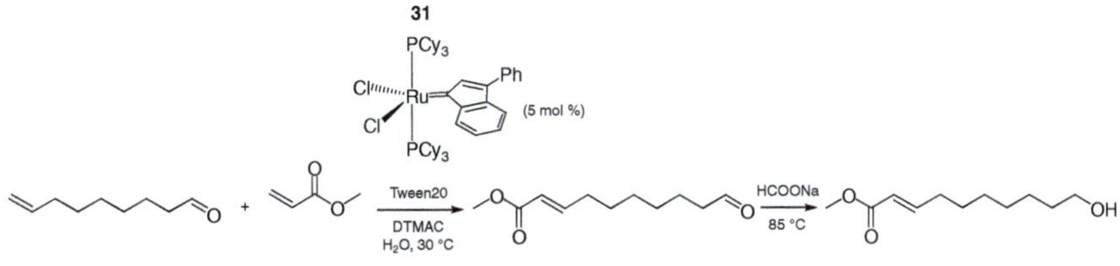

Scheme 8.5 Sequential cross-metathesis and transfer hydrogenation in water.

8.3 Biopolymer Integrated Olefin Metathesis Catalysts

Biopolymers such as β-cyclodextrin and alginate are used in metathesis chemistry for different purposes (Figure 8.8). For example, Chung et al. address the separation problem of water-soluble Hoveyda–Grubbs catalysts (**32**) from aqueous media by using host-guest interactions of adamantly substituted HG2 catalysts

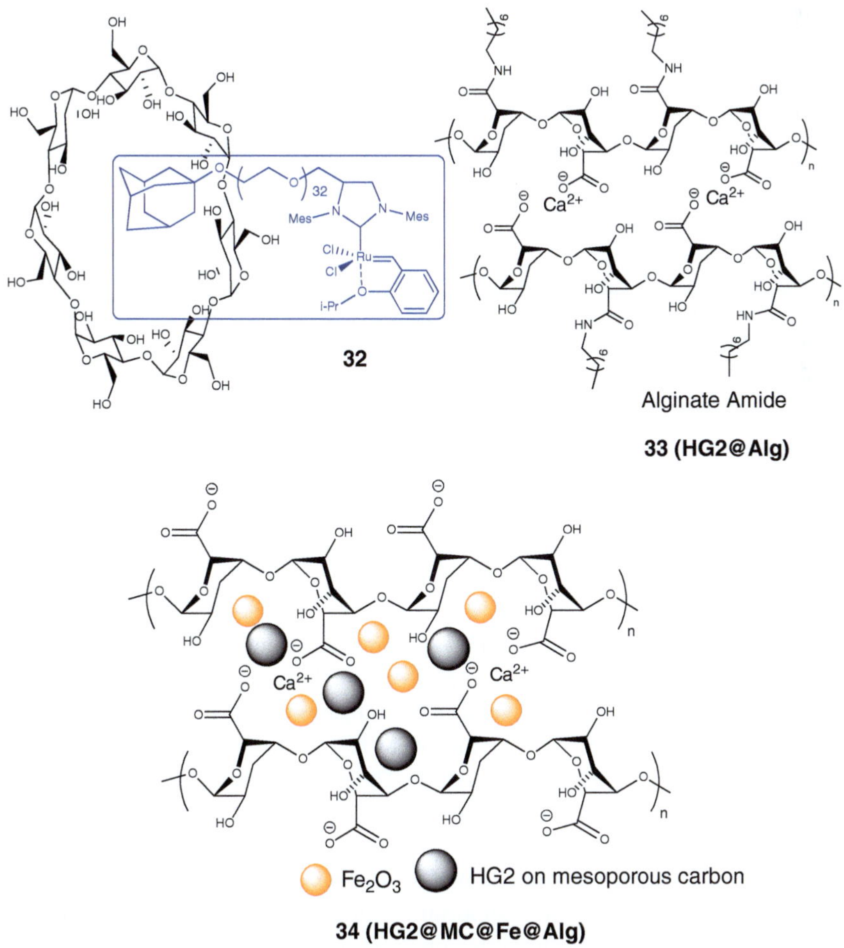

Figure 8.8 Dextrin and alginate-based olefin metathesis catalyst systems.

with β-cyclodextrin [38, 39]. The ruthenium content of the final product can be decreased to 53 ppm after filtration, indicating the efficiency of host-guest interaction between two molecules. A hydrophilic natural polymer, alginate, was used as an encapsulation agent for Hoveyda–Grubbs second (HG2)-generation catalysts for aqueous metathesis applications [40, 41]. Pauly et al. have modified alginate beads with an n-octyl group to improve the diffusion of hydrophobic olefinic substrates into the alginate beads through creating hydrophobic domains [40]. Although the performance of **33** (HG2@Alg) is quite comparable to HG2 in neat water, the encapsulation allowed the efficient recycling of the catalyst after the metathesis reactions. In 2022, HG2 loaded on hydrophobic nanomesoporous carbon and magnetic γ-Fe_2O_3 nanoparticles were encapsulated within calcium alginate beads. The mesoporous carbon allowed the homogeneous distribution of HG2 catalysts throughout the alginate matrix and also provided hydrophobic surfaces for olefinic substrates [41]. The catalyst 34 (HG2@MC@Fe@Alg) showed high performance in the RCM of diethyldiallyl malonate, and the catalyst can be easily separated from the reaction mixture with the help of a magnet. The performance comparison of **33** and **34** was given in Scheme 8.6. Another encapsulation strategy is based on the kinetic protection of quaternary ammonium-tagged Hoveyda–Grubbs catalysts, namely, Aquamet (**13**), within metal-organic cages with the general formula of $[Pt_6L_4](NO_3)_{12}$ in aqueous medium [42]. Although Aquamet was effectively encapsulated within the void of the cage structure, the encapsulation of G1, G2, and HG2 failed due to the low polarity of the complexes.

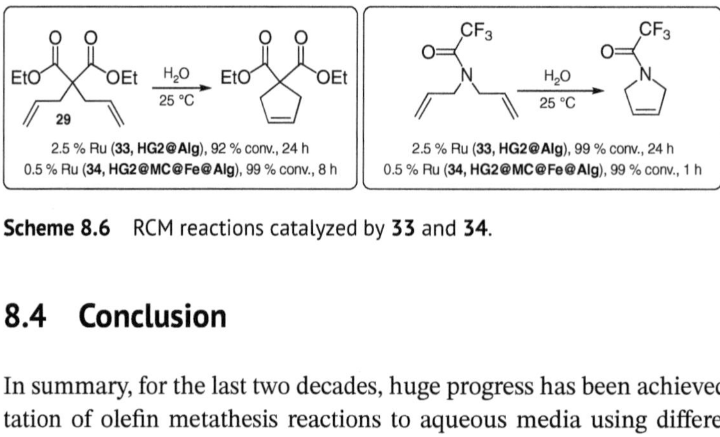

Scheme 8.6 RCM reactions catalyzed by **33** and **34**.

8.4 Conclusion

In summary, for the last two decades, huge progress has been achieved in the adaptation of olefin metathesis reactions to aqueous media using different strategies. Water-soluble ruthenium catalysts, as well as emulsion-based ruthenium catalytic systems, can catalyze the ROMP, CM, and RCM reactions of both hydrophilic and hydrophobic olefinic substrates to an extent. The elimination of organic solvents in metathesis reactions not only improves the sustainability of the process but also reduces production costs. In addition, the recycling of water-compatible catalysts in the aqueous phase allows the production of olefinic products with minimum ruthenium contamination, which is very important for the pharmaceutical industry [43]. Recently, protein- and enzyme-supported metathesis catalyst systems have emerged as an alternative to the current strategies to conduct olefin metathesis in pre-defined confinements in water [44]. However, the intolerance of biomolecules to pH and the salt environment of the aqueous media is an ongoing challenge in the field. Considering the recent evolvement shows us that aqueous olefin metathesis can be an alternative to traditional industrial metathesis protocols.

Acknowledgments

Dr. B. Ö. Öztürk would like to thank the BAGEP Award of the Science Academy with funding supplied by the 2023 BAGEP program.

References

1 Schuster, M. and Blechert, S. (1997). *Angew. Chem. Int. Ed.* 36: 2036–2056.

2 Ogba, O.M., Warner, N.C., O'Leary, D.J., and Grubbs, R.H. (2018). *Chem. Soc. Rev.* 47: 4510–4544.

3 Barteczko, N., Grymel, M., and Chrobok, A. (2023). *Catal. Commun.* 177: 106662–110681.

4 Grau, B.W., Neuhauser, A., Aghazada, S. et al. (2022). *Chem. Eur. J.* 7: 202201414.

5 Mol, J.C. (2004). *J. Mol. Catal. A. Chem.* 213: 39–45.

6 Phillips, J.H. (2020). Latest industrial uses of olefin metathesis. In: *Organometallic Chemistry in Industry: A Practical Approach* (ed. T.J. Colacot and C.C.C.J. Seechurni), 259–282. Weinheim-Germany: Wiley-VCH.

7 Hughes, D., Wheeler, P., and Ene, D. (2017). *Org. Proc. Res. Dev.* 21: 1938–1962.

8 Smoleń, M., Marczyk, A., Kośnik, W. et al. (2019). *Eur. J. Org. Chem.* 2019: 640–646.

9 Tyszka-Gumkowska, A., Purohit, V.B., Nienałtowski, T. et al. (2022). *iScience* 25: 104131–104159.

10 (a) Mol, J.C. (2002). *Green Chem.* 4: 5–13. (b) Skowerski, K., Białecki, J., Tracz, A., and Olszewski, T.K. (2014). *Green Chem.* 16: 1125–1130.

11 (a) Jordan, J.P. and Grubbs, R.H. (2007). *Angew. Chem. Int. Ed.* 46: 5152–5155. (b) Mohr, B., Lynn, D.M., and Grubbs, R.H. (1996). *Organometallics* 15: 4317–4325.

12 Robers, A.N., Cochran, A.C., Rankin, D.A. et al. (2007). *Organometallics* 26: 6515–6518.

13 (a) Skowerski, K., Szczepaniak, G., Wierzbickai, C. et al. (2012). *Catal. Sci. Technol.* 2: 2424–2427. (b) Skowerski, K., Wierzbicka, C., Szczepaniak, G. et al. (2012). *Catal. Sci. Technol.* 14: 3264–3268.

14 Michrowska, A., Gulajski, L., Kaczmarska, Z. et al. (2006). *Green Chem.* 8: 685–688.

15 Gulajski, L., Michrowska, A., Naroznik, J. et al. (2008). *ChemSusChem* 22: 103–109.

16 P'Poo, S.J. and Schanz, H.J. (2007). *J. Am. Chem. Soc.* 129: 14200–14212.

17 Öztürk, B.Ö., Sariaslan, B., Bayramgil, N.P., and Sehitoglu Karabulut, S. (2014). *App. Catal. A-Gen.* 483: 19–24.

18 Öztürk, B.Ö., Şehitoğlu Karabulut, S., and Meier, M.A.R. (2015). *Eur. J. Polym. Chem.* 62: 116–123.
19 Öztürk, B.Ö., Kolberg, A., and Şehitoğlu Karabulut, S. (2017). *Macromol. Chem. Phys.* 218: 1600594.
20 Suzuki, N., Watanabe, K., Takahashi, C. et al. (2023). *Curr. Org. Chem.* 27: 1347–1356.
21 Lynn, D.M., Kanaoka, S., and Grubbs, R.H. (1996). *J. Am. Chem. Soc.* 118: 784–790.
22 Connon, S.J. and Blechert, S. (2002). *Bioorg. Med. Chem. Lett.* 12: 1873–1876.
23 Zaman, S., Chen, H., and Abell, A.D. (2011). *Tetrahedron Lett.* 52: 878–880.
24 Yao, Q. (2000). *Angew. Chem. Int. Ed.* 39: 3896–3898.
25 Gallivan, J.P., Jordan, J.P., and Grubbs, R.H. (2005). *Tetrahedron Lett.* 46: 2577–2580.
26 Hong, S.H. and Grubbs, R.H. (2006). *J. Am. Chem. Soc.* 128: 3508–3509.
27 Tomasek, J. and Schatz, J. (2013). *Green Chem.* 15: 2317–2338.
28 Zarka, M.T., Nuyken, O., and Weberskirch, R. (2004). *Macromol. Rapid Commun.* 25: 858–862.
29 Öztürk, B.Ö., Durmuş, B., and Şehitoğlu Karabulut, S. (2018). *Catal. Sci. Technol.* 8: 5807–5815.
30 Lipshutz, B.H. and Ghorai, S. (2009). *Org. Lett.* 11: 705–708.
31 Krause, J.O., Zarka, M.T., Anders, U. et al. (2003). *Angew. Chem. Int. Ed.* 42: 5965–5969.
32 Laville, L., Charnay, C., Lamaty, F. et al. (2012). *Chem. Eur. J.* 18: 760–764.
33 Davis, K.J. and Sinou, D. (2002). *J. Mol. Catal. A. Chem.* 177: 173–178.
34 Lipshutz, B.H., Aguinaldo, G.T., Ghorai, S., and Voigtritter, K. (2008). *Org. Lett.* 10: 1325–1328.
35 Lipshutz, B.H., Ghorai, S., and Aguinaldo, G.T. (2008). *Adv. Synth. Catal.* 350: 953–956.
36 Foster, J.C., Grocott, M.C., Arkinstall, L.A. et al. (2020). *J. Am. Chem. Soc.* 142: 13878–13885.
37 Öztürk, B.Ö. and Öztürk, S. (2020). *Mol. Catal.* 480: 110640–110646.
38 Wright, D.B., Touve, M.A., Thompson, M.P., and Gianneschi, N.C. (2018). *ACS Macro Lett.* 7: 401–405.
39 Kim, C., Ondrusek, B.A., and Chung, H. (2018). *Org. Lett.* 20: 736–739.
40 Pauly, J., Gröger, H., and Patel, A.V. (2018). *Green Chem.* 20: 5179–5187.
41 Tunalı, Z., Sagdic, K., Inci, F., and Öztürk, B.Ö. (2022). *React. Chem. Eng.* 7: 1617–1625.
42 James, C.C., Laan, P.C.M., de Bruin, B., and Reek, J.N.H. (2023). *ChemCatChem* 15: e202201272.
43 Mayer, C., Gillingham, D.G., Ward, T.R., and Hilvert, D. (2011). *Chem. Commun.* 47: 12068–12070.
44 (a) Sauer, D.F., Schiffels, J., Hayashi, T. et al. (2018). *Beilstein J. Org. Chem.* 14: 2861–2871. (b) Lo, C., Ringenberg, M.R., Gnandt, D. et al. (2011). *Chem. Commun.* 47: 12065–12067.

9

Oxidation and Reduction Reactions in Water

Rose Mary Philip[1], *Padinjare Veetil Saranya*[1], *and Gopinathan Anilkumar*[1,2]

[1] *Mahatma Gandhi University, School of Chemical Sciences, Priyadarsini Hills PO, Kottayam, Kerala 686560, India*
[2] *Mahatma Gandhi University, Institute for Integrated programs and Research in Basic Sciences (IIRBS), Priyadarsini Hills PO, Kottayam, Kerala 686560, India*

9.1 Introduction

Oxidation reactions are fundamental reactions in synthetic organic chemistry and have been traditionally conducted using stoichiometric quantities of inorganic or organic oxidants or metal-based reagents. The deployment of water as the solvent in oxidations is of great importance from the sustainable point of view. Oxidations in an aqueous medium have been developed by employing water-compatible complexes for metal-based catalytic oxidations and the use of oxidants like H_2O_2 or molecular oxygen [1]. Either the reaction follows true catalysis in water or a phase transfer catalyst is used to carry the oxidant to the organic phase [2]. Recent developments in alkene epoxidation and the oxidation of alcohols, aldehydes, and ketones in water are included in this chapter, while the reports covered in previous reviews are excluded [3]. The current discussion on oxidation reactions will be limited to those catalyzed by water-soluble metal complexes, organocatalysts, and metal nanoparticle systems under true catalysis in water without the use of any organic solvent.

Reduction reactions in water are not much explored. The available methodologies are those in which water compatible catalytic systems or reducing agents are used. Water's pH-sensitive behavior and specific ligand characteristics make it a distinctive solvent that can be exploited to increase the selectivity and effectiveness of hydrogenation processes. We will deal with a number of reduction reactions in this chapter, such as the hydrogenation of C=O, C=C, and C=N bonds and aromatic nitro compounds in water as the medium, and the reactions are classified based on the catalyst used.

Organic Transformations in Water: Principles and Applications, First Edition.
Edited by Gopinathan Anilkumar, Nissy Ann Harry, and Sankuviruthiyil M. Ujwaldev.

9.2 Oxidation Reactions in Water

9.2.1 Epoxidation of Alkenes in Water

The synthesis of epoxides via the oxidation of C=C bonds is a major conversion in organic chemistry, and sustainable methods of utilizing water-compatible catalysts and oxidants have been developed over the years.

Tan and coworkers reported the synthesis and catalytic activity of thermo-responsive catalytic nanoreactors (chiral salen-based Cr(III) complexes encapsulated in an amphiphilic copolymer) for the asymmetric epoxidation of alkenes in 2019 [4]. Compared to the conventional Cr(salen) systems, the catalytic nanoreactors offered higher yields (up to 95%) and selectivity (*ee* up to 99%) for the asymmetric epoxidation of varied alkenes in water. Moreover, the catalytic system hindered the undesired hydrolysis of products, and the catalytic nanoreactor can be easily recovered and reused.

A Mn-salen micellar nanoreactor was developed for the enantioselective epoxidation of alkenes in water by Sfrazzetto and coworkers [5]. The catalytic nanoreactor was formed from a Mn-salen catalyst functionalized with a choline group and a long-chain aliphatic group, forming a surfactant catalyst which was used with cetyltrimethylammonium bromide (CTABr) as the surfactant to form a nanoreactor for the epoxidation. The epoxidation of 1,2-dihydronaphthalene and 6-CN-2,2-dimethylchromene afforded high conversions and enantioselectivity in the presence of NaClO as an oxidant in water (Scheme 9.1).

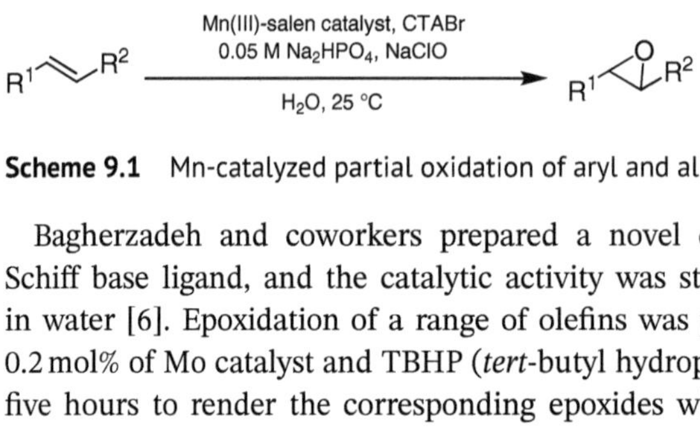

Scheme 9.1 Mn-catalyzed partial oxidation of aryl and aliphatic alcohols.

Bagherzadeh and coworkers prepared a novel dioxo Mo^{VI} complex with a Schiff base ligand, and the catalytic activity was studied for alkene epoxidation in water [6]. Epoxidation of a range of olefins was performed in the presence of 0.2 mol% of Mo catalyst and TBHP (*tert*-butyl hydroperoxide) in water at 80 °C for five hours to render the corresponding epoxides with higher turnover numbers (Scheme 9.2). Notably, epoxidation of cyclooctene was achieved with a turnover number greater than 1400. In the following year, Pisk and coworkers successfully employed molybdenum hydrazonato complexes as catalysts for the epoxidation of cyclooctene using TBHP as the oxidant with good activity and selectivity in water [7].

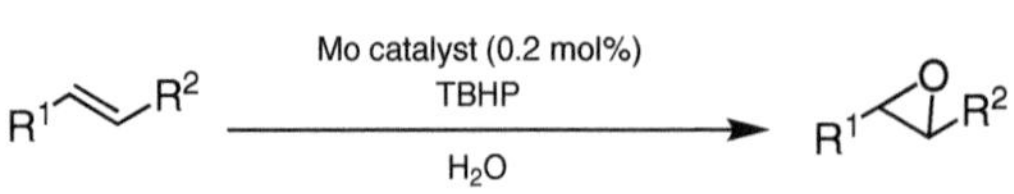

Scheme 9.2 Mo^{VI} complex-catalyzed epoxidation of alkenes in water.

The cobalt bis(dicarbollide) complex (Na[3,3-Co(η^5-1,2-$C_2B_9H_{11}$) was utilized as a photoredox catalyst for the olefin oxidation to access diverse epoxides in water, by Teixidor and coworkers in 2021 [8]. This catalyst exhibited high solubility in water and the high oxidizing ability of the $Co^{4+/3+}$ couple ensured effective olefin epoxidation. Importantly, epoxidation of varied aryl and aliphatic olefins was performed with a low catalyst loading of 0.01 mol% and offered excellent selectivity and a high turnover number of the corresponding epoxides in 15 minutes.

In 2022, Liu and coworkers put forward a green epoxidation protocol using a titanosilicate/H_2O_2 catalytic system in water as the solvent (Scheme 9.3) [9]. Their studies on propylene epoxidation showed that the catalytic activity of the system is enhanced when the optimal H_2O structure is achieved in the reaction. They also concluded that the introduction of alkali metal ions like K^+ ions and organic solvents can break the hydrogen bonds and therefore attain the favored H_2O structure, improving the efficiency of the system.

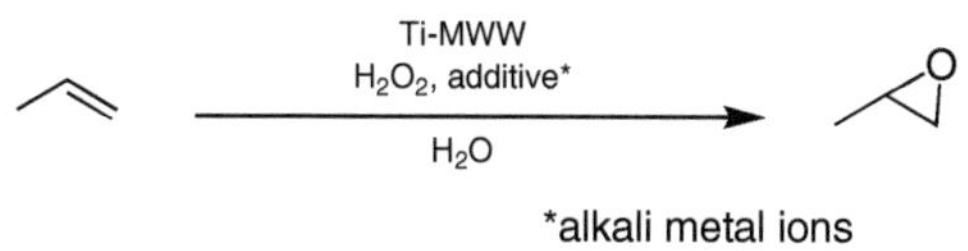

Scheme 9.3 Titanosilicate/H_2O_2 catalyzed epoxidation of propylene in water.

An Fe(III)-bis(phenol) diamine complex immobilized upon silica-coated magnetic nanoparticles bearing amine functional groups was prepared, and its catalytic activity for the epoxidation of varied olefins was studied [10]. This heterogeneous catalyst carried out efficient olefin epoxidation in the presence of SO_3^{2-} as a co-reductant at room temperature in water, and the catalyst could be magnetically recovered. They also proposed a novel TEMPO (2,2,6,6-tetramethyl-1-piperidinyloxy)-assisted epoxidation pathway which furnished different epoxides in moderate to good yields.

9.2.2 Oxidation of Alcohols to Carbonyl Compounds in Water

9.2.2.1 Oxidation of Alcohols to Aldehydes or Ketones

Oxidation of primary and secondary alcohols selectively to the corresponding aldehydes and ketones forms a key reaction in organic synthesis. The deployment of water as an eco-friendly solvent for this transformation is achieved using water-soluble homogeneous catalysts, supported heterogeneous catalysts or hybrid catalysts that work at the interface of homogeneous and heterogeneous catalysis. Along with metal- or metal-free reactions, in water, oxidations are developed under photocatalytic conditions.

Fujita and coworkers developed a water-soluble iridium-based catalyst for the dehydrogenative oxidation of secondary alcohols to access ketones [11]. This approach to acceptorless dehydrogenation of alcohols in aqueous media presented a greener alternative to other known methods, as molecular hydrogen is the sole byproduct formed. A range of aliphatic and aromatic secondary alcohols underwent successful oxidation in the presence of 3 mol% of Ir-catalyst in water under reflux conditions for 20 hours to afford good yields of the corresponding ketones (Scheme 9.4).

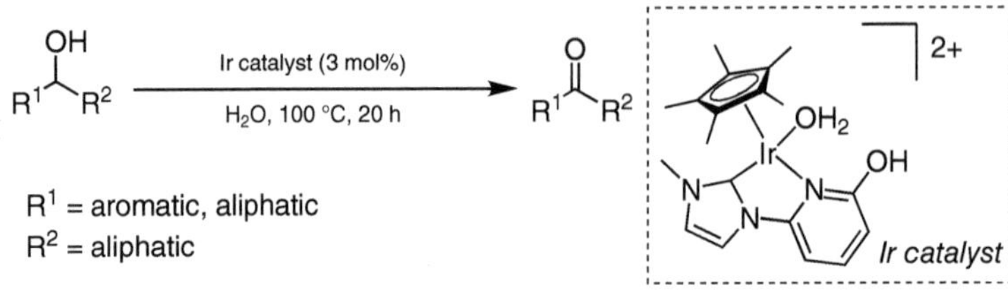

Scheme 9.4 Ir-catalyzed partial oxidation of aryl and aliphatic alcohols.

In the same year, Lee and coworkers put forward a method for the acceptorless dehydrogenation of alcohols to the respective carbonyl compounds with the use of rhodium based nanoparticles in water as the solvent [12].

A novel transition metal-free method for the oxidation of aromatic and aliphatic secondary alcohols to access diverse ketones in moderate to excellent yields was achieved by Shen and coworkers using 9-azabicyclo[3.3.1]nonane *N*-oxyl (ABNO, 6 mol%)/*tert*-butyl nitrite (5 mol%)/KPF_6 (4 mol%) as the catalytic system in water [13]. This green and efficient reaction was conducted in the presence of O_2 to enable the aerobic oxidation of a variety of secondary alcohols in water.

In 2023, Wang and coworkers prepared a mesoporous carbon-supported metal-based catalyst (WO_3–WS_2 anchored on mesoporous S, N co-doped carbon) and its catalytic performance for the oxidation of alcohols using H_2O_2 as an oxidant in water under mild conditions was studied [14].

A sustainable approach in alcohol oxidation is photocatalyzed oxidation using different photocatalysts. Recent reports on photocatalyzed alcohol oxidation to aldehydes, ketones, or acids in water medium are summarized in Scheme 9.5.

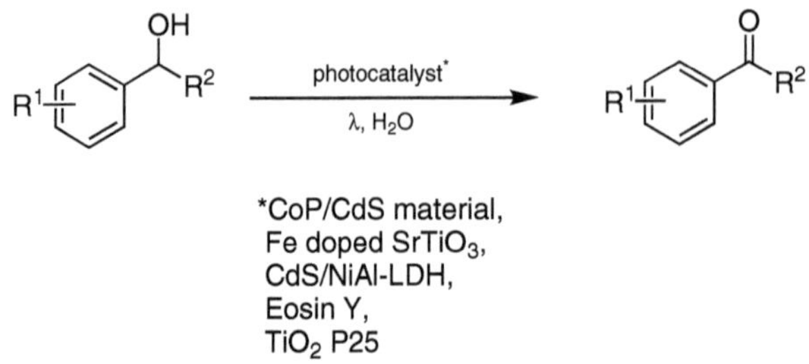

Scheme 9.5 Photocatalyzed oxidation of alcohol to the corresponding aldehydes or ketones using different photocatalysts.

In 2018, Fu and coworkers described the oxidation of benzyl alcohol derivatives to the corresponding aryl aldehydes or ketones using a hybrid photocatalytic system composed of CoP, cobalt phosphide combined with CdS [15]. Here, the CoP/CdS material (5 wt%) was employed as a visible-light catalyst (LED ($\lambda > 420$ nm)) for the photoinduced oxidation at room temperature under a nitrogen atmosphere. This reaction in water is accompanied by the evolution of hydrogen and the catalytic reaction is atom economical from a sustainable point of view.

In the subsequent year, Muhler and coworkers reported a strategy for the oxidation of alcohols in the presence of Fe-doped $SrTiO_3$ as the photocatalyst [16].

Here, the anaerobic photooxidation was carried out under visible light (300W Xe lamp (>400 nm)) and was accomplished with simultaneous H_2 generation over Fe-doped $SrTiO_3$ with Pt nanoparticles in water. The activation of alpha-C—H bonds in alcohols, accompanied by the avoidance of over-oxidation of the carbonyl compounds is attributed to the Fe-doped photocatalytic surface. The method provided an eco-friendly route to access carbonyl compounds using an economical photocatalyst under mild conditions.

In 2021, a method employing CdS/NiAl-LDH, a combination of CdS quantum dots and ultrathin NiAl-LDH nanosheets, was developed to ensure high selectivity and yields for alcohol oxidation reactions under visible light [17]. Herein, CdS/NiAl-LDH acted as a type-II heterojunction photocatalyst and exhibited better stability, efficient oxidation, and an enhanced rate of hydrogen evolution in a clean water medium. With the use of the catalyst, they achieved highly selective oxidation to synthesize anisaldehyde, preventing possible overoxidation.

Organo-photocatalysts are also known to conduct the selective oxidation of alcohols to carbonyl compounds in simple and eco-friendly conditions. A visible-light-induced (green LED, 525 nm, 3W) oxidation of differently substituted benzylic alcohols to the respective aldehydes or ketones was accomplished by Chen and coworkers in the presence of a Ni-based dual catalyst system comprising 1 mol% of eosin Y as the photocatalyst and 7.5 mol% of Ni-MPA (Ni^{II}-3-mercaptopropionic acid) [18]. The strategy presented an eco-friendly and mild route with H_2 as the sole byproduct.

In 2023, the selective photocatalytic oxidation of 2- and 4-hydroxybenzyl alcohols in water using TiO_2 P25 as the photocatalyst was reported [19]. Herein, partial oxidation of 2-hydroxybenzyl alcohol to aldehyde and oxidation of 4-hydroxybenzyl alcohol to the respective acid were achieved, with the latter exhibiting higher selectivity. The differential selectivity of the alcohols was studied by monitoring the surface processes and different adsorption mechanisms were suggested.

9.2.2.2 Oxidation of Alcohols or Aldehydes to Carboxylic Acids

The direct synthesis of carboxylic acids *via* sustainable methods, including reactions employing water as the solvent, is of great attention due to their applicability in both the laboratory and industries.

Bera and coworkers described $[RuH(CO)(py\text{-}NP)(PPh_3)_2]Cl$ (5 mol%) as the catalyst for the conversion of aryl and aliphatic primary alcohols to the respective carboxylic acid salts with the generation of H_2 [20]. This reaction was performed in alkaline water and occurred through acceptorless dehydrogenation of alcohol, followed by a rapid OH^-/H_2O attack on the Ru-bound aldehyde formed in the initial step. Interestingly, amino alcohols were also reactive, suggesting the utility of this dehydrogenative protocol toward the direct synthesis of amino acids (Scheme 9.6).

Ru(III) complexes composed of 3-(*p*-substituted phenyl azo)-4-hydroxy-1,2-naphthoquinones-based ligands were employed for the selective oxidation of benzyl alcohols to the corresponding benzoic acid derivatives by Almalki in 2021. The conversion was accomplished in an aqueous KOH medium and $K_2S_2O_8$ as the co-oxidant at ambient temperature [21].

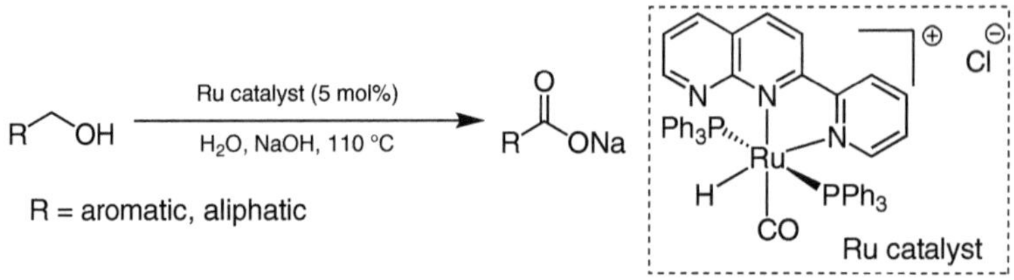

Scheme 9.6 Ru-catalyzed oxidation of aryl and aliphatic primary alcohols to the corresponding sodium carboxylates.

In the same year, a transition metal-free I_2/NaOH-catalyzed oxidation of alcohols and aldehydes toward the direct synthesis of varied carboxylic acids with water as the reaction solvent was described by Elias and coworkers (Scheme 9.7) [22]. The reaction proceeds through an initial dehydrogenation of alcohol and a sequential attack of H_2O on the aldehyde. The key highlights of the method include large-scale applicability, circumvention of purification, enlarged substrate scope, and use of green solvent.

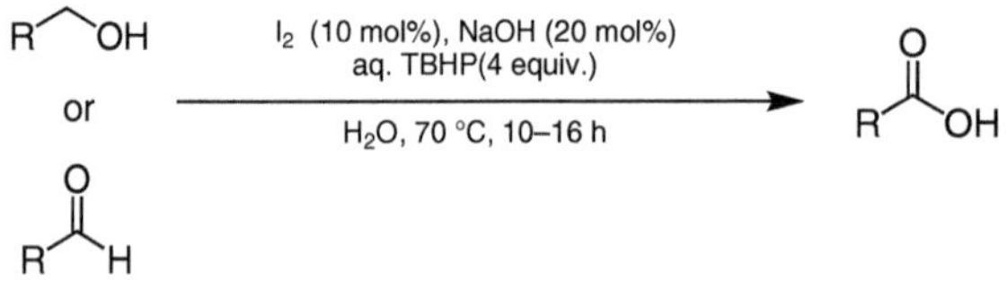

Scheme 9.7 I_2/NaOH/TBHP-catalyzed oxidation of alcohols and aldehydes to carboxylic acids.

The same group of researchers reported the first-time oxidation of aryl or heteroaryl alcohols to carboxylic acids or ketones in water catalyzed by Cl^- ion [23]. A range of electron-donating and electron-withdrawing group substituted aryl alcohols efficiently reacted under the optimized conditions of 20 mol% of NaCl, 50 mol% of aqueous NaOH and TBHP (4 equiv.) at 70 °C for 10–16 hours to furnish the carboxylic acids in high yields (Scheme 9.8). They even successfully demonstrated the use of filtered sea water as the catalyst and solvent for the reaction.

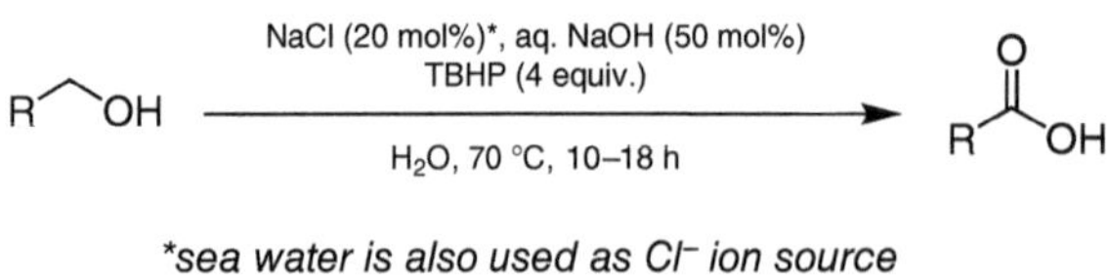

Scheme 9.8 Cl^- ion-catalyzed oxidation of alcohols to carboxylic acids.

Sashidhara and coworkers developed a method for the oxidation of aromatic ketones and benzyl nitrile derivatives to carboxylic acids using the NaOH/TBHP catalytic system [24]. Herein, the oxidative cleavage of the C—C bond of aromatic ketones and benzyl nitriles was carried out in the presence of NaOH as the base and aqueous TBHP (70%) as the oxidant (Scheme 9.9). The key highlights of the

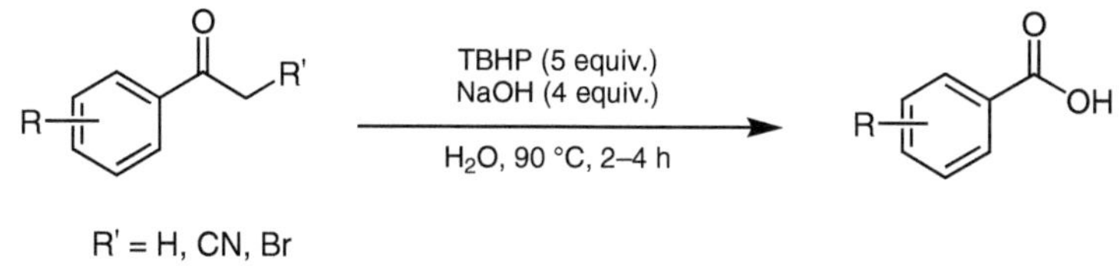

Scheme 9.9 Oxidation of aryl ketones and benzyl nitrile derivatives to carboxylic acids in water.

work include the wide substrate scope, additive-free conditions, and gram-scale applicability. They also accomplished the synthesis of industrially important acids, including salicylic acid, anthranilic acid, terephthalic acid, and 2,5-furan dicarboxylic acid (FDCA).

Sun and coworkers established the oxidative transformation of furfural to furancarboxylic acid utilizing a commercially available Mn-Cu bimetallic oxide as the catalyst [25]. The furfural oxidation was successfully performed using $Mn_2Cu_1O_x$ in water, and an excellent conversion rate of 99% was obtained (Scheme 9.10). Notably, the heterogeneous catalyst, $Mn_2Cu_1O_x$, could be recovered and reused for five cycles. The authors suggested the importance of the doping of CuO which enhances the catalytic activity of the transformation.

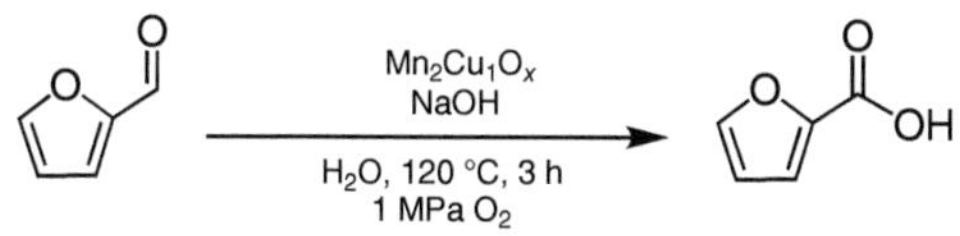

Scheme 9.10 Oxidation of furfural to furancarboxylic acid in water.

A novel method to synthesize varied carboxylic acids from aldehydes through iodoarene-mediated aerobic oxidation was reported by Zhang and coworkers [26]. The protocol proceeded through the formation of hypervalent iodine intermediates from aldehydes and their sequential interception by aldehyde hydrate in the presence of water (Scheme 9.11). Under the optimal conditions of 10 mol% of PhI at 50 °C under an O_2 atmosphere, aryl aldehydes and linear/cyclic aliphatic aldehydes were transformed into differently functionalized carboxylic acids.

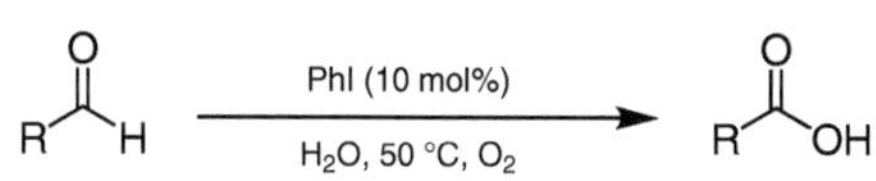

Scheme 9.11 Oxidation of aldehydes to carboxylic acids in water.

9.3 Reduction Reactions in Water

9.3.1 Iridium-Catalyzed Reduction

Xu and coworkers achieved a green, chemoselective, and efficient iridium-catalyzed reduction of ketones with low catalyst loading by employing formic acid as the

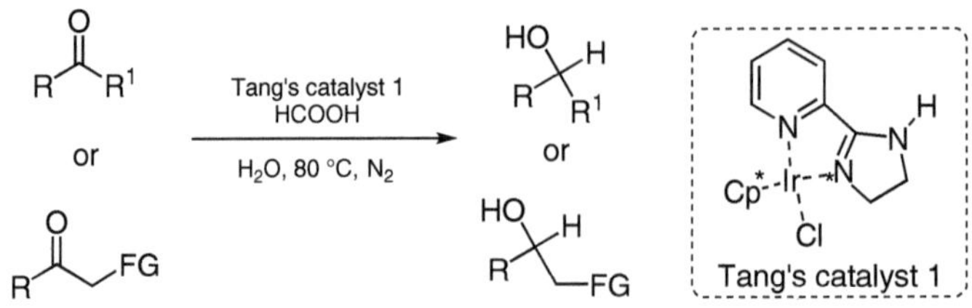

Scheme 9.12 Transfer hydrogenation of ketones utilizing formic acid as the hydride source.

hydrogen source in water as the solvent (Scheme 9.12) [27]. Not only can simple ketones be easily reduced, but also ketones with α- or β-functionalization. Another characteristic of the reduction is that it tolerates a wide range of functional groups well, including alkyloxy, aryloxy, cyano, halogen, ester, and phenol.

In 2019, Fujita and coworkers disclosed a strategy for the transfer hydrogenation of aldehydes and ketones to the corresponding alcohols using 0.1–1 mol% of an iridium complex as the catalyst and glucose as the hydrogen donor [28]. In this protocol, further investigations showed that hydrogen was transferred from the hydroxyl group at the C1 position in glucose to the substrates. This paper's findings were the first instance of transfer hydrogenation performed in water with glucose functioning as a hydrogen donor.

Chemoselective, pH-dependent and mild transfer hydrogenation of α,β-unsaturated aldehydes using water as reaction medium was put forward by Luo and team (Scheme 9.13) [29]. This Ir-catalyzed reaction yielded unsaturated alcohols with HCOOH (pH ≈ 2) as the hydride source and saturated alcohols with HCOONa (pH ≈ 8) as the hydride source. A broad variety of functional groups, both electron-withdrawing and electron-donating, were compatible with the reaction condition. The pH value was found to be essential for the chemoselectivity of the reaction.

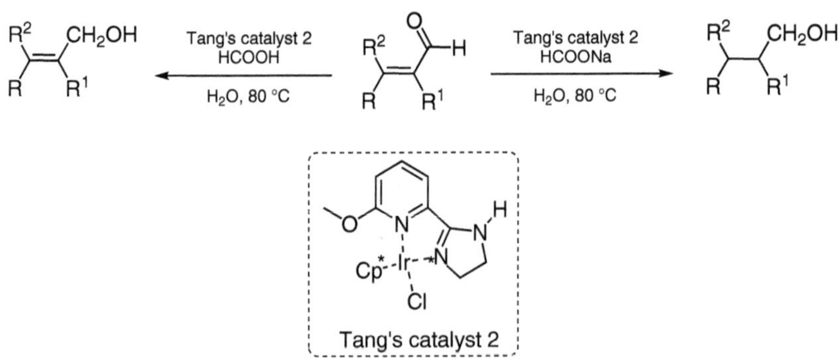

Scheme 9.13 Chemoselective reduction of α,β-unsaturated aldehydes through pH-dependence.

Sun and coworkers developed an iridium-catalyzed protocol for the transfer hydrogenation of quinolines and *N*-heteroaryl compounds in an asymmetric

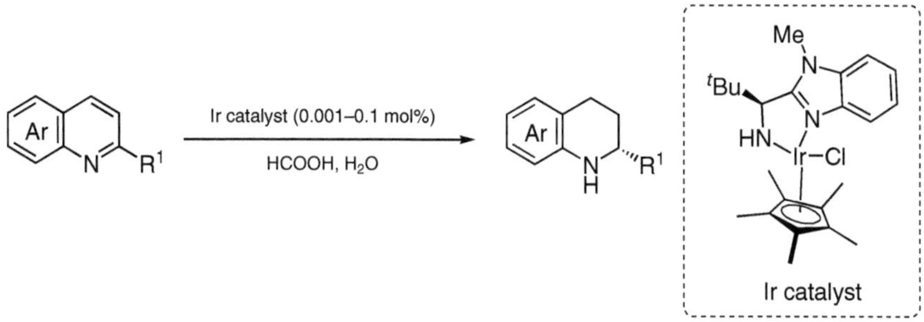

Scheme 9.14 Asymmetric transfer hydrogenation of quinolines and *N*-heteroaryl compounds.

manner (Scheme 9.14) [30]. They have selected formic acid as the source of hydrogen. A readily available aminobenzimidazole ligand was present in the iridium catalyst, which afforded enantioselectivity up to 99%.

Similarly, a reaction was developed for the iridium/imidazoline-catalyzed asymmetric transfer hydrogenation of quinoline in water (Scheme 9.15) [31]. With mild conditions, it is possible to successfully convert a variety of quinolines and *N*-heteroaryl structures into the corresponding products with moderate to good enantioselectivities.

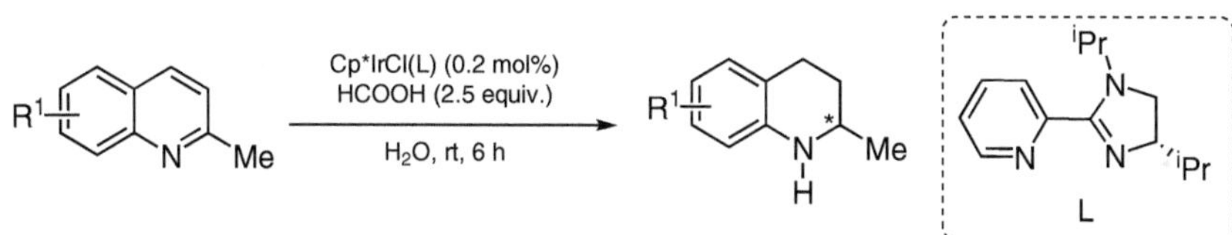

Scheme 9.15 Asymmetric transfer hydrogenation of quinolone derivatives.

Albrecht and coworkers put forward a mild, iridium-catalyzed protocol for the reduction of quinoline with hydrogen (5 atm) [32]. This strategy provided the corresponding product, tetrahydroquinoline, utilizing water as the medium.

9.3.2 Ruthenium-Catalyzed Reduction

Ruthenium-catalyzed synthesis of furfuryl alcohol in water through the hydrogenation of furfural was developed by Rodríguez-Ramos and coworkers [33].

Öztürk and Öztürk disclosed the transfer hydrogenation of aldehydes and ketones using water as the medium [34]. Here, the first-generation ruthenium-indenylidene complex (dichloro(3- phenyl-1H-inden-1-ylidene)bis(tricyclohexylphosphine)ruthenium(II) was employed as the catalyst along with dodecyltrimethylammonium chloride and Tween 20 as the surfactants, and sodium formate as the reducing agent. Strong functional group compatibility was demonstrated by the catalyst, and there was not a noticeable difference in reactivity upon varying the position of the substituents. Olefins were employed and sequential cross-metathesis-transfer hydrogenation reactions were also performed.

Yoon and coworkers reported the transfer hydrogenation of carbonyl compounds to alcohols in water via Ru catalysis [35]. A Ru catalyst based on a covalent triazine framework containing bipyridyl units was used as the recyclable, heterogeneous catalyst and environmentally friendly HCOONa as the hydrogen donor (Scheme 9.16). The catalyst exhibited high activity in the case of several aldehydes and ketones. Excellent yields were achieved when aromatic ketones containing both electron-rich and electron-deficient functional groups were transformed into alcohols. Furfuryl alcohol was obtained from furfuraldehyde. Throughout the recycling tests, the catalyst exhibited high activity, and it continued to do so until the fifth cycle.

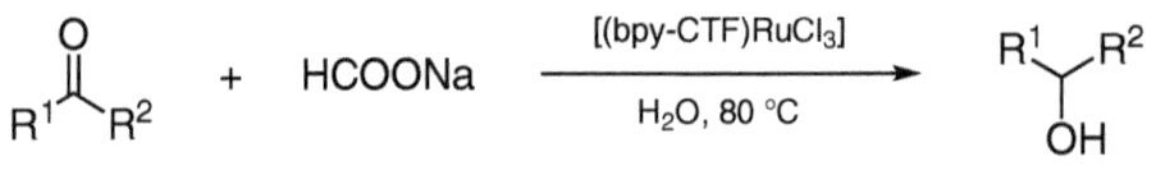

R^1, R^2 = H, Aryl, Alkyl

Scheme 9.16 Ruthenium-catalyzed transfer hydrogenation of carbonyl compounds.

Jia and coworkers established a green protocol for the polymer ionic liquid (PIL)-functionalized ruthenium complex-catalyzed asymmetric hydrogenation of aliphatic ketones in water with high selectivity [36]. Without enduring an important decrease in enantioselectivity, the catalyst could potentially be retrieved and recycled up to nine cycles with ease and efficiency. It was established that the enantioselectivity increased with increasing the alkyl chain length. Under the optimized condition, aromatic ketone, acetophenone was also examined, and excellent enantioselectivity was achieved.

9.3.3 Palladium-Catalyzed Reduction

Lipshutz and coworkers reported the hydrogenation of olefins at room temperature in water (Scheme 9.17) [37]. The protocol used only ppm levels of heterogeneous Pd/C along with 1 atm H_2 gas, and the reaction proceeded in a safe and chemoselective manner. Reduction was possible for a wide range of alkenes including differently conjugated, terminal, and highly substituted internal. Several natural products that include olefins have also been subjected to this approach. The fact that they can be effectively hydrogenated without affecting other functional groups points to the extremely mild conditions entailed.

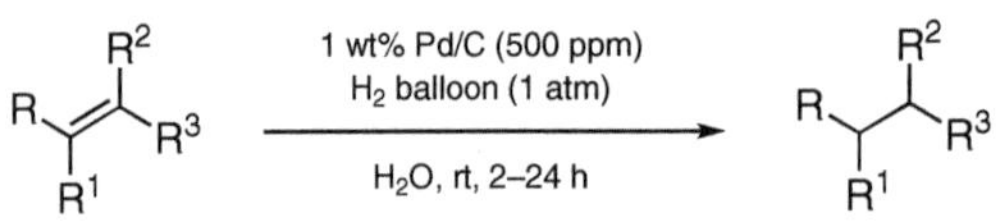

Scheme 9.17 Hydrogenation of olefins.

Liu and coworkers discussed the utilization of Pd/UiO-66 in the reduction of furfural to tetrahydrofurfuryl alcohol which is a green solvent [38]. Under mild conditions, Pd/UiO-66 showed excellent activity, by fully converting furfural to tetrahydrofurfuryl alcohol (100% selectivity) with a pressure of H_2 as 1.0 MPa at a temperature greater than 50 °C.

Palladium nanoparticle-catalyzed reduction of epoxides under mild conditions was explored by Roucoux and coworkers (Scheme 9.18) [39]. The reduction involved the selective conversion of diverse epoxides to the corresponding alcohols under basic conditions. The reaction proceeded in the presence of the heterogeneous Pd@HEA16Cl colloidal suspension as the catalyst.

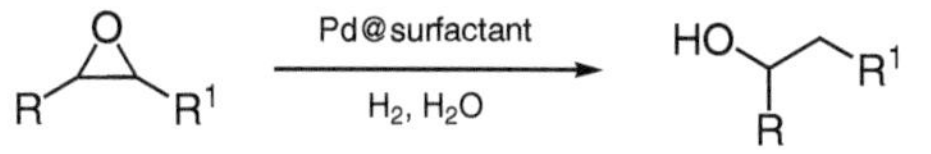

Scheme 9.18 Palladium(0)nanoparticle-catalyzed reduction of epoxides.

Varma and coworkers put forth a green approach for the Pd nanoparticle (supported on hydroxyapatite)-catalyzed reduction of nitroarenes (Scheme 9.19) [40]. This heterogeneous protocol was found to occur in water at room temperature, using $NaBH_4$ to reduce nitroarenes to the corresponding amines with high selectivities and yields. Six runs of the reaction were conducted without a discernible decrease in catalytic activity, confirming the catalyst reusability.

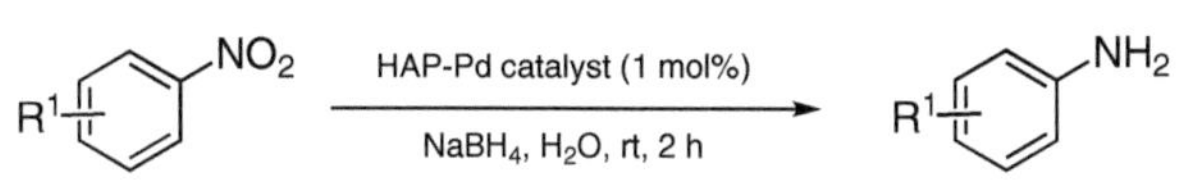

Scheme 9.19 Reduction of nitroarenes via heterogeneous Pd catalysis.

9.3.4 Rhodium-Catalyzed Reduction

Ratovelomanana-Vidal and coworkers achieved an environmentally sustainable reaction toward the asymmetric transfer hydrogenation of α-methoxy β-ketoesters [41]. The reaction was performed using a new rhodium(III) complex as the catalyst. When using sodium formate in water, the reaction was effective and through a dynamic kinetic resolution procedure, the *syn* α-methoxy β-hydroxyesters were produced with excellent enantioselectivities and high diastereoselectivities. A scale-up reaction also revealed the same high degree of stereoselectivity.

An investigation of influence factors on asymmetric transfer hydrogenation of α,β-unsaturated ketones catalyzed by Rh along with surfactant-type ligand was developed by Li and co-workers [42]. To investigate the impact of the length of the aliphatic substrates on the asymmetric transfer hydrogenation, the researchers in this work conducted experiments at 5 °C utilizing L1 as the ligand and various liner aliphatic ketones as the substrates, and the highest *ee* of 93% was obtained when the alkyl chain had nine carbons (Scheme 9.20). Up to that point, the *ee* value increased as the alkyl chain's length increased.

9.3.5 Nickel-Catalyzed Reduction

In 2021, work was reported on the reduction of ketones, aldehydes, nitroarenes, and nitriles utilizing air-stable, heterogeneous nano-Ni_2P as the catalyst and water as the

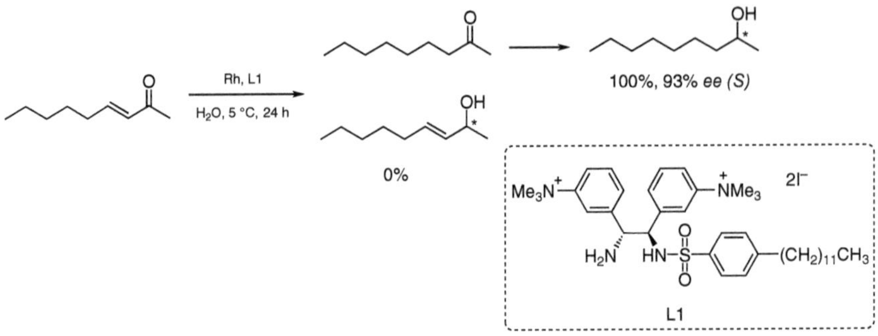

Scheme 9.20 Surfactant-type catalyst for the asymmetric transfer hydrogenation of α,β-unsaturated ketones.

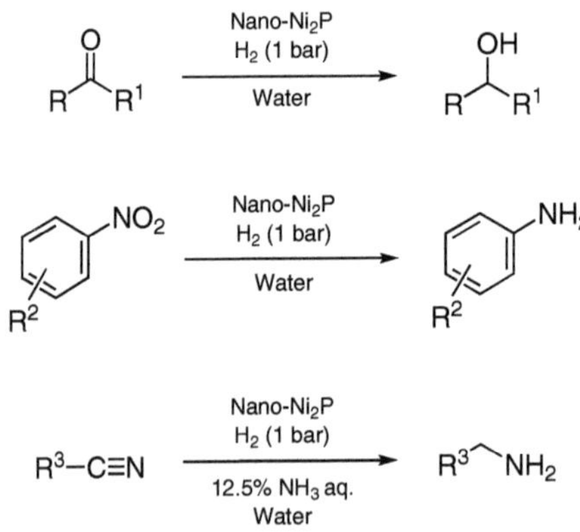

Scheme 9.21 Nano-Ni_2P as hydrogenation catalyst.

solvent (Scheme 9.21) [43]. Centrifugation made it simple to recover the catalyst, which could then be recycled up to six times without losing its activity.

A chemoselective, DNA-supported nickel nanoparticle-catalyzed hydrogenation of nitroarene was developed by Niakan and Asadi [44]. This sustainable method involved the hydrogenation of diversely substituted nitroarenes at room temperature in water using sodium borohydride as the hydride source. This heterogeneous procedure allowed the recycling and reuse of the catalyst for multiple reactions. The reaction efficiency was not significantly affected by the steric nature of the substrates.

9.3.6 Platinum-Catalyzed Reduction

Li and coworkers used an effective 3 wt% Pt/SiC-C catalyst (Pt catalyst supported on SiC-C composite) for the reduction of furfural, utilizing water as the solvent at room temperature (Scheme 9.22) [45]. At 25 °C and 1 MPa hydrogen, the catalyst provided full furfural conversion with more than 99% selectivity to furfuryl alcohol. The good recyclability of the catalyst was demonstrated when the right solvent was employed for washing the spent catalyst.

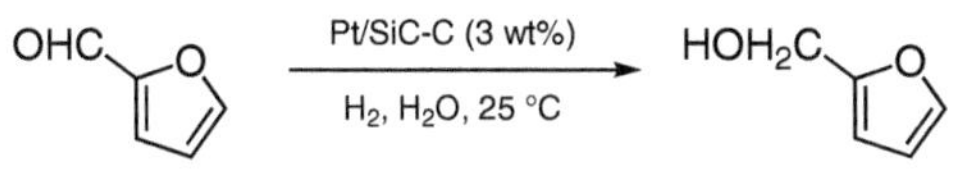

Scheme 9.22 Selective reduction of furfural to furfuryl alcohol catalyzed by Pt/SiC-C catalyst.

N-Heterocyclic carbene (NHC)-stabilized platinum nanoparticle-catalyzed strategy toward the hydrogenation of alkynes, ketones, and nitroarenes was reported [46]. The catalyst was recovered and reused up to 10 cycles.

9.3.7 Iron-Catalyzed Reduction

A phosphine-free, iron-catalyzed hydrogenation of carbon dioxide and carbonates to formate was reported [47]. Here, a bifunctional cyclopentadienone-iron tricarbonyl complex was employed as the catalyst. A base was needed for the hydrogenation of CO_2, and triethanolamine turned out to be the most suitable one. A combination of CO_2 and H_2 was used to perform the reduction in an aqueous solution for 20 hours at 100 °C.

Electrocatalytic reduction of nitrite was designed by Bren and coworkers [48]. An Fe complex was selected as the catalyst for this process and could convert NO_2^- to NH_2OH and NH_4^+ (Scheme 9.23).

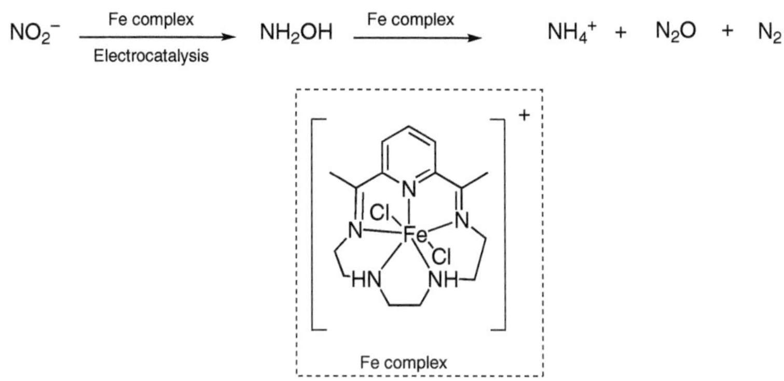

Scheme 9.23 Electrocatalytic reduction of nitrite catalyzed by iron complex.

9.3.8 Cobalt-Catalyzed Reduction

Cobalt nanocluster-catalyzed hydrogenation of aromatic ketones and aldehydes in water was developed by Fu and coworkers [49]. This reaction employed $Co/ZrLa_{0.2}O_x$ as the catalyst at 40 °C under 2 MPa of H_2. Furfural was hydrogenated in 10 hours, with a maximum yield of 95% (Scheme 9.24). Throughout the recycling experiments, the catalyst stayed stable and displayed minimal deactivation. Acetophenone and benzaldehyde were also examined as substrates and thus proved the efficiency of the catalyst.

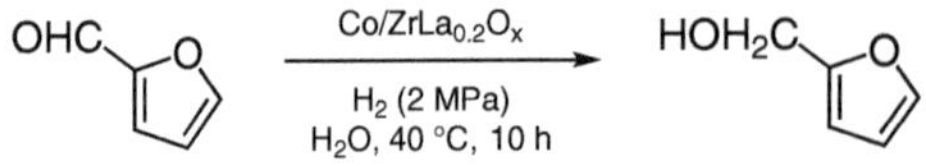

Scheme 9.24 Co nanocluster-catalyzed hydrogenation of furfural.

An efficient, chemoselective procedure for the reduction of nitroarenes was designed through another Co nanocluster-catalyzed approach [50]. Here, the Co nanocluster was coated with N-doped carbon. The optimized conditions were applied to diversely substituted nitroarenes with electron-deficient and electron-rich functional groups at various ring positions.

9.3.9 Copper-Catalyzed Reduction

Arsalani and coworkers mentioned a procedure for the reduction of 2- and 4-nitroanilines to the corresponding diamines through copper ferrite nanoparticle catalysis [51]. This heterogeneous process used $NaBH_4$ as the reducing agent and the reaction was carried out at room temperature (Scheme 9.25). With no appreciable loss of activity, the catalyst was recovered using magnetic separation and utilized up to six cycles. This eco-friendly and cost-effective reaction was performed with water as the solvent.

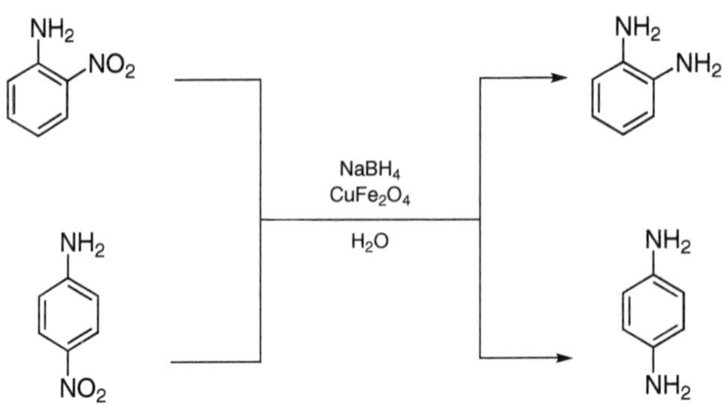

Scheme 9.25 Reduction of 2- and 4-nitroanilines.

9.3.10 Bimetal-Catalyzed Reduction

Zhao and coworkers disclosed a strategy for the selective reduction of cinnamaldehyde to cinnamyl alcohol using Fe–Co alloy nanoparticles supported on N-doped carbon as the catalyst (Scheme 9.26) [52]. The catalyst was able to attain an astounding 91.7% selectivity toward cinnamyl alcohol.

Chen and coworkers used Ni–Co alloy supported on ZrO_2 as a catalyst for the reduction of phenol to cyclohexanol in water [53]. The results showed that the catalyst exhibited good hydrothermal stability, with a phenol conversion of 96.5%. The current catalyst has potential in comparison to earlier ones since it needs a shorter time and less catalyst-reactant ratios to achieve high conversion rates.

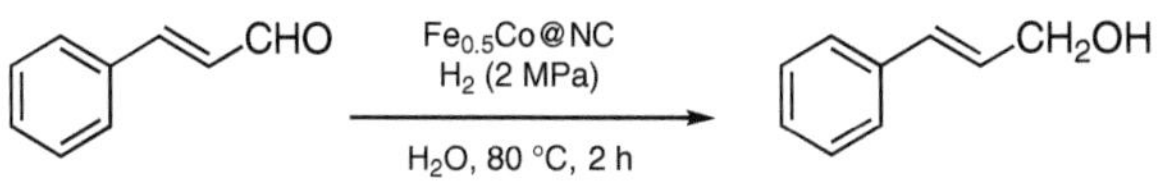

Scheme 9.26 Selective reduction of cinnamaldehyde to cinnamyl alcohol using Fe-Co alloy nanoparticles.

9.3.11 Miscellaneous Reactions

A protocol was reported for the reduction of unsaturated aromatics using visible light and Au@ZrO_2 nanoparticles [54]. This mild reaction disclosed the combination of formic acid with water as a source of hydrogen.

Salmain and coworkers reported a synthetic procedure for asymmetric transfer hydrogenation of ketones via the metal-catalyzed reaction with proteins as macromolecular ligands in water [55].

A scheme for the reduction of carbonyl groups to alcohol via a biomimetic method using hydrogen sulfide in water was achieved by Chatgilialoglu and coworkers (Scheme 9.27) [56].

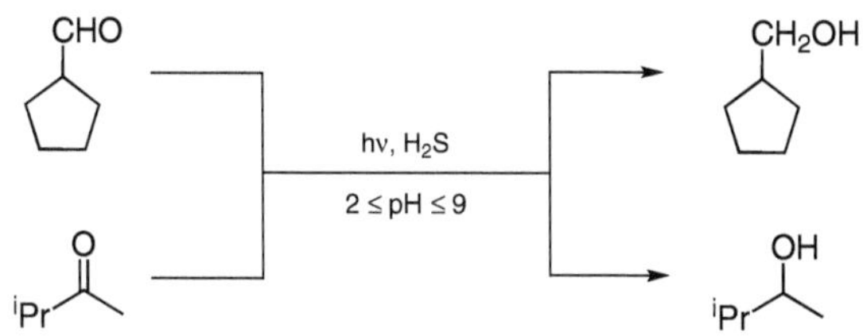

Scheme 9.27 Reduction of carbonyl compounds in water utilizing hydrogen sulfide.

Daasbjerg and coworkers utilized nickel, cobalt, or iron and N-doped carbon electrocatalysts for the selective reduction of CO_2 to CO [57].

9.4 Summary and Conclusions

Oxidation reactions in water have developed in the last few years with respect to environment-friendly oxidants in aqueous media, and utilizing photocatalytic and metal-free conditions in the absence of any organic solvents. Even though low-cost and earth abundant 3d transition metal-catalyzed epoxidations are widely studied, efficient catalytic systems in water are not largely explored and are anticipated. In the case of the catalytic oxidation of alcohols, novel methods utilizing metal-free, organocatalytic, and photocatalytic dehydrogenative oxidations presented sustainable alternatives to traditional catalytic systems. The use of nanoparticles as efficient catalysts, with the benefits of catalyst recovery and reusability, is receiving increasing attention. We anticipate the development of novel protocols for efficient and selective oxidation of different functionalities in aqueous media in the future.

Reduction reactions discussed above encompass the hydrogenation of diverse compounds such as aldehydes, ketones, quinolines, furfurals, olefins, nitroarenes etc. in water using different reductants. Catalysts such as iridium, ruthenium, rhodium, palladium, nickel, and so on are employed. Low catalyst loading, wide functional group tolerance, mild and green reaction conditions are some of the features of these reactions. The reusability of the catalyst is evident in majority of the reactions. Most of the reducing agents, like hydrogen, formic acid, glucose, sodium formate, and $NaBH_4$ underwent reaction, producing moderate to excellent yields of the desired products. The results obtained are frequently on par with, and occasionally even better than, those produced in conventional organic solvents in terms of yields and selectivities. Furthermore, using water in place of typical organic solvents ought to have a positive impact on the environment and human health.

References

1 Li, C.J. and Chen, L. (2006). *Chem. Soc. Rev.* 35: 68–82.
2 Sheldon, R.A. (2015). *Catal. Today* 247: 4–13.
3 Kitanosono, T., Masuda, K., Xu, P., and Kobayashi, S. (2018). *Chem. Rev.* 118: 679–746.
4 Wang, W., Li, C., Pi, Y. et al. (2019). *Catal. Sci. Technol.* 9: 5626–5635.
5 Ballistreri, F.P., Toscano, R.M., Amato, M.E. et al. (2018). *Catalysts* 8: 129.
6 Mahmoudi, H., Bagherzadeh, M., Ataie, S. et al. (2020). *Inorg. Chim. Acta.* 511: 119775.
7 Mrkonja, S., Topić, E., Mandarić, M. et al. (2021). *Catalysts* 11: 756.
8 Guerrero, I., Viñas, C., Romero, I., and Teixidor, F. (2021). *Green Chem.* 23: 10123–10131.
9 Yu, Y., Fang, N., Chen, Z. et al. (2022). *ACS Sustain. Chem. Eng.* 10: 11641–11654.
10 Mohammadpour, P., Safaei, E., Mazarei, E., and Zeinalipour-Yazdi, C.D. (2023). *Phys. Chem. Chem. Phys.* 25: 26588–26603.
11 Fujita, K.I., Tamura, R., Tanaka, Y. et al. (2017). *ACS Catal.* 7: 7226–7230.
12 Yao, L., Zhao, J., and Lee, J.M. (2017). *ACS Sustain. Chem. Eng.* 5: 2056–2060.
13 Ma, J., Hong, C., Wan, Y. et al. (2017). *Tetrahedron Lett.* 58: 652–657.
14 Xu, C., Dai, L., Chen, Y. et al. (2023). *Appl. Surf. Sci.* 609: 155296.
15 Xu, Y., Zeng, L.Z., Fu, Z.C. et al. (2018). *Catal. Sci. Technol.* 8: 2540–2545.
16 Hu, Y., Zhao, G., Pan, Q. et al. (2019). *ChemCatChem* 11: 5139–5144.
17 Jiao, L., Zhang, D., Hao, Z. et al. (2021). *ACS Catal.* 11: 8727–8735.
18 Yang, X.J., Zheng, Y.W., Zheng, L.Q. et al. (2019). *Green Chem.* 21: 1401–1405.
19 Bellardita, M., Escolano-Casado, G., Palmisano, L., and Mino, L. (2023). *Catal. Today* 413: 113983.
20 Sarbajna, A., Dutta, I., Daw, P. et al. (2017). *ACS Catal.* 7: 2786–2790.
21 Almalki, A.S. (2021). *ChemistrySelect* 6: 7525–7536.
22 Hazra, S., Deb, M., and Elias, A.J. (2017). *Green Chem.* 19: 5548–5552.
23 Hazra, S., Kushawaha, A.K., Yadav, D. et al. (2019). *Green Chem.* 21: 1929–1934.
24 Jaiswal, A.K., Kushawaha, A.K., Pandey, S. et al. (2023). *Tetrahedron* 136: 133359.
25 Yu, X., Jin, T., Wang, H. et al. (2023). *Green Energy Environ.* 8: 1683–1692.

26 Zeng, K., Stückl, A.C., Qin, J. et al. (2023). *Mol. Catal.* 537: 112919.
27 Liu, J., Yang, S., Tang, W. et al. (2018). *Green Chem.* 20: 2118–2124.
28 Yoshida, M., Hirahata, R., Inoue, T. et al. (2019). *Catalysts* 9: 503–515.
29 Luo, N., Liao, J., Ouyang, L. et al. (2019). *Organometallics* 38: 3025–3031.
30 Wang, L., Lin, J., Xia, C., and Sun, W. (2021). *J. Org. Chem.* 86: 16641–16651.
31 Qi, H., Wang, L., Sun, Q., and Sun, W. (2022). *Mol. Catal.* 532: 112715.
32 Vivancos, Á., Beller, M., and Albrecht, M. (2018). *ACS Catal.* 8: 17–21.
33 Ramirez-Barria, C., Isaacs, M., Wilson, K. et al. (2018). *Appl. Catal. A Gen.* 563: 177–184.
34 Öztürk, B.Ö. and Öztürk, S. (2020). *Mol. Catal.* 480: 110640.
35 Padmanaban, S., Gunasekar, G.H., Lee, M., and Yoon, S. (2019). *ACS Sustain. Chem. Eng.* 7: 8893–8899.
36 Li, X., Sun, Y., Wang, S., and Jia, X. (2020). *ACS Appl. Polym. Mater.* 2: 1268–1275.
37 Takale, B.S., Thakore, R.R., Gao, E.S. et al. (2020). *Green Chem.* 22: 6055–6061.
38 Wang, C., Wang, A., Yu, Z. et al. (2021). *Catal. Commun.* 148: 106178.
39 Duval, M., Deboos, V., Hallonet, A. et al. (2021). *J. Catal.* 396: 261–268.
40 Shokouhimehr, M., Yek, S.M.-G., Nasrollahzadeh, M. et al. (2019). *Appl. Sci.* 9: 4183–4195.
41 He, B., Zheng, L.-S., Phansavath, P., and Ratovelomanana-Vidal, V. (2019). *ChemSusChem* 12: 3032–3036.
42 Li, J., Han, J., Lin, Z. et al. (2019). *Tetrahedron* 75: 422–428.
43 Fujita, S., Yamaguchi, S., Yamasaki, J. et al. (2021). *Chem. Eur. J.* 27: 4439–4446.
44 Niakan, M. and Asadi, Z. (2019). *Catal. Lett.* 149: 2234–2246.
45 Wang, G., Yao, R., Xin, H. et al. (2018). *RSC Adv.* 8: 37243–37253.
46 Ruiz-Varilla, A.M., Baquero, E.A., Chaudret, B. et al. (2020). *Catal. Sci. Technol.* 10: 2874–2881.
47 Coufourier, S., Gaillard, Q.G., Lohier, J.-F. et al. (2020). *ACS Catal.* 10: 2108–2116.
48 Stroka, J.R., Kandemir, B., Matson, E.M., and Bren, K.L. (2020). *ACS Catal.* 10: 13968–13972.
49 Ma, Y., Xu, G., Wang, H. et al. (2018). *ACS Catal.* 8: 1268–1277.
50 Gutiérrez-Tarriño, S., Rojas-Buzo, S., Lopes, C.W. et al. (2021). *Green Chem.* 23: 4490–4501.
51 Naghash-Hamed, S., Arsalani, N., and Mousavi, S.B. (2022). *ChemistryOpen* 11: e202200156.
52 Lv, Y., Han, M., Gong, W. et al. (2020). *Angew. Chem. Int. Ed.* 59: 23521–23526.
53 Zhang, X., Wang, Z., Shu, S. et al. (2023). *React. Kinet. Mech. Catal.* 136: 937–952.
54 Huang, Y., Liu, Z., Gao, G. et al. (2018). *Catal. Sci. Technol.* 8: 726–734.
55 Cázares-Marinero, J.J., Przybylski, C., and Salmain, M. (2018). *Eur. J. Inorg. Chem.* 2018: 1383–1393.
56 Barata-Vallejo, S., Ferreri, C., Golding, B.T., and Chatgilialoglu, C. (2018). *Org. Lett.* 20: 4290–4294.
57 Hu, X.-M., Hval, H.H., Bjerglund, E.T. et al. (2018). *ACS Catal.* 8: 6255–6264.

10

Radical Reactions in Water

Xiaoye Yu[1,2], Ying Zhang[1], and Huan-Ming Huang[1]

[1] *ShanghaiTech University, School of Physical Science and Technology, 393 Middle Huaxia Road, Pudong, Shanghai 201210, China*
[2] *Universität Münster, Organisch-Chemisches Institut, Corrensstrasse 40, Münster 48149, Germany*

10.1 Introduction

Chemists have embarked on a prolonged quest to discover environmentally friendly reaction media that can facilitate organic synthetic transformations. This pursuit aligns seamlessly with the inevitable trajectory of green chemistry, marking a crucial trend that is expected to shape the future of chemical research and innovation [1]. Water, being a naturally abundant and eco-friendly reaction medium, stands out for its affordability, safety, and non-flammability, garnering substantial attention in contrast to traditional organic solvents. The adoption of water as a solvent holds immense potential for environmental benefits when utilized judiciously. This aligns seamlessly with the principles of "green chemistry," offering a pathway to effectively curtail waste production in chemical processes [2]. Considering the potential rewards of water as a green solvent, chemists face great challenge of developing new organic transformations that tolerate well in water [3]. In recent years, the field of radical reactions in water has grown steadily, which is mainly attributed to the bloom of new catalytic systems and newly developed precursors. Here, water plays a vital role in a favorable promotion in the reaction rate and reaction outcome. However, an overview to emphasize the recent advantages of radical reaction in water is extremely rare. In this chapter, our primary focus lies on synthetic transformations that involve radical intermediates in water. It is important to note that we specifically exclude those reactions conducted in mixed solutions of organic solvents and water. In recent decades, radical reactions have garnered significant attention from chemists owing to their substantial synthetic potential [4]. These highly reactive radical species could be integrated with diverse catalytic systems such as transition metal catalysis, photoredox chemistry, and electrosynthesis, providing a robust synthetic toolkit to access complex molecules that are otherwise difficult to achieve by classical ionic

Huan-Ming Huang and Xiaoye Yu are corresponding authors to this chapter.

Organic Transformations in Water: Principles and Applications, First Edition.
Edited by Gopinathan Anilkumar, Nissy Ann Harry, and Sankuviruthiyil M. Ujwaldev.

reactions. Water is frequently chosen as an ideal solvent in radical reactions, primarily because a significant number of organic radicals exhibit stability in aqueous environments [5]. Furthermore, significant efforts have been dedicated to developing an extensive toolbox of catalysts and reagents to facilitate efficient and selective radical reactions in water. In this section, we aim to summarize the recent advances in radical reactions in water, categorizing them according to different catalytic modes.

10.2 Transition Metal-Catalyzed Radical Reaction in Water

10.2.1 Transition Metal-Catalyzed Oxidation

In 2017, Liu developed a copper-catalyzed ligand-free oxidation of the benzylic sp^3 C—H bond in the presence of 70% *tert*-butyl hydroperoxide (TBHP) using water as a solvent (Scheme 10.1a) [6a].

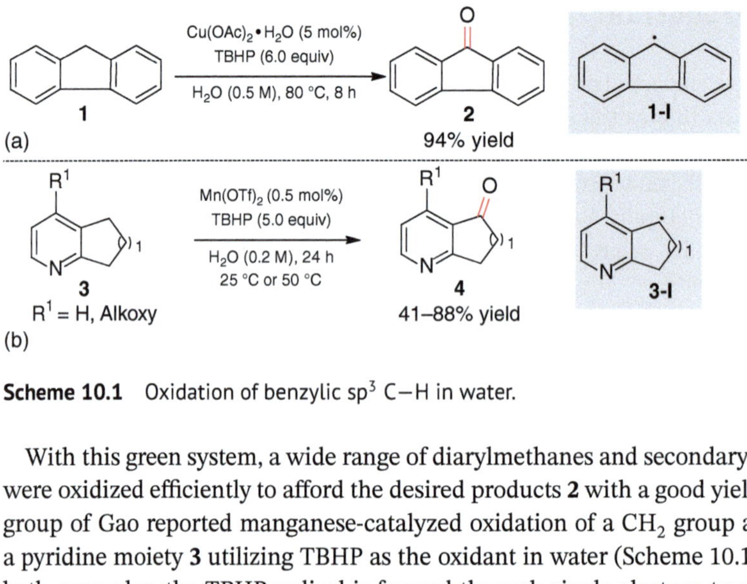

Scheme 10.1 Oxidation of benzylic sp^3 C—H in water.

With this green system, a wide range of diarylmethanes and secondary alcohols **1** were oxidized efficiently to afford the desired products **2** with a good yield. And the group of Gao reported manganese-catalyzed oxidation of a CH_2 group adjacent to a pyridine moiety **3** utilizing TBHP as the oxidant in water (Scheme 10.1b) [6b]. In both examples, the TBHP radical is formed through single electron transfer (SET) to cleave the benzylic C—H bond, leading to the subsequent formation of the key benzylic intermediate **1-I** or **3-I**. The Zhao group described a direct oxidative amidation between methylarenes and free amines by employing TBHP as an oxidant with co-catalysis of TBAI and $FeCl_3$ in water [7]. The substrate scope was quite broad, and the corresponding coupling product could be obtained in moderate to good yield. In addition to that, cobalt-catalyzed C—H activation and peroxidation of 2-oxindoles **5** in water were also reported by Cheng and coworkers (Scheme 10.2) [8]. They proposed that the oxindole radical **5-I** was formed through cobalt-catalyzed homolytic decomposition of peroxides, and then coupled with the longer-lived peroxyl radical, the final product **6** could be obtained with good to excellent yields.

In 2011, Taniguchi et al. developed a mild oxyarylation of alkenes **7** by oxidative radical addition in the presence of Fe^{III} (Scheme 10.3) [9]. The reaction was

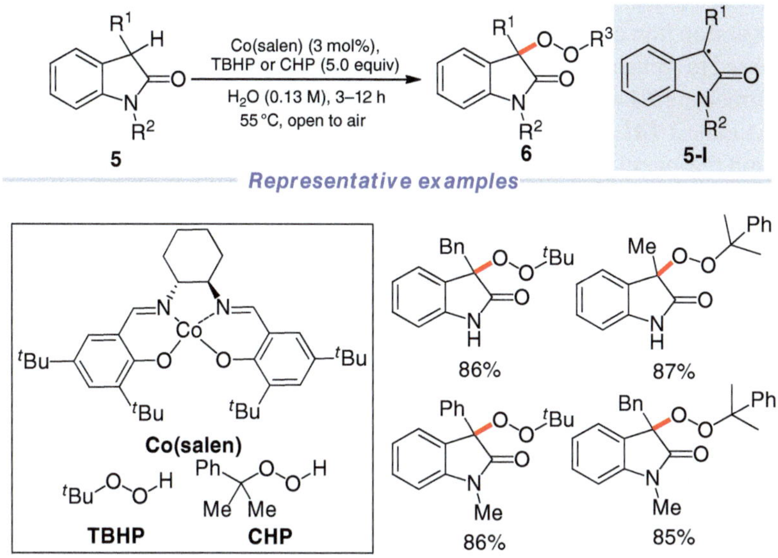

Scheme 10.2 Cobalt-catalyzed peroxidation of 2-oxindoles in water.

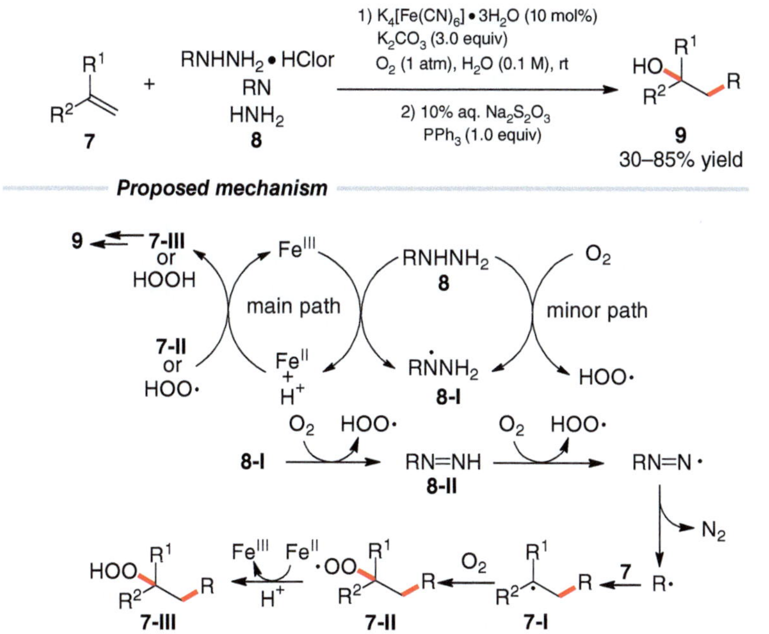

Scheme 10.3 Oxidative radical addition of aryl hydrazines in water.

conducted well in water at room temperature due to the hydrophobic effect. The authors proposed that phenylhydrazyl radical **8-I** is formed via SET between phenylhydrazine **8** and an iron catalyst. After undergoing twice oxidation by O_2, the radical **8-I** is converted to a phenyl radical. Subsequent addition to alkenes and coupling of radical intermediate **7-I** with O_2 gave peroxyl radical **7-II**. Finally, the redox reaction between the radical **7-II** or hydrogen peroxide radical, and the Fe^{II} species leads to peroxides and regeneration of the Fe^{III} species. By simple workup, a wide variety of alcohols **9** could be obtained in moderate to good yields.

10.2.2 Transition Metal-Catalyzed Fluorination

Recently, researchers have successfully combined "untamed" fluorine and "mild" water in chemical reactions, thus opening up new prospects for green fluorine chemistry [10]. For example, Lipshutz and coworkers studied the reaction of arylacrylamide **10** with Langlois' reagent (CF_3SO_2Na) in water (Scheme 10.4) [11].

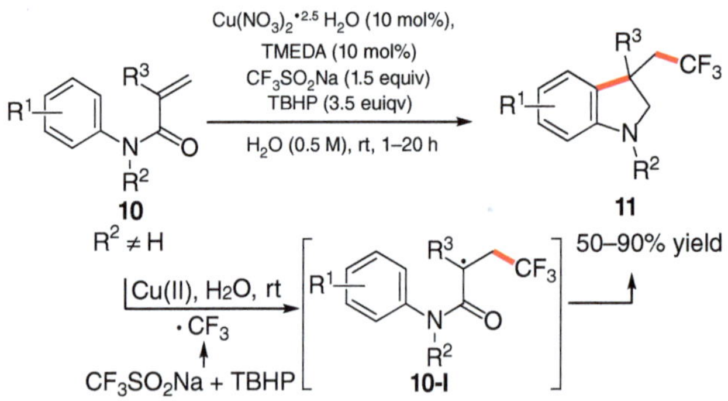

Scheme 10.4 Copper(II)-catalyzed trifluoromethylation of *N*-aryl acrylamides in water.

They found the combination of CF_3SO_2Na and TBHP could effectively generate trifluoromethyl radicals in the presence of catalytic $Cu(NO_3)_2$. Langlois' reagent could also be applied for regioselective C—H trifluoromethylation of imidazoheterocycles **12**. This reaction proceeded in the presence of a catalytic amount of $Cu(OAc)_2$ in a recyclable mixed medium of 1-butyl-3-methylimidazoliumtetrafluoroborate ([Bmim]BF_4) and water (Scheme 10.5a) [12a]. Subsequently, Zhang developed nickel(II)-catalyzed site-selective C—H bond trifluoromethylation of arylamine **14** in water by the utilization of picolinamide as a directing group (Scheme 10.5b) [12b].

Difluoromethylene and trifluoromethylthio groups are two highly valuable fluorinated moieties that have been used to modify the property of a molecule [13]. In 2020, Li et al. developed a silver-catalyzed decarboxylative allylation of α,α-difluoroarylacetic acids **16** with allyl sulfones **17** in water (Scheme 10.6) [14].

Various aryl difluoroacetic acids **16** could react with allylsulfone **17** to afford the desired products **18** in moderate to good yields. The mechanism demonstrated that difluoroalkyl radical **16-I** was formed through the Ag-catalyzed SET process. And Silver-catalyzed decarboxylative trifluoromethylthiolation of secondary and tertiary carboxylic acids **19** in water was also reported (Scheme 10.7) [15].

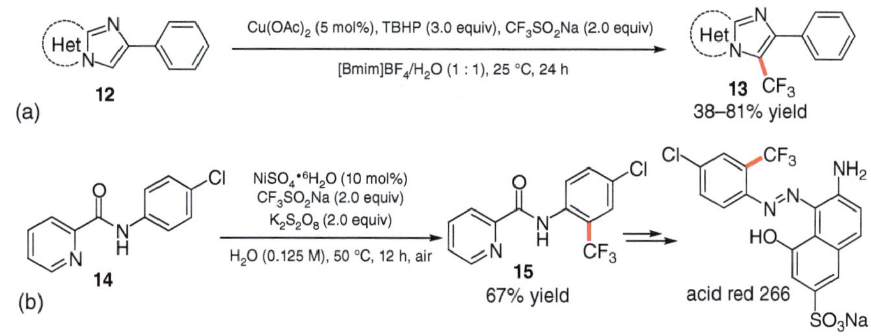

Scheme 10.5 Metal-catalyzed C–H bond trifluoromethylation in water.

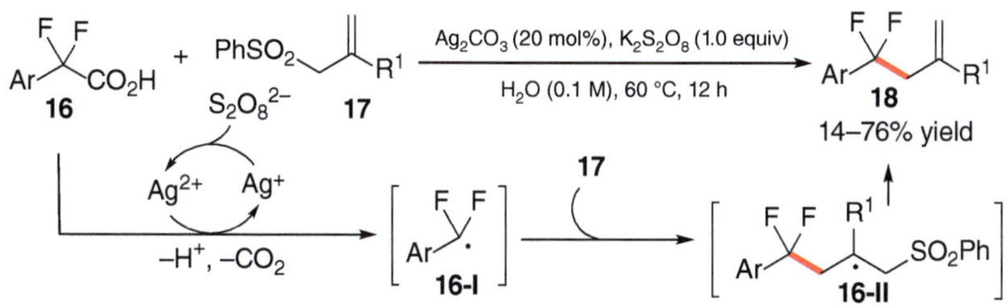

Scheme 10.6 Silver-catalyzed decarboxylative allylation in water.

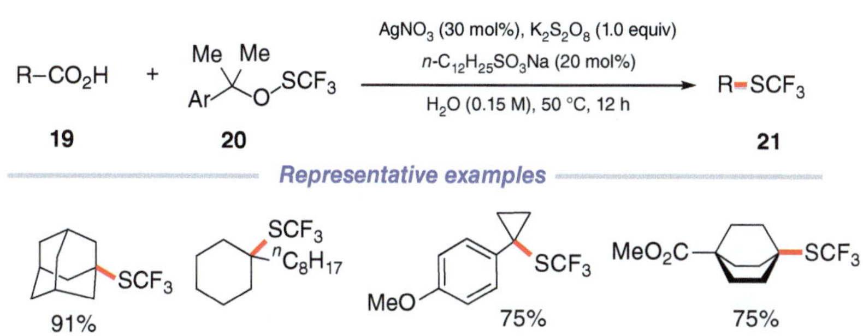

Scheme 10.7 Silver-catalyzed decarboxylative trifluoromethylthiolation in water.

10.2.3 Transition Metal-Catalyzed Difunctionalization of Alkenes

Recently, an indium-mediated radical iodoperfluorobutylation of but-3-yn-1-ol in water was reported by Takagi and Kanamori [16]. A mechanistic study demonstrated that R_f radical was formed by spontaneous ET from In metal to R_fI. In 2002, the Oshima group discovered that water plays a considerably important role in the Pd(0)-catalyzed Kharasch reaction of alkenes and alkynes [17]. Moreover, the reaction of alkenes with n-$C_6F_{13}I$ afforded better yields than those with $BrCCl_3$. Besides that, Bao developed copper-catalyzed radical oxyazidation and diazidation in water (Scheme 10.8) [18]. A wide range of styrenes, 1,3-dienes, and unactivated alkenes **22** were well tolerated.

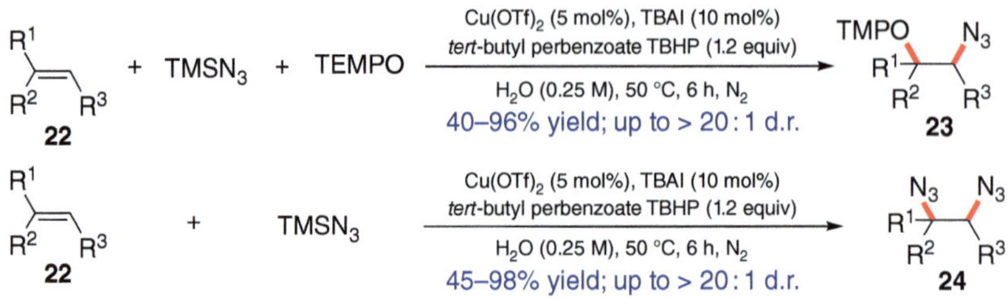

Scheme 10.8 Copper-catalyzed difunctionalization of alkene in water.

10.2.4 Miscellaneous

Methyl sulfones **26**, as important synthetic targets, have been prepared by Wang and coworkers from sulfonyl hydrazides using TBHP as the radical initiator and water as solvent in the presence of $Cu(OAc)_2$ (Scheme 10.9) [19]. Substrates examined in this study were highly converted, regardless of whether they are aryl-substituted sulfonyl hydrazide or alkyl-substituted one. A facile and efficient synthesis of α-naphthols, binaphthols, and anthracenol through Ag(I)-catalyzed C(sp^3)–H/C(sp)–H and C(sp^2)–H/C(sp)–H functionalization of β-ketoesters and alkynes in water was developed [20]. The reaction exhibited high regioselectivity and environmental benignancy.

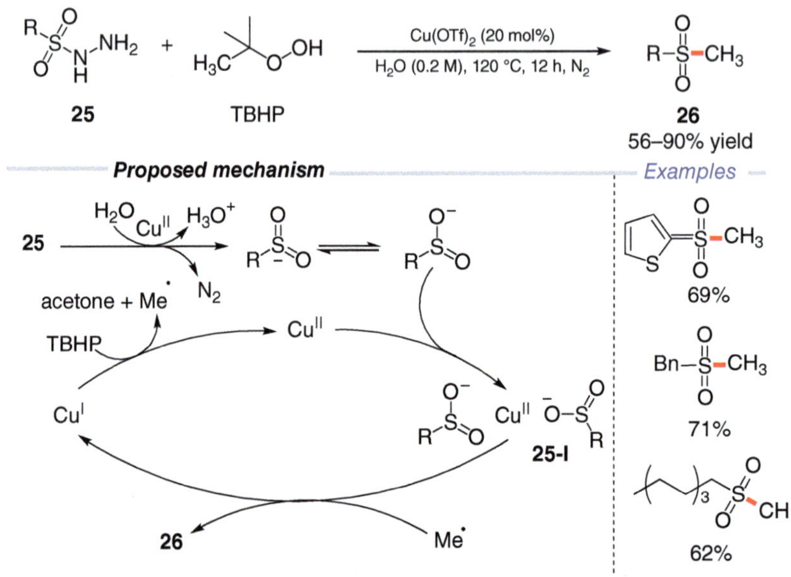

Scheme 10.9 Copper-catalyzed S-methylation of sulfonyl hydrazidesin water.

10.3 Transition Metal-Free Radical Reaction in Water

Over the past decades, there has been a rapid and significant development in transition metal-free radical reactions conducted in water. Under the well-defined condition, various radicals could be generated by the addition of stoichiometric amounts of initiator [21]. In response to this, numerous methods have been devised employing different kinds of conditions and reagents. In this section, we mainly describe recent advances in metal-free radical reactions in water based on the different oxidant systems.

10.3.1 Azo-Type Initiators

In 1998, Oshima reported the atom-transfer radical cyclization of *N*-allyl-2-iodoacetamide **29** using a water-soluble azo initiator in water (Scheme 10.10) [22, 23]. This innovative approach resulted in the synthesis of the desired γ-lactam derivatives **30** with a good yield. Of note, this reaction could be extended to the intermolecular radical cyclization of 2-iodoacetamide or 2-iodoacetic acid **31** with

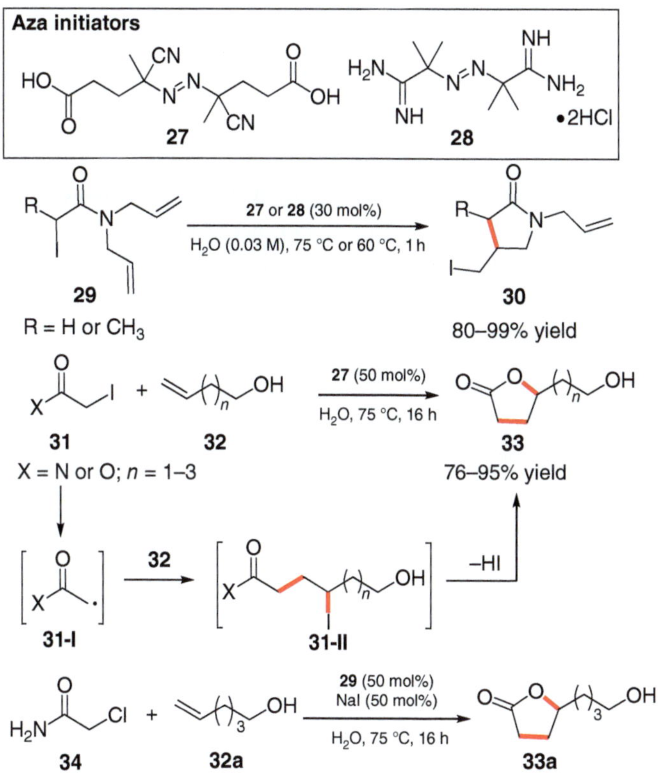

Scheme 10.10 Atom transfer radical cyclization of 2-iodoacetamidein water.

alkenol **32**. A wide range of γ-lactones **33** were obtained with favorable to excellent yields. In this transformation, iodide was liberated via ionic lactonization. Hence, the radical-ionic tandem reaction of 2-chloroacetamide **34** proceeds very smoothly by adding catalytic amount of sodium iodide. Kita and coworkers developed a novel intramolecular radical cyclization for a wide range of hydrophobic substrates in water [24]. In 2003, Murphy and coworkers reported a simple and highly efficient radical reaction for the preparation of indolones and found that the best condition entailed the use of the reagent diethylphosphine oxide and radical initiator **35** in water with no need of additional additives (Scheme 10.11) [25].

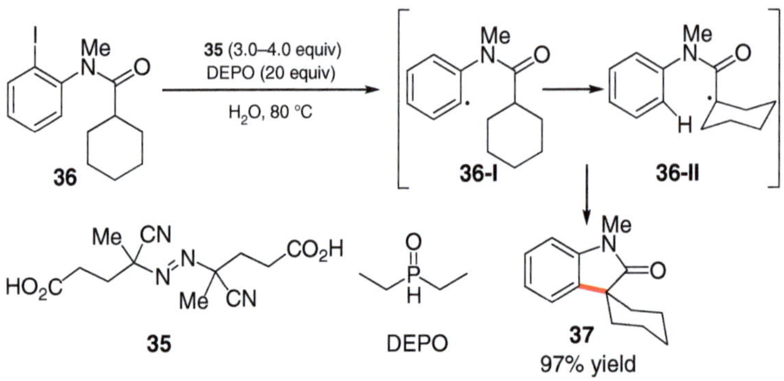

Scheme 10.11 DEPO-mediated arylation of lactamsin water.

Recently, Nudelman and coworker reported perfluoroalkylation of alkenes and alkynes using thermal initiation by 1,1′-azobis(cyclohexanecarbonitrile) in water [26]. This strategy proves versatile for constructing consecutive C—C/C—I bond, exhibiting high functional group tolerance. In addition, radical deoxygenation of alcohols and formation of carbon–carbon bonds in water were achieved by the Jang group (Scheme 10.12) [27]. In this case, additional additives, such as surfactants, are

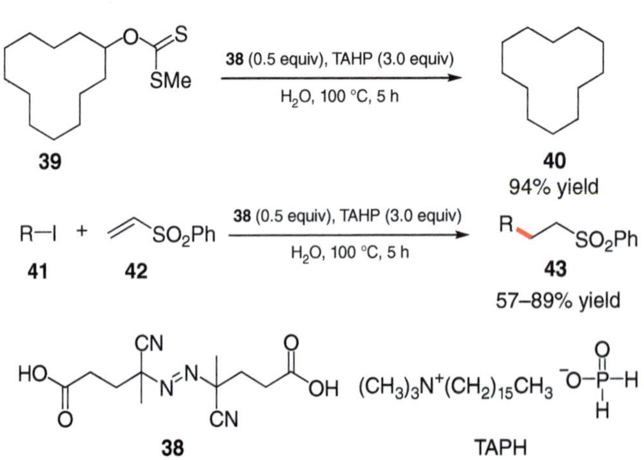

Scheme 10.12 Radical deoxygenation of alcohols and intermolecular C–C bond formation in water.

not necessary. By employing water-soluble initiators **38** and surfactant-type radical chain carriers TAPH, deoxygenation of alcohols and radical addition reactions could proceed smoothly.

This kind of water-soluble initiator could also be used for C—S bond formation. Kita developed a general intermolecular radical reaction for the synthesis of thioesters in water using a radical initiator and the cationic surfactant cetyltrimethylammonium bromide CTAB (Scheme 10.13) [28, 29]. A wide range of disulfides were investigated and pentafluorodiphenyl sulfide gave the best yield.

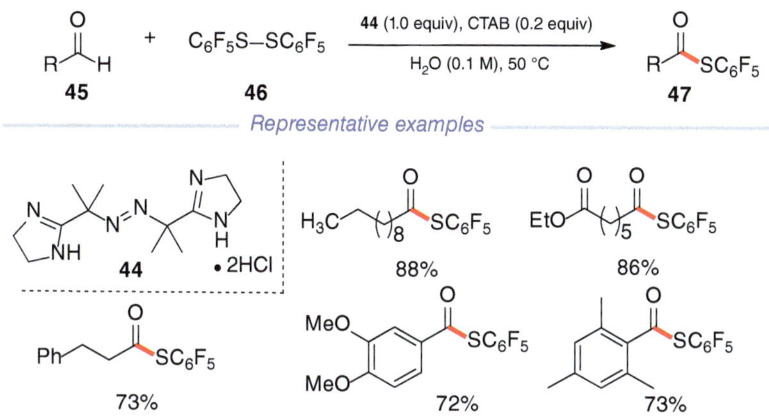

Scheme 10.13 C–S bond formation in water.

10.3.2 Triethylborane as Initiator

In 2000, Oshima and coworkers reported the radical cyclization of *N*-allyl-2-iodo amide [30]. This reaction exhibited excellent performance in water, affording the corresponding γ-lactam adducts in good yield. They also accomplished the radical cyclization of iodo acetals and iodoacetates in water [31]. It was found that the conversion is significantly higher in water for the radical cyclization of allyl iodoacetate than that in nonpolar solvents such as benzene and hexane (Table 10.1). Interestingly, water as a reaction solvent also promoted the cyclization reaction of large-membered rings. Subsequently, Renaud and coworker developed tandem radical annulation-azidation of iodomalonate with 1-octene [32]. This process involved radical addition, successive cyclization, and azidation, and gave the desired tertiary azide with a good yield. Oshima reported the intermolecular addition of halogenated compounds to carbon–carbon multiple bonds [33, 34].

10.3.3 Persulfate Salts as Oxidants

Persulfate salts are stable and non toxic, and easy to handle which have been extensively involved in a wide array of organic transformations. In 2017, Natarajan et al. developed an intramolecular oxidative radical cyclization of *N*-substituted 2-aminobiaryls **50** in water (Scheme 10.14) [35]. A wide range of substituted 2-aminobiaryls were all compatible with this transformation to give the desired

Table 10.1 Effect of the solvent on radical cyclization of allyl iodoacetate.

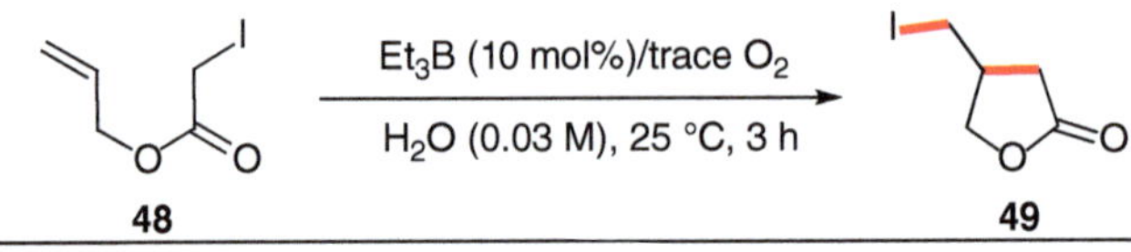

Solvent	Yield (%)	Solvent	Yield (%)	Solvent	Yield (%)
H_2O	78	Benzene	<1	CH_3CN	13
Hexane	<1	Methanol	6	DMF	13

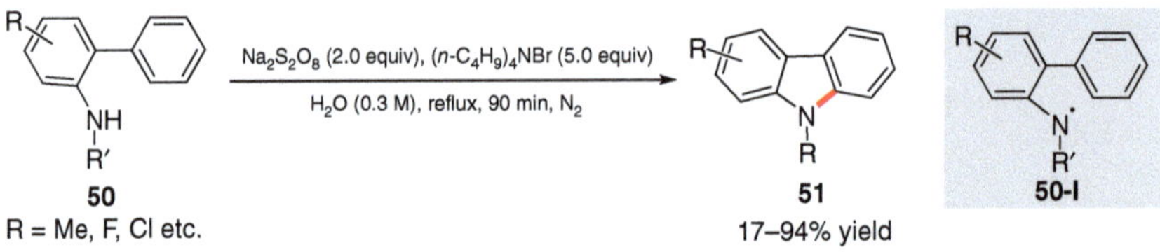

Scheme 10.14 Intramolecular oxidative radical cyclization of *N*-substituted 2-aminobiaryls in water.

N-substituted carbazoles **51** in good yields. The mechanistic study involves the formation of the nitrogen-centered amidyl radical **50-I**. Later, $K_2S_2O_8$-promoted radical cyclization of enynes/dienes with alcohols in water was developed [36].

Recently, Xu and coworkers utilized $Na_2S_2O_8$ as the green oxidant to develop an efficient and practical protocol to synthesize isothiocyanates **54** from amines and carbon disulfide with water as a solvent (Scheme 10.15) [37]. This protocol can proceed smoothly through either a stepwise or one-pot procedure under the basic conditions, giving the desired isothiocyanates **54** in good yields. A wide range of linear and branched alkyl amines and aryl amines **52** were readily converted by these two procedures. Moreover, chiral isothiocyanates can also be synthesized

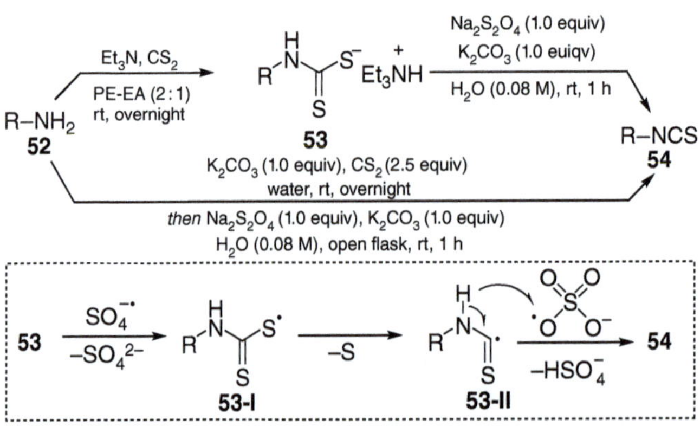

Scheme 10.15 $Na_2S_2O_8$-mediated efficient synthesis of isothiocyanates in water.

by chiral amines following the one-pot procedure in water. The mechanism study demonstrated that carbamothionyl radical **53-II** was involved.

The use of inexpensive $K_2S_2O_8$ in water for the synthesis and functionalization of heterocycles from arylglyoxylic acid was well reported by the group of Laha (Scheme 10.16) [38]. This transformation is very innovative because glucose was used as a $K_2S_2O_8$ activator, facilitating the reaction to occur at room temperature. Under the optimized condition, a series of heterocycles like pyrrole, benzothiazole, thiophene, and indole etc. could all be acylated smoothly. Also, the tandem reactions of various *ortho*-substituted anilines **58** with aryl glyoxylates **55** were achieved. Sharma and coworkers presented $K_2S_2O_8$-glucose-mediated oxidative coupling for the synthesis of bis(indolyl)methanes (BIMs) from simple indoles and alcohols in water [39]. The key step of this reaction is to oxidize alcohols to carbonyl compounds by *in situ* generating radicals from $K_2S_2O_8$. Most recently, functionalized 3-(*S*)-arylthioindoles were synthesized from indoles and aryl thiols in good to excellent yields by utilizing the $K_2S_2O_8$-glucose system in water [40].

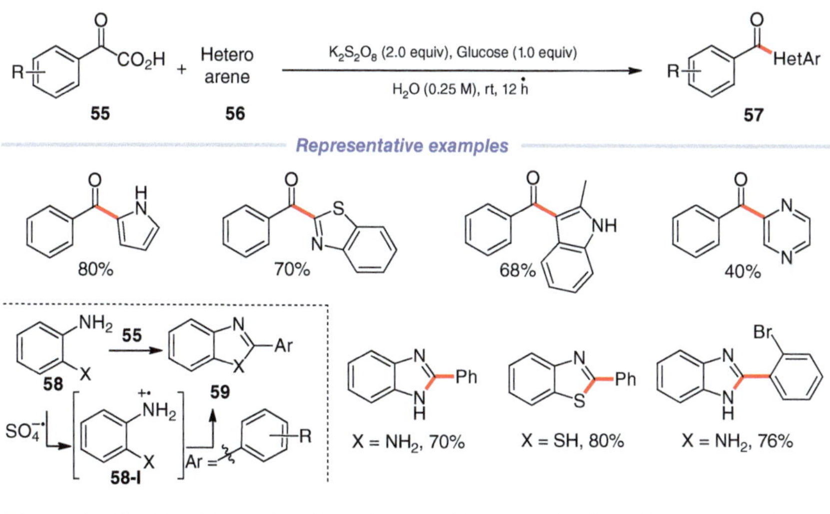

Scheme 10.16 $K_2S_2O_8$-mediated synthesis and functionalization of heterocycles in water.

10.3.4 Peroxides as Oxidants

Recently, organic chemists combined a catalytic amount of I_2 with TBHP to develop several interesting radical cyclization reactions. For example, Liang and coworkers utilized the I_2/TBHP catalytic oxidation system to achieve a one-pot radical cyclization reaction of 1,6-enynes **60** with sulfonyl hydrazides in water (Scheme 10.17) [41]. This reaction proceeded smoothly, giving the five-membered or six-membered sulfonylated products **61** or **62** with good yields. Two types of cyclization products were formed due to the control of the steric effect. Afterwards, cyclization/hydrolysis of 1,5-enenitriles with sulfonyl hydrazides has been established for the preparation of valuable pyrrolidine-2,4-diones [42]. In the presence

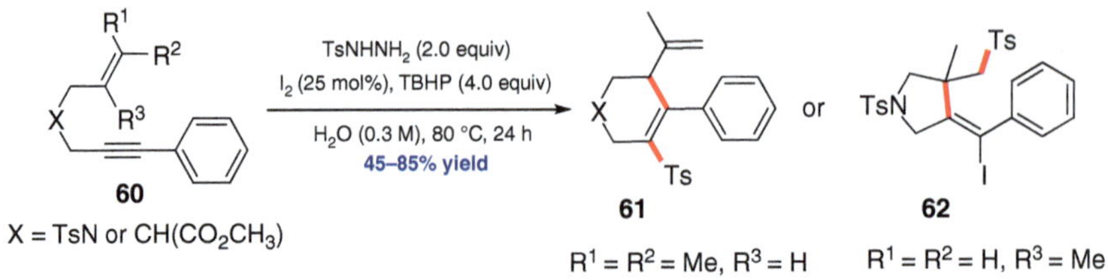

Scheme 10.17 Radical cyclization reaction of 1,6-enynes with sulfonyl hydrazides in water.

of I_2/TBHP system with water as a solvent, sulfonyl hydrazides could generate sulfonyl radicals, which underwent radical addition, intramolecular cyclization, hydrogen abstraction, and hydrolysis to afford the final products.

In 2022, Wei and coworkers developed three types of radical cyclization based on 1,6-enynes **63** to afford five-membered and six-membered cyclic lactams **64–66** (Scheme 10.18) [43]. This protocol showed noteworthy features such as mild reaction conditions and the employment of water as a reaction medium. In addition, radical C(sp^3)-H functionalization triggered cyclization of olefinic 1,3-dicarbonyls and olefinic amides was also developed [44].

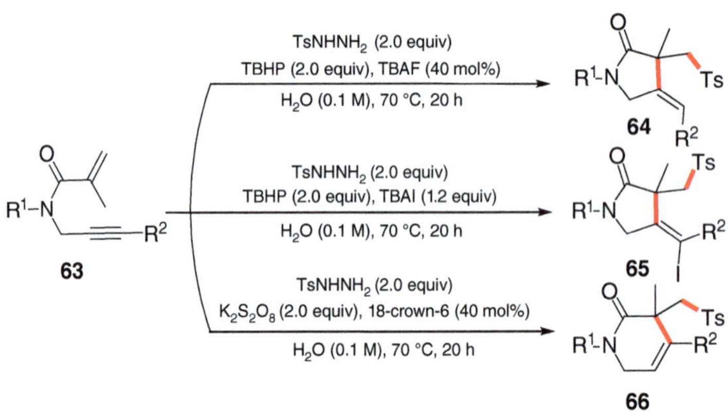

Scheme 10.18 Three types of radical cyclization of 1,6-enynes with sulfonyl hydrazides in water.

C(sp^3)–H functionalization of inert alkanes was also accomplished, allowing for access to various benzimidazo[2,1-*a*]isoquinolin-6(5*H*)-ones **69** (Scheme 10.19) [45]. Notably, using di-*tert*-butyl peroxide (DTBP) as the oxidant and sodium dodecyl sulfate (SDS) as the phase transfer agent in a water-based medium proved highly efficient for this conversion. Moreover, a wide range of cyclic alkanes, long-chain alkanes, toluene derivatives could be compatible, and demonstrating a broad substrate scope. Remarkably, this methodology could be extended for the radical cyclization of unactivated alkanes with alkene-substituted quinazolinones, showcasing the great potential of the C(sp^3)–H functionalization strategy employing DTBP in water [46].

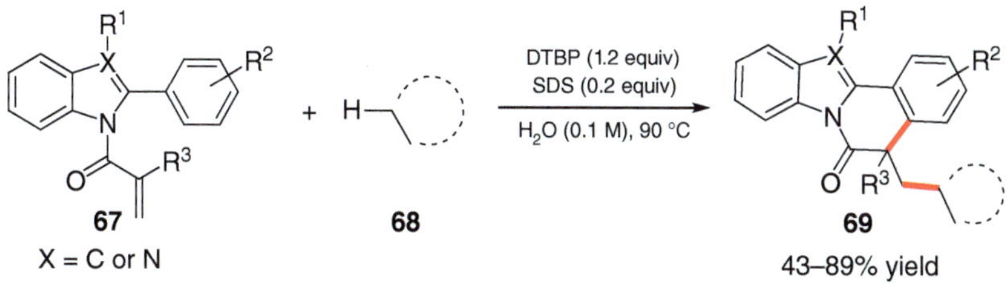

Scheme 10.19 Radical cyclization of 2-arylbenzoimidazoles with unactivated alkanes in water.

In 2018, Wei et al. developed a new method for the C(sp^3)–H hydroxylation of 2-oxindoles **70** via a radical cross-coupling reaction exposed to air in water, affording various 3-hydroxy-2-oxindoles **71** in good yields (Scheme 10.20) [47]. In this reaction, *t*-BuOOH was used as a hydroxylation reagent, and the reaction was promoted by the addition of SDS as a surfactant. Additionally, a temperature-controlled sulfonylation of DTBP in water was also reported by the Yuan group, which was used for the formation of diverse aryl methyl sulfones and β-hydroxysulfones [48].

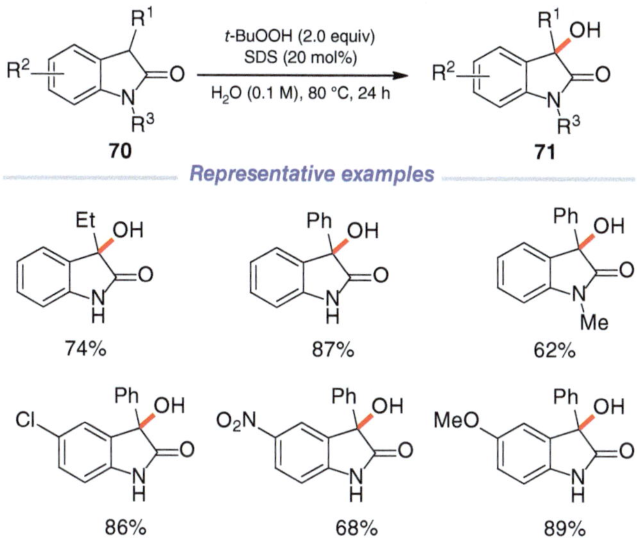

Scheme 10.20 C(sp^3)–H hydroxylation of 2-oxindoles in water.

Recently, Wang and coworkers discovered the oxidative radical addition of arylhydrazines with α-aryl vinyl azides in the presence of H_2O_2 as the oxidant and PEG-800 as the phase-transfer catalyst [49]. This reaction took place in water, providing various 2-arylacetophenones under conditions compatible with various functional groups. Liu and coworkersreported the ionic liquid promoted a C—H bond oxidant cross-coupling reaction for the synthesis of heterocyclic chromeno[2,3-*c*]pyrazol-4(1*H*)-ones **74** (Scheme 10.21) [50]. The optimal yield was

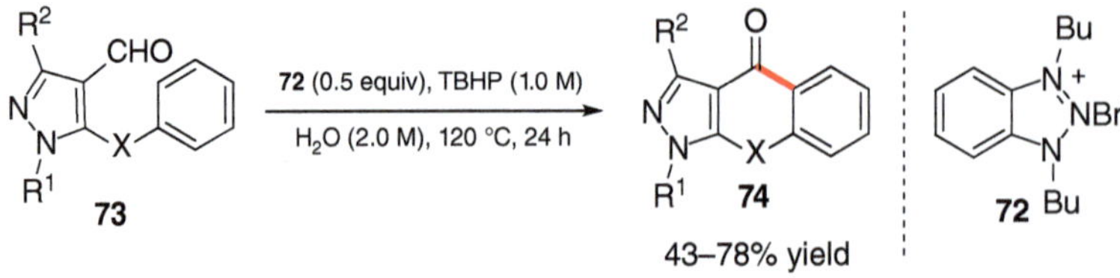

Scheme 10.21 Ionic liquid promoted C—H bond oxidant cross-coupling reaction in water.

obtained by employing TBHP as the oxidant and 1,3-dibutyl-1*H*-benzo[*d*][1,2,3]-triazol-3-ium bromide **72** as the promotor in the presence of water.

Recently, the combination of MI-TBHP as an oxidation system (M=Na, K, and NH_4) in aqueous media has emerged as a promising avenue for transition-metal-free oxidative transformations [51]. In 2015, Li and coworkers developed I^-/TBHP catalyzed oxidative amination with benzophenone imine in water [52]. A series of substituted benzaldehydes furnished the corresponding amides successfully. Afterwards, oxidative cross-coupling of C(sp^3)—H bonds adjacent to nitrogen with alkyl O—H and N—H bonds was reported by the Punniyamurthy group (Scheme 10.22) [53]. Under the optimal conditions, a wide range of functionalized oxazolidines and imidazolidines **76** were obtained with good yield.

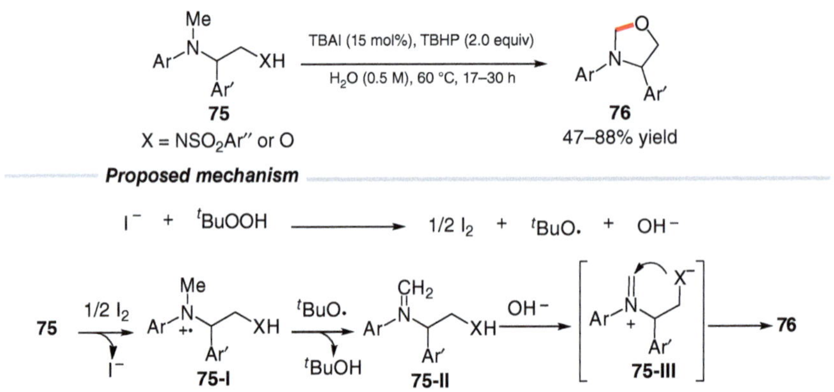

Scheme 10.22 KI/TBHP-catalyzed C(sp^3)–H functionalization/C—O/C—N bond formation in water.

10.3.5 O_2 as the Oxidant

Given that O_2 is both inexpensive and environmentally benign. For example, dioxygen radical initiation has recently been successfully applied for the hydrosilylation of alkynes **77** in water, leading to high yields of hydrosilylated alkene products **78** with excellent *Z*:*E* stereoselective ratios (Scheme 10.23) [54]. Thus, a tentative mechanism for this process was suggested. The dioxygen initiation triggers radical events, where $(Me_3Si)_3Si\bullet$ radical derived from the $(Me_3Si)_3SiH$ adds to the C—C triple bond. Then, this radical **77-I** abstracts a hydrogen atom from $(Me_3Si)_3SiH$ to give

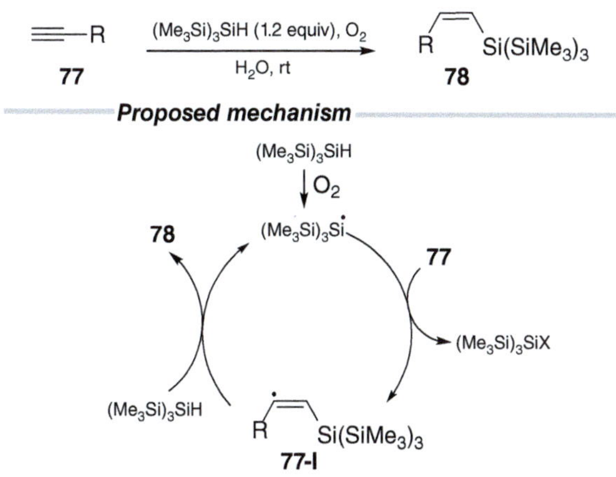

Scheme 10.23 Hydrosilylation of triple-bonded substrates in water.

the final product **78** with regeneration of $(Me_3Si)_3Si\bullet$ radical to propagate a radical chain. The Chatgilialoglu group examined different initiation methods for hydrosilylation and indeed found that dioxygen initiation exhibited superior performance for the hydrosilylation reaction of C—C multiple bonds compared to that of C—C double bonds [55]. Besides that, $(Me_3Si)_3Si\bullet$ radical is also used to generate alkyl radicals. The Postigo group developed the $(Me_3Si)_3SiH$-mediated intermolecular addition of perfluoroalkyl radicals to electron-rich or -poor alkenes in water [56]. In this transformation, the azo compound is found to be a more efficient radical initiator than dioxygen.

Recently, Baruah and coworkers utilized O_2 as the oxidant to develop three-component reactions for the synthesis of 1,3-oxazines (Scheme 10.24) [57]. This reaction proceeded very smoothly in water, enabling the assembly of naphthol **79**, aldehyde **81** and tetrahydroisoquinoline **80** into 1,3-oxazines **82** with moderate to good yield.

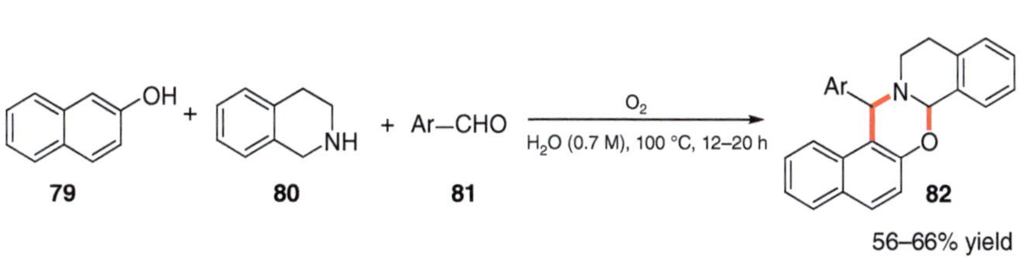

Scheme 10.24 Three-component reactions of naphthol, aldehyde, and tetrahydroisoquinoline in water.

In 2017, Wei et al. reported radical coupling reactions of oxindoles with *t*-BuONO in water (Scheme 10.25) [58]. The reaction proceeded under air, and a variety of oxindoles were successfully transformed into isatins with good yields. Moreover, this reaction was operationally simple and showed good functional group tolerance [59].

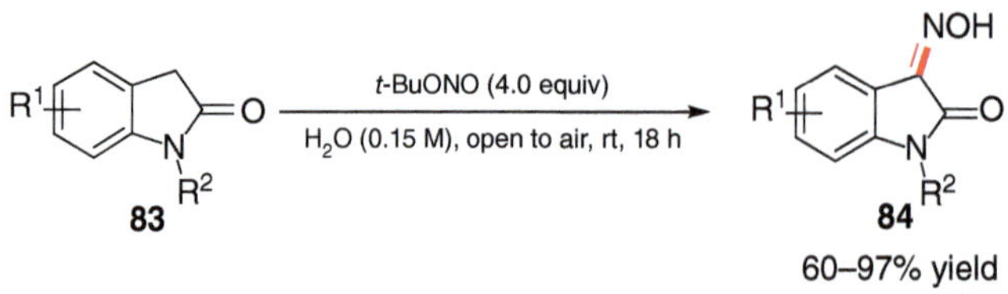

Scheme 10.25 Radical coupling reactions of oxindoles with *t*-BuONO.

10.3.6 Miscellaneous

In 2016, Wang and coworkers reported catalyst-free fluorination of sulfonyl hydrazides through a free radical pathway (Scheme 10.26) [60]. Under the optimal conditions, both arylsulfonyl hydrazides and aliphatic sulfonyl hydrazides **85** were well tolerated.

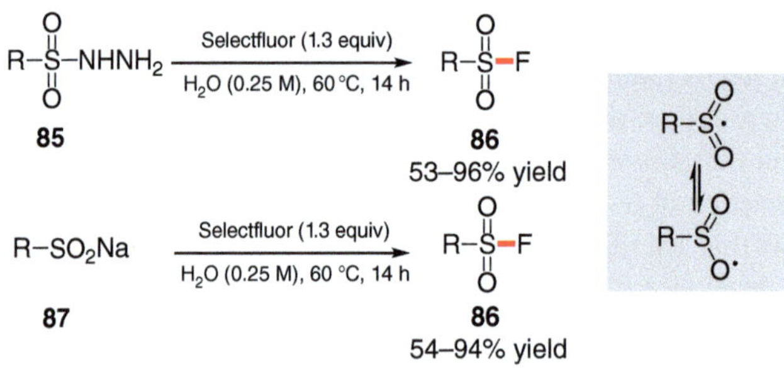

Scheme 10.26 Catalyst-free fluorination in water.

In addition, iodine-catalyzed amination of benzothiazoles **88** with KSeCN in water was described by Song and coworkers (Scheme 10.27) [61]. This method provided efficient access to primary 2-benzothiazolamines **90** in the presence of catalytic amounts of I_2 using (diacet oxyiodo)benzene (PIDA) as the oxidant and Tf_2O as the additive.

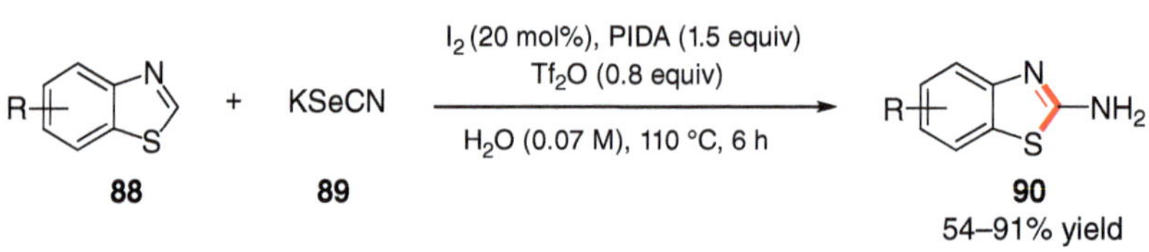

Scheme 10.27 Iodine-catalyzed amination of benzothiazolesin water.

10.4 Photoredox Reactions Excited by Visible Light in Water

Visible light photoredox reactions have been esteemed as green chemistry tools for various organic transformations [62]. However, visible-light catalytic reactions in water have historically received little attention due to the poor solubility of most organic substrates and catalysts. In the recent decades, merging the inherently green properties of visible light catalysis with the use of water as a solvent has experienced a significant boost in order to diminish the environmental impact of photocatalytic processes [63]. In this section, we will present the recent advances in visible light catalytic reactions in water and emphasize their potential advantages according to the different synthetic transformations.

10.4.1 Radical Addition to α,β-Unsaturated Systems

In 2012, Postigo and coworkers reported light-induced iodoperfluoroalkylation reactions of alkenes and alkynes in water [64]. Recently, Lei and coworkers developed a visible light-induced radical addition to alkynes **91** in water for the direct synthesis of *Z*-alkenylphosphine oxides **93** (Scheme 10.28) [65]. A wide range of terminal and internal alkynes **91** were well tolerated, affording the products **93** with high *Z* stereoselectivity. The mechanism study demonstrated that high *Z*-selectivity came from π–π stacking interaction between the aromatic rings of the aryl acetylene and diphenylphosphine oxide.

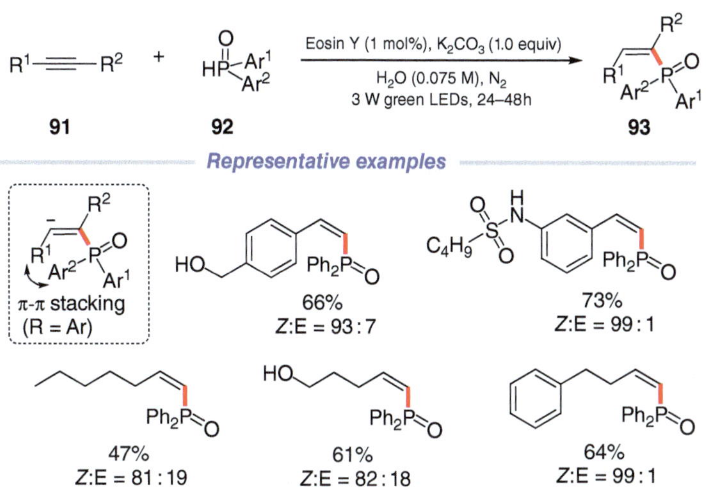

Scheme 10.28 Addition of diaryl phosphine oxide to alkyne in water.

Besides, a visible light-driven hydroacylation and epoxyacylation of olefins **94** or **97** in water have been presented (Scheme 10.29) [66]. The Salles et al. found that optimum reaction conditions consist of the use of methylene blue as a photoredox catalyst, persulfate as an oxidant, and potassium carbonate as a base in water as a solvent in the presence of visible light. Moreover, these two transformations could be accomplished using only one set of reagents. Under the optimized condition, this reaction performed very well for aromatic and aliphatic aldehydes as well as conjugated and nonconjugated olefins, affording the ketones **98** and epoxyketones **96** in excellent yields. Furthermore, the reaction was carried out on a gram scale with no loss of reactivity.

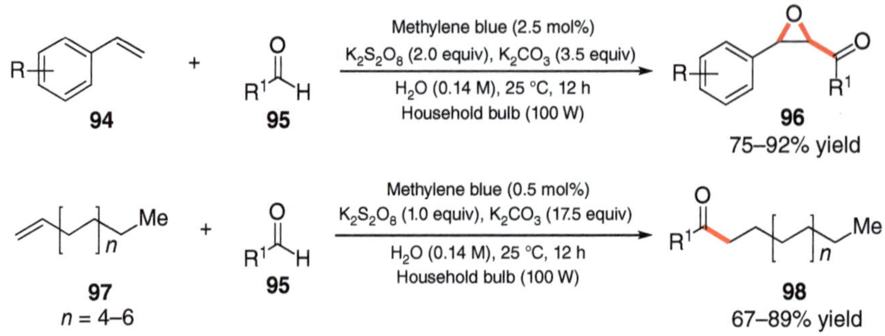

Scheme 10.29 Visible-light-driven epoxyacylation and hydroacylation of olefins in water.

More recently, Xu and coworkers showed water could have a key role in promote iremote 1,5-trifluoromethylthio-sulfonylation of vinylcyclopropane **99** with $ArSO_2SCF_3$ reagents **100** (Scheme 10.30) [67]. Moreover, SDS was also necessary to increase the solubility of organic substrates in water. This green and mild approach allowed a wide range of aryl vinylcyclopropane and alkyl vinylcyclopropane to be available in this transformation. According to the reaction mechanism, the reaction was initiated by a SET process between eosin Y* and $ArSO_2SCF_3$ **100** to generate sulfonyl radicals and the eosin Y/SCF_3 complex. Then, arylsulfonyl radical was added to vinylcyclopropane **99**, followed by a rearrangement of cyclopropane to form intermediate **99-I**, which sequentially reacted with the eosin Y/SCF_3 complex to furnish the corresponding product **101**.

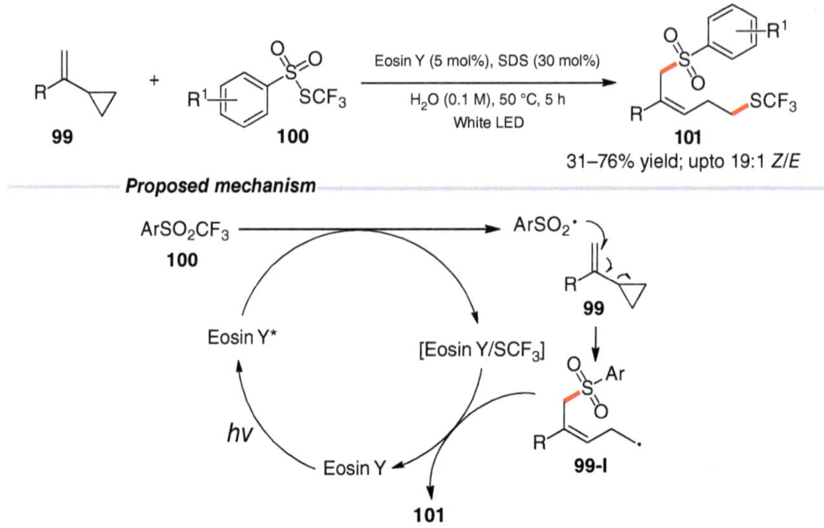

Scheme 10.30 Remote 1,5-trifluoromethylthio-sulfonylation of vinylcyclopropane in water.

Visible light-induced iodosulfonyl reaction of alkenes in water using an EDA complex strategy has been developed by Li and coworkers (Scheme 10.31) [68]. This reaction afforded β-iodo-substituted sulfone derivatives **104** in the presence of a cationic surfactant, CTAB with good yields.

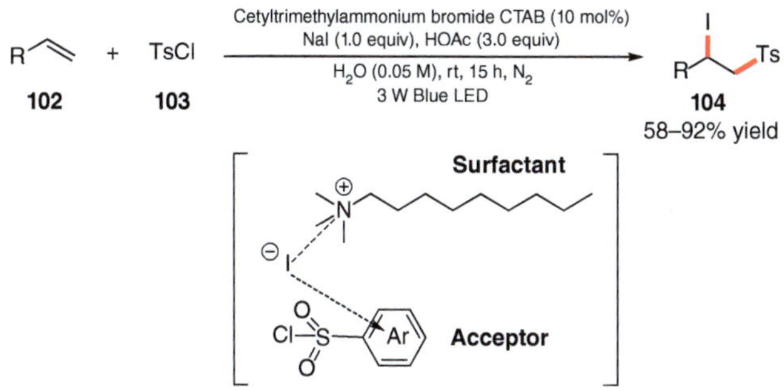

Scheme 10.31 Visible-light-induced sulfonylation of alkenes with sulfonyl chloride in water.

In 2018, Lipshutz and coworkers utilized task-specific amphiphilic iridium as a photocatalyst to realize difunctionalization of alkenes and sulfonylation of enol acetates in water (Scheme 10.32) [69]. The amphiphilic photocatalyst is composed of a hydrophilic component (MPEG), a lipophilic side chain, and an Ir center. Using this catalyst, the reaction of both alkenes **105** or enol acetates **108** with sulfonyl

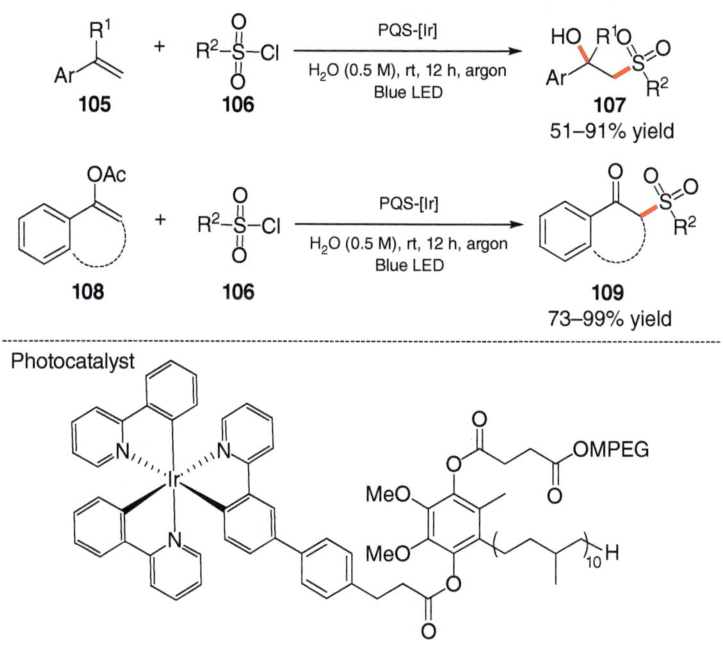

Scheme 10.32 An amphiphilic iridium catalyst mediated difunctionalization of alkenes in water.

chloride **106** proceeded smoothly to give the desired products with a moderate to good yield. Of note, this aqueous medium could be recycled several times with no loss of reaction efficiency.

Recently, Niu and coworkers reported a visible light-induced Meerwein hydration reaction of alkenes **111** in water using mpg-C_3N_4 as a recyclabe photocatalyst (Scheme 10.33) [70]. A wide range of diazonium salts **110** and alkenes **111** with different functional groups were all compatible in this transformation, furnishing the racemic alcohols **112** in moderate to high yields.

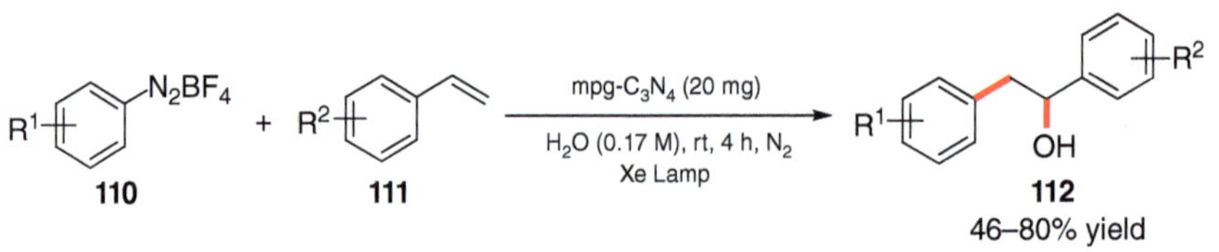

Scheme 10.33 Visible light induced Meerwein hydration reaction in water.

10.4.2 Cross-Dehydrogenative Coupling

In 2023, Kundu and coworkers developed visible light-mediated organocatalytic dehydrogenative aza-coupling of 1,3-diones **113** with aryldiazonium salts **114** in water (Scheme 10.34) [71]. By employing water as a reaction medium, a wide range of alkyl diazenes **115** could be obtained in acceptable yields. Moreover, mechanistic studies using cyclic voltammetry suggested differential reactivities of various substituted aryldiazonium cations according to their reduction potentials.

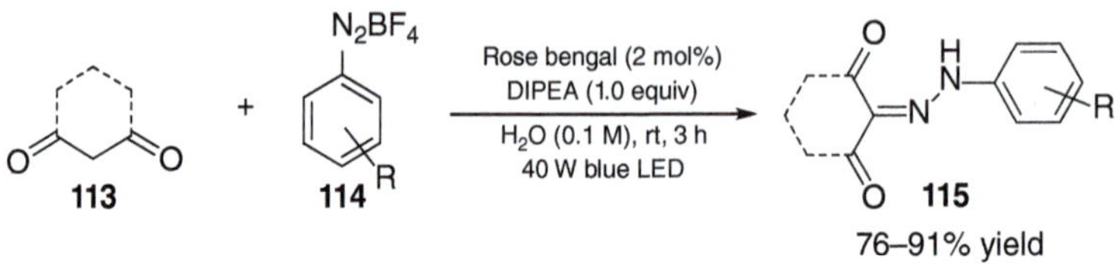

Scheme 10.34 Visible light induced dehydrogenative aza-coupling of 1,3-diones in water.

Recently, 1,2,3,5-tetrakis(carbazol-9-yl)-4,6-dicyanobenzene (4CzIPN)-based porous organic polymers which were prepared and used in cross-dehydrogenative coupling of *N*-aryl tetrahydroisoquinolines **116** with diarylphosphine oxides **117** for the construction of the C—P bond in water (Scheme 10.35) [72]. A wide range of substrates, including various functionalized diarylphosphine oxides **117** and *N*-substituted tetrahydroisoquinolines **116**, were presented, and the desired products **118** were formed in moderate to good yields. In addition, Zhang, and coworkers also reported efficient cross-dehydrogenative coupling of *N*-aryl tetrahydroisoquinolines **116** with ketones **119** in the presence of a sponge photocatalyst in water (Scheme 10.35) [73]. The preparation of this sponge catalyst is based on the modification of the polymer surface and classical solid-phase peptide synthesis.

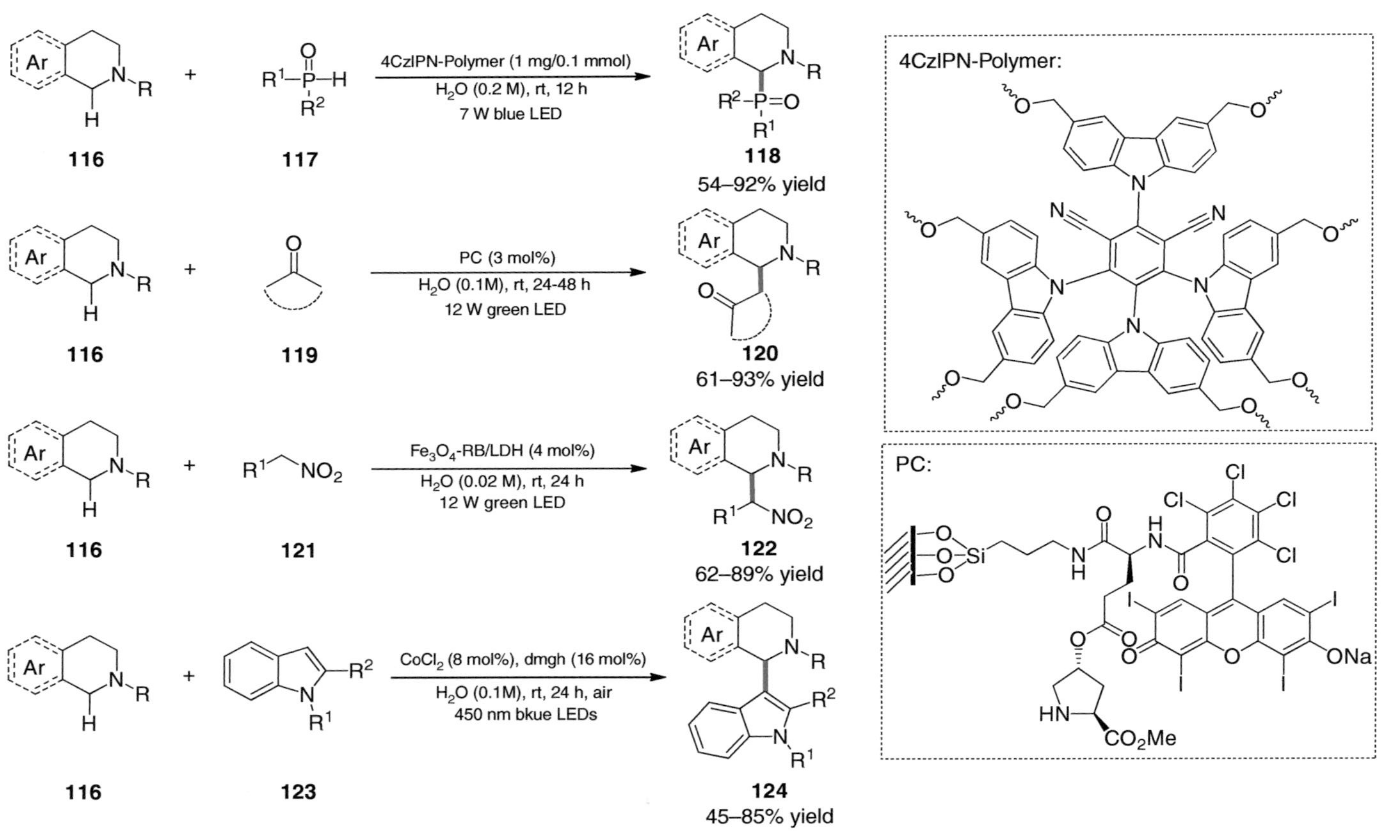

Scheme 10.35 Visible light induced cross-dehydrogenative coupling in water.

By employing this photocatalyst, *N*-phenyltetrahydroisoquinoline **116** with various substituents and various ketones **119** were all compatible with this reaction (Scheme 10.35) [74]. This catalyst was synthesized through ultrasound-assisted self-assembly among the anionic rose bengal dye, negatively charged Fe_3O_4 NPs, and positively charged MgAl-LDH. Under the optimal conditions, this reaction with nitroalkanes proceeded very smoothly in water to give the corresponding products **122** in moderate to good yields. In addition to this, Wu and coworkers successfully employed *in situ*-generated photosensitizer derived from a catalytic amount of $CoCl_2$ with dmgH (dimethylglyoxime) as a ligand for the development of the cross-dehydrogenative coupling reaction of *N*-aryl tetrahydroisoquinolines **116** and indoles **123** (Scheme 10.35) [75].

10.5 Conclusions

In conclusion, this chapter provides a comprehensive summary of radical reactions in water using traditional and modern radical methods. Simultaneously, the opportunities and the challenges of utilizing water as a solvent have been highlighted. And the related mechanisms are also described in detail to guide further developments in radical reactions. In the past few years, a growing number of new reactions accommodated to water have appeared, opening a new prospect for green chemistry. They feature several advantages, such as mild reaction conditions, broad substrate scope, and simple operation.

Despite these advantages, significant improvement is still necessary. For example, addressing challenges related to the reuse of aqueous systems and the treatment of waste water is crucial. Moreover, asymmetric catalysis in this field still requires further development. For example, more environmentally friendly strategies need to be explored, especially in areas such as photoredox chemistry, electrosynthesis, and photobiocatalytic transformations. In addition, the development of new water-soluble mediators to broaden the scope of reaction types is also necessary. With the increasing interest in green chemistry, the pursuit of developing milder and more sustainable processes for the synthesis of natural products and complex molecules has never stopped. We hope this chapter inspires more chemists to explore more novel, relevant strategies.

References

1 (a) Li, C.-J. (2016). *Green Chem.* 18: 1836–1838. (b) Sheldon, R.A. (2018). *ACS Sustain. Chem. Eng.* 6: 32–48. (c) Li, C.-J. and Anastas, P.-T. (2012). *Chem. Soc. Rev.* 41: 1413–1414. (d) Simon, M.-O. and Li, C.-J. (2012). *Chem. Soc. Rev.* 41: 1415–1427.

2 Anastas, P. and Eghbali, N. (2010). *Chem. Soc. Rev.* 39: 301–312.

3 (a) Kitanosono, T., Masuda, K., Xu, P., and Kobayashi, S. (2018). *Chem. Rev.* 118: 679–746. (b) Li, C.-J. and Chen, L. (2006). *Chem. Soc. Rev.* 35: 68–82.

4 (a) Sibi, M.P., Li, C.-Z., and Jiao, N. (2019). *Sci. China Chem.* 62: 1423–1424. (b) Studer, A. and Curran, D.P. (2016). *Angew. Chem. Int. Ed.* 55: 58–102.

5 (a) Huang, Y. and Wei, W. (2018). *Prog. Chem.* 30: 1819–1826. (b) Liu, Y., Zuo, L., Lv, Q., and Yu, B. (2023). *Curr. Opin. Green Sustain. Chem.* 40: 100754. (c) Ouyang, W.-T., Xiao, F., Ou, L.-J., and He, W.-M. (2023). *Curr. Opin. Green Sustainable Chem.* 40: 100760. (d) Barata-Vallejo, S., Yerien, D.E., and Postigo, A. (2021). *ACS Sustain. Chem. Eng.* 9: 10016–10047. (e) Sun, K., Lv, Q.-Y., Chen, X.-L. et al. (2021). *Green Chem.* 23: 232–248. (f) Chen, D., Liu, J., Zhang, X. et al. (2019). *Chin. J. Org. Chem.* 39: 3353–3362.

6 (a) Wu, J., Liu, Y., Ma, X. et al. (2017). *Chin. J. Chem.* 35: 1391–1395. (b) Ren, L., Wang, L., Lv, Y. et al. (2015). *Green Chem.* 17: 2369–2372.

7 Wang, T., Yuan, L., Zhao, Z. et al. (2015). *Green Chem.* 17: 2741–2744.

8 Kong, D.-L., Cheng, L., Yue, T. et al. (2016). *J. Org. Chem.* 81: 5337–5344.

9 Taniguchi, T., Zaimoku, H., and Ishibashi, H. (2011). *Chem. Eur. J.* 17: 4307–4312.

10 (a) Barata-Vallejo, S. and Postigo, A. (2012). *Eur. J. Org. Chem.* 2012: 1889–1899. (b) Yang, L., Dong, T., Revankar, H.M., and Zhang, C.-P. (2017). *Green Chem.* 19: 3951–3992. (c) Song, H.-X., Han, Q.-Y., Zhao, C.-L., and Zhang, C.-P. (2018). *Green Chem.* 20: 1662–1731.

11 Yang, F., Klumphu, P., Liang, Y.-M., and Lipshutz, B.H. (2014). *Chem. Commun.* 50: 936–938.

12 (a) Ji, X.-M., Wei, L., Chen, F., and Tang, R.-Y. (2015). *RSC Adv.* 5: 29766–29773. (b) Xu, J., Qiao, L., Shen, J. et al. (2017). *Org. Lett.* 19: 5661–5664.

13 (a) Leroux, F., Jeschke, P., and Schlosser, M. (2005). *Chem. Rev.* 105: 827–856. (b) Ni, C., Hu, M., and Hu, J. (2015). *Chem. Rev.* 115: 765–825.

14 Li, X., Zhang, R., Zhang, X. et al. (2020). *Chem. Asian J.* 15: 1175–1179.

15 Hu, F., Shao, X., Zhu, D. et al. (2014). *Angew. Chem. Int. Ed.* 53: 6105–6109.

16 Takagi, T. and Kanamori, T. (2011). *J. Fluorine Chem.* 132: 427–429.

17 Motoda, D., Kinoshita, H., Shinokubo, H., and Oshima, K. (2002). *Adv. Synth. Catal.* 344: 261–265.

18 (a) Cui, J., Zhou, H., Li, Y., and Bao, H. (2022). *Synthesis* 54: 3262–3270. (b) Zhou, H., Jian, W., Qian, B. et al. (2017). *Org. Lett.* 19: 6120–6123.

19 Yang, Y., Bao, Y., Guan, Q. et al. (2017). *Green Chem.* 19: 112–116.

20 Naresh, G., Kant, R., and Narender, T. (2015). *Org. Lett.* 17: 3446–3449.

21 (a) Székely, A. and Klussmann, M. (2019). *Chem. Asian J.* 14: 105–115. (b) Postigo, A. (2011). *RSC Adv.* 1: 14–32.

22 Yorimitsu, H., Wakabayashi, K., Shinokubo, H., and Oshima, K. (1999). *Tetrahedron Lett.* 40: 519–522.

23 Yorimitsu, H., Wakabayashi, K., Shinokubo, H., and Oshima, K. (2001). *Bull. Chem. Soc. Jpn.* 74: 1963–1970.

24 Kita, Y., Nambu, H., Ramesh, N.G. et al. (2001). *Org. Lett.* 3: 1157–1160.

25 Khan, T.A., Tripoli, R., Crawford, J.J. et al. (2003). *Org. Lett.* 5: 2971–2974.

26 Vázquez, A.J. and Nudelman, N.S. (2015). *ARKIVOC* 5: 190–201.

27 Cho, D.H. and Jang, D.O. (2005). *Tetrahedron Lett.* 46: 1799–1802.

28 Nambu, H., Hata, K., Matsugi, M., and Kita, Y. (2005). *Chem. Eur. J.* 11: 719–727.

29 Nambu, H., Hata, K., Matsugi, M., and Kita, Y. (2002). *Chem. Commun.* 1082–1083.

30 Wakabayashi, K., Yorimitsu, H., Shinokubo, H., and Oshima, K. (2000). *Bull. Chem. Soc. Jpn.* 73: 2377–2378.

31 Yorimitsu, H., Nakamura, T., Shinokubo, H. et al. (2000). *J. Am. Chem. Soc.* 122: 11041–11047.

32 Panchaud, P. and Renaud, P. (2004). *J. Org. Chem.* 69: 3205–3207.

33 Nakamura, T., Yorimitsu, H., Shinokubo, H., and Oshima, K. (1998). *Synlett* 1998: 1351–1352.

34 Yorimitsu, H., Shinokubo, H., Matsubara, S. et al. (2001). *J. Org. Chem.* 66: 7776–7785.

35 Natarajan, P., Priya, P., and Chuskit, D. (2017). *Green Chem.* 19: 5854–5861.

36 Wang, D.-K., Fang, Y.-L., Zhang, J. et al. (2020). *Org. Biomol. Chem.* 18: 8491–8495.

37 Fu, Z., Yuan, W., Chen, N. et al. (2018). *Green Chem.* 20: 4484–4491.

38 Laha, J.K. and Hunjan, M.K. (2021). *Chem. Commun.* 57: 8437–8440.

39 Indurthi, H.K., Das, S., Kumari, A., and Sharma, D.K. (2022). *New J. Chem.* 46: 13924–13930.

40 Kumar, N., Venkatesh, R., Singh, S., and Kandasamy, J. (2023). *Eur. J. Org. Chem.* 26: e202300679.

41 Zheng, L., Zhou, Z.-Z., He, Y.-T. et al. (2016). *J. Org. Chem.* 81: 66–76.

42 Hu, S.-J., Jiang, L.-L., Qiu, H. et al. (2022). *Org. Biomol. Chem.* 20: 6418–6422.

43 Yu, X.-C., Zheng, Y.-N., Zhang, J.-H. et al. (2022). *ACS Sustain. Chem. Eng.* 10: 6057–6062.

44 Qin, F.-H., Wang, H.-Z., Cao, T.-T. et al. (2021). *Asian J. Org. Chem.* 10: 3380–3383.

45 Li, J.-Z., Mei, L., Cai, X.-E. et al. (2022). *Adv. Synth. Catal.* 364: 2080–2085.

46 Zhang, W.-K., Li, J.-Z., Zhang, C.-C. et al. (2022). *Eur. J. Org. Chem.* 2022: e202200523.

47 Wei, W.-T., Zhu, W.-M., Shao, Q. et al. (2018). *ACS Sustainable Chem. Eng.* 6: 8029–8033.

48 Lai, J. and Yuan, G. (2018). *Tetrahedron Lett.* 59: 524–527.

49 Luo, M., Zhang, Y., Fang, P. et al. (2022). *Org. Biomol. Chem.* 20: 630–635.

50 Li, H., Liu, C., Zhang, Y. et al. (2015). *Org. Lett.* 17: 932–935.

51 Ghosh, P., Ganguly, B., and Das, S. (2021). *Org. Biomol. Chem.* 19: 2146–2167.

52 Huang, H., Chen, W., Xu, Y., and Li, J. (2015). *Green Chem.* 17: 4715–4719.

53 Satheesh, V., Sengoden, M., and Punniyamurthy, T. (2016). *J. Org. Chem.* 81: 9792–9801.

54 Postigo, A. and Sbarbati Nudelman, N. (2010). *J. Phys. Org. Chem.* 23: 910–914.

55 Postigo, A., Kopsov, S., Zlotsky, S.S. et al. (2009). *Organometallics* 28: 3282–3287.

56 Barata-Vallejo, S. and Postigo, A. (2010). *J. Org. Chem.* 75: 6141–6148.

57 Deb, M.L., Pegu, C.D., Borpatra, P.J. et al. (2017). *Green Chem.* 19: 4036–4042.

58 Wei, W.-T., Zhu, W.-M., Ying, W.-W. et al. (2017). *Org. Biomol. Chem.* 15: 5254–5257.

59 Rodrigues, C.A.B., de Matos, M.N., Guerreiro, B.M.H. et al. (2011). *Tetrahedron Lett.* 52: 2803–2807.

60 Tang, L., Yang, Y., Wen, L. et al. (2016). *Green Chem.* 18: 1224–1228.

61 Zhu, Y.-S., Shi, L., Fu, L. et al. (2022). *Chin. Chem. Lett.* 33: 1497–1500.

62 Chanda, A. and Fokin, V.V. (2009). *Chem. Rev.* 109: 725–748.

63 (a) Russo, C., Brunelli, F., Tron, G.C., and Giustiniano, M. (2023). *J. Org. Chem.* 88: 6284–6293. (b) Fukuzumi, S., Lee, Y.-M., and Nam, W. (2019). *ChemSusChem* 12: 3931–3940. *Curr. Org. Chem.* **2012**, *16*, 2354-2364.

64 Slodowicz, M., Barata-Vallejo, S., Vázquez, A. et al. (2012). *J. Fluorine Chem.* 135: 137–143.

65 Wang, H., Li, Y., Tang, Z. et al. (2018). *ACS Catal.* 8: 10599–10605.

66 de Souza, G.F.P., Bonacin, J.A., and Salles, A.G. Jr., (2018). *J. Org. Chem.* 83: 8331–8340.

67 Liu, J., Yao, H., Li, X. et al. (2020). *Org. Chem. Front.* 7: 1314–1320.

68 Lin, L., Yang, Z., Liu, J. et al. (2021). *Green Chem.* 23: 5467–5473.

69 Bu, M.-J., Cai, C., Gallou, F., and Lipshutz, B.H. (2018). *Green Chem.* 20: 1233–1237.

70 Wang, J., Xue, L., Hong, M. et al. (2020). *Green Chem.* 22: 411–416.

71 Das, R., Kundu, T., and Basumatary, J. (2023). *RSC Adv.* 13: 3147–3154.

72 Zhu, S.-S., Zuo, L., Liu, Y., and Yu, B. (2022). *Green Chem.* 24: 8725–8732.

73 Zhang, T., Liang, W., Huang, Y. et al. (2017). *Chem. Commun.* 53: 12536–12539.

74 Huang, Y., Xin, Z., Yao, W. et al. (2018). *Chem. Commun.* 54: 13587–13590.

75 Wu, C.-J., Zhong, J.-J., Meng, Q.-Y. et al. (2015). *Org. Lett.* 17: 884–887.

11

Carbene Reactions in Water

Xiaoye Yu[1,2], Ying Zhang[1], and Huan-Ming Huang[1]

[1] *ShanghaiTech University, School of Physical Science and Technology, 393 Middle Huaxia Road, Pudong, Shanghai 201210, China*
[2] *Universität Münster, Organisch-Chemisches Institut, Corrensstrasse 40, Münster 48149, Germany*

11.1 Introduction

Scientists have embarked on a persistent exploration to identify environmentally sustainable reaction media that can facilitate organic synthetic transformations. This endeavor harmoniously aligns with the inexorable progression of green chemistry, representing a pivotal trend expected to significantly influence the future of chemical research and innovation [1]. Water, as a naturally abundant and environmentally friendly reaction medium, distinguishes itself with qualities such as affordability, safety, and nonflammability, attracting considerable attention in contrast to traditional organic solvents [2]. Despite the potential advantages of water as a green solvent, chemists are faced with the formidable challenge of developing new organic transformations that thrive in water [3]. In recent years, the field of carbene reactions in water has experienced steady growth, propelled by the emergence of new catalytic systems and innovative precursors. Water assumes a crucial role in enhancing reaction rates and outcomes favorably. However, a comprehensive overview highlighting recent advancements in carbene reactions in water remains notably scarce. This chapter primarily delves into synthetic transformations involving carbene intermediates in water. It is important to emphasize that we specifically exclude reactions conducted in mixed solutions of organic solvents and water.

11.2 Carbene Reaction in Water

Water, as an environmentally friendly green solvent, possesses numerous attractive physicochemical properties. Over the past decades, water-promoted approaches in carbene chemistry have gained significant attention [4]. With the increasing number of studies on carbene reactions with water as co-solvent or additive, one recent

Huan-Ming Huang and Xiaoye Yu are corresponding authors to this chapter.

Organic Transformations in Water: Principles and Applications, First Edition.
Edited by Gopinathan Anilkumar, Nissy Ann Harry, and Sankuviruthiyil M. Ujwaldev.

review has systematically summarized the recent developments in this emerging field [4]. Hence, in this chapter, we mainly focus on the chemistry of carbene reactions in pure water and simultaneously highlight the most important breakthroughs in this field, providing a comprehensive guide for reaction design. In this chapter, the organization of reactions is based on the different reaction types, and the last part deals with organocatalytic *N*-heterocyclic carbene (NHC) reactions in water.

11.2.1 Cyclopropanation and Cyclopropenization

The cyclopropanation of diazo compounds with alkenes is a general and powerful protocol for the synthesis of three-membered carbocycles [5]. For example, Carreira and coworker reported iron-catalyzed cyclopropanation with trifluoroethylamine hydrochloride 1 and olefins 2 in the presence of sodium nitrite in water (Scheme 11.1) [6]. When commercially available tetraphenylporphyriniron(III) complex Fe(TPP)Cl was employed as the catalyst, the cyclopropanation of styrene derivatives took place via *in situ* diazotation, affording chemo- and stereoselectively *trans*-cyclopropanes 3 in high yield. Later, the same group extended this methodology to dienes and enynes [7]. The reaction of trifluoroethylamine hydrochloride with 1-arylbuta-1,3-dienes or 1-aryl-but-3-ene-1-ynes in water gave the trifluoromethyl-substituted vinyl- and alkynylcyclopropanes in good yields with complete *trans* selectivity.

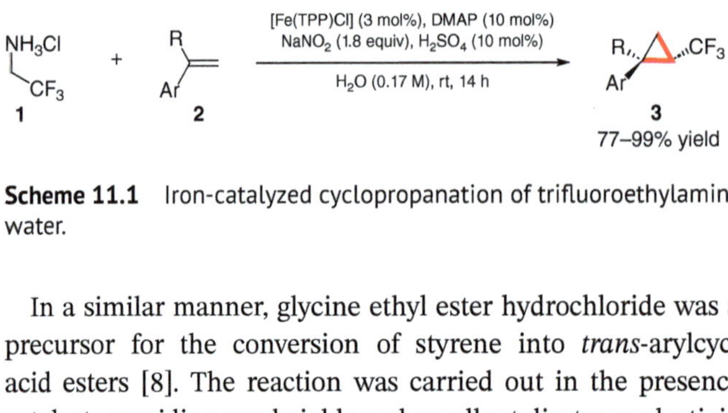

Scheme 11.1 Iron-catalyzed cyclopropanation of trifluoroethylamine hydrochloride in water.

In a similar manner, glycine ethyl ester hydrochloride was also used as carbene precursor for the conversion of styrene into *trans*-arylcyclopropanecarboxylic acid esters [8]. The reaction was carried out in the presence of Fe(TPP)Cl as a catalyst, providing good yields and excellent diastereoselectivity of target products. In addition, cyclopropanation of styrenes, dienes, and enynes by *in situ* formed diazomethane was achieved by Carreira (Scheme 11.2) [9]. In 2008, Simonneaux

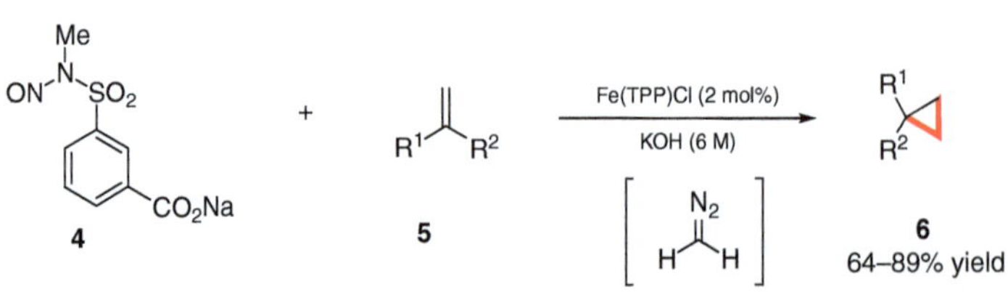

Scheme 11.2 Iron-catalyzed cyclopropanation with *in situ* generation of diazomethane in water.

and coworkers developed asymmetric cyclopropanation of styrenes with ethyldiazoacetate in water catalyzed by a water-soluble *para*-tetrasulfonated Halterman porphyriniron complex [10].

In 2010, Carreira and coworker disclosed rhodium-catalyzed cyclopropenation of trifluoroethylamine hydrochloride 7 with alkynes 8 in water medium (Scheme 11.3) [11]. In this transformation, trifluoroethylamine hydrochloride 7 was converted into the reactive diazoalkane in the presence of H_2SO_4 and $NaNO_2$, ultimately leading to a series of trifluoromethyl-substituted cyclopropenes 9 in 67–78% yields. Crucial to the success of this transformation was compatibility of rhodium catalyst [$Rh_2(esp)_2$] under the harsh condition.

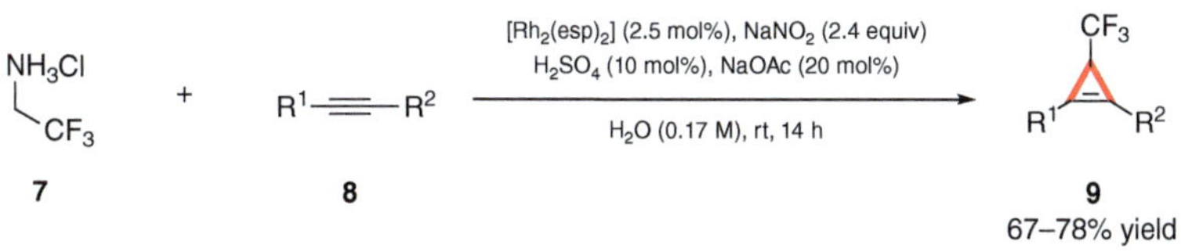

Scheme 11.3 Rhodium-catalyzed cyclopropenation in water.

Subsequently, visible light-promoted cyclopropenizations of *N*-tosylhydrazones 10 in water were investigated by Wu group (Scheme 11.4) [12]. The key step of the mechanism was the formation of key singlet carbene species under blue light irradiation. Furthermore, this protocol was also applied for X—H insertion reactions (X=N, S) and cyclopropanations.

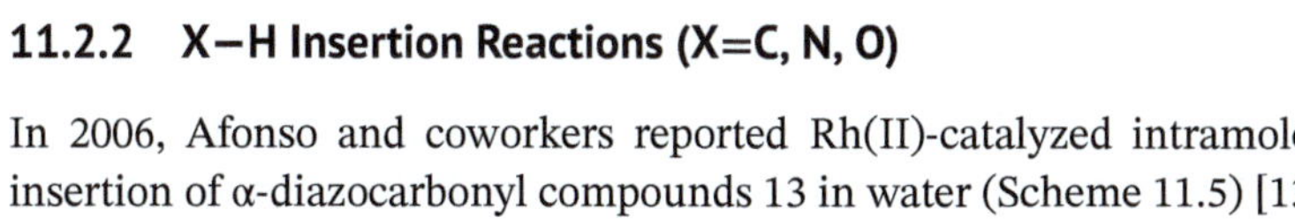

Scheme 11.4 Blue light-promoted cyclopropenizations in water.

11.2.2 X—H Insertion Reactions (X=C, N, O)

In 2006, Afonso and coworkers reported Rh(II)-catalyzed intramolecular C—H insertion of α-diazocarbonyl compounds 13 in water (Scheme 11.5) [13]. The study provided evidence that the selectivity of the C—H insertion relied on the structure

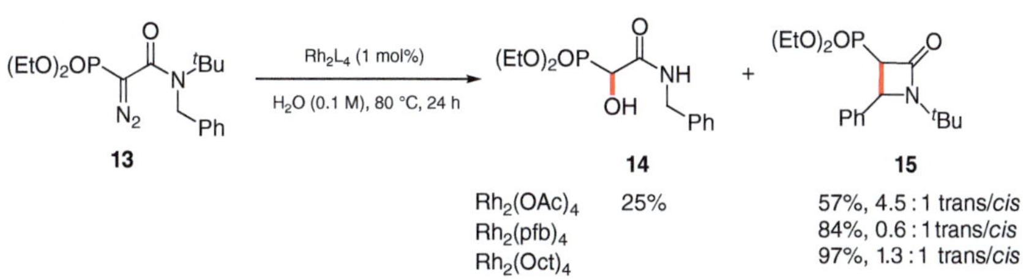

Scheme 11.5 Rh(II)-catalyzed intramolecular C—H insertion of diazo substrates in water.

of the catalyst and the hydrophobic nature of the amide substituents. Recently, Pérez and coworkers discovered a silver catalyst for the intermolecular insertion of carbene into alkane C—H bonds in water [14].

Sivasankar and coworker achieved an iridium-catalyzed acceptor/acceptor carbene insertion into N—H bonds of various anilines 17 (Scheme 11.6) [15]. Various α-amino diesters 16 could be obtained in good-to-excellent yields employing water as the solvent. And noncationic $[(COD)IrCl]_2$ was found to be an effective catalyst for carbene insertion of these less-reactive acceptor/acceptor precursors.

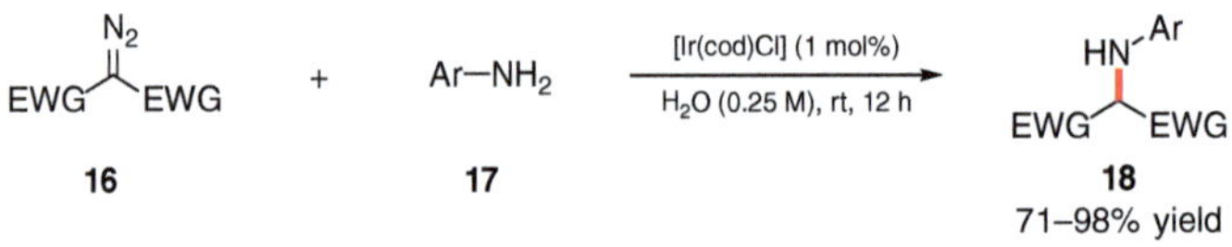

Scheme 11.6 Iridium-catalyzed carbene insertion into N–H bonds in water.

In addition, difluoromethylation of alcohols 20 using $TMSCF_2Br$ as the difluorocarbene reagent and KHF_2 as a mild activator in water was accomplished by Hu and coworkers (Scheme 11.7) [16]. This reaction with primary, secondary, and tertiary alcohols 20 performed very well, affording various difluoromethyl ethers 21 in moderate-to-good yields. In 2007, Wang and coworker demonstrated [2,3]-sigmatropic rearrangement of sulfonium ylides generated from Rh(II) carbene and sulfides in water [17].

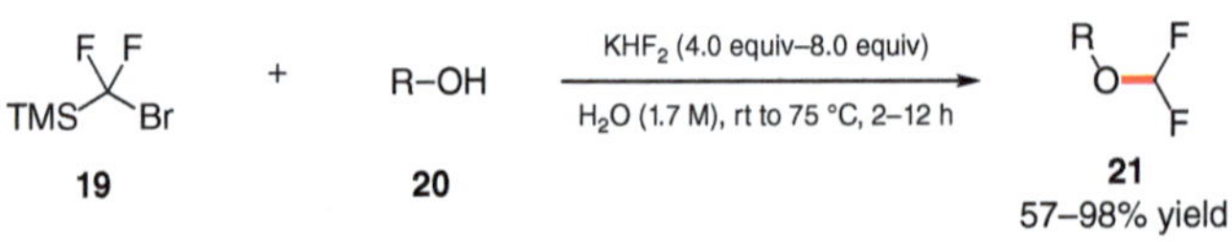

Scheme 11.7 Difluoromethylation of alcohols with $TMSCF_2Br$ in water.

11.2.3 Organocatalytic NHC Reactions

NHCs are not only used as ligands in the catalysis and organometallic chemistry but as powerful neutral nucleophiles to react with diverse electrophiles. In 2010, García-Tellado and coworkers developed a three-component protocol for the C2-functionalization of *N*-alkylated imidazoles in water (Scheme 11.8) [18]. The optimal order of addition was determined to be alkynoate 23, followed

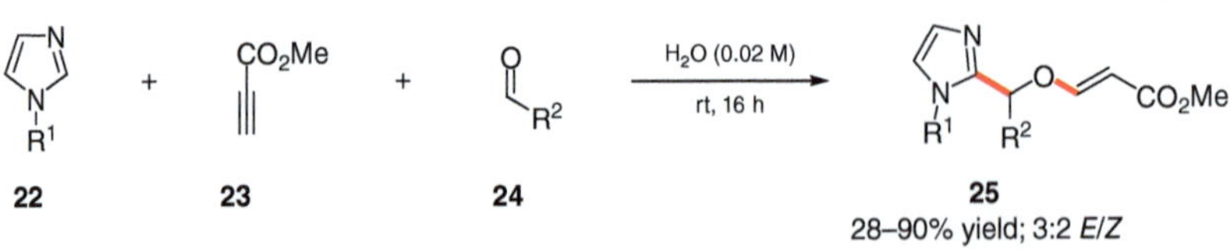

Scheme 11.8 C2-Functionalization of 1-substituted imidazoles in water.

by N-alkylated imidazole 22 and aldehyde 24, which was identified to be most optimal. The reaction underwent carbene–aldehyde addition, alkoxide–alkynoate addition, and protonation and hydrolysis to afford the desired products 25 with moderate-to-good yields.

Later, Chen and coworkers reported silver-promoted intramolecular addition reaction between NHC and unsaturated esters in water (Scheme 11.9) [19]. A wide range of 1H-imidazo[1,2-a]indole derivatives 27 could be obtained with moderate-to-excellent yields.

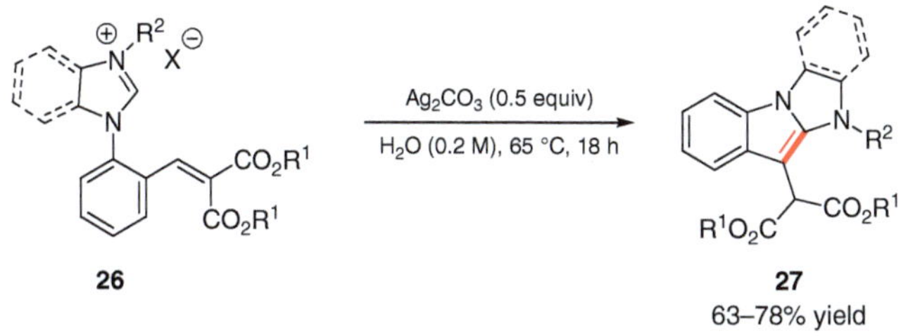

Scheme 11.9 Silver(I)-promoted intramolecular addition in water.

In 2018, Zhong and coworkers developed a chiral NHC-catalyzed benzoin condensation reaction in water (Scheme 11.10) [20]. A reasonable reaction mechanism involved the formation of an enol-type Breslow intermediate followed by addition of another benzaldehyde and a 1,4-H shift process with water as the proton shuttle.

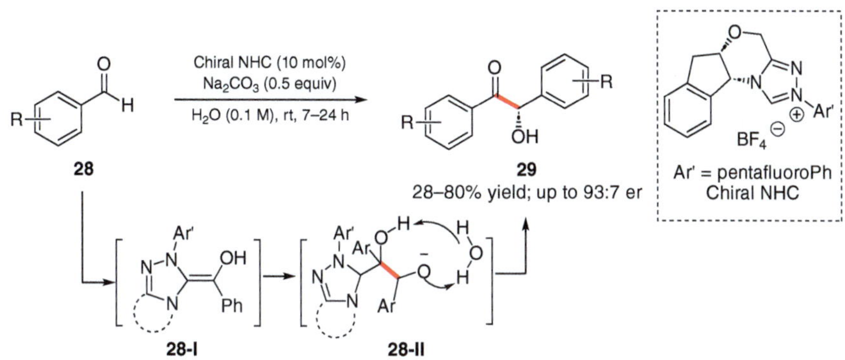

Scheme 11.10 N-Heterocyclic carbene-catalyzed asymmetric benzoin reaction in water.

In 2020, Desvergnes and coworkers reported NHC-catalyzed Stetter reaction of benzaldehyde with chalcone in an aqueous medium [21]. Recently, NHC-catalyzed aza-Michael addition reaction of β-aryl ethenesulfonyl fluorides 30 with non-activated aniline-type amines 31 was investigated by Bae group to access β-aminosulfonyl fluorides 32 (Scheme 11.11) [22]. The NMR analyses revealed NHC-activated amines 238 into a nucleophilic amine species.

Chi and coworkers demonstrated NHC-catalyzed reactions of enals 33 and enone 34 with water as the solvent (Scheme 11.12) [23]. By using 5 mol% of azolium

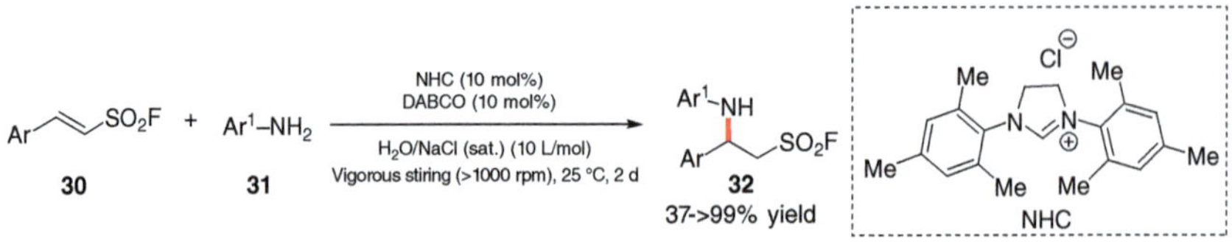

Scheme 11.11 Aza-Michael addition reaction in water.

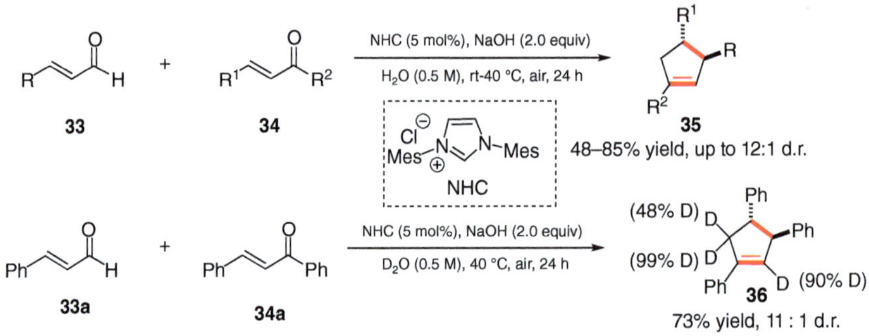

Scheme 11.12 NHC-catalyzed reactions of enalsin water.

NHC precatalyst and NaOH as the base, the reaction performed very well to afford cyclopentene products 35 in generally good yields and high diastereoselectivities.

Two years later, Snider disclosed a similar procedure using air-stable NHC precatalyst and NaOH as the base, resulting in the synthesis of 1,3,4-triphenylcyclopentene [24]. Afterwards, Ganesan and coworkers found that the reaction of isatins 37 with enals 38 as the electrophiles performed well in water medium with NaCl as an additive (Scheme 11.13) [25]. Numerous experiments demonstrated the use of brine as reaction medium greatly increased the reaction rate by hydrophobic hydration. This methodology could be further extended to synthesis of a variety of heterocycles and carbocycles.

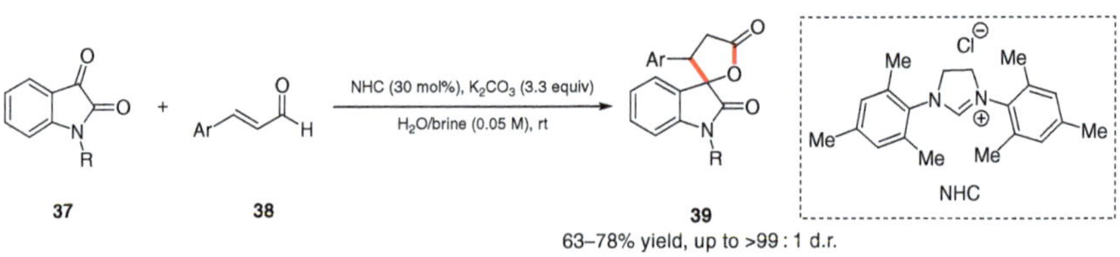

Scheme 11.13 Reaction of enals with isatins with the use of brine as reaction medium.

11.3 Conclusions

In summary, this chapter offers a comprehensive overview of carbene reactions in water, encompassing processes like cyclopropanation and cyclopropenization. Additionally, X—H insertion reactions and NHC-catalyzed reactions are thoroughly discussed, providing a thorough exploration of the advancements and intricacies within this domain. Simultaneously, this chapter emphasizes both the opportunities and challenges associated with utilizing water as a solvent. Detailed descriptions of the related mechanisms are provided to guide further developments in carbene reactions. Over the past few years, an increasing number of new reactions compatible with water have emerged, presenting a new prospect for green chemistry. These carbene reactions offer several advantages, including mild reaction conditions, a broad substrate scope, and straightforward operation.

Despite these advantages, there is still a need for significant improvement. Addressing challenges related to the reuse of aqueous systems and the treatment of wastewater is crucial. Furthermore, the development of asymmetric catalysis in this field requires further attention. The creation of new water-soluble mediators to broaden the scope of reaction types is also necessary. In the context of carbene reactions in water, the development of new carbene catalysts and the exploration of diverse reaction types are evident.

With the growing interest in green chemistry, the pursuit of developing milder and more sustainable processes for the synthesis of natural products and complex molecules remains ongoing. This chapter aims to inspire more chemists to explore novel and relevant strategies for advancing the field.

References

1 (a) Li, C.-J. (2016). *Green Chem.* 18: 1836–1838. (b) Sheldon, R.A. (2018). *ACS Sustain. Chem. Eng.* 6: 32–48. (c) Li, C.-J. and Anastas, P.-T. (2012). *Chem. Soc. Rev.* 41: 1413–1414. (d) Simon, M.-O. and Li, C.-J. (2012). *Chem. Soc. Rev.* 41: 1415–1427.

2 Anastas, P. and Eghbali, N. (2010). *Chem. Soc. Rev.* 39: 301–312.

3 (a) Kitanosono, T., Masuda, K., Xu, P., and Kobayashi, S. (2018). *Chem. Rev.* 118: 679–746. (b) Li, C.-J. and Chen, L. (2006). *Chem. Soc. Rev.* 35: 68–82.

4 Levin, E., Ivry, E., Diesendruck, C.E., and Lemcoff, N.G. (2015). *Chem. Rev.* 115: 4607–4692.

5 Ebner, C. and Carreira, E.M. (2017). *Chem. Rev.* 117: 11651–11679.

6 Morandi, B. and Carreira, E.M. (2010). *Angew. Chem. Int. Ed.* 49: 938–941.

7 Morandi, B., Cheang, J., and Carreira, E.M. (2011). *Org. Lett.* 13: 3080–3081.

8 Morandi, B., Dolva, A., and Carreira, E.M. (2012). *Org. Lett.* 14: 2162–2163.

9 Morandi, B. and Carreira, E.M. (2012). *Science* 335: 1471–1474.

10 Nicolas, I., Maux, P.L., and Simonneaux, G. (2008). *Tetrahedron Lett.* 49: 5793–5795.

11 Morandi, B. and Carreira, E.M. (2010). *Angew. Chem. Int. Ed.* 49: 4294–4296.

12 Yan, K., He, H., Li, J. et al. (2021). *Chin. Chem. Lett.* 32: 3984–3987.

13 Candeias, N.R., Gois, P.M., and Afonso, C.A.M. (2006). *J. Org. Chem.* 71: 5489–5497.

14 Álvarez, M., Gava, R., Rodríguez, M.R. et al. (2017). *ACS Catal.* 7: 3707–3711.

15 Ramakrishna, K. and Sivasankar, C. (2017). *Org. Biomol. Chem.* 15: 2392–2396.

16 Zhang, R., Ni, C., Xie, Q., and Hu, J. (2020). *Tetrahedron* 76: 131676–131681.

17 Liao, M. and Wang, J. (2007). *Green Chem.* 9: 184–188.

18 Cruz-Acosta, F., Armas, P., and García-Tellado, F. (2010). *Synlett* 16: 2421–2424.

19 Shang, X., Chen, C., Qiu, H., and Chen, W. (2014). *Tetrahedron* 70: 3073–3077.

20 Yan, J., Sun, R., Shi, K. et al. (2018). *J. Org. Chem.* 83: 7547–7552.

21 Debiais, M., Hamoud, A., Drain, R. et al. (2020). *RSC Adv.* 10: 40709–40718.

22 Park, J.H., Lee, S.B., Koo, B.J., and Bae, H.Y. (2022). *ChemSusChem* 15: e202201000.

23 Leong, W.W.Y., Chen, X., and Chi, Y.R. (2013). *Green Chem.* 15: 1505–1508.

24 Snider, B.B. (2015). *J. Chem. Educ.* 92: 1394–1397.

25 Suresh, P., Thamotharan, S., and Ganesan, S.S. (2018). *Catal. Commun.* 111: 47–51.

12

Nucleophilic Addition and Substitution Reactions In-Water

Panackalchirayil S. Devi[1], *Pulluparambil X. T. Rinu*[1], *and Gopinathan Anilkumar*[1,2]

[1] *Mahatma Gandhi University, School of Chemical Sciences, Priyadarsini Hills PO, Kottayam, Kerala 686560, India*
[2] *Mahatma Gandhi University, Institute for Integrated programs and Research in Basic Sciences (IIRBS), Priyadarsini Hills PO, Kottayam, Kerala 686560, India*

12.1 Introduction

As the global economy continues to burgeon, the spotlight on environmental issues has intensified. This has sparked considerable interest in utilizing water-based protocols for synthetic chemistry. Water emerges as a prime candidate in an organic chemist's solvent kit, not only due to its widespread availability but also because of its exceptional environmental credentials. It is inexpensive, nontoxic, nonflammable, highly stable, and easy to handle, making it an inherently quintessential "green" solvent choice. Both heterogeneous "on-water" and homogeneous "in-water" reactions and micellar catalysis driven by the hydrophobic effect have been successfully developed, showcasing the versatility of water as a solvent.

Recent advancements in nucleophilic addition and substitution reactions conducted in water have revolutionized the field of organic synthesis. These developments have been driven by the increasing demand for sustainable and environmentally friendly synthetic methodologies. The key advancements in this area include the design of novel catalysts and reaction conditions optimized for aqueous environments, allowing for the efficient and selective functionalization of a wide range of substrates. Moreover, innovative strategies such as micellar catalysis and flow chemistry have further enhanced the efficiency and scalability of nucleophilic reactions in water, enabling their application in industrial-scale synthesis. Furthermore, these advancements have not only expanded the synthetic toolbox but also contributed to the development of biocompatible reactions for applications in medicinal chemistry and drug discovery. Overall, recent progress in nucleophilic addition and substitution reactions in water holds immense promise for advancing sustainable chemistry and addressing environmental challenges in the field of organic synthesis.

Organic Transformations in Water: Principles and Applications, First Edition.
Edited by Gopinathan Anilkumar, Nissy Ann Harry, and Sankuviruthiyil M. Ujwaldev.

12.2 Nucleophilic Addition Reactions

Nucleophilic addition reactions entail the interaction between a compound containing an electrophilic triple or double bond and a nucleophile, resulting in the cleavage of the bond. Additionally, when nucleophiles are added to triple or double bonds involving carbon and a heteroatom, diverse types of polar bonds, such as —C≡N or —C=O, display significant variation in nucleophilic addition reactions. The introduction of electron-withdrawing groups onto multiple-bonded carbon atoms reduces the electron density of carbon, thereby facilitating nucleophilic addition. These reactions are viable in both aqueous-organic and solely organic solvents, with an increasing emphasis on conducting reactions in-water or water–organic mixtures. This discussion delves into the realm of nucleophilic addition reactions, particularly focusing on those executed in-water or water–organic environments.

12.2.1 Knoevenagel Reactions

The Knoevenagel reaction is the modified version of Aldol reaction. Here, C–C bonds are formed by the nucleophilic addition of a compound having active hydrogen to a ketone or aldehyde by utilizing a basic catalyst [1]. This reaction proceeded in water media and produced adducts in reasonable yields. The recyclability of catalysts, the short reaction time, and the requirement of only low temperature are the main peculiarities of this approach.

Morpholine-catalyzed synthesis of (*E*)-ethyl 2-cyano compounds in water was developed by Menegatti and coworkers [2]. This green protocol proceeded with the Knoevenagel reaction of aryl aldehydes with ethyl cyanoacetate. This approach rendered the expected products in moderate to excellent yields without the aid of any organic solvent.

Chang and coworkers achieved an effective green protocol for the synthesis of novel varieties of Knoevenagel adducts using water as the solvent (Scheme 12.1a) [3]. Here, the Knoevenagel reaction was successfully carried out between aldehydes or ketones with activated methylene compounds in the presence of [Nmm-PDO][OAc] in water at room temperature and delivered the expected products in good to excellent yields. The activated methylene compounds, like 2-arylacetonitrile, β-diketone, alkyl cyanoacetate, cyanoacetamide, benzothiazole-2-acetonitrile, malononitrile, and thiazolidinedione, were found to be tolerated well in the reaction. However, thiazolidinedione underwent this reaction under a solvent-free environment at 60 °C utilizing the [Nmm-PDO][Gly] catalyst. The product separation via simple filtration techniques and the recyclability of the catalyst are the main benefits of this protocol.

In 2018, Xu and coworkers developed *N*-methyl piperidine-based ionic liquids ([Nmp-PDO][OAc]) as an efficient recyclable catalyst that can be used for both the Knoevenagel reaction and the hetero-Michael addition reaction in H_2O as the solvent (Scheme 12.1b,c) [4]. Interestingly, this is the first report on the Michael addition of α,β-unsaturated amides in H_2O using [Nmp-PDO][OAc] as catalyst to

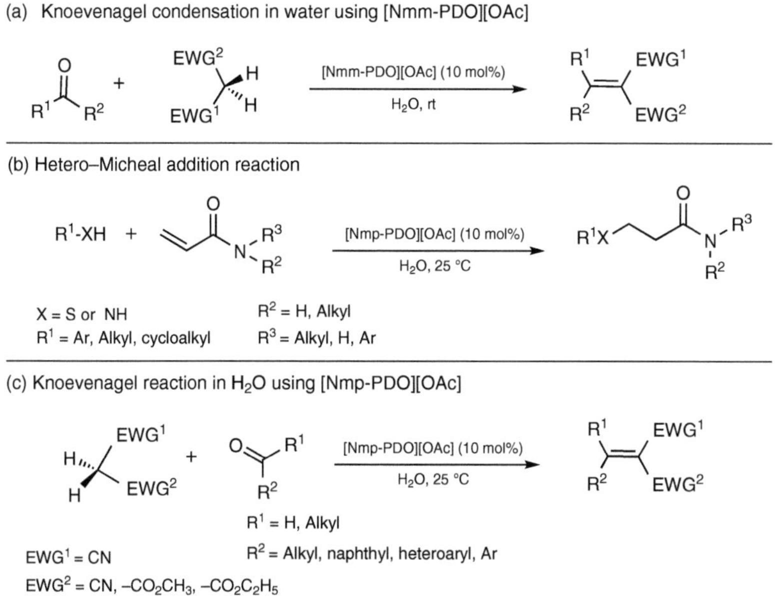

Scheme 12.1 [Nmm-PDO][OAc] and [Nmp-PDO][OAc]-catalyzed reactions in-water.

afford the corresponding expected products in high yields. Various *N*-substituted acrylamides were tolerated well in this protocol. In Knoevenagel condensation, several α,β-unsaturated, heteroaryl, aryl aldehydes, and aliphatic ketones were found compatible with the reaction.

Xi and coworkers explored a Knoevenagel condensation reaction of aryl aldehydes with ethyl cyanoacetate using N-(2-hydroxy-ethyl)-DBU bromide ([HyEtDBU]Br)-H_2O-DABCO catalytic system at room temperature [5]. Substituted aryl aldehydes with electron-rich or electron-deficient groups, furaldehyde, naphthaldehyde, and anthracene carboxaldehyde were tolerated well in this protocol. The reaction proceeded at a short reaction time, imparting excellent yields with high catalytic recyclability.

Recently, Das and coworkers developed a synthetic approach toward electrophilic olefins from the Knoevenagel condensation reaction [6]. The reaction between organocyano compounds with carbonyl compounds catalyzed by 20 mol% Mg powder in water afforded the expected products in good yields (Scheme 12.2). The versatility of this protocol was demonstrated through the execution of a one-pot Knoevenagel–Michael reaction and a sequential Knoevenagel reaction followed by chemoselective reduction in alkenes. These methodologies facilitated the efficient synthesis of pharmaceutically relevant cyano compounds, highlighting the protocol's practical utility in medicinal chemistry applications. The major benefits of the protocol include a solvent-free process, eco-friendly conditions, together with a wide substrate scope, and the recyclability of the catalyst.

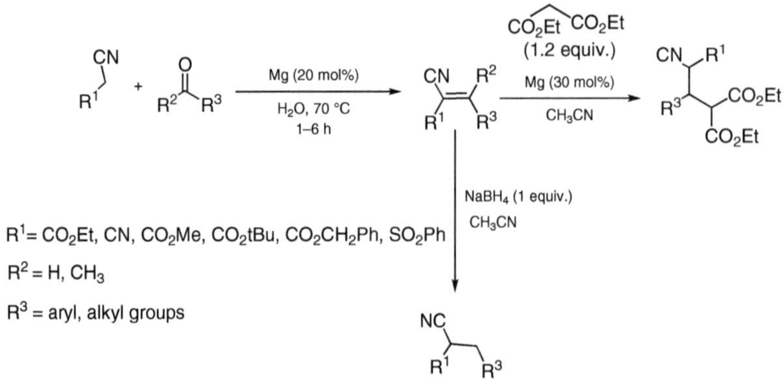

Scheme 12.2 Pharmaceutically relevant cyano compounds were synthesized *via* the one-pot Knoevenagel–Michael reaction as well as Knoevenagel-chemoselective reduction.

12.2.2 Michael Addition Reaction

A Michael donor, such as an enolate or other nucleophile, and a Michael acceptor, typically an α,β-unsaturated carbonyl compound, react to form a Michael adduct by forming a C–C bond at the β-carbon of the acceptor. This reaction is called the Michael 1,4-addition or Michael reaction [7]. Recently, chiral surfactant-type, metal enzyme, lipopeptides, and amino acid-based catalysts have been used in water-mediated Michael reactions. This protocol was performed in water or a mixture of organic solvent/water media. There are some advantages to this approach, such as good yield, mild reaction conditions, and good stereoselectivity.

Metallomicelles, novel chiral surfactant-type catalysts, were synthesized utilizing Schiff-based ligands, and it was developed by Wang and coworkers [8]. This catalyst showed high catalytic performance in the Michael addition of 2-enoylpyridine 1-oxides and indoles in-water as the solvent. This protocol was found successful for a wide range of indole and 2-enoylpyridine 1-oxides to the corresponding Michael adducts with high enantioselectivity and yields. The heterocyclic, polyclic, aryl *ortho*-, and *meta*-substituted 2-enoylpyridine 1-oxides underwent the Michael reaction smoothly. Also, this methodology was found highly effective for indoles with electron-donating and halogen substituents, whereas the strategy was entirely inert in electron-poor indoles.

Wang and coworkers established nitro-bearing pyrrolidines *via* the enantioselective Michael addition reaction (Scheme 12.3a) [9]. In this synthetic methodology, nitroalkanes underwent Michael addition with 2, 3-dioxopyrrolidines in the presence of 10 mol% $Cu(OTf)_2$, 10 mol% ligand L_1, 30 mol% *N*-ethylmorpholine, 10 mol% SLS in $H_2O/CHCl_3$ at 25 °C, which provided the desired products in good yield and diastereoselectivity. *N*-Aryl and *N*-alkyl 2,3-dioxopyrrolidines were well tolerated in this strategy.

In 2022, a novel route for spirooxindole synthesis *via* the Michael/hemiketalization reaction in $CHCl_3$:H_2O media was reported by Wang and coworkers [10]. Here, 4-hydroxycoumarins reacted with unsaturated α-ketoesters (derived from isatin) in the presence of triethylamine, $Cu(OTf)_2$, ligand L_2, and SDS in $CHCl_3$: H_2O

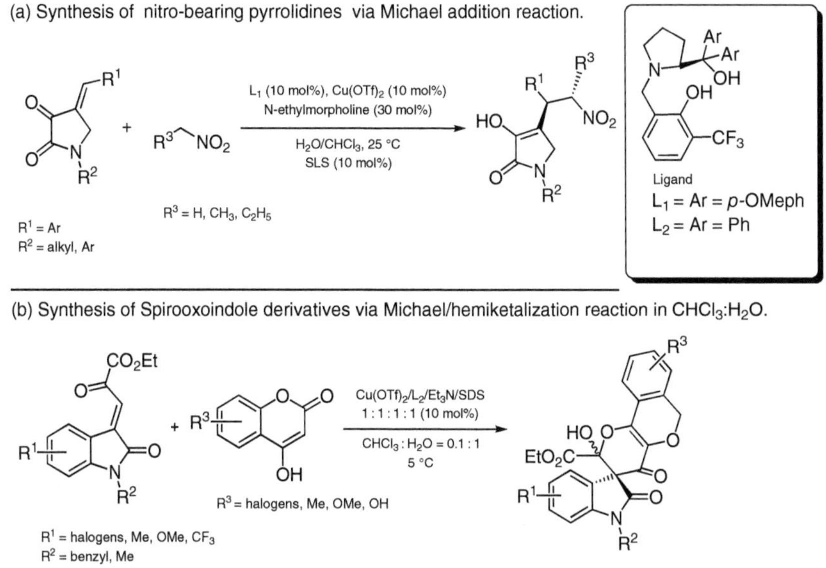

Scheme 12.3 Cu-catalyzed Michael addition reaction in aqueous media.

at 5 °C and delivered spirooxindoles in excellent yield and enantioselectivity. Various *N*-substituted isatin-derived α-ketoesters and 4-hydroxycoumarins bearing electron-poor and electron-rich substituents performed well in this protocol. The strategy showed many advantages, including the eco-friendly nature of the reaction, a wide substrate scope, and high yields of the product (Scheme 12.3b).

Itoh and coworkers synthesized highly lipophilic-substituted 3-decyl-β-proline and also studied the catalytic activity of this compound in the Michael addition reaction in aqueous media [11]. This catalyst was found to be efficient when one equivalent of nitrostyrene was reacted with two equivalents of carbonyl compound in the presence of a saturated aqueous KCl solution at room temperature for five hours, which afforded the Michael adducts. Low catalyst loading, mild reaction conditions, excellent yield, and diastereoselectivity were the main highlights of this protocol.

Paixão and coworkers developed an *N*-alkylated lipopeptide and studied its catalytic activity in the Michael addition reaction of several aldehydes to *E*-nitrostyrene in aqueous media [12]. The nitroalkenes having alkyl, heterocyclic, and also aryl substituents with electron-rich and electron-deficient moieties resulted in moderate to excellent yields of the expected products. The aldehydes like hexanal, isovaleraldehyde, and (*R*)-(+)-citronellal tolerated this strategy. The positive aspects of this protocol include low catalyst loadings, free of additives, high yield, good recyclability of catalyst, excellent diastereoselectivity, and enantioselectivity.

Ryu and coworkers developed a novel *L*-proline-derived organocatalyst, 1-(3,5-Bis (trifluoromethyl)phenyl)-3-((*R*)-phenyl((*S*)-pyrrolidin-2-yl)methyl)urea (PHU) for asymmetric Michael addition of various nitroolefins and carbonyl compounds [13]. This catalyst was also found efficient for this protocol in the presence of 10 mol%

of PHU and 2.5 mol% trifluoroacetic acid (TFA) in H_2O at 25 °C that afforded the expected products in excellent yield and stereoselectivity. Under these reaction conditions, various aldehydes, cycloketones, and arylnitroalkenes reacted well. The significance of this protocol was established by the synthesis of a new histamine receptor agonist, Sch 50971.

In the Aza-Michael reaction, nitrogen-based nucleophiles react with electron-deficient C=C bonds (Michael acceptor) to form C–N bonds. Under the influence of a suitable catalyst, this reaction proceeds efficiently in water media. For example, Tyagi and coworkers demonstrated the first aza-Michael addition of substituted aryl and heteroaryl amines to methyl vinyl ketone, catalyzed by α-amylase in H_2O [14]. They also synthesized an efficient and reusable catalyst, α-amylase@CuNPs that was used for this protocol. Additionally, the research team illustrated the synthetic applicability of a cascade aza-Michael/aldol/aromatization process by producing biologically relevant 3-acetyl quinoline.

In the thia-Michael reaction, a variant of Michael addition, a thiol molecule acts as the Michael donor (nucleophile) and reacts with an electron-poor C=C bond. Here also, by choosing the appropriate catalyst, the reaction could be performed in water.

Hoven and coworkers devised poly(*N*-(benzyl acrylamide))-ran-poly(*N*-(2-hydroxypropyl)acrylamide) (PBAM-ran-PHPAM) that was utilized for the sulfa-Michael reaction in H_2O as the solvent [15]. This protocol was successfully achieved by reacting thiol derivatives with various β-nitrostyrenes in the presence of $PBAM_{26}$-ran-$PHPAM_{74}$ micelles at room temperature for 24 hours. Excellent yields, mild reaction environments, and the reusability of polymeric micelles up to 10 times were the main attractions of this strategy.

12.2.3 Mannich Reaction

The Mannich reaction, often referred to as the amino alkylation reaction, is the condensation of an enolizable carbonyl molecule with a 1° or 2° amine, a non-enolizable aldehyde, and NH_3 to produce an amino carbonyl compound, which is also known as the Mannich base [16]. This compound plays a major role in the synthesis of natural products. Recently, metals, anionic surfactants, and acids like calix[4]arene tetracarboxylic acids and polyacrylic acids (PAAs) were used as catalysts in water-mediated Mannich reactions and provided adducts in reasonable yields.

Mukhopadhyay and coworker reported a calix[4]arene tetracarboxylic acid (C4A4)-catalyzed one-pot synthetic methodology for the preparation of spiro [dihydropyridine-oxindoles] utilizing water as the reaction media through a multi-component Mannich-type reaction [17]. This green protocol proceeds by reacting anilines, *N*-substituted isatins, and 1,3-diketones in the presence of a 5 mol% C4A4 catalyst in H_2O at 80 °C for 12 hours to afford the expected products in high yields. A range of anilines with electron-withdrawing and electron-donating substituents underwent the reaction successfully. This catalytic system showed advantages such as catalyst recyclability, eco-friendliness, and enhanced product yield.

Chakraborti and coworkers established an eco-friendly multicomponent reaction for the preparation of tetrahydropyridines using sodium dioctyl sulfosuccinate

(SDOSS) as the anionic surfactant [18]. Substrates like amine, β-ketoester, and benzaldehyde in the presence of 10 mol% SDOSS in H_2O at room temperature for 4–10 hours provided the desired products in good yields. The proposed mechanism suggested that the protocol proceeded *via* tandem intra- and intermolecular Mannich-type reaction routes. The strategy showed many advantages, like increased product yield, short reaction time, eco-friendly process, room temperature reaction, and needed only simple purification techniques like crystallization to purify the product.

Gao and coworkers developed a PAA-catalyzed multicomponent Mannich reaction for the synthesis of β-amino ketones in water media [19]. This protocol was achieved by reacting substrates such as aldehydes, anilines, and ketones under optimized conditions [(PAA) (7.5 mol%), H_2O, 25 °C, 6–12 hours] to afford the required products in high yields. Aryl amines and aryl aldehydes containing either electron-rich or electron-deficient substituents were well tolerated in the reaction. Furthermore, this methodology was also found effective for aliphatic and heteroaldehydes.

Recently, Naseri and coworkers disclosed the synthesis of two novel catalystssuch as HybPOM/HPW_{12} and Go/Fe_3O_4/HybPOM/HPW_{12} via the immobilization of heteropolyacids supported on polyoxometalate organic–inorganic nanoparticles [20]. At ambient temperature, these recyclable catalysts were found to have good catalytic activity in one-pot Mannich reactions between aldehydes, ketones, and amine in water, affording a good yield of β-amino carbonyl compounds (Scheme 12.4).

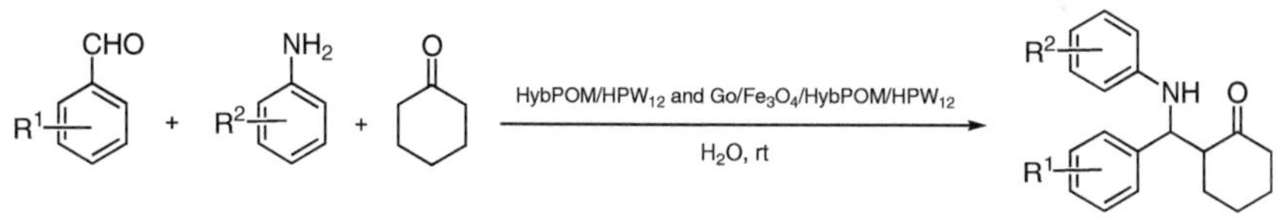

Scheme 12.4 Synthesis of β-amino carbonyl compounds via HybPOM/HPW_{12} and Go/Fe_3O_4/HybPOM/HPW_{12}-catalyzed Mannich reaction.

The piperdin-2-yl-indoles were synthesized via an eco-friendly Mannich reaction mediated by MsOH and were devised by Lu and coworkers [21]. The desired products have the core skeleton of certain alkaloids. The protocol was carried out between indoles and aliphatic cyclic imines having no α-substitution, using 100 mol% MsOH and 30 mol% tetra-*n*-butylammonium iodide (TBAI) in H_2O at 60 °C for 18 hours. 5-Fluoroindole, 5-chloroindole, and isoquinoline derived and five-membered imines are compatible with this protocol, affording reasonable yields.

A wide range of bioactive [22] and naturally occurring compounds [23] have the isoindolinone framework. The one-pot multicomponent Mannich-type reactions were reported with 1°-amines, 2-naphthols, and 2-formyl benzoic acids, affording isoindolinones in moderate to good yields [24]. This catalyst-free strategy was conducted in water at 120 °C. Amines like benzyl amines, 4-phenylbutyl, 1-phenylethyl, and furfuryl amines were compatible with this strategy, whereas the sterically hindered *tert*-butylamine did not give the desired products. Besides,

2-naphthol-bearing $-CO_2Me$ and Br groups at its sixth position, as well as −CN and $-NO_2$ substituents at various positions (6-nitro, 4-cyano) on the aryl ring of phthalaldehydic acid, were tolerated well. Transition metal-free strategy, eco-friendly approach, one-pot multicomponent reaction, and water as solvent are the major features of this protocol.

Zarea and coworkers reported on the novel *N*-alkylation of indoles in aqueous microdroplets [25]. The alkylated indoles were formed by a multicomponent Mannich-type approach in both bulk as well as microdroplet phases. Aldehydes, indole, and amines underwent the conventional bulk method, which provided C-alkylated products, whereas a microdroplet Mannich reaction with the same starting materials offered N-alkylated indoles in moderate yields. The reaction was found to be highly chemoselective and worked well with functionalized aryl aldehydes and indoles.

Ye and coworkers devised a highly selective and efficient method for producing 3,4-dihydro-2-quinolones through Cu-catalyzed alkyne hydration followed by intramolecular Mannich reaction, and this protocol was applied to carbonyl-ynamides [26]. Moreover, when carbonyl-ynamides underwent Zn-catalyzed aldol reactions in water, they also afforded 3,4-dihydro-2-quinolones (Scheme 12.5). The formal synthesis of (±)-martinellic acid demonstrated the synthetic relevance of this protocol.

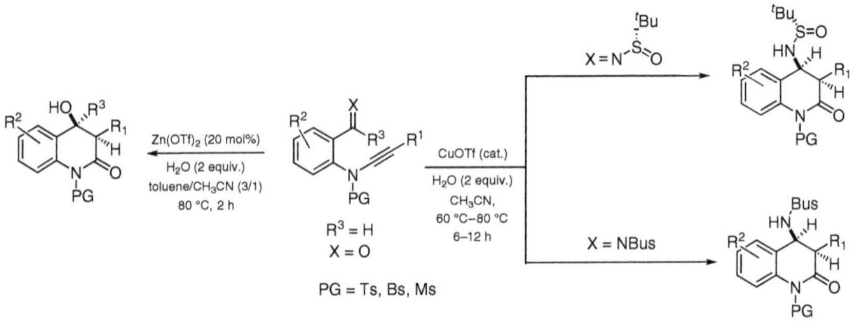

Scheme 12.5 Synthesis of 3,4-dihydro-2-quinolones using Cu-catalyzed alkyne hydration/intramolecular Mannich reaction or Zn-catalyzed aldol reaction in H_2O media.

12.2.4 Aldol Reaction

The aldol reaction involves the formation of a C–C bond by the reaction of an aldehyde or ketone having an α-hydrogen in the presence of a base to produce the corresponding β-hydroxyaldehyde or β-hydroxyketone [27]. This reaction was also carried out in aqueous media, and the corresponding aldol adducts were obtained in good yield with high enantio- and diastereoselectivity. Here, high temperature is not required in most of the water-mediated aldol reactions, which is the main merit of this strategy.

At room temperature, a semi-heterogenous polydiacetylene-proline (pDApro)-catalyzed aldol reaction between several aldehydes and cyclohexanone using water as a solvent was devised by Doris and coworkers [28]. Here, aryl aldehydes bearing electron-withdrawing groups afforded good yields of the aldol adducts, whereas benzaldehyde and para-methoxy benzaldehyde provided low yields. The low catalyst loading, use of water as the media, reusability of the catalyst, good yields, as well as good enantio- and diastereoselectivity of the products are the main peculiarities of this protocol.

In 2019, Wang and coworkers discussed the vinylogous aldol reaction of α,α-dicyanoolefins with silyl glyoxylates catalyzed by DABCO in water media [29]. This strategy tolerates silyl glyoxylates bearing *t*-butyl ester, cyclohexyl-based ester, and electron-rich groups, whereas both electron-poor and electron-rich substituents in the aryl rings of α,α-dicyanoolefins smoothly reacted well and afforded α-hydroxysilanes in high yields.

The water-mediated aldol protocol is an inevitable part of the construction of C–C bonds in various natural products and bioactive compounds. For example, 3-hydroxy oxindole, a well-known bioactive compound, was synthesized using WERSA (water extract of rice straw ash) by Chowhan and coworkers [30]. The aldol reaction of isatin with β-ketoacid was performed via WERSA-catalyzed aldol reaction. A wide range of substrates were explored, which revealed that the C-5 positions of isatin-bearing electron-rich and electron-poor substituents were compatible with this methodology. In the case of β-ketoacid, the presence of electron-donating moieties at the C-4 and C-3 positions of its aryl ring afforded satisfactory yields of the expected products in comparison with the electron-deficient ones. It was noticed that the electronic effects offered by functional groups in the various positions of both isatin and β-ketoacid played a significant role in determining the yield of the reaction.

Recently, a cobalt nanocomposite (CoOx@NC-800) catalyst was found to be effective in the chemoselective hydrogenation of α,β-unsaturated carbonyls in water [31]. This nanocomposite catalyzed one-pot cross-aldol reaction followed by hydrogenation has been utilized for the synthesis of saturated carbonyls from aldehydes and ketones, and the strategy was applied in the synthesis of Loureirin A, a well-known bioactive compound.

Sumathi and coworker reacted 2-ethoxy-2H-chromene chalcones with barbituric acid and pyrazolin-5-one in water media, catalyzed by *L*-proline, to produce 4-hydroxy-4H-chromene barbiturates and 4-hydroxy-4H-chromene-pyrazoles in good yields [32]. This protocol involves the removal of the $-OC_2H_5$ group, Michael addition, aldol condensation, and hydrolysis reactions. High yields and a short reaction time are the merits of this eco-friendly approach (Scheme 12.6).

Wang and coworkers synthesized highly electron-poor molecular cage catalysts that were used in the aldol decarboxylative reactions [33]. Here, the reaction was carried out between half thioesters of malonic acid and aryl aldehydes in the presence of a 5 mol% cage catalyst in water at room temperature that afforded the expected products in good yields.

Kim and coworkers synthesized a variety of α-hydroxy-α-nitromethylamides through a water-mediated nitroaldol reaction (the Henry reaction) [34]. The reaction between nitromethane and α-keto amide was catalyzed by *L*-arginine in H_2O at 25 °C for six hours, which afforded the expected products in high yields.

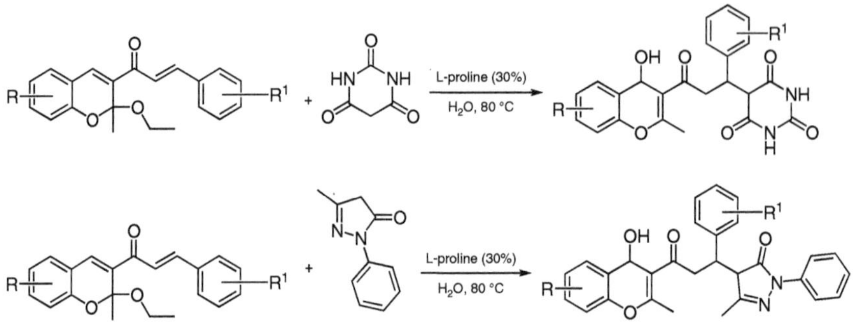

Scheme 12.6 Synthesis of 4-hydroxy-4H-chromene barbiturates and 4-hydroxy-4H-chromene-pyrazoles using *L*-proline as a catalyst.

The highlights of this protocol include the wide substrate scope, eco-friendliness, as well as mild reaction environments.

The asymmetric aldol reaction of ketones with isatins, described by Da and coworkers is one of the best routes for the synthesis of 3-hydroxy-2-oxindoles [35]. This reaction was found to take place at room temperature in the presence of a bifunctional chiral amine catalyst in water media. The wide substrate scope, recyclability of the catalyst, cost-effectiveness, and eco-friendly nature of the protocol are the main attractions of this strategy. The position and electronic effects of the substituents on isatin have a great influence on the stereoselectivity of the products. Cyclic ketones like tetrahydro-4*H*-thio-pyran-4-one, 3-methyl, and 4-methyl cyclohexanone were tolerated in this protocol. A gram-scale reaction was also done to ensure the scalability of the reaction, using cyclohexanone and isatin as the starting materials.

12.3 Nucleophilic Substitution Reactions

Nucleophilic substitution reactions represent fundamental processes in organic chemistry, wherein a nucleophile replaces a leaving group in a molecule. These reactions are crucial for the construction of new bonds and for the modification of organic compounds. Nucleophilic substitution reactions find widespread applications in organic synthesis, pharmaceuticals, and materials science, making them one of the standing pillars of organic synthesis. Nucleophilic substitution reactions are mainly classified into S_N1 (substitution nucleophilic unimolecular), S_N2 (substitution nucleophilic bimolecular), and S_NAr (nucleophilic aromatic substitution) reactions. Performing these reactions in aqueous media faces many challenges, and the scientific community is constantly working to overcome them. The past few years have witnessed a dramatic evolution in this field by designing new catalysts, using surfactants, phase transfer catalysts, flow chemistry, catalyst-free reflux protocols, etc.

12.3.1 Nucleophilic Aromatic Substitution Reactions (S_NAr)

Nucleophilic aromatic substitution reactions (S_NAr) can modify aromatic systems with precision and play a pivotal role in organic synthesis. In these reactions, an aromatic ring bearing an electron-withdrawing group acts as the stage for a nucleophile's attack, leading to the substitution of the aromatic ring. Unlike the conventional nucleophilic substitutions occurring at saturated carbon centers, S_NAr reactions occur at sp^2-hybridized carbon atoms within aromatic systems. Performing S_NAr reactions in aqueous environments poses significant challenges due to the comparatively low reactivity of many nucleophiles and substrates in water. Innovative approaches such as micellar catalysis, biocatalysis, and polymer-based hydrophobic pockets have been harnessed to enhance the reactivity and selectivity of S_NAr reactions in water. This transformative shift toward conducting S_NAr reactions in water not only overcomes environmental concerns associated with traditional solvents but also expands the synthetic toolbox to embrace sustainable methodologies.

A highly greener S_NAr reaction in micellar medium using oxygen, nitrogen, and sulfur nucleophiles was reported by Lipshutz and coworkers at near room temperature conditions [36]. Heteroaromatic and aromatic substrates smoothly underwent this micellar catalysis using 2 wt % TPGS-750-M as the surfactant and K_3PO_4 base (Scheme 12.7). The key advantages of this protocol include a 1 : 1 ratio of reaction partners, easy recyclability of the catalytic system, replacing aprotic solvents using nanomicellar reactors in an aqueous medium, a low E factor, etc.

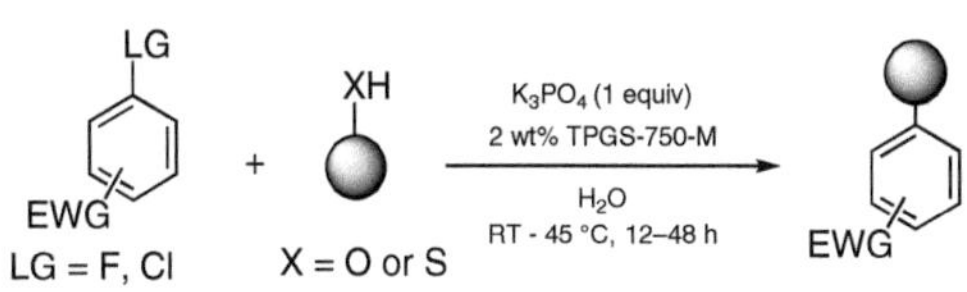

Scheme 12.7 S_NAr reaction using TPGS-750-M and K_3PO_4 in water.

A base-free and metal-free aqueous protocol for the construction of pharmaceutically relevant tetrahydroquinoline spiro motifs was achieved *via* a cascade reaction involving S_NAr [37]. Various active methylene compounds and 2-(pyrrolidin-1-yl) benzaldehydes were the building blocks of this protocol, which reacted on water at room temperature for 48 hours. It is a four-step cascade protocol involving an initial S_NAr, followed by Knoevenagel condensation and [1,5]-hydride transfer, and a final cyclization. This innovative protocol features high atom economy and step economy, a simple reaction set-up, and mild conditions.

Later, Verma and coworkers designed an innovative metal-free and base-free multicomponent cascade synthesis of benzo[*d*]thiazol-2(3*H*)-ylidene benzamide on water [38]. They treated acrylates, aroyl isothiocyanates, and *ortho*-iodoanilines using triethyl amine as a catalyst on water at 90 °C for 20 hours. The initial thiourea intermediate formation and subsequent S_NAr and Michael additions gave the desired products in good yields. Operationally simple cascade approaches, excellent yields and regioselectivity, and mild reaction conditions are the attractive features of this one-pot protocol.

Shortly thereafter, another catalyst-free and base-free reaction in water was put forward by Al-Harrasi and coworkers for the nucleophilic substitution of nitroimidazoles with enolizable 1,3-dicarbonyl compounds [39]. This flexible S_NAr reaction was conducted at a refluxing temperature without the aid of a catalyst or additives.

The S_NAr substitution of perfluorobenzocycloalkenones was successfully achieved in alkaline water by Mezhenkova and coworkers in 2021 [40]. When polyfluorinated aromatic ketones were treated with K_2CO_3 under aqueous conditions, the fluorine atom *ortho-* or *para-* to the carbonyl group underwent S_NAr substitution by a hydroxide ion. Interestingly, the dicarbonyl derivatives were more susceptible to transformation than the corresponding monoketones.

Subsequently, an efficient and sustainable S_NAr reaction between a variety of nucleophiles and electrophiles was achieved using the environmentally benign polymeric substance hydroxypropyl methylcellulose (HPMC) in aqueous medium [41]. This newly designed HMPC–water system offers S_NAr reactions of various N, O, and S nucleophiles in the presence of KOH (Scheme 12.8). The corresponding products were obtained in good yields with broad functional group tolerance. The main attractive features of this interesting protocol include a short reaction time, excellent regio- and chemoselectivities, and ambient reaction temperature.

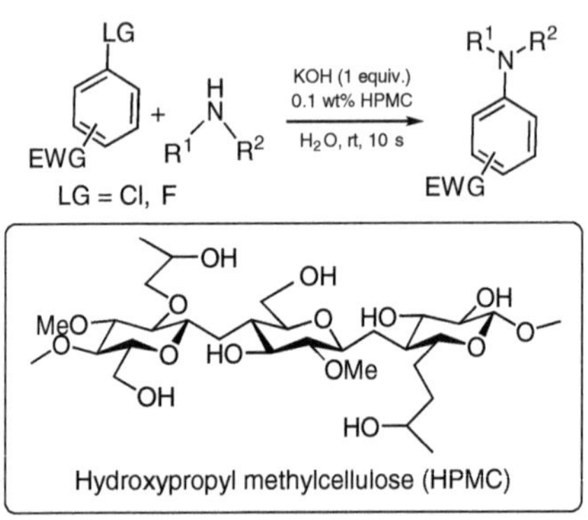

Scheme 12.8 S_NAr reactions of various N, O, and S nucleophiles in the HMPC–water system using KOH.

Very recently, a biomass-derived polymeric surface-active additive, HPMC, directed aromatic nucleophilic substitution reactions in water via hydrophobic effects was reported [42]. Kappe and his research team carried out these S_NAr reactions of various nucleophiles with electrophiles in an oscillatory plug flow reactor (OFR). Notably, this aqueous reaction is scalable and robust.

12.3.2 S_N1 and S_N2 Reactions

Aqueous S_N1 and S_N2 reactions have propelled the field of organic synthesis toward a more sustainable and greener phase. S_N1 and S_N2 reactions are fundamental transformations in organic chemistry, involving the substitution of a leaving group with a nucleophile at a saturated carbon center. Traditionally, these reactions were predominantly carried out in organic solvents. Carrying out S_N1 and S_N2 reactions in aqueous media presents several challenges, primarily due to the unique properties of water and the nature of the reactions themselves. Substrate solubility issues, nucleophile solvation and reduction of its reactivity, instability of intermediates due to nucleophilic attack by water molecules, and less compatibility of many catalytic systems in water are the major hurdles in this arena. Researchers have developed innovative catalytic systems to promote S_N1 and S_N2 reactions in water. Transition metal catalysis, phase-transfer catalysis, and micellar catalysis have been optimized to enhance the reaction rates, selectivity, and efficiency in aqueous media. Furthermore, deep eutectic solvents have been investigated for their potential to enhance S_N1 and S_N2 reactions in water.

Azumaya and team designed an aqueous *S*-benzylation with the aid of π-benzylpalladium(II) [43]. In this synthetic strategy, benzylic alcohols underwent nucleophilic substitution with benzenethiols bearing electron-deficient groups in aqueous media in the presence of 5 mol% $Pd(OAc)_2$ and 10 mol% sodium diphenylphosphinobenzene-3-sulfonate (TPPMS) at 120 °C for 16 hours (Scheme 12.9a). The *S*-benzylated products were obtained in good yields. Surprisingly, water plays an important role in the benzylic alcohol activation and the formation of crucial π-benzylpalladium(II) complexes. This is an efficient atom economical pathway, rendering water the only byproduct.

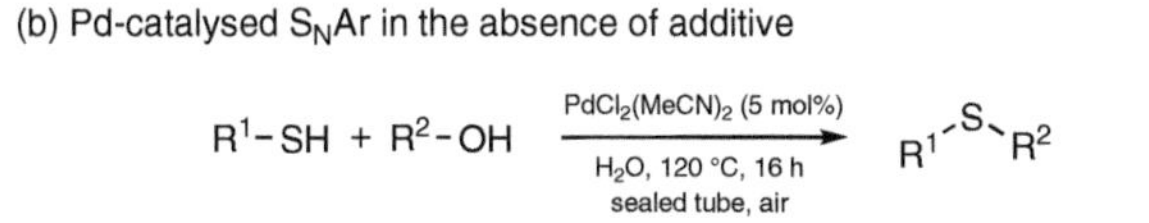

Scheme 12.9 Palladium-catalyzed nucleophilic substitution of alcohols and thiols.

Soon after, the same research group came up with a cationic palladium(II) additive-free catalytic system for the aqueous *S*-benzylation of benzhydryl alcohols [44]. Benzhydryl alcohol underwent nucleophilic substitution with benzenethiol having electron-deficient groups in the presence of 5 mol% $PdCl_2(MeCN)_2$ in water at 120 °C for 16 hours under air (Scheme 12.9b). This additive-free nucleophilic substitution is an atom economic process with water as the only byproduct.

Subsequently, Vigalok and coworkers explored hydrotropic NaOT solution as a green reaction medium for S_N1 and S_N2 reactions [45]. They carried out *exo*-2-norbornyl sulfonate hydrolysis via the S_N1 reaction using 3.8M NaOTs at room temperature. The hydrotropic NaOT solution was also proven to be the best reaction medium for benzylic bromide cyanation *via* the S_N2 reaction using NaCN at 30 °C. This hydrotropic media can be easily recycled and reused, and it opens a new arena for organic reactions in an environmentally benign medium.

Continuing the exploration into aqueous protocols, Azumaya and coworkers reported a cobalt(II)/TPPMS (sodium diphenylphosphinobenzene-3-sulfonate) catalyzed conversion of allylic and benzylic anilines to the corresponding amines in water [46]. The alcohols underwent dehydrative amination in the presence of cobalt(II)/TPPMS catalytic system in an aqueous medium at 120 °C for 16 hours. The atom economical nature, simple reaction set-up, and additive- and base-free conditions elevate the greener face of this nucleophilic substitution protocol.

A novel dicataonic PEG ionic liquid-associated $MnCl_4^{2-}$ (PEG-DIL-$MnCl_4$) catalyst for the nucleophilic substitution of benzyl halides was put forward by Davarpanah and team [47]. Using this environment-friendly catalyst, they synthesized a variety of benzyl azides and thiocyanates in water under reflux conditions.

(E)-1,3-Diphenylallyl acetate underwent allylic substitution using various nucleophiles in an aqueous medium. This metal-free nucleophilic substitution was carried out using nucleophiles such as benzylamine, phenol, acetylacetone, sodium azide, and indole at 50 °C for 24 hours in water. Notably, for phenol and acetylacetone, the presence of 20 mol% NaO^tBu was needed for the smooth conversion into allylated products. Moreover, the amines needed only 16 hours for the complete conversion into products. This green protocol follows an S_N1 mechanism.

Concurrently, with the aid of the NaOH-water system, biologically important Ataluren analogues and 3,5-disubstituted 1,2,4-oxadiazoles were synthesized by Rangappa and coworkers [48] The amidoxime was reacted with dithioesters *via* an interesting nucleophilic substitution–dehydration–intramolecular ring formation under transition metal-free conditions at 90 °C for 7–10 hours. This green protocol offers excellent yields and wide functional group compatibility.

An interesting P–S coupling via nucleophilic substitution in water was developed by Zeng and coworkers in 2019 [49]. This metal-free reaction between diarylphosphines, H-phosphonate esters, and H-phosphine oxides with *N*-(aryl/alkylthio)succinimides rendered thiophosphates, thiophosphinites, and thiophosphinates in aqueous medium at 40 °C in 120 minutes under an air atmosphere. The mild reaction conditions, metal-free and base-free conditions, environmentally benign nature, scalability, and wide substrate scope are the appealing features of this protocol.

In the same year, a transition-metal-free aqueous $C(sp^3)$–S coupling was developed by Dong and coworker [50]. The S_N2 reaction between tetraalkylthiuram disulfide with benzyl halide using one equivalents of K_2CO_3 in water at 100 °C for 0.5–3 hours delivered the benzyl dithiocarbamates in excellent yields. Environmental friendliness, simplicity in performing the reaction, excellent product yields, and functional group tolerance are the attractive features of this protocol.

Active methylene compounds were selectively C-arylalkylated in an aqueous medium with the aid of CTAB [51]. This environmentally friendly nucleophilic substitution of arylalkyl halides with active methylene compounds was accomplished using K_2CO_3 in the presence of the phase transfer catalyst CTAB in water at 60 °C for 16 hours. The research group also demonstrated the gram-scale synthesis of this highly regioselective transformation.

Herrera and coworkers effectively functionalized π-activated alcohols *via* nucleophilic substitution in water [52]. They successfully utilized C-, S-, and P-centered nucleophiles such as 1,3-dicarbonyl compounds, pyrroles, indoles, anilines, pyridine-2-thiol, and diphenyl phosphite to substitute π-activated alcohols in water in the presence of 20 mol% AcOH at room temperature for 3–10 hours. Operational simplicity, mild reaction conditions, regioselectivity, recyclability of the catalytic system, and high atom economy are the attractive features of this green protocol.

An excellent atom economic strategy to synthesize allylic amines in an aqueous medium via a metal-free catalytic route was established by Miao and coworkers in 2021[53]. They used Bu_4NHSO_4 as the non-metal catalyst for this direct pyrazole N-allylation. Allylic alcohols were reacted with pyrazoles at 100 °C for 12 hours in water in the presence of Bu_4NHSO_4 (Scheme 12.10). This protocol offered excellent recyclability and was found to be suitable for gram-scale synthesis. Interestingly, the intramolecular H-bonding interaction between water and allylic alcohols plays an inevitable role in the successful proceeding of this nucleophilic substitution reaction.

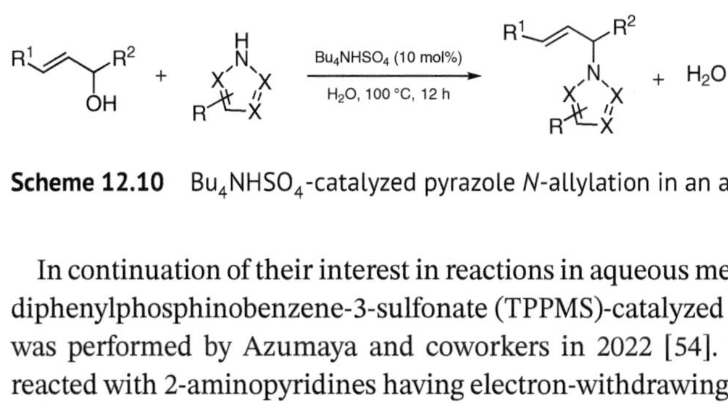

Scheme 12.10 Bu_4NHSO_4-catalyzed pyrazole *N*-allylation in an aqueous medium.

In continuation of their interest in reactions in aqueous media, an Au(III)/sodium diphenylphosphinobenzene-3-sulfonate (TPPMS)-catalyzed dehydrative amination was performed by Azumaya and coworkers in 2022 [54]. The benzhydrols were reacted with 2-aminopyridines having electron-withdrawing groups in the presence of 5 mol% $HAuCl_4$ and 10 mol% TPPMS in water at 120 °C for 16 hours under an air atmosphere. The current gold-catalyzed water protocol offers excellent yields and high functional group tolerance

In 2023, Yu and coworkers developed an interesting photocatalyst-free, visible-light-driven α-substitution of amines with various nucleophiles in an organo-aqua ($MeCN/THF/H_2O = 2:1:2$) solvent system [55]. Water is essential to the *in situ* production of substrate aggregates in this catalyst-free air-promoted reaction, which is a crucial stage in this transformation. The attractive features of the protocol include green reaction set-up, high functional group compatibility, and easy operation.

Biologically potent selenocyanates and thiocyanites were constructed in a water-dimethyl carbonate (DMC) (20 : 1) solvent system using selenonium salt as the effective catalyst [56]. Alberto and his team reacted organic bromides with seleno- or thiocyanates at ambient temperature for 10 minutes, which delivered the anticipated products in good yields. The effectiveness of this protocol lies in the key chalcogen bond interactions between the organic bromides and selenonium catalyst.

Later, Mukherjee and coworkers successfully established vanadium-catalyzed S- and C-alkylation in an aqueous medium *via* nucleophilic substitution [57]. This dehydrative alkylation was achieved by reacting benzhydrols with thiols and arenes in the presence of a 4 mol% divanadium oxoperoxo system $[K_3(V^{+5})_2(O_2^{2-})_4(O^{2-})_2(\mu\text{-OH})]$ at 40 °C in water for four hours under air. Interestingly, the catalytic system retains its activity over seven runs. Excellent substrate scope, mild reaction conditions, base- and additive-free nature, and water as the only co-product are the key advantages of this protocol.

During the same year, Acocella and coworkers synthesized a novel phosphonium-carbon-based phase transfer catalyst oCB/TPP (oxidized carbon black-tetraphenyl phosphonium filler) [58]. This newly fabricated heterogeneous catalyst successfully catalyzed the S_N2 reaction between NaSCN and various benzyl bromides, employing water as the reaction medium. Only 5 mol% of the catalyst was needed for this transformation. A short reaction span and excellent recyclability are the other attractive features of this protocol.

The introduction of carbon black intercalated compounds as phase transfer catalysts for the S_N2 reaction of 4-nitrobenzyl bromide with NaSCN was reported by Acocella and coworkers in 2023 [59]. This nucleophilic substitution was carried out in a biphasic $CHCl_3/H_2O$ liquid system using 5 mol% oCB/2HT (oxidized carbon black intercalated with quaternary ammonium salt). This heterogeneous catalyst shows excellent recyclability without losing its catalytic activity many times.

12.4 Summary and Conclusions

Recent years have seen remarkable progress in nucleophilic addition and substitution reactions conducted in water. The development of novel water-active catalytic systems, the engineering of dual solvent systems with water as a key component, the evolution of surfactants capable of facilitating reactions within hydrophobic pockets, and the implementation of flow chemistry have been notable advancements in this field. These innovations tackle several challenges previously encountered in carrying out nucleophilic addition and substitution reactions in water, such as substrate solubility issues, nucleophile solvation and reduction of its reactivity, the instability of intermediates due to nucleophilic attack by water molecules, and the limited compatibility of many catalytic systems with aqueous environments. By addressing these obstacles, the recent advancements pave the way for more efficient and sustainable nucleophilic reactions in water. Overall, these advancements represent significant strides toward overcoming limitations and expanding the scope of nucleophilic addition and substitution reactions in aqueous environments.

So, nucleophilic substitution and addition reactions in-water offer numerous advantages, including their alignment with green chemistry principles, abundance and accessibility of water, safety considerations, versatility in facilitating hydrophilic and hydrophobic interactions, potential for reaction acceleration, and relevance to biological systems. Persisting research inquiries surrounding nucleophilic addition and substitution methodologies in water, alongside the proliferation of catalytic systems and methodologies effective in aqueous environments, indicate that the field is ripe with untapped potential. This chapter seeks to underscore the profound promise of water as a reaction medium, capable of facilitating both nucleophilic addition and substitution. By simplifying synthetic routes and driving advancements toward sustainable chemistry, these water protocols emerge as a transformative force in the realm of organic synthesis.

References

1 Knoevengel, E. (1898). *Ber. Dtsch. Chem. Ges.* 31: 2596–2619.
2 Gomes, M.N., de Oliveira, C.M.A., Garrote, C.F.D. et al. (2011). *Synth. Commun.* 41: 52–57.
3 Xu, H., Pan, L., Fang, X. et al. (2017). *Tetrahedron Lett.* 58: 2360–2365.
4 Yang, P., Liu, Y., Chai, L. et al. (2018). *Synth. Commun.* 48: 1060–1067.
5 Zhao, S.D., Meng, D., Wei, L. et al. (2019). *Green Chem. Lett.* 12: 271–277.
6 Kalar, P.L., Jain, K., Agrawal, S. et al. (2023). *J. Org. Chem.* 88: 16829–16844.
7 Little, R.D., Masjedizadeh, M.R., Wallquist, O., and McLoughlin, J.I. (1995). *Org. React.* 47: 315–552.
8 Liang, X., Gui, Y., Li, K. et al. (2020). *Chem. Commun.* 56: 11118–11121.
9 Huang, Y., Zha, Z., and Wang, Z. (2020). *Org. Lett.* 22: 2512–2516.
10 Li, N., Lu, W., Gu, W. et al. (2022). *Chem. Commun.* 58: 10957–10960.
11 Nagata, K., Nakagawa, C., Yokoyama, W. et al. (2021). *ACS Omega* 6: 19642–19646.
12 Delgado, J.A.C., Vicente, F.E.M., de la Torre, A.F. et al. (2021). *New J. Chem.* 45: 14050–14057.
13 Bae, D., Lee, J.W., and Ryu, D.H. (2022). *J. Org. Chem.* 87: 16532–16541.
14 Dutt, S., Goel, V., Garg, N. et al. (2019). *Adv. Synth. Catal.* 362: 858–866.
15 Sombat, W., Authai, P., Padungros, P., and Hoven, V.P. (2023). *ACS Appl. Polym. Mater.* 5: 7288–7297.
16 Mannich, C. and Krosche, W. (1912). *Archiv der Pharmazie* 250: 647–667.
17 Sarkar, P. and Mukhopadhyay, C. (2016). *Tetrahedron Lett.* 57: 4306–4310.
18 Parikh, N., Roy, S.R., Seth, K. et al. (2016). *Synthesis* 48: 547–556.
19 Chen, X., Sun, H., Luo, Y. et al. (2016). *RSC Adv.* 6: 39343–39347.
20 Khoshnavazi, R., Bahrami, L., Havasi, F., and Naseri, E. (2017). *RSC Adv.* 7: 11510–11521.
21 Xie, L.-H., Cheng, J., Luo, Z.-W., and Lu, G. (2018). *Tetrahedron Lett.* 59: 457–461.
22 (a) Ferland, J.M., Demerson, C.A., and Humber, L.G. (1985). *Can. J. Chem.* 63: 361–365. (b) Zhuang, Z.P., Kung, M.P., Mu, M., and Kung, H.F. (1998). *J. Med. Chem.* 41: 157–166. (c) Anzini, M., Cappelli, A., Vomero, S. et al. (1996).

J. Med. Chem. 39: 4275–4284. (d) Augner, D., Gerbino, D.C., Slavov, N. et al. (2011). *Org. Lett.* 13: 5374–5377.

23 (a) Heaney, H. and Shuhaibar, K.F. (1995). *Synlett* 1995: 47–48. (b) Othman, M., Pigeon, P., and Decroix, B. (1997). *Tetrahedron* 53: 2495–2504. (c) Boger, D.L., Lee, J.K., Goldberg, J., and Jin, Q. (2000). *J. Org. Chem.* 65: 1467–1474. (d) Comins, D.L., Schilling, S., and Zhang, Y. (2005). *Org. Lett.* 7: 95–98. (e) Speck, K. and Magauer, T. (2013). *Beilstein J. Org. Chem.* 9: 2048–2078.

24 Tian, Y., Liu, Q., Liu, Y. et al. (2018). *Tetrahedron Lett.* 59: 1454–1457.

25 Gnanamani, E., Yana, X., and Zarea, R.N. (2019). *Angew. Chem. Int. Ed.* 59: 3069–3072.

26 Zhu, B.-H., Zhang, Y.-Q., Xu, H.-J. et al. (2021). *ACS Catal.* 11: 1706–1713.

27 Wurtz, C.A. (1872). *J. Prakt. Chem.* 5: 457–461.

28 Hoang, M.-D., Kumar, R.A., Buisson, D.A. et al. (2019). *ChemCatChem* 12: 1156–1160.

29 Pan, H., Han, M.-Y., Li, P., and Wang, L. (2019). *J. Org. Chem.* 84 (21): 14281–14290.

30 Dwivedi, K.D., Reddy, M.S., Kumar, N.S., and Chowhan, L.R. (2019). *ChemistryS-elect* 4: 8602–8605.

31 Song, T., Ma, Z., and Yang, Y. (2019). *ChemCatChem* 11: 1313–1319.

32 Reddy, C.V.S. and Sumathi, S. (2020). *ChemistrySelect* 5: 8729–8734.

33 Luo, N., Ao, Y.-F., Wang, D.-X., and Wang, Q.-Q. (2021). *Chem. Asian J.* 16: 1–6.

34 Wu, C., Hu, B., Liu, H. et al. (2022). *ChemistrySelect* 7: e202104433.

35 Lv, X.-X., Liu, N., Chen, F. et al. (2023). *Org. Biomol. Chem.* 21: 8695–8701.

36 Isley, N.A., Linstadt, R.T.H., Kelly, S.M. et al. (2015). *Org. Lett.* 17: 4734–4737.

37 Zhu, S., Chen, C., Xiao, M. et al. (2017). *Green Chem.* 19: 5653–5658.

38 Saini, K.M., Saunthwal, R.K., Kumar, S., and Verma, A.K. (2019). *J. Org. Chem.* 84: 2689–2698.

39 Shah, S.R., Ali, M., Anwar, M.U., and Al-Harrasi, A. (2020). *RSC Adv.* 10: 21656.

40 Zonov, Y.V., Wang, S., Karpov, V.M., and Mezhenkova, T.V. (2021). *J. Fluor. Chem.* 249: 109851.

41 Borlinghaus, N., Ansari, T.N., Braje, L.H. et al. (2021). *Green Chem.* 23: 3955–3962.

42 Hammer, S., Nanto, F., Canu, P. et al. (2024). *ChemSusChem.* 17: e202301149.

43 Hikawa, H., Toyomoto, M., Kikkawa, S., and Azumaya, I. (2015). *Org. Biomol. Chem.* 13: 11459–11465.

44 Hikawa, H., Machino, Y., Toyomoto, M. et al. (2016). *Org. Biomol. Chem.* 14: 7038–7045.

45 Sela, T., Lin, X., and Vigalok, A. (2017). *J. Org. Chem.* 82: 11609–11612.

46 Hikawa, H., Ijichi, Y., Kikkawa, S., and Azumaya, I. (2017). *Eur. J. Org. Chem.* 2017: 465–468.

47 Goodajdar, B.M., Akbari, F., and Davarpanah, J. (2018). *Appl. Organomet. Chem.* 33: e4647.

48 Swarup, H.A., Chaithra, N., Mantelingu, K., and Rangappa, K.S. (2018). *Chem-istrySelect* 3: 539053–539094.

49 Wang, G., Zhou, L., Li, N., and Zeng, Q. (2019). *ChemistrySelect* 4: 13899–13903.

50 Peng, H.-Y. and Dong, Z.-B. (2019). *Eur. J. Org. Chem.* 2019: 949–956.
51 Godha, A.K., Thiruvengadam, J., Abhilash, V. et al. (2019). *New J. Chem.* 43: 16041–16045.
52 Ortiz, R., Koukouras, A., Marqués-López, E., and Herrera, R.P. (2020). *Arab. J. Chem.* 13: 1866–1873.
53 Zhuang, H., Lu, N., Ji, N. et al. (2021). *Adv. Synth. Catal.* 363: 5461–5472.
54 Hikawa, H., Nakayama, T., Nakamura, S. et al. (2022). *Org. Biomol. Chem.* 20: 4183–4188.
55 Zhao, Y., Hou, X., He, M. et al. (2023). *Org. Lett.* 25: 7344–7348.
56 Ángel, A.Y.B., Campos, P.R.O., and Alberto, E.E. (2023). *Molecules* 28: 3056.
57 Singh, P., Chowdhury, D., and Mukherjee, A. (2023). *Eur. J. Org. Chem.* 26: e202201300.
58 Kiani, A., Coscia, D., Schettini, R., and Acocella, M.R. (2023). *Mol. Catal.* 550: 113558.
59 Melillo, A., Kiani, A., Schettini, R., and Acocella, M.R. (2023). *Mol. Catal.* 537: 112951.

13

Asymmetric Organic Reactions in Water

Han-Yong Bae, Woo-Hee Kim, and Seok-Ju Hong

Department of Chemistry, Sungkyunkwan University, 2066 Seobu-ro, Jangan-gu, Suwon, Republic of Korea

13.1 Introduction

Asymmetric synthesis in water has been extensively explored over the past several decades [1]. Merck's asymmetric alkylation using cinchona-based chiral quaternary ammonium salts (phase-transfer catalysts) in 1984 probably represents a significant starting point for modern asymmetric aqueous catalytic processes [2]. Subsequently, reactions using water-compatible catalysts gained considerable attention, and there is a growing anticipation that sustainable chemistry can be applied to synthesizing natural products and myriad pharmaceuticals [1, 3]. Water-compatible asymmetric synthesis in aqueous environments was initially explored enzymatically, inspired by biological processes [4]. However, the limited reaction scope of enzymatic processes led to ongoing efforts to create new substances efficiently and selectively using precisely designed molecular catalyst systems. These molecular catalytic systems include combinations of metals with organic chiral ligands and, more recently, chiral organocatalysts in finely designed homogeneous catalytic systems [5]. The reported asymmetric reactions in water can be classified based on the key catalytic activation mechanisms as (i) *asymmetric metal (Lewis acid) catalysis,* (ii) *asymmetric Lewis base organocatalysis*, (iii) *asymmetric Brønsted acid organocatalysis,* and (iv) *asymmetric Brønsted base organocatalysis.*

In this chapter, we focus on the recent trend of asymmetric catalytic reactions in water and emphasize significant findings from past decades that have not been covered in previous publications [6]. More specifically, we highlight reactions that involve the use of water as the reaction medium and showcase outcomes characterized by positive effects in reactivity and selectivity attributable to bulk-water environments.

Organic Transformations in Water: Principles and Applications, First Edition.
Edited by Gopinathan Anilkumar, Nissy Ann Harry, and Sankuviruthiyil M. Ujwaldev.

13.2 Asymmetric Metal (Lewis Acid) Catalysis in Water

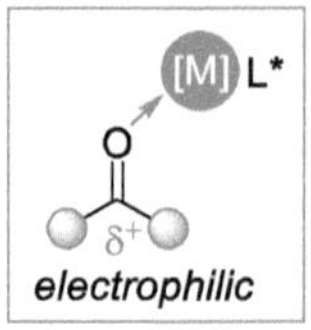

13.2.1 Asymmetric Reduction

13.2.1.1 Carbonyl (C=O Bond) Reduction

Deng and coworkers reported an asymmetric transfer hydrogenation of aliphatic ketones in water using a rhodium (Rh) catalyst [7]. High enantioselectivities resulted from a synergistic interaction between the catalytic center and the lipophilic microenvironment within a metallomicelle core. Chiral surfactant-type ligand **1** complexed with rhodium was successfully utilized for the asymmetric reduction of diverse aliphatic ketones. The proposed transition state was based on the considerations of steric interactions between the alkyl groups of aliphatic ketones and the cyclopentadienyl (Cp*) ligand of an amphiphilic catalyst, during which hydrophobic interactions in the metallomicelle contributed to the high enantioselectivity of ketone reduction (Scheme 13.1).

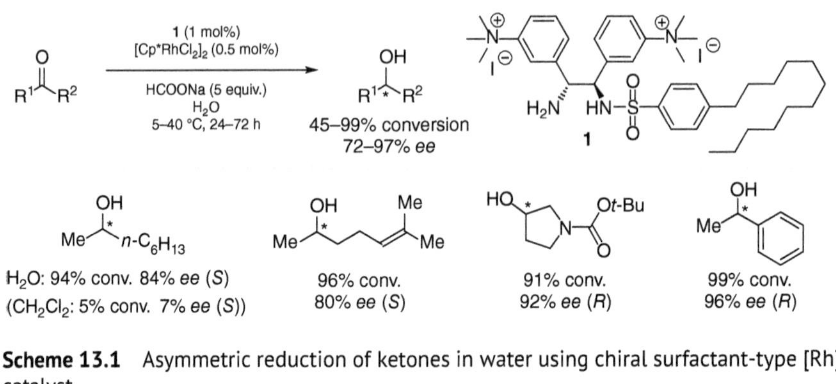

Scheme 13.1 Asymmetric reduction of ketones in water using chiral surfactant-type [Rh] catalyst.

13.2.1.2 Imine (C=N Bond) Reduction

The Deng group presented an asymmetric reduction of cyclic imines and iminiums in water using a ruthenium (Ru) catalyst (Scheme 13.2) [8]. By employing sodium formate as the hydrogen source and cetrimonium bromide (CTAB) as a surfactant additive, a water-soluble Ru(II) complex of the (*R*,*R*)-ligand **2** was used as a recyclable catalyst in an aqueous reaction. Unnatural isoquinoline and β-carboline alkaloids with potential biological (psychotropic) properties were catalytically prepared with excellent enantioselectivity (up to 99% *ee*).

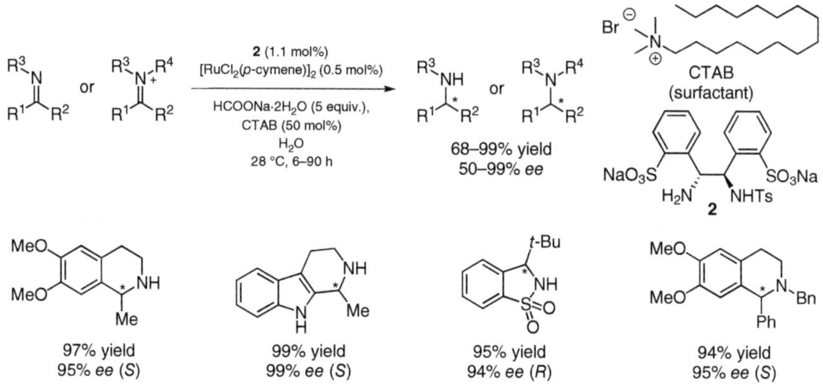

Scheme 13.2 Asymmetric transfer hydrogenation of imines and iminiums in water using chiral [Ru] catalyst.

13.2.2 Asymmetric Oxidation

13.2.2.1 Epoxidation

The Malkov group reported asymmetric vanadium (V)-catalyzed epoxidation of allylic alcohols in water [9]. The hydroxamic acid functional group at chiral ligand **3** was crucial for this reaction. Various allylic alcohols as starting materials were successfully converted to the desired enantio-enriched epoxides. Well known for its ligand-deceleration effect under conventional organic solvents, it was transformed into a ligand-accelerated process under an aqueous reaction medium. The aqueous environment enabled the V-catalyzed epoxidation of allylic alcohols at up to 94% *ee* (Scheme 13.3).

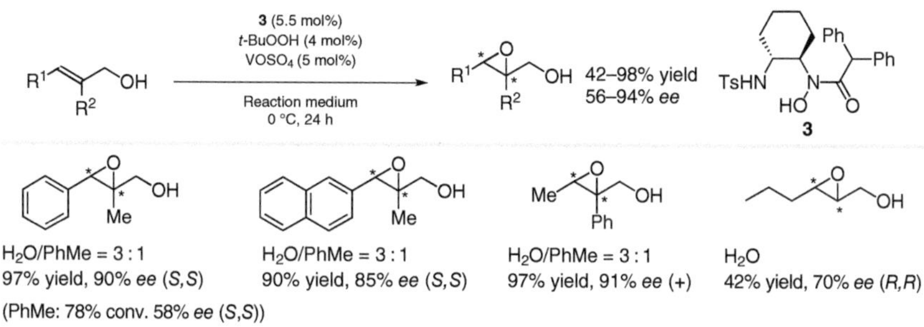

Scheme 13.3 Asymmetric epoxidation of allylic alcohols in water using chiral [V] catalyst.

13.2.2.2 Sulfoxidation

Katsuki and coworker reported highly enantioselective sulfoxidation in water using a novel iron complex (**4**, Fe(salan), salan = *N,N′*-disubstituted bis(aminophenoxide))/ aqueous hydrogen peroxide system without surfactant (Scheme 13.4) [10]. Substrates from alkyl-aryl sulfides to methyl-alkyl sulfides were successfully oxidized to the corresponding chiral sulfoxides at enantioselectivities of up to 96% *ee*. Notably, the Fe-**4** catalyst achieved a turnover number (TON) of 8000.

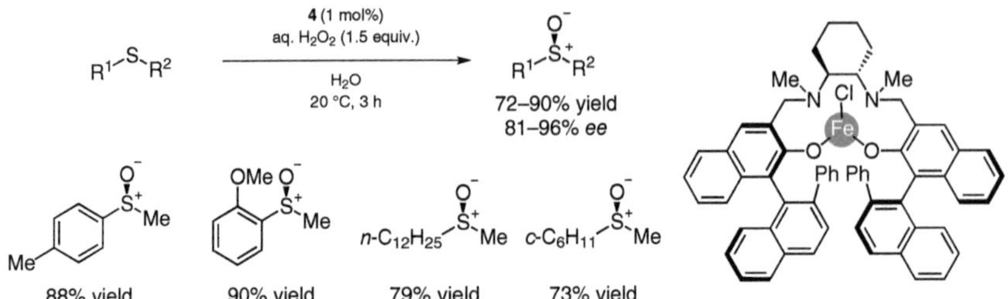

4

Scheme 13.4 Asymmetric sulfoxidation of sulfides in water using [Fe(salan)] catalyst.

13.2.3 Asymmetric Carbon–Carbon Bond Formation

13.2.3.1 Allylation

The Nakamura group reported catalytic asymmetric Barbier-type allylation using indium (In) powder and chiral bis(imidazoline) ligand **5** in water via a six-membered transition state involving In(III) (Scheme 13.5) [11]. The reaction in organic solvents (e.g. THF: tetrahydrofuran) resulted in a significant decrease in enantioselectivity. The inclusion of sodium iodide as an additive enhanced both yield and enantioselectivity. In the transition state, the resulting octahedral structure allowed the allyl group to approach the *Re*-face of ketones, ensuring (*R*)-homoallylic alcohols and avoiding steric repulsion between the phenyl group in ligand **5** and ketone substituents.

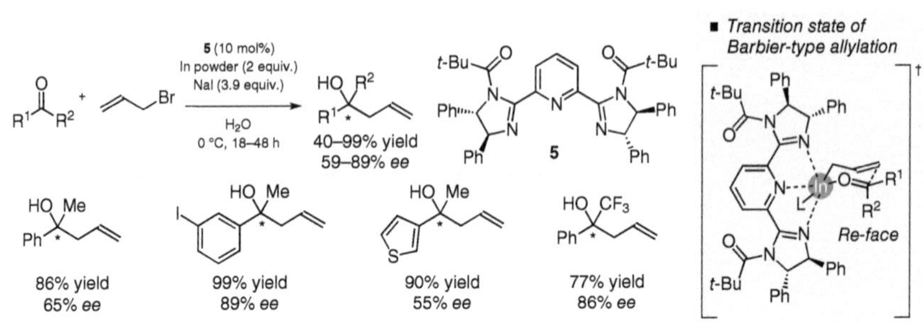

Scheme 13.5 Asymmetric Barbier-type allylation of ketones in water using chiral [In] catalyst.

13.2.3.2 Alkynylation

Nakamura and coworkers reported enantioselective copper (Cu)-catalyzed three-component alkynylation involving aldehydes, amines, and alkynes in water (Scheme 13.6) [12]. Cu(I) catalyst with hydrophobic bis(imidazoline) ligand **5** and sodium dodecyl sulfate (SDS) surfactant were used as efficient catalytic components. The reaction demonstrated broad substrate scope across various aldehydes and alkynes, yielding optically active propargyl amines with outstanding yields (up to 99%) and high enantioselectivities (up to 99% *ee*).

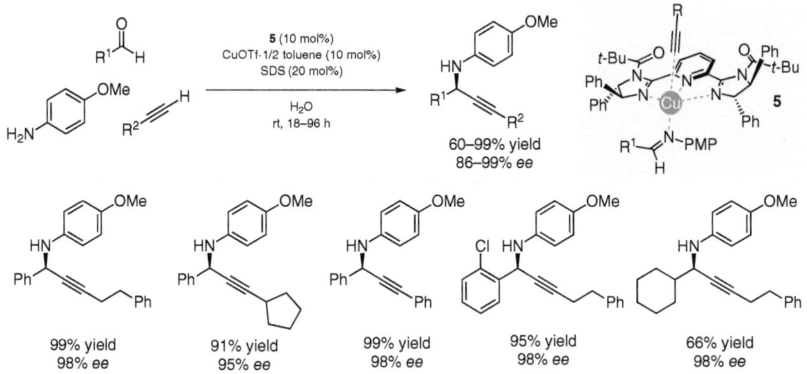

Scheme 13.6 Asymmetric three-component alkynylation of aldehydes in water using a chiral [Cu] catalyst.

13.2.3.3 Friedel–Crafts Reaction

Kobayashi and coworkers described a chiral palladium (Pd) aqua complex catalyzed by Friedel–Crafts-type alkylation of indoles in water (Scheme 13.7) [13]. Pd-coordinated water molecules have recently emerged as novel soft Lewis acid catalysts. Employing a chiral 2,2′-bipyridine (Bolm's ligand) [14] **6** as both ligand and hydrogen bond donor facilitated the creation of a robust, highly Lewis acidic, chiral aqua complex in water. On the other hand, enantioselectivity was difficult to control in organic solvents or under solvent-free conditions.

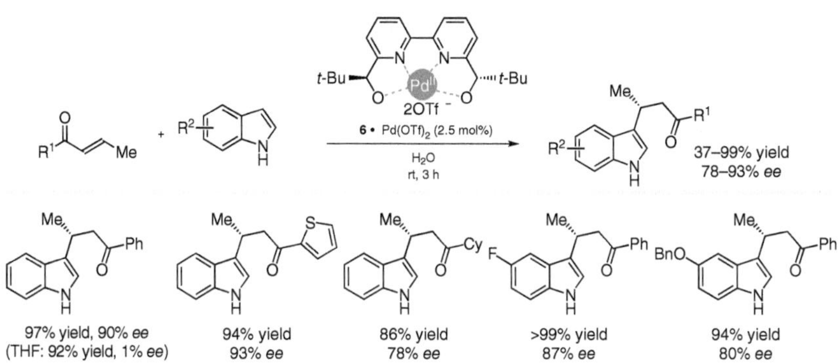

Scheme 13.7 Asymmetric Friedel–Crafts-type reaction of indoles in water using a chiral [Pd] catalyst.

13.2.3.4 Aldol-Type Reaction

The Kobayashi group presented chiral scandium (Sc) catalysts that serve as simple metalloenzyme-like catalysts in water that mimic enzyme structures (Scheme 13.8) [15]. These artificial organometallic catalysts (Sc with **7**) efficiently facilitated enantioselective direct aldol-type reactions of enolates using aqueous formaldehyde as an acceptor. The aqueous catalytic system follows Michaelis–Menten kinetics and possesses heat-resistant properties.

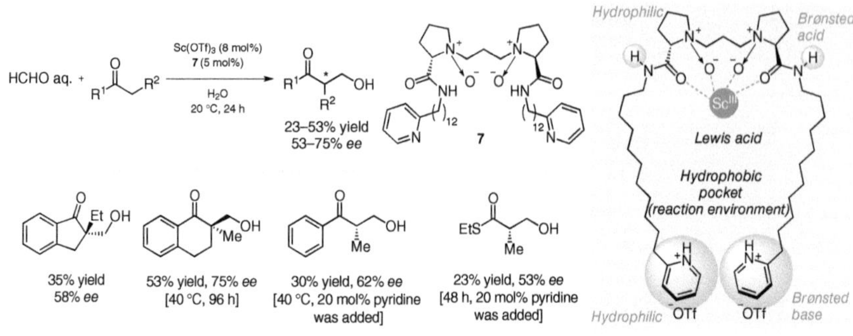

Scheme 13.8 Asymmetric aldol reaction of formaldehyde in water using chiral [Sc] catalysts.

13.2.4 Asymmetric Carbon-Heteroatom Bond Formation

13.2.4.1 Carbon–Nitrogen (C–N) Bond Formation

The Kobayashi group reported a scandium-chiral 2,2′-bipyridine (**6**)-complex-catalyzed asymmetric ring-opening of *meso*-epoxides using amines in water (Scheme 13.9) [16]. Chiral β-amino alcohols were produced at high yields and enantioselectivities using catalytic amounts of scandium tris(dodecyl sulfate) ($Sc(DS)_3$) as a Lewis acid surfactant, without diol formation. The authors commented that the use of organic components at greater concentrations might enhance hydrophobic interactions and improve results.

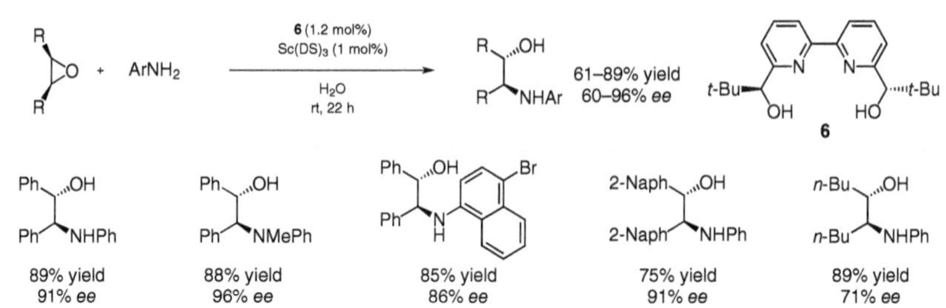

Scheme 13.9 Asymmetric ring opening of *meso*-epoxides using amines in water using a [Sc] catalyst.

13.2.4.2 Carbon–Sulfur (C–S) Bond Formation

The Kobayashi group reported asymmetric thia-Michael/protonation reactions of thiols to enones catalyzed by a scandium-chiral 2,2′-bipyridine (**6**)-complex in water [17]. These reactions proceeded in water and organic solvents, though superior selectivity was achieved in water. As regards protonation, lower yields and enantioselectivities were observed in organic solvents, and notably, highly enantioselective protonation occurred in water (Scheme 13.10).

13.2.4.3 Carbon–Boron (C–B) Bond Formation

The Kobayashi group developed an asymmetric 1,4-addition reaction of diboron to a wide variety of α,β-unsaturated carbonyls and a nitrile, catalyzed by copper(II)

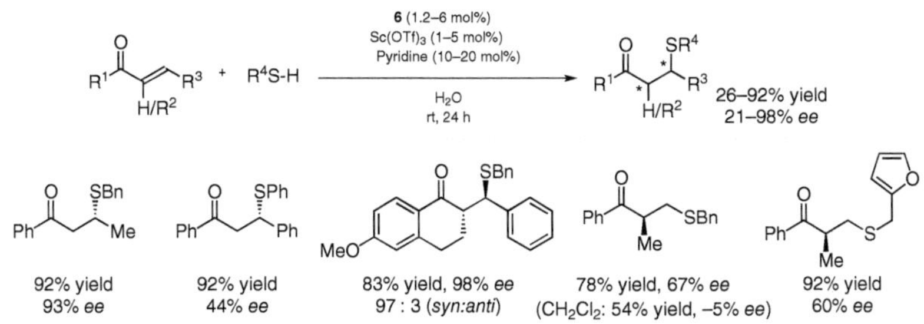

Scheme 13.10 Asymmetric thia-Michael addition/protonation of thiols in water using a [Sc] catalyst.

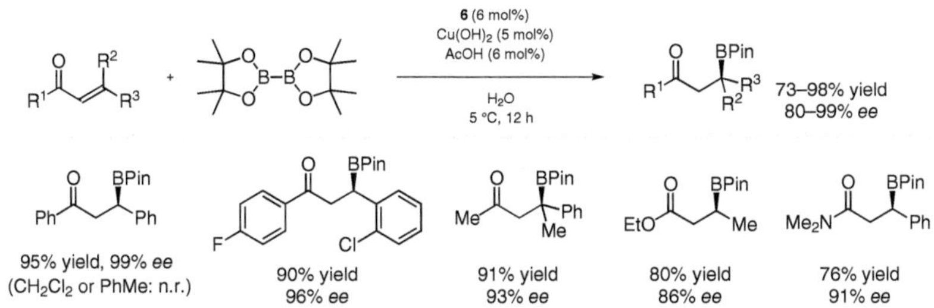

Scheme 13.11 Asymmetric boron conjugate additions in water using a [Cu] catalyst.

hydroxide/chiral 2,2′-bipyridine ligand **6** in water (Scheme 13.11) [18]. Excellent enantioselectivities (up to 99% *ee*) and high turnover frequencies (TOF: up to 17 600 h^{-1}) were obtained. Similar reactions using a Cu(I) catalyst in organic solvents were attempted, though a Cu(II) catalyst provided exceptional C–B bond-forming activity under aqueous reaction conditions.

13.2.4.4 Carbon–Silicon (C–Si) Bond Formation

The Kobayashi group reported that an insoluble copper(II)–chiral 2,2′-bipyridine (**6**) complex enabled silyl conjugate addition to electron-deficient olefins in water (Scheme 13.12) [19]. Notably, water was the only efficient reaction medium and was considered crucial in establishing and stabilizing confined transition states. The chiral solid crystals of Cu(II) functioned as heterogeneous catalysts. From the green

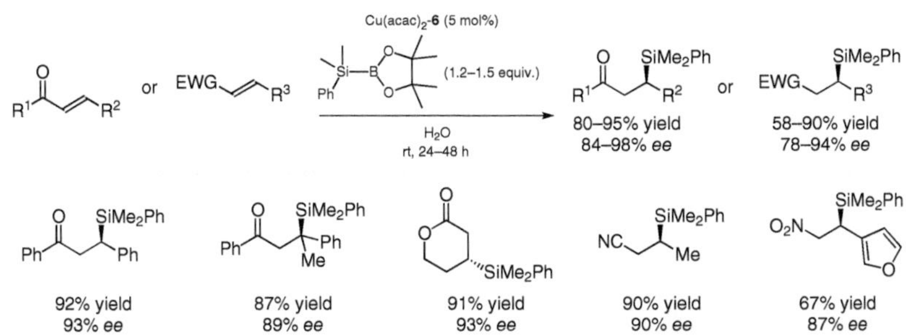

Scheme 13.12 Asymmetric silyl conjugate additions in water using a [Cu] catalyst.

chemistry perspective, this heterogeneous system might reduce organic solvent use, enabling catalyst recovery and recycling.

13.3 Asymmetric Lewis Base Organocatalysis in Water

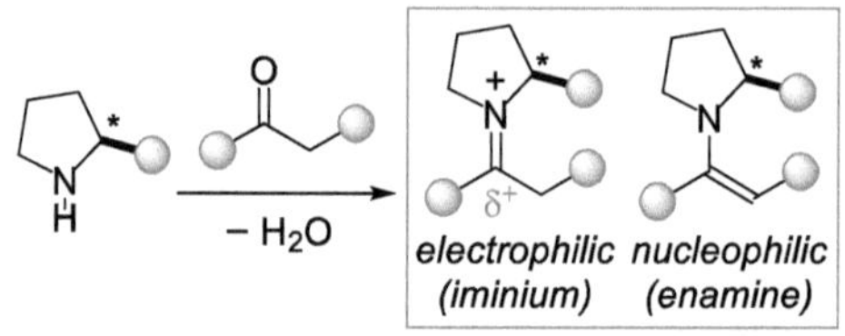

13.3.1 Asymmetric Aldol Reaction

13.3.1.1 Using Hydrophobic Diamine Catalyst

The Barbas group disclosed a seminal result on direct asymmetric aldol reactions in water (Scheme 13.13) [20]. A combination of chiral-diamine-organocatalyst-incorporating hydrophobic alkyl chains (**8**) with a catalytic amount of trifluoroacetic acid (TFA) efficiently promoted reactions and provided the desired aldol products in excellent yield with high enantioselectivity. The transformation did not occur in water when using the other catalyst without a hydrophobic alkyl chain.

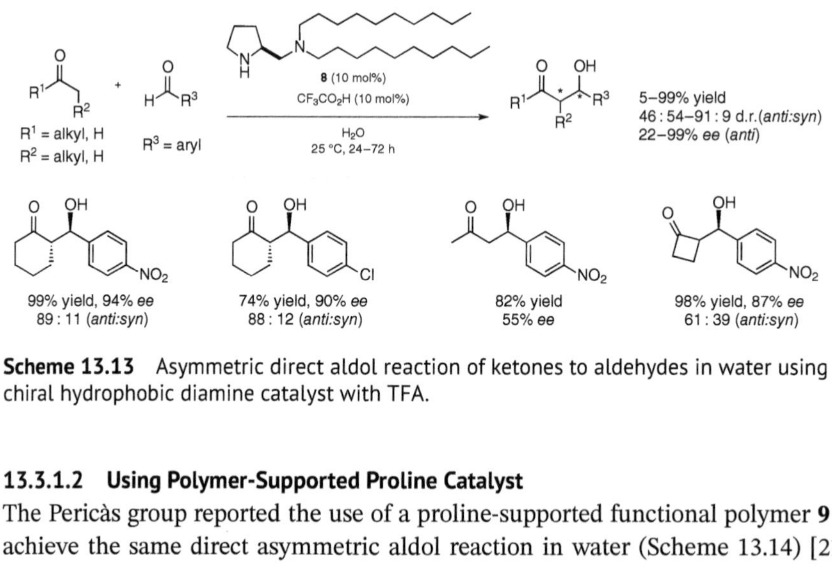

Scheme 13.13 Asymmetric direct aldol reaction of ketones to aldehydes in water using a chiral hydrophobic diamine catalyst with TFA.

13.3.1.2 Using Polymer-Supported Proline Catalyst

The Pericàs group reported the use of a proline-supported functional polymer **9** to achieve the same direct asymmetric aldol reaction in water (Scheme 13.14) [21]. Proline bound to polystyrene with a 1,2,3-triazole linker exhibited catalytic properties tailored for 'artificial aldolase'. Despite the hydrophobic nature of the polymeric backbone, the resin swelled in water, creating an aqueous microenvironment. The exceptionally high reactivity and enantioselectivity observed were presumed to be due to the confinement effect of the hydrogen-bond networks produced by proline and 1,2,3-triazole fragments.

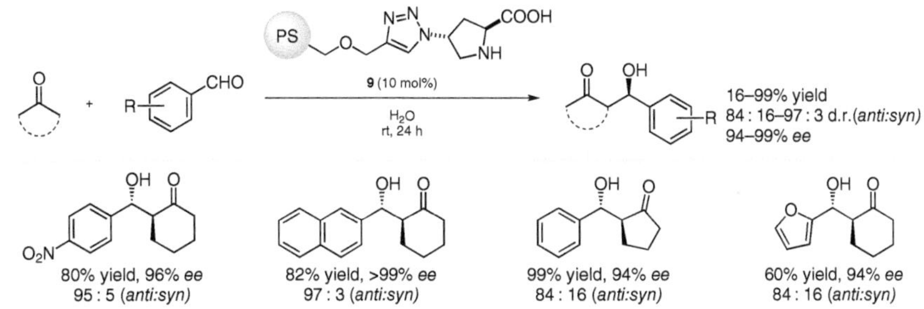

Scheme 13.14 Asymmetric direct aldol reaction in water using a polymer-supported proline catalyst.

13.3.2 Asymmetric Michael Reaction

13.3.2.1 Using Hydrophobic Diamine Catalyst

The Barbas group reported an asymmetric Michael addition reaction of carbonyl compounds to β-nitroolefins in saturated NaCl and seawater [22]. A bifunctional hydrophobic organocatalyst **8** was used and efficiently afforded the desired Michael products in excellent yields with high enantioselectivity. The aggregation of organic components led to the exclusion of water from the oil phase and shifted the equilibrium toward enamine formation. The process was also possibly enhanced by a salting-out effect. The rapid Michael addition reaction took place within the highly concentrated organic phase via a transition state similar to that observed in homogeneous organic solvents (Scheme 13.15).

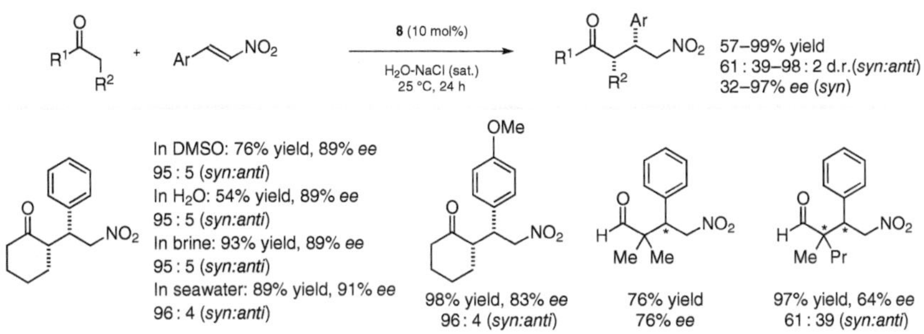

Scheme 13.15 Asymmetric direct Michael reaction of ketones or aldehydes to β-nitroolefins in brine using a chiral hydrophobic diamine catalyst.

13.3.2.2 Using Superparamagnetic Nanoparticle-Supported Diarylprolinol Silyl Ether Catalyst

The Wang group developed a superparamagnetic nanoparticle-supported Jørgensen–Hayashi catalyst ((*S*)-diphenylprolinol trimethylsilyl ether) **10** and used it for the asymmetric Michael addition of aldehydes to nitroolefins in water [23]. The desired Michael addition products were obtained in moderate-to-good yields (up to 96%) and good stereoselectivities (up to 90% *ee*, and up to 99 : 1 d.r.). Due to its paramagnetic nature, the immobilized chiral catalyst was extracted from

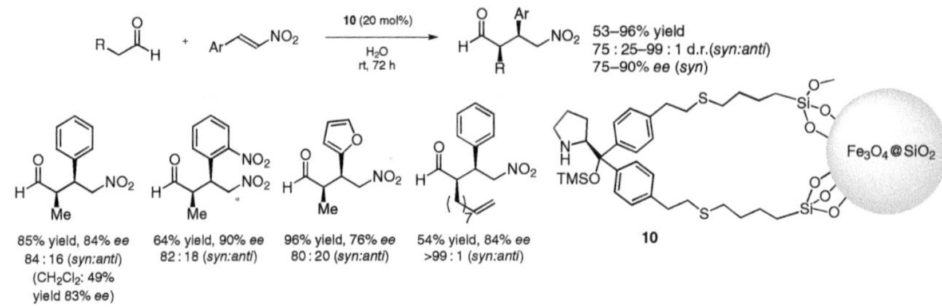

Scheme 13.16 Asymmetric direct Michael reaction in water using a superparamagnetic nanoparticle-supported chiral diarylprolinol silyl ether catalyst.

reaction mixtures using an external magnet. In addition, the catalyst was recycled four times without significant degradation of the activity (Scheme 13.16).

13.3.3 Asymmetric Diels–Alder Reaction

13.3.3.1 Using Imidazolidinone Catalyst

MacMillan and coworker reported an asymmetric Diels–Alder reaction between dienes and ketone dienophiles in water (Scheme 13.17) [24]. Chiral imidazolidinone **11** used in iminium catalysis enabled the enantioselective synthesis of Diels–Alder products, previously inaccessible in the presence of metal Lewis acid catalysts. A large number of cyclic and acyclic dienes were reacted with α,β-unsaturated ketones via asymmetric [4+2] cycloaddition. A computational study provided theoretical evidence for the enantioinduction observed during the catalytic transformation.

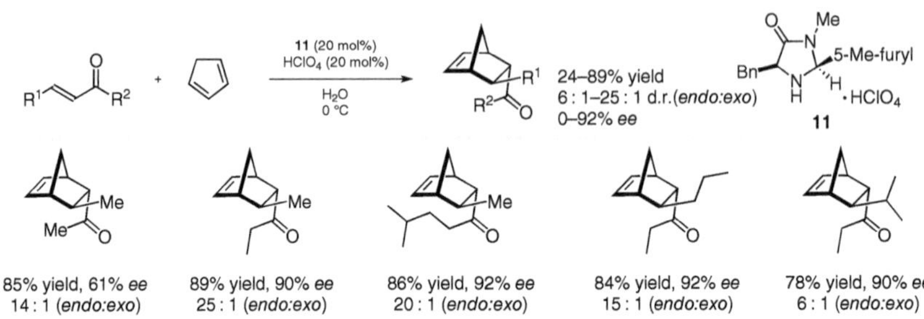

Scheme 13.17 Asymmetric Diels–Alder reaction in water using a chiral imidazolidinone catalyst.

13.3.3.2 Using Diarylprolinol Silyl Ether Catalyst

The Hayashi group reported catalytic asymmetric Diels–Alder reactions between a diene and α,β-unsaturated aldehydes in water (Scheme 13.18) [25]. A diarylprolinol silyl ether salt organocatalyst **12** provided highly enantioselective [4+2] cycloaddition products without organic solvents. According to the authors, the reaction occurred in the organic phase (not at the water/oil interface). Notably, a tiny amount of water dissolved in the organic reaction mixture phase was found to affect the transition state, accelerate the reaction, and enhance enantioselectivity.

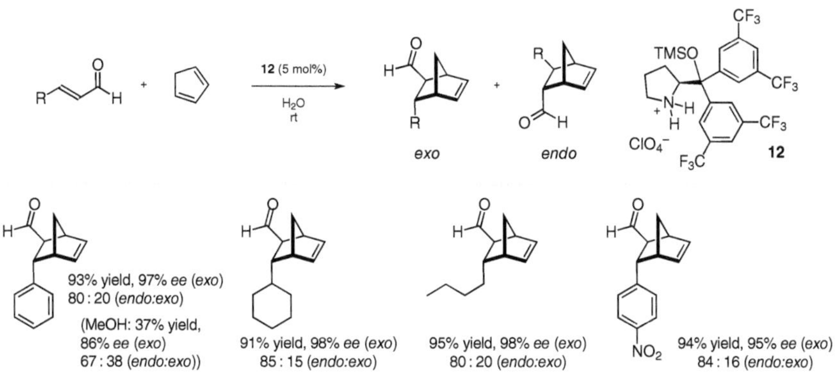

Scheme 13.18 Asymmetric Diels–Alder reaction in water using a chiral diarylprolinol silyl ether catalyst.

13.4 Asymmetric Brønsted Acid Organocatalysis in Water

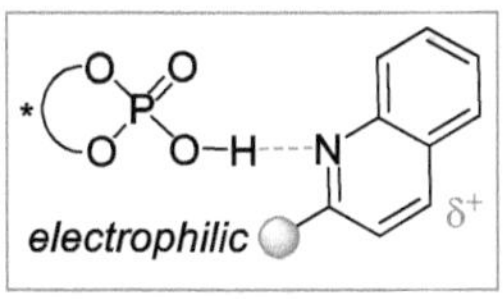

13.4.1 Asymmetric Transfer Hydrogenation

13.4.1.1 Using Chiral Phosphoric Acid Catalyst

Recent advancements in asymmetric synthesis utilizing Brønsted acid catalysts have been remarkable [5, 26]; however, there is a still rare example of such a reaction operating in an aqueous environment. Rueping and Theissmann developed a biomimetic Brønsted acid-catalyzed transfer hydrogenation process in water (Scheme 13.19) [27]. A (*R*)-3,3′-Bis(2,4,6-triisopropylphenyl)-1,1′-binaphthyl-2,2′-diyl hydrogenphosphate (TRIP) catalyst **13** enabled the reduction of quinolines employing

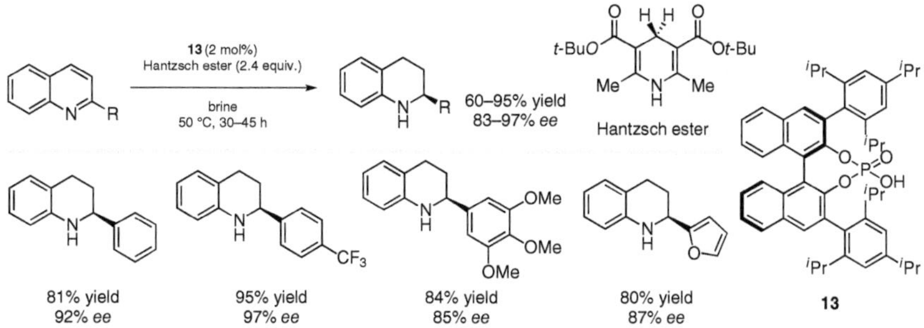

Scheme 13.19 Asymmetric transfer hydrogenation in water using a chiral phosphoric acid catalyst.

Hantzsch ester as an organic hydride donor. According to a mechanistic investigation, the reaction proceeded stepwise with (i) 1,4-hydride addition, (ii) protonation, and (iii) 1,2-hydride addition. Novel "non-covalent asymmetric induction" and the hydrophobic hydration effect afforded chiral tetrahydroquinolines in aqueous environments.

13.5 Asymmetric Brønsted Base Organocatalysis in Water

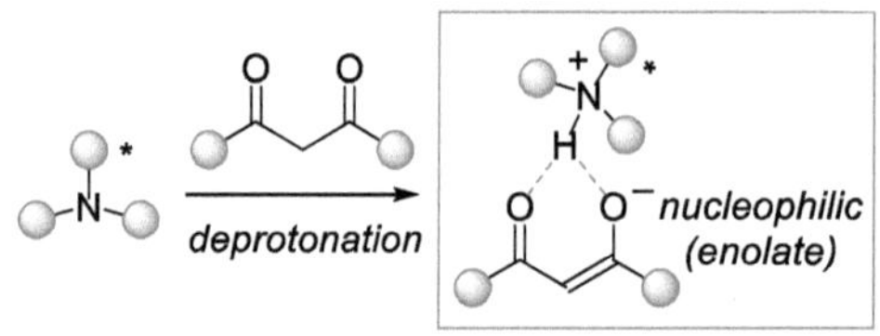

13.5.1 Asymmetric Michael Addition

13.5.1.1 Using Cinchona-Based Squaramide Catalyst: Amplification of Reactivity

Bae and Song reported biomimetic Michael addition "on-water" [28] via Brønsted base catalysis (Scheme 13.20) [29]. The quinuclidine moiety in cinchona alkaloids (e.g. quinine, quinidine, cinchonidine, and cinchonine) [30] serves as highly efficient tertiary amine bases to deprotonate enolate precursors. Water proved to be an exceptionally efficient medium for the non-covalent, hydrogen bonding-mediated asymmetric Michael addition of malonates (and β-ketoesters) to diverse aryl- and alkyl-nitroolefins using dihydroquinine-derived squaramide bifunctional (weak acid–Brønsted base) organocatalyst **14**. The developed aquatic organocatalysis was applied to the multigram scalable syntheses of (*S*)-rolipram (an antidepressant) and (*S*)-pregabalin (an anticonvulsant).

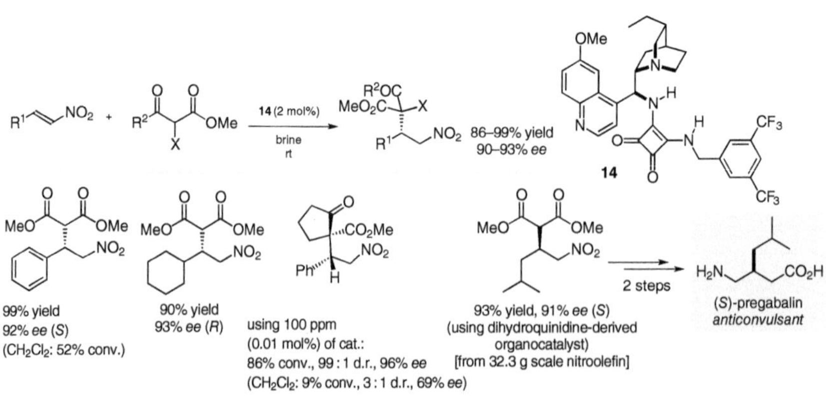

Scheme 13.20 Asymmetric Michael addition of malonates to β-nitroolefins "on-water" using cinchona-based squaramide catalysts.

Unreactive acceptors worked during similar Michael addition by taking advantage of the hydrophobic hydration effects of water [31]. The Song group reported dithiomalonates (DTMs), which proved to be superior enolate donors rather than simple malonates for the Michael addition using the same bifunctional organocatalyst **14** [32]. Complex substrates such as β,β-disubstituted nitroolefins successfully participated in asymmetric Michael addition reactions to provide products with chiral all-carbon quaternary centers (Scheme 13.21).

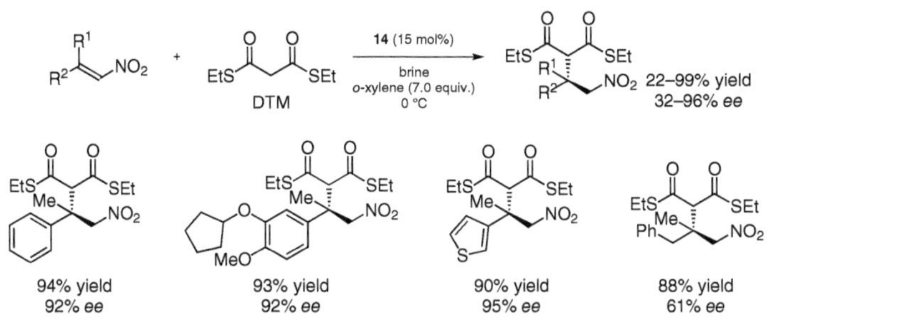

Scheme 13.21 Asymmetric Michael addition of DTMs to β-nitroolefins "on-water" using chiral cinchona-based squaramide catalyst.

13.5.2 Asymmetric Mannich Reaction

13.5.2.1 Using Cinchona-Based Squaramide Catalyst: Amplification of Reactivity

The Song group employed DTMs for asymmetric Mannich reactions using diverse α-amidosulfones as imine precursors in water using quinidine-derived squaramide catalyst **15** [33]. The enhanced reactivity of DTMs proved crucial for facilitating reactions with several challenging substrates incorporating primary alkyl substituents; these reactions did not proceed with less-reactive simple malonates and monothiomalonates. The synthetic utility of the enantio-enriched Mannich adduct was utilized for the first organocatalytic preparation of the antidiabetic drug (*R*)-sitagliptin without the use of coupling reagent (Scheme 13.22).

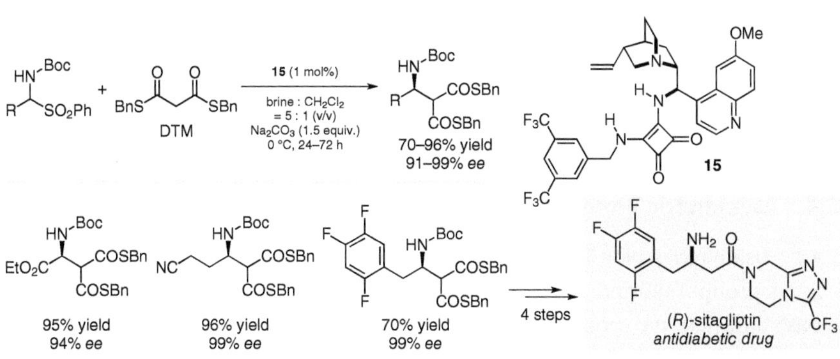

Scheme 13.22 Asymmetric Mannich reaction of DTMs to primary alkyl α-amidosulfones in water using chiral cinchona-based squaramide catalyst.

13.5.2.2 Using Hydrophobic Cinchona Alkaloids: Amplification of Enantioselectivity

Song et al. reported remarkable "chirality amplification" for an asymmetric Mannich reaction within a confined water cage "on-water" [34]. Hydrophobic hydration effects increased proximity between the organocatalyst and substrates, resulting in better enantioselectivity than that obtained using conventional organic solvents. Control experiments showed that reducing droplet size could further enhance chirality amplification. Furthermore, simply prepared hydrophobic cinchona alkaloid catalysts (such as **16**) were more enantioselective than natural cinchona alkaloids (cinchonine or cinchonidine). The authors commented on a plausible and interesting scenario for the homochiral state, derived from the aqueous environment of the primordial earth that included phenomena, such as mist, clouds, and spray (Scheme 13.23).

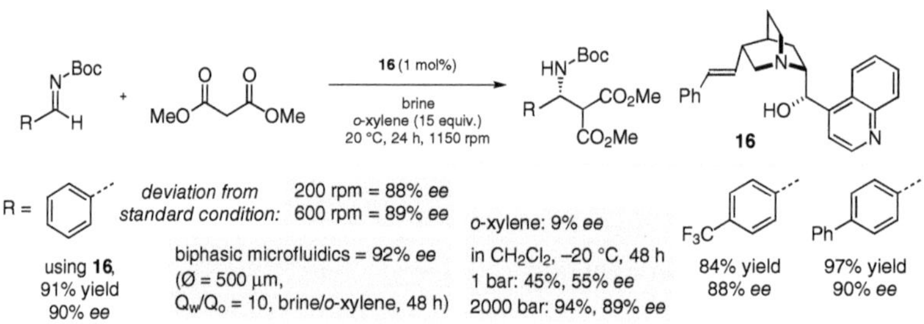

Scheme 13.23 Asymmetric Mannich reaction of malonates to aryl *N*-Boc imines on water using chiral hydrophobic cinchona alkaloid catalyst.

13.5.3 Asymmetric Thiolation

13.5.3.1 Using *tert*-Amine Squaramide Catalyst

Li and coworkers reported the asymmetric conjugated addition of triphenylmethanethiol to directly generated ortho-quinone methides (*o*-QMs). The water/oil biphasic system was crucial for achieving high reactivity and stereoselectivity [35]. In the organic phase, *o*-QMs are generated by a Brønsted basic tertiary amine at the catalyst **17**; the resulting protonated catalyst can be regenerated by an inorganic base (Na_2CO_3) in the aqueous phase. Tertiary amine activates thiol as a nucleophile, and squaramide hydrogen bond donors facilitate Michael-type addition to *o*-QMs (Scheme 13.24).

13.5.4 Asymmetric Protonation

13.5.4.1 Using *tert*-Amine Thiourea Catalyst

The Song group reported catalytic asymmetric protonation in water using an acid–base bifunctional organocatalyst **18** [36]. Due to the extreme mobility of protons in aqueous media, enantioselective protonation of a prochiral carbon ion is challenging. Inspired by enzymatic systems, such as glyoxalase I, the authors established enantioselective isomerization of *in situ* generated hemithioacetals

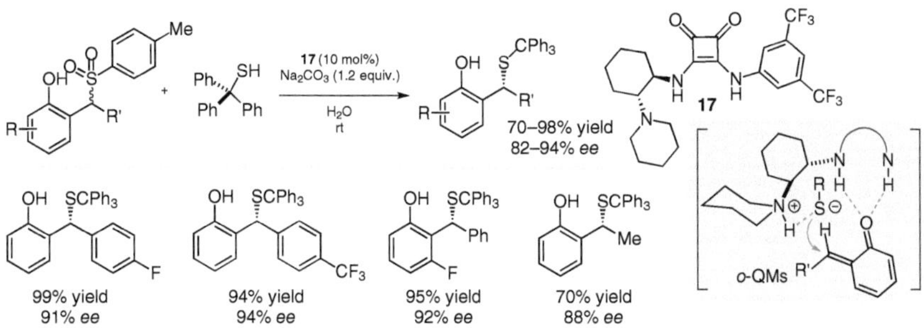

Scheme 13.24 Asymmetric thiolation in water using chiral *tert*-amine squaramide catalyst.

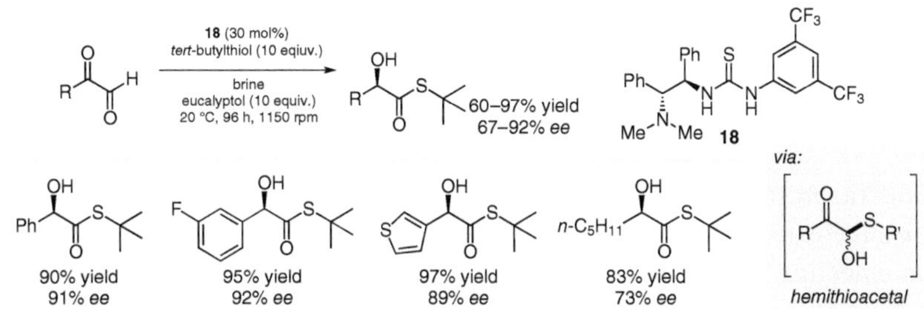

Scheme 13.25 Asymmetric protonation in water using a chiral *tert*-amine thiourea catalyst.

and produced a transient enediol intermediate using a chiral *tert*-amine thiourea catalyst. Experiments using a biphasic microfluidic system revealed that bulk water is essential for the reaction (Scheme 13.25).

13.6 Summary and Conclusions

Recent trends in enantioselective aqueous organic synthesis primarily involve the use of water-compatible metal catalysts or organocatalysts. Breakthrough examples of asymmetric syntheses in water based on different key catalytic activation modes are provided in this chapter. Catalytic activation via ionic (electron pair-involved) acid–base interactions broadly represents the current state of the art. Therefore, the mechanisms can be categorized as Lewis acid, Lewis base, Brønsted acid, and Brønsted base catalyses. Despite significant advancements in organocatalysis including Nobel Prize-winning achievements, [5, 37] a water-compatible asymmetric Brønsted acid catalysis requires further development. One possibly interesting example of non-asymmetric synergistic Brønsted acid-hydrogen bonding donor (DBSA [4-dodecylbenzenesulfonic acid] with **19**) catalysis in water by the Bae group is displayed in Scheme 13.26 [38]. An asymmetric version of this process can be a useful challenge for future investigations.

Additionally, water-accelerated catalytic reactions targeting the syntheses of functionalized alkyl sulfonyl fluorides, which facilitate sulfur(VI) fluoride exchange

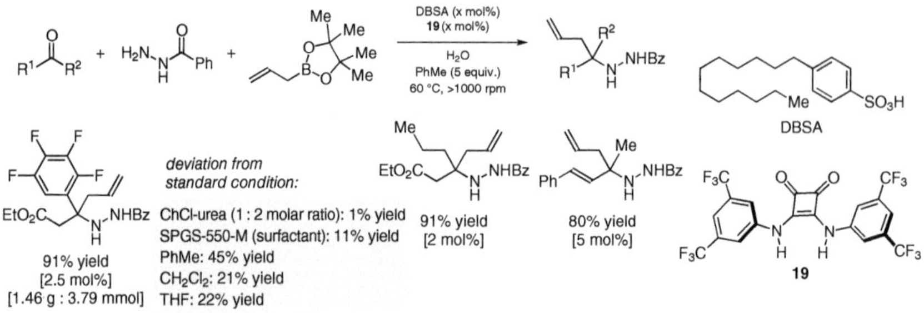

Scheme 13.26 Synergistic Brønsted acid-hydrogen bonding donor catalyzed multicomponent allylation reaction.

(SuFEx) click reactions, [39] were reported very recently [40]. The development of such related examples showcasing asymmetric processes utilizing emerging trends in phosphazene organosuperbase [41] or *N*-heterocyclic carbene [42] remains elusive.

Asymmetric reactions of photochemistry [43] or electrochemistry [44] in bulk water are anticipated in the near future. Some attractive examples of new reactions based on emerging radical mechanisms (e.g. single-electron and energy transfer) in water have been described [45], though enantioselective versions remain in the early developmental stage. These fascinating catalytic processes will undoubtedly provide an efficient means of generating enantio-enriched products in aqueous environments.

Further, water-promoted asymmetric synthetic methodologies utilizing tailored biocatalysts (directed evolution) are of considerable interest [46]. Although not discussed in the main text, technical advances have led to the development of biotechnologies that enable the design of new functional enzymes. For example, chiral pharmaceutical, such as sitagliptin, is now produced industrially using asymmetric reductive amination reactions using designed enzymes [47]. Numerous research groups are actively exploring this field and developing novel biocatalysts.

Acknowledgments

The generous supports of the Ministry of Science, ICT, and Future Planning of Korea (RS-2023-00259659, RS-2023-00219859, 2020R1C1C1006440, 2019R1A6A1A10073079, and 2021R1A2C2093597), and Ministry of Education (2022R1A6C101A751 and 2022R1A6C102A913) are gratefully acknowledged.

References

1 (a) Kobayashi, S. (ed.) (2012). *Science of Synthesis: Water in Organic Synthesis*, 1–867. Stuttgart, Germany: Thieme. (b) Kobayashi, S. (2007). *Pure Appl. Chem.* 79: 235–245. (c) Butler, R.N. and Coyne, A.G. (2010). *Chem. Rev.* 110: 6302–6337.

(d) Li, C.J. and Chen, L. (2006). *Chem. Soc. Rev.* 35: 68–82. (e) Kitanosono, T., Masuda, K., Xu, P., and Kobayashi, S. (2018). *Chem. Rev.* 118: 679–746.
2 Dolling, U.H., Davis, P., and Grabowski, E.J.J. (1984). *J. Am. Chem. Soc.* 106: 446–447.
3 Hayler, J.D., Leahy, D.K., and Simmons, E.M. (2019). *Organometallics* 38: 36–46.
4 Hall, M. (2021). *RSC Chem. Biol.* 2: 958–989.
5 Akiyama, T. and Ojima, I. (ed.) (2022). *Catalytic Asymmetric Synthesis*, 4e, 1–890. Weinheim, Germany: Wiley-VCH.
6 Lindström, U.M. (2007). *Organic Reactions in Water: Principles, Strategies and Applications*, 1–405. Weinheim, Germany: Wiley-Blackwell.
7 Li, J., Tang, Y., Wang, Q. et al. (2012). *J. Am. Chem. Soc.* 134: 18522–18525.
8 Wu, J., Wang, F., Ma, Y. et al. (2006). *Chem. Commun.* 37: 1766–1768.
9 Malkov, A.V., Czemerys, L., and Malyshev, D.A. (2009). *J. Org. Chem.* 74: 3350–3355.
10 Egami, H. and Katsuki, T. (2007). *J. Am. Chem. Soc.* 129: 8940–8941.
11 Nakamura, S., Hara, Y., Furukawa, T., and Hirashita, T. (2017). *RSC Adv.* 7: 15582–15585.
12 Ohara, M., Hara, Y., Ohnuki, T., and Nakamura, S. (2014). *Chem. Eur. J.* 20: 8848–8851.
13 Kitanosono, T., Hisada, T., Yamashita, Y., and Kobayashi, S. (2021). *Angew. Chem. Int. Ed.* 60: 3407–3411.
14 Bednářová, E., Malatinec, Š., and Kotora, M. (2020). *Molecules* 25: 958.
15 Kitanosono, T. and Kobayashi, S. (2015). *Chem. Asian J.* 10: 133–138.
16 Azoulay, S., Manabe, K., and Kobayashi, S. (2005). *Org. Lett.* 7: 4593–4595.
17 Kitanosono, T., Sakai, M., Ueno, M., and Kobayashi, S. (2012). *Org. Biomol. Chem.* 10: 7134–7147.
18 Kobayashi, S., Xu, P., Endo, T. et al. (2012). *Angew. Chem. Int. Ed.* 51: 12763–12766.
19 Kitanosono, T., Zhu, L., Liu, C. et al. (2015). *J. Am. Chem. Soc.* 137: 15422–15425.
20 Mase, N., Nakai, Y., Ohara, N. et al. (2006). *J. Am. Chem. Soc.* 128: 734–735.
21 Font, D., Sayalero, S., Bastero, A. et al. (2008). *Org. Lett.* 10: 337–340.
22 Mase, N., Watanabe, K., Yoda, H. et al. (2006). *J. Am. Chem. Soc.* 128: 4966–4967.
23 Wang, B.G., Ma, B.C., Wang, Q., and Wang, W. (2010). *Adv. Synth. Catal.* 352: 2923–2928.
24 Northrup, A.B. and MacMillan, D.W.C. (2002). *J. Am. Chem. Soc.* 124: 2458–2460.
25 Hayashi, Y., Samanta, S., Gotoh, H., and Ishikawa, H. (2008). *Angew. Chem. Int. Ed.* 47: 6634–6637.
26 (a) Akiyama, T. (2007). *Chem. Rev.* 107: 5744–5758. (b) Akiyama, T. and Mori, K. (2015). *Chem. Rev.* 115: 9277–9306.
27 Rueping, M. and Theissmann, T. (2010). *Chem. Sci.* 1: 473–476.
28 Narayan, S., Muldoon, J., Finn, M.G. et al. (2005). *Angew. Chem. Int. Ed.* 44: 3275–3279.
29 (a) Bae, H.Y. and Song, C.E. (2015). *ACS Catal.* 5: 3613–3619. (b) Bae, H.Y., Some, S., Oh, J.S. et al. (2011). *Chem. Commun.* 47: 9621–9623.

30 Song, C.E. (ed.) (2009). *Cinchona Alkaloids in Synthesis and Catalysis: Ligands, Immobilization and Organocatalysis*, 1–525. Weinheim, Germany: Wiley–VCH.

31 Kleiner, C.M. and Schreiner, P.R. (2006). *Chem. Commun.* 41: 4315–4317.

32 Sim, J.H. and Song, C.E. (2017). *Angew. Chem. Int. Ed.* 56: 1835–1839.

33 Bae, H.Y., Kim, M.J., Sim, J.H., and Song, C.E. (2016). *Angew. Chem. Int. Ed.* 55: 10825–10829.

34 Song, C.E., Park, S.J., Hwang, I.S. et al. (2019). *Nat. Commun.* 10: 851.

35 Guo, W., Wu, B., Zhou, X. et al. (2015). *Angew. Chem. Int. Ed.* 54: 4522–4526.

36 Park, S.J., Hwang, I.S., Chang, Y.J., and Song, C.E. (2021). *J. Am. Chem. Soc.* 143: 2552–2557.

37 (a) https://nobelprize.org/prizes/chemistry/2021/press-release/; (b) MacMillan, D.W.C. (2008). *Nature* 455: 304–308. (c) List, B. and Yang, J.W. (2006). *Science* 313: 1584–1586. (d) Xiang, S.-H. and Tan, B. (2020). *Nat. Commun.* 11: 3786. (e) Albrecht, L., Albrecht, A., and Dell'Amico, L. (ed.) (2022). *Asymmetric Organocatalysis: New Strategies, Catalysts, and Opportunities*, 1–724. Weinheim, Germany: Wiley–VCH.

38 Goswami, P., Cho, S.Y., Park, J.H. et al. (2022). *Nat. Commun.* 13: 2702.

39 Dong, J., Sharpless, K.B., Kwisnek, L. et al. (2014). *Angew. Chem. Int. Ed.* 53: 9466–9470.

40 Koo, B.J., Lee, S.B., Kim, W.H. et al. (2023). *Aldrichimica Acta* 56: 35–46.

41 (a) Park, J.H., González-Montiel, G.A., Cheong, P.H.-Y., and Bae, H.Y. (2023). *Org. Lett.* 25: 1056–1060. (b) Lee, S.B., Park, J.H., and Bae, H.Y. (2022). *ChemSusChem* 15: e202200634. (c) Kim, S., Lee, S.B., Park, J.H. et al. (2024). *Org. Lett.* 26: 7352–7356.

42 Park, J.H., Lee, S.B., Koo, B.J., and Bae, H.Y. (2022). *ChemSusChem* 15: e202201000.

43 Genzink, M.J., Kidd, J.B., Swords, W.B., and Yoon, T.P. (2022). *Chem. Rev.* 122: 1654–1716.

44 Rein, J., Zacate, S.B., Mao, K., and Lin, S. (2023). *Chem. Soc. Rev.* 52: 8106–8125.

45 Kim, S.B., Kim, D.H., and Bae, H.Y. (2024). *Nat. Commun.* 15: 3876.

46 (a) Bell, E.L., Finnigan, W., France, S.P. et al. (2021). *Nat. Rev. Methods Primers* 1: 46. (b) Wang, Y., Xue, P., Cao, M. et al. (2021). *Chem. Rev.* 121: 12384–12444. (c) Reetz, M.T. (ed.) (2017). *Directed Evolution of Selective Enzymes: Catalysts for Organic Chemistry and Biotechnology*, 1–308. Weinheim, Germany: Wiley–VCH.

47 Savile, C.K., Janey, J.M., Mundorff, E.C. et al. (2010). *Science* 329: 305–309.

14

Organocatalytic Reactions in Water

Bubun Banerjee[1], Anu Priya[1], Zsuzsanna Szalai[2], Aditi Sharma[1], Manmeet Kaur[1], Arvind Singh[1], and György Keglevich[2]

[1]*Akal University, Department of Chemistry, Talwandi Sabo, Bathinda, Punjab 151302, India*
[2]*Budapest University of Technology and Economics, Faculty of Chemical Technology and Biotechnology, Department of Organic Chemistry and Technology, Műegyetem rkp. 3, 1111 Budapest, Hungary*

14.1 Introduction

Nowadays, sustainable organic transformations are in high demand both in academia and in the chemical industries to prevent the side effects of the hazardous substances associated or generated in chemical processes [1–4]. Scientists are trying to take all necessary steps to protect our "*Mother Nature*" from the ever-increasing chemical pollution. Thus, special emphasis has been given to design the chemical transformations more environmentally friendly and sustainable by using non-hazardous reagents, catalysts, and solvents [5–8]. As a result, the last decade has witnessed a tremendous outburst in the modification of chemical processes to make them more sustainable and environmentally benign. Catalysts play an important role in organic synthesis among other parameters. Though a large number of various metal catalysts have been applied for diverse organic transformations, these days scientists are reluctant to use metal catalysts due to their toxic nature and to avoid metal contamination. A large number of research groups aimed at developing suitable, as well as sustainable alternatives to avoid the use of toxic metal catalysts. The use of metal-free organocatalysis in chemical transformations fulfills this goal on many occasions due to their nontoxicity and biodegradable nature [9–13]. Realizing the importance of organocatalysis, in 2021, List and MacMillan were awarded the Nobel Prize in Chemistry for carrying out diverse asymmetric reactions by using chiral organic substances as catalysts.

On the other hand, water is considered the safest solvent to carry out organic transformations as it is cheap, environmentally benign, nonflammable, and abundantly available. Water behaves differently than other organic solvents, as it can activate the functional groups by forming hydrogen bonds [14, 15]. Additionally, the high surface tension and hydrophobic nature of organic reactants bound to aggregate with each other in aqueous medium to decrease the exposed organic surface area. This helps reactants to get closer, thereby increasing the rate of the reaction [16–18].

Organic Transformations in Water: Principles and Applications, First Edition.
Edited by Gopinathan Anilkumar, Nissy Ann Harry, and Sankuviruthiyil M. Ujwaldev.

Therefore, a "strong collaboration" between metal-free organocatalyst and water as the solvent holds the key to achieving the sustainability goal. The beneficial features in terms of sustainability have motivated organic chemists to explore the feasibility of water-mediated organocatalyzed reactions in more depth. As a result, the catalytic activity of a wide variety of metal-free organocatalysts has been screened in aqueous medium for diverse organic transformations. A series of amines, amino acids, surfactants, and other organic compounds with reactive functional groups are frequently used as efficient organocatalysts. In 2016, Jimeno [19] compiled a large number of literature related to water-mediated asymmetric synthesis. In 2019, Eelkema and coworkers[20] published an interesting review on water-mediated organocatalysis. In this chapter, we have summarized the latest developments on water-mediated organocatalyzed reactions reported during the last four years.

14.2 Organocatalyzed Synthesis of Biologically Promising *N*-Heterocycles in Water

14.2.1 Synthesis of Decahydroacridine-1,8-Diones and Dihydropyrido[2,3-d:6,5-*d′*]Dipyrimidines

In 2020, Bhosle et al. [21] developed a simple, facile, and water-mediated protocol for the synthesis of a series of decahydroacridine-1,8-diones (**5**), as well as dihydropyrido[2,3-d:6,5-*d′*]dipyrimidine derivatives (**6**), in excellent yields *via* the one-pot three-component reaction of substituted anilines (**1**), 4-((2-phenylthiazol-4-yl)methoxy)benzaldehyde (**2**), and dimedone (**3**) or barbituric acid (**4**), respectively, in the presence of 30 mol% taurine (2-aminoethanesulfonic acid) as an efficient organocatalyst under reflux conditions (Scheme 14.1).

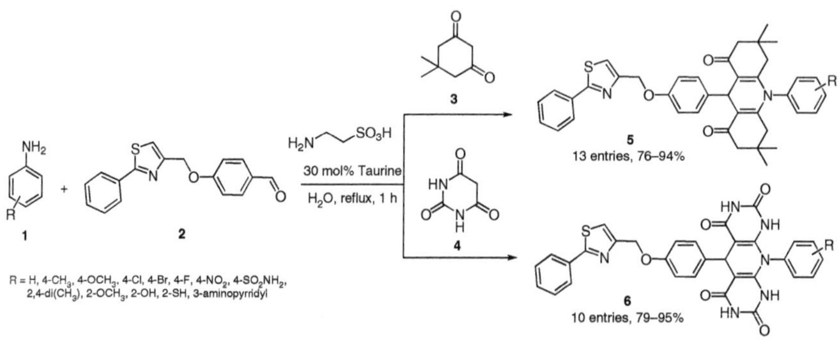

Scheme 14.1 Synthesis of decahydroacridine-1,8-diones (**5**) and dihydropyrido-[2,3-*d*:6,5-*d′*] dipyrimidines (**6**) catalyzed by taurine as an organocatalyst in water.

14.2.2 Synthesis of Aminocyanopyridines

In 2020, Tamaddon and Tadayonfar [22] prepared a diester cationic gemini surfactant (DCGS) under microwave conditions according to the pathway depicted in Scheme 14.2. Using just 5 mol% of this freshly prepared DCGS as the catalyst, they synthesized a series of 2-amino-4-substituted-6-phenylnicotinonitriles (**11**) *via* one-pot four-component reactions of substituted benzaldehydes (**7**), acetophenone (**8**), malononitrile (**9**), and ammonium carbonate (**10**) under reflux conditions in water (Scheme 14.3). It was proposed that the reaction occurred in the micellar medium formed by the ionic catalyst in water. After the completion of the reaction, the entire micellar medium was recovered and recycled for four successive runs with practically similar efficiency.

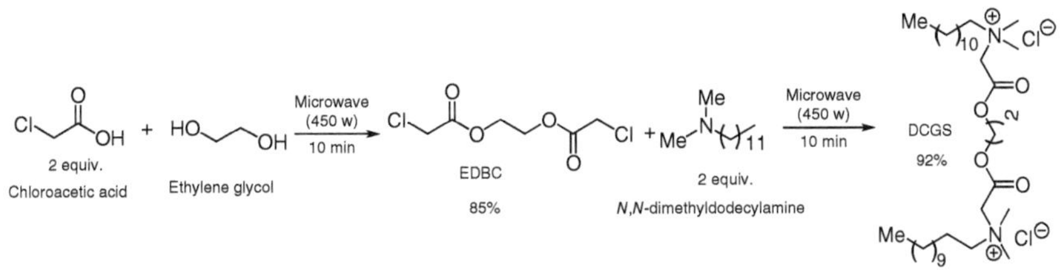

Scheme 14.2 Microwave-assisted synthesis of the DCGS.

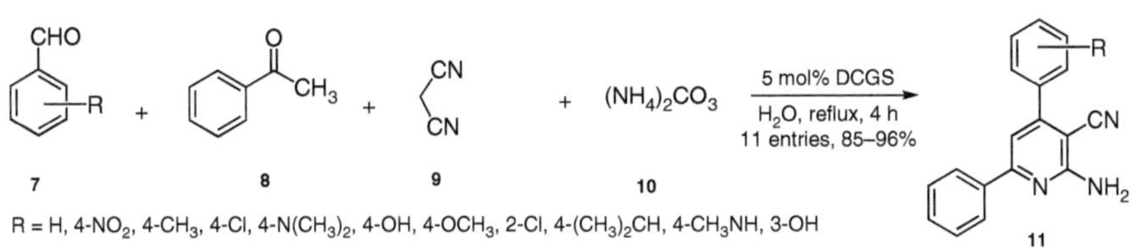

Scheme 14.3 DCGS-catalyzed one-pot four-component synthesis of 2-amino-4-substituted-6-phenylnicotinonitrile (**11**) in water.

Recently, in 2023, Bhadke et al. [23]synthesized another series of 2-amino-4-substituted-6-phenylnicotinonitrile derivatives (**11**) using 10 mol% of *L*-proline as the catalyst at 60 °C in water.

14.2.3 Synthesis of 1,2,6-Triaryl-4-Arylamino-Piperidine-3-Ene-3-Carboxylates

1,2,6-Triaryl-4-arylamino-piperidine-3-ene-3-carboxylate derivatives (**13**) possess immense biological activities [24]. A series of 1,2,6-triaryl-4-arylamino-piperidine-3-ene-3-carboxylate derivatives (**13**) was synthesized by Dogra et al. [25] *via* a one-pot pseudofive-component reaction of two molecules of substituted anilines (**1**), two molecules of aromatic aldehydes (**7**), and one equivalent of β-keto esters (**12**) in the presence of a catalytic amount of citric acid as an efficient organocatalyst at room temperature in water (Scheme 14.4). Majority of the synthesized compounds showed excellent antioxidant activities.

14.2.4 Synthesis of 2-Substituted-2,3-Dihydroquinazolin-4(1*H*)-ones

In 2020, Pariyar et al. [26] synthesized a series of 2-substituted-2,3-dihydroquinazolin-4(1*H*)-one derivatives (**15**) by the reaction of various aldehydes (**7**) and 2-aminobenzamide (**14**) at 90 °C in the presence of a catalytic amount of ascorbic

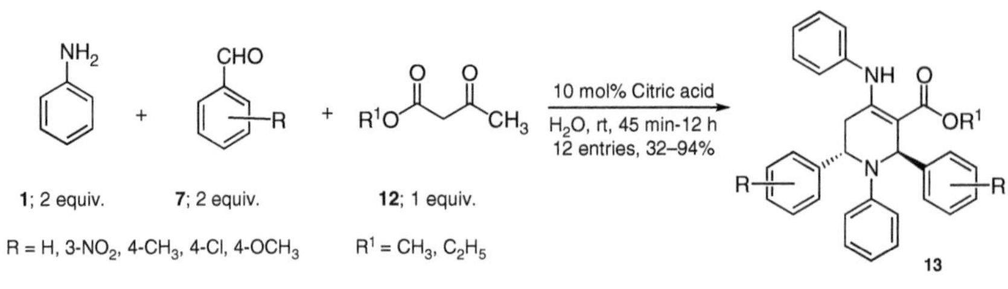

Scheme 14.4 Citric acid-catalyzed one-pot pseudo five-component synthesis of 1,2,6-triaryl-4-arylamino-piperidine-3-ene-3-carboxylates (**13**) in water.

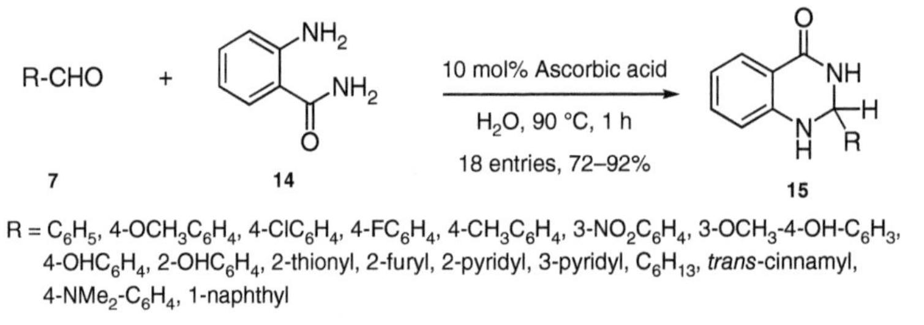

Scheme 14.5 Ascorbic acid-catalyzed synthesis of 2-substituted-2,3-dihydroquinazolin-4(1*H*)-one 2,3-dihydroquinazolin-4(1*H*)-one (**15**) in water.

acid in water (Scheme 14.5). All reactions were completed within one hour and afforded product **15** in excellent yields (72–92%).

In 2019, Chate et al. [27] reported a facile organocatalyzed water-mediated protocol for the synthesis of a series of 1,2-(dihydroquinazolin-3(4*H*)-isonicotinamides (**19**) by the reaction of isatoic anhydride (**16**), isoniazid (**17**), and aromatic aldehydes (**7**) in the presence of 15 mol% taurine as catalyst under reflux conditions (Scheme 14.6). Under similar conditions, a series of spirooxindole dihydroquinazolinones (**20**) was also prepared in excellent yields *via* the one-pot three-component reaction of isatoic anhydride (**16**), aromatic amines (**1**), and isatins (**18**). Recently, the same reaction was realized using *L*-proline as an efficient organocatalyst in the presence of Triton X-100 as a surfactant in water [28].

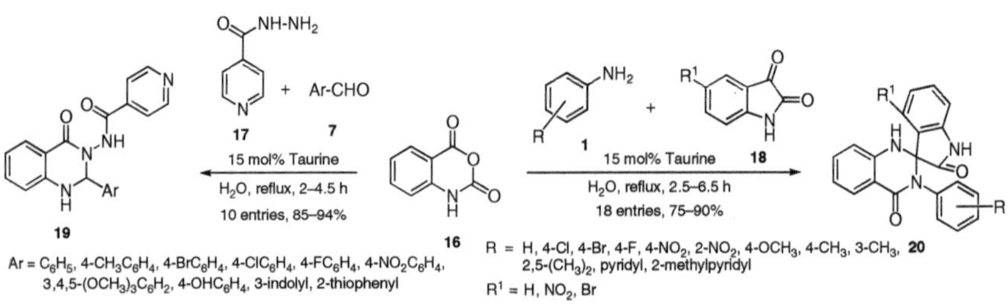

Scheme 14.6 Taurine-catalyzed synthesis of novel 1,2-(dihydroquinazolin-3(4*H*)-isonicotinamides (**19**) and spirooxindole dihydroquinazolinones (**20**) in water.

14.2.5 Synthesis of 2,3-Dihydro-1*H*-Perimidines

In 2020, Khopkar and Shankarling [29] employed catalytic amount of squaric acid as an efficient organocatalyst for the synthesis of a series of 2,3-dihydro-1*H*-perimidine derivatives (**23**) by the reaction of 1,8-diamino naphthalene (**21**) and structurally diverse cyclic or acyclic ketones (**22**) at 80 °C in water (Scheme 14.7).

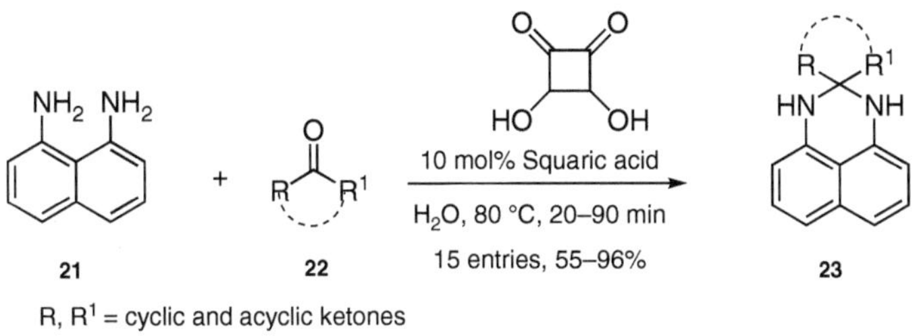

Scheme 14.7 Squaric acid-catalyzed synthesis of 2,3-dihydro-1*H*-perimidines (**23**) in water.

14.2.6 Synthesis of Imidazopyridines

In 2020, Bhutia et al. [30] synthesized a number of 2-arylimidazo[1,2-*a*]pyridine derivatives (**25**) by the reaction of 2-aminopyridines (**24**) and various arylmethyl ketones (**8a**) at 40 °C in the presence of 30 mol% iodine and 10 mol% sodium dodecyl sulphate (SDS) as a cocatalyst (Scheme 14.8).

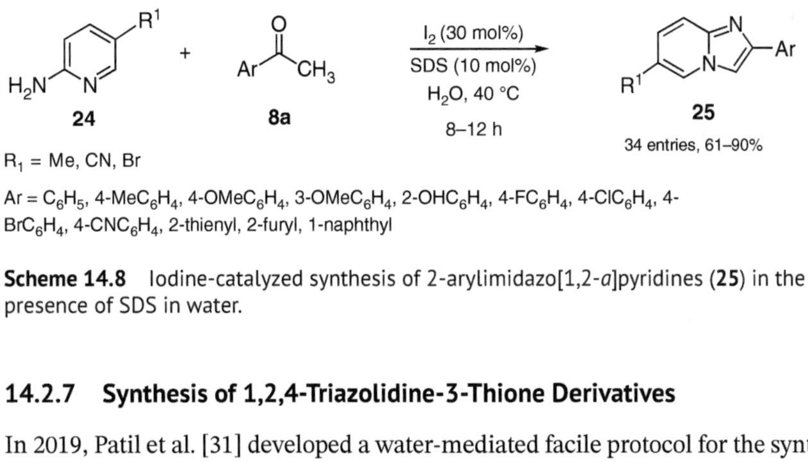

Scheme 14.8 Iodine-catalyzed synthesis of 2-arylimidazo[1,2-*a*]pyridines (**25**) in the presence of SDS in water.

14.2.7 Synthesis of 1,2,4-Triazolidine-3-Thione Derivatives

In 2019, Patil et al. [31] developed a water-mediated facile protocol for the synthesis of a series of structurally diverse 1,2,4-triazolidine-3-thiones (**28**, **28a**, **29**, and **30**) by the reaction of thiosemicarbazides (**26**) and various carbonyl compounds, such as aromatic aldehydes (**7**), substituted acetophenones (**8a**)/isatins (**18**)/cyclic ketones (**27a–e**) at room temperature using a catalytic amount of thiamine hydrochloride as an efficient organocatalyst (Scheme 14.9). The catalyst was recovered and recycled for five successive runs with similar efficiency.

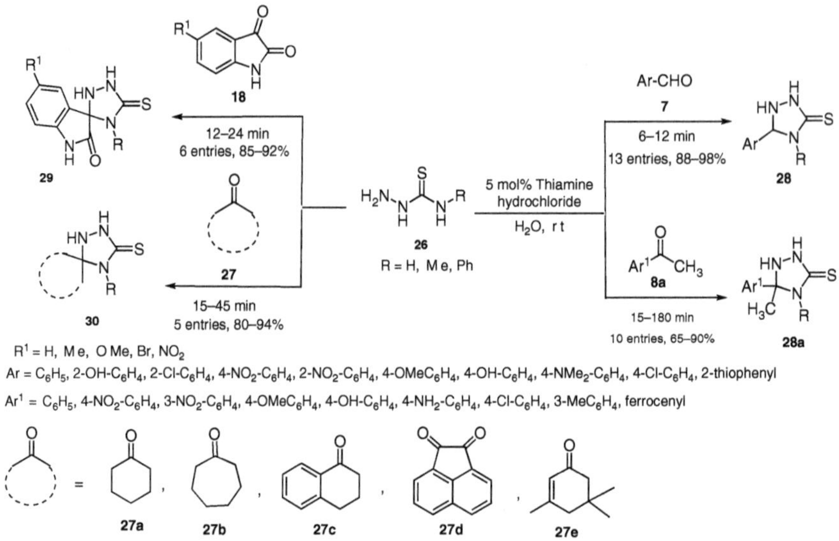

Scheme 14.9 Thiamine hydrochloride-catalyzed synthesis of 1,2,4-triazolidine-3-thione derivatives (**28** and **28a**).

14.3 Organocatalyzed Synthesis of Biologically Promising *O*-Heterocycles in Water

14.3.1 Synthesis of 4*H*-Pyrans, Pyran-Annulated Heterocycles, and Spiro-Pyrans

4*H*-pyrans and pyran-annulated heterocycles were found to possess a wide range of biological efficacies including anticancer, antitumor, antifungal, antibacterial, anti-HIV etc activities [32–35]. In 2020, Auria-Luna et al. [36] reported a simple, ultrasound-assisted water-mediated protocol for the synthesis of a series of diethyl 6-amino-5-cyano-4-aryl-4*H*-pyran-2,3-dicarboxylate derivatives (**32**) *via* the one-pot three-component reaction of aromatic aldehydes (**7**), malononitrile (**9**), and diethyl 2-oxosuccinate (**31**) at room temperature in the presence of a catalytic amount of triethyl amine (Scheme 14.10). A few products showed excellent DNA binding ability.

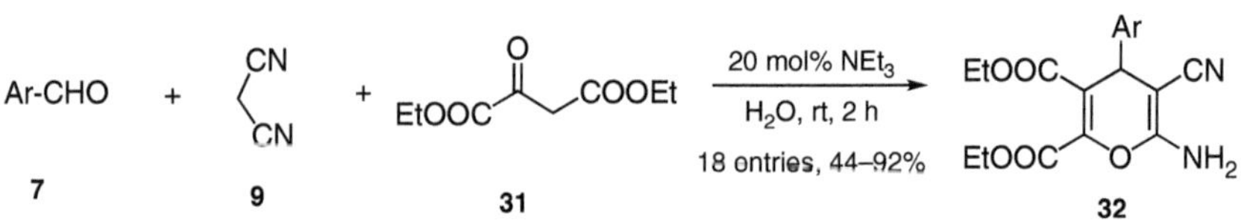

Ar = C_6H_5, 4-CNC_6H_4, 4-ClC_6H_4, 3-ClC_6H_4, 4-$CF_3C_6H_4$, 4-BrC_6H_4, 4-$NO_2C_6H_4$, 3-$NO_2C_6H_4$, 4-$OMeC_6H_4$, 4-MeC_6H_4, 3,5-$(CF_3)_2C_6H_4$, benzyl, furyl, thienyl, pyridyl

Scheme 14.10 Ultrasound-assisted, triethylamine-catalyzed synthesis of diethyl 6-amino-5-cyano-4-aryl-4*H*-pyran-2,3-dicarboxylates in water.

Recently, our group has synthesized 2-amino-3-cyano-4-aryltetrahydrobenzo[*b*] pyrans (**33**) *via* the one-pot three-component reaction of aromatic aldehydes (**7**), malononitrile (**9**), and dimedone (**3**) or 1,3-cyclohexanedione (**3a**) at room temperature using SDS as an efficient surfactant type catalyst in water (Scheme 14.11) [37]. A plausible mechanism and the role of SDS as the catalyst are shown in Scheme 14.12. It was proposed that SDS in water formed micelles, which helped the organic reactants to get closer to the hydrophobic interior. Under similar reaction conditions, a series of spiropyrans (**35** and **36**) was synthesized in excellent yields by the reaction of dimedone (**3**) or 1,3-cyclohexanedione (**3a**), malononitrile (**9**), and isatin (**18**) or ninhydrin (**34**), respectively (Scheme 14.13).

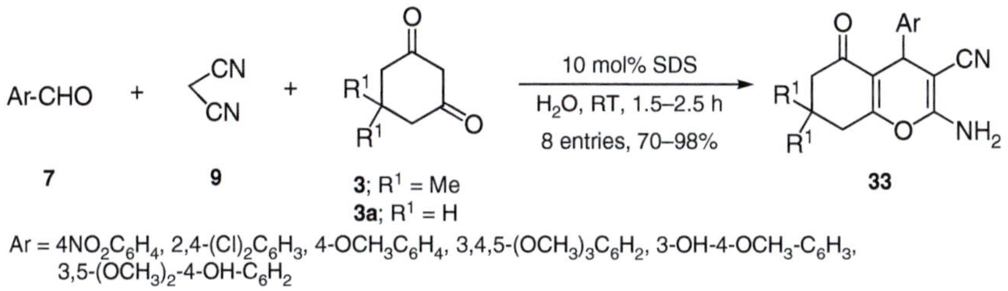

Scheme 14.11 SDS-catalyzed synthesis of 2-amino-3-cyano-4-aryltetrahydrobenzo[*b*]pyrans in water.

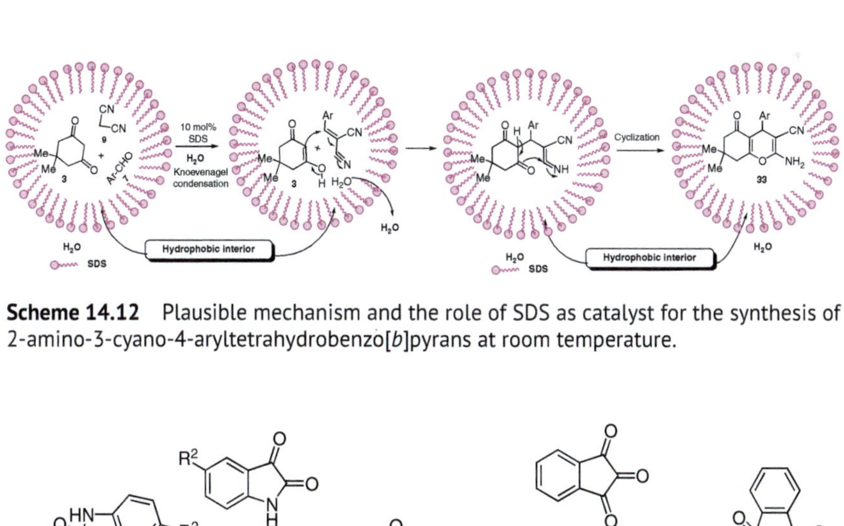

Scheme 14.12 Plausible mechanism and the role of SDS as catalyst for the synthesis of 2-amino-3-cyano-4-aryltetrahydrobenzo[*b*]pyrans at room temperature.

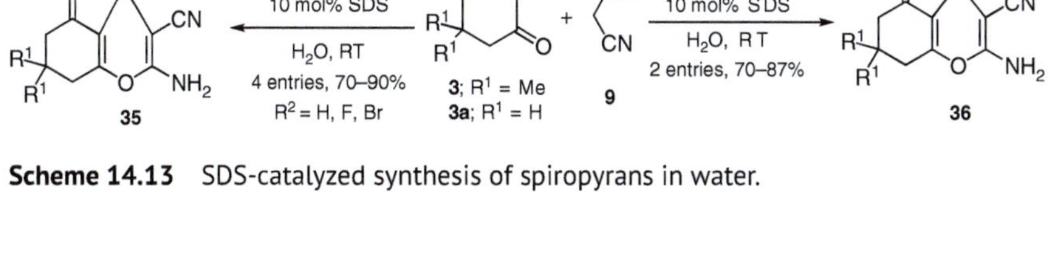

Scheme 14.13 SDS-catalyzed synthesis of spiropyrans in water.

14.3.2 Synthesis of Chromenes

In 2022, Rani et al. [38] reported a water-mediated facile protocol for the efficient synthesis of 2-aryl-3-nitro-2*H*-chromene derivatives (**38**) in excellent yields *via* tandem oxa-Michael Henry reactions of substituted salicylaldehydes (**7a**) and β-nitrostyrenes (**37**) in the presence of a catalytic amount of *L*-prolinamide and benzoic acid as a cocatalyst at 70 °C (Scheme 14.14). A number of other organic solvents, chloroform, acetonitrile, toluene, tetrahydrofuran, DMSO, 1,4-dioxane, EtOH, etc., were also tried out, but the best results in terms of yields as well as reaction times were obtained in aqueous medium.

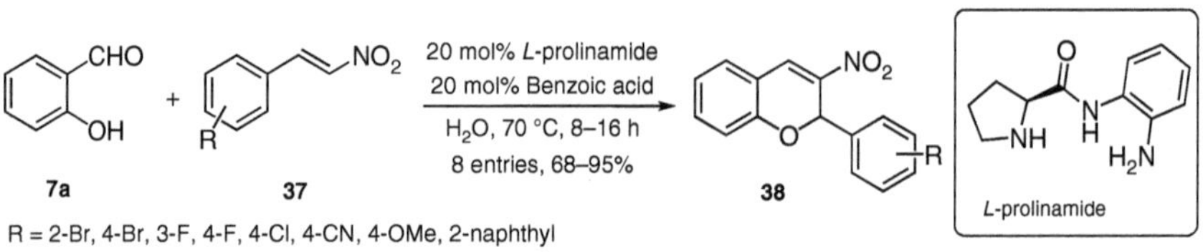

Scheme 14.14 *L*-prolinamide-catalyzed synthesis of 2-aryl-3-nitro-2*H*-chromene derivatives in water.

Recently, in 2023, Thangalipalli et al. [39] synthesized a series of 2-amino-4-aryl-4*H*-chromenes (**40**) in high yields (88–97%) *via* the one-pot three-component reaction of aromatic aldehydes (**7**), malononitrile (**9**)/ethyl 2-cyanoacetate (**9a**), and resorcinols (**39**) using 4-dimethylaminopyridine (DMAP) as an organocatalyst at room temperature in water (Scheme 14.15). The catalyst applied was recovered and recycled for six subsequent runs without the loss of its catalytic activities. Earlier, in 2012, the same reaction was performed in water using glycine as the catalyst under the effect of ultrasonic irradiation [40].

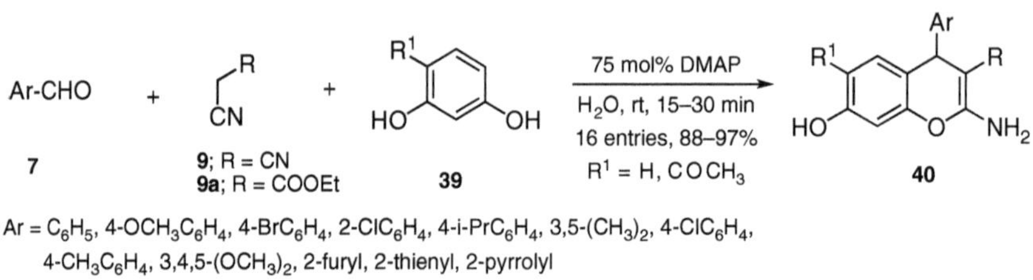

Scheme 14.15 4-Dimethylaminopyridine-catalyzed synthesis of 2-amino-4-aryl-4*H*-chromenes in water.

Last year, our group [41] developed one-pot three-component protocol for the synthesis of 2-amino-3-cyano-4-substitued-4*H*-chromenes (**46a–j**) by the reaction of substituted salicyaldehydes (**7a**); malononitrile (**9**); and a variety of nucleophiles for example another molecule of malononitrile (**9**), indoles (**41**), thiophenol (**42**), nitromethane (**43**), 2-napthalenethiol (**44**), and cyanoacetamide (**45**) in the presence of SDS as an efficient surfactant type organocatalyst at room temperature in water (Scheme 14.16). Proposed mechanism of this reaction and the role of the catalyst are shown in Scheme 14.17.

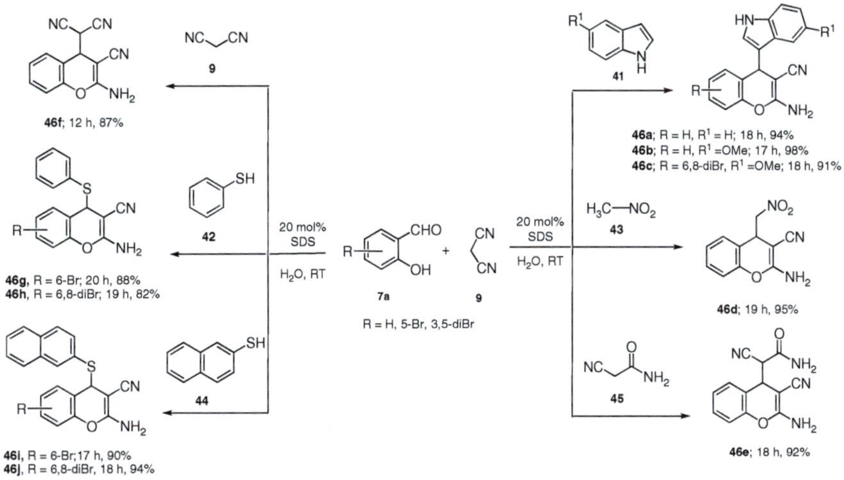

Scheme 14.16 SDS-catalyzed synthesis of 2-amino-3-cyano-4-substitued-4*H*-chromenes in water.

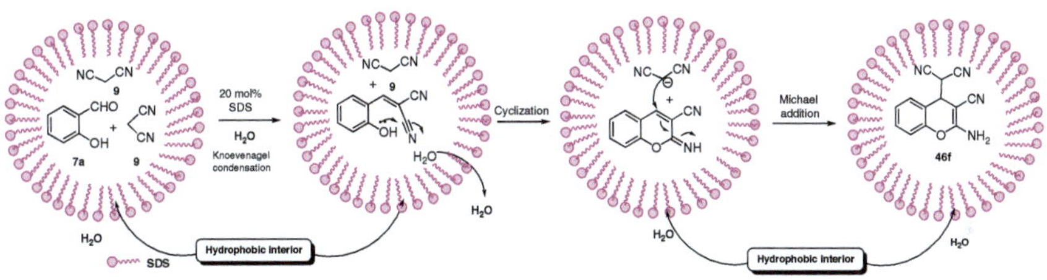

Scheme 14.17 Proposed mechanism for the synthesis of 2-amino-3-cyano-4-substitued-4*H*-chromenes.

14.3.3 Synthesis of Tetrahydrobenzo[a]xanthene-11-ones

In 2021, Chavan et al. [42] utilized Baker's yeast as an efficient catalyst for the water-mediated synthesis of a series of tetrahydrobenzo[*a*]xanthene-11-ones (**48**) *via* one-pot three-component reactions of aromatic aldehydes (**7**), dimedone (**3**), and β-naphthol (**47**) at room temperature (Scheme 14.18). All reactions were complete within 45 minutes and afforded the products (**48**) in good yields (79–90%).

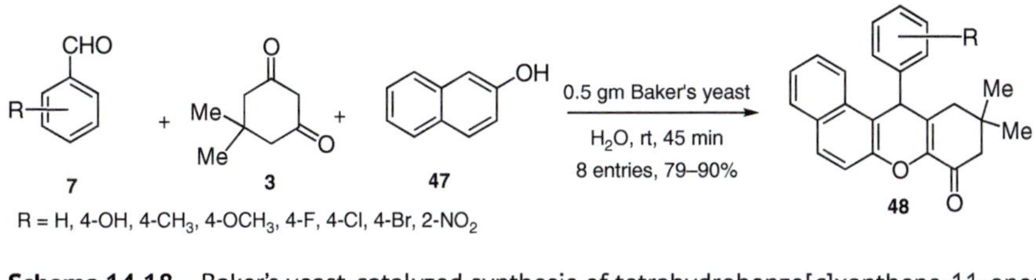

Scheme 14.18 Baker's yeast-catalyzed synthesis of tetrahydrobenzo[*a*]xanthene-11-ones in water.

14.4 Organocatalyzed Synthesis of Biologically Promising *N*,*O*-Heterocycles in Water

14.4.1 Synthesis of Pyrano[3,2-*c*]Quinolone Derivatives

A series of pyrano[3,2-*c*]quinolone derivatives (**50**) was synthesized using a catalytic amount of taurine as an efficient and green organocatalyst under reflux conditions in water in the one-pot three-component reaction of 4-hydroxy-1-methyl-2(1*H*)-quinolone (**49**), substituted benzaldehydes (**7**), and malononitrile (**9**)/ethylcyanoacetate (**9a**) (Scheme 14.19) [43]. The catalyst was recovered and recycled for three subsequent runs without loss of its catalytic activity.

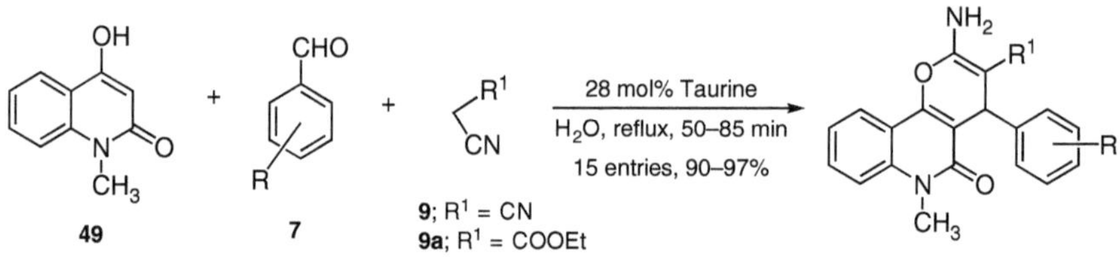

50

R = H, 3,5-$(OCH_3)_2$, 2-OH, 3,4,5-$(OCH_3)_3$, 4-NO_2, 2-NO_2, 4-Cl, 4-OMe, trans-cinnamyl

R^1 = CN, $COOC_2H_5$

Scheme 14.19 Taurine-catalyzed synthesis of pyrano[3,2-*c*]quinolone derivatives in water.

14.4.2 Synthesis of Chromeno[4,3-*d*]Pyrimidines

In 2019, Kaurav et al. [44] synthesized a series of chromeno[4,3-*d*]pyrimidine derivatives (**23**) by the one-pot three-component reactions of various aromatic aldehydes (**7**), urea (**51**)/thiourea (**51a**), and 4-hydroxycoumarin (**52**) in the presence of 10 mol% *L*-proline as an efficient organocatalyst at 70 °C in water (Scheme 14.20). Under conventional heating, completion of the reaction required 3–5 hours, whereas the same transformations were complete within 10 minutes under microwave conditions using the same amount of catalyst.

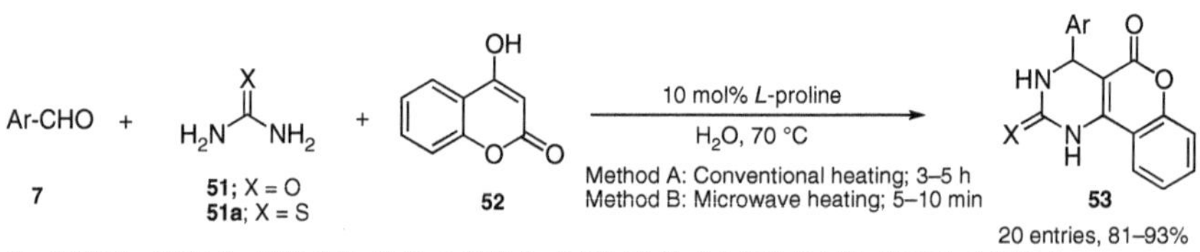

Ar = 4-ClC_6H_4, 2-ClC_6H_4, 3-$NO_2C_6H_4$, C_6H_5, 2-OHC_6H_4, 4-$OCH_3C_6H_4$, 3,4-$(OCH_3)_2C_6H_3$, 3-OCH_3-4-OHC_6H_3, 4-$N(CH_3)_2C_6H_4$, indolyl, furyl

Scheme 14.20 *L*-proline-catalyzed synthesis of chromeno[4,3-*d*]pyrimidines in water.

14.4.3 Synthesis of 3,4-Disubstituted Isoxazol-5(4*H*)-ones

In 2021, Ghogare et al. [45] reported a simple and efficient one-pot three-component protocol for the synthesis of 3,4-disubstituted isoxazol-5(4*H*)-one derivative (**55**) by

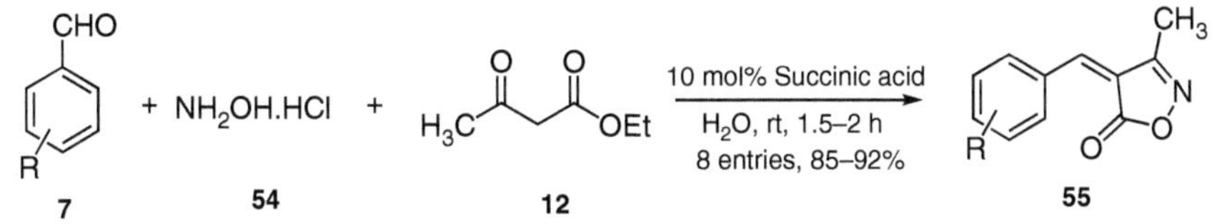

R= H, 4-OH, 4-CH$_3$, 4-OCH$_3$, 3,4-(OCH$_3$)$_2$, 4-OH-3-OMe, 3,4,5-(OCH$_3$)$_3$, 4-NMe$_2$

Scheme 14.21 Succinic acid-catalyzed synthesis of 3,4-disubstituted isoxazol-5(4*H*)-ones in water.

the reaction of aromatic aldehydes (**7**), hydroxylamine hydrochloride (**54**), and ethyl acetoacetate (**12**) in the presence of a catalytic amount of succinic acid as the catalyst at room temperature in water (Scheme 14.21).

14.4.4 Synthesis of Chromeno[4,3-*b*]Pyrano[3,4-*e*]Pyridines

In 2021, a catalytic amount of *L*-proline (10 mol%) was used as an efficient organocatalyst for the synthesis of a series of chromeno[4,3-*b*]pyrano[3,4-*e*]pyridine derivatives (**56**) *via* a one-pot four-component reaction of substituted benzaldehydes (**7**), ammonium acetate (**56**), 4-hydroxycoumarin (**52**), and 6-methyl-2*H*-pyran-2,4(3*H*)-dione (**57**) under reflux conditions (Scheme 14.22) in water [46].

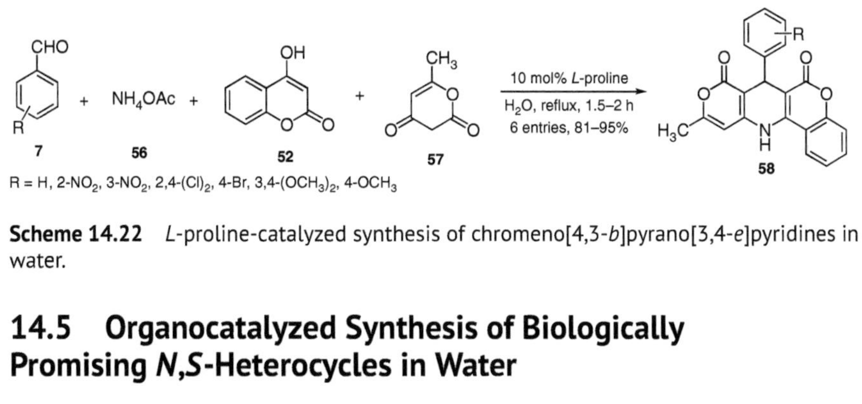

R = H, 2-NO$_2$, 3-NO$_2$, 2,4-(Cl)$_2$, 4-Br, 3,4-(OCH$_3$)$_2$, 4-OCH$_3$

Scheme 14.22 *L*-proline-catalyzed synthesis of chromeno[4,3-*b*]pyrano[3,4-*e*]pyridines in water.

14.5 Organocatalyzed Synthesis of Biologically Promising *N*,*S*-Heterocycles in Water

14.5.1 Synthesis of Pyrimido[2,1-*b*]Benzothiazole Derivatives

Singh and Lal [47] synthesized a series of pyrimido[2,1-*b*]benzothiazolederivatives (**60**) *via* a one-pot three-component cyclo-condensation of various aromatic aldehydes (**7**), 2-aminobenzothiazole (**59**), and ethyl acetoacetate (**12**) in the presence of a catalytic amount of 1,8-diazabicyclo[5.4.0]undec-7-ene (DBU) in a micellar medium at 60 °C (Scheme 14.23). The micellar was prepared by applying the surfactant, cetylpyridinium bromide, in water.

14.5.2 Synthesis of Spiro[Benzothiazolo[3,2-*a*]Pyrimidine-4,3′-Indolines

The catalytic activity of *L*-proline was investigated in water in the synthesis of a series of spiro[benzothiazolo[3,2-*a*]pyrimidine-4,3′-indoline derivatives (**61**)

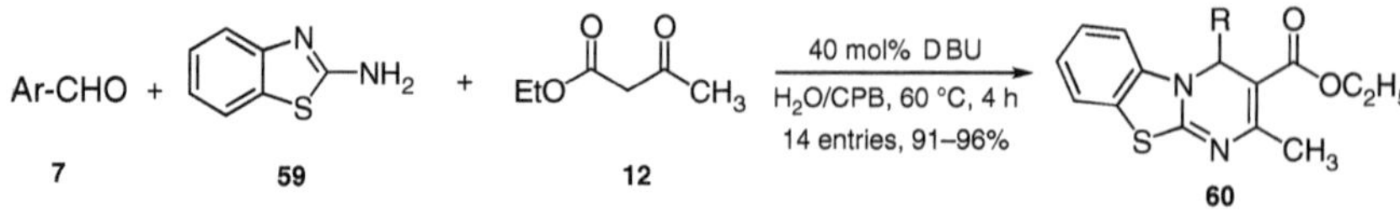

Ar = C_6H_5, 4-$NO_2C_6H_4$, 3-$CH_3C_6H_4$, 3-Cl-C_6H_4, 4-$N(CH_3)_2C_6H_4$, 4-OH-3-$OCH_3C_6H_3$, 2-ClC_6H_4, 3-CH_3-4-$OCH_3C_6H_3$, 3-$(CH_3)_2C_6H_4$, 4-$OCH_3C_6H_4$, 2-Cl-6-ClC_6H_3, 2-OHC_6H_4, 3-$OCH_3C_6H_4$, 4-OHC_6H_4, 3-$NO2C_6H_4$, 2,4-$Cl_2C_6H_3$

Scheme 14.23 DBU-catalyzed 4*H*-pyrimido[2,1-*b*]benzothiazole derivatives in micellar medium.

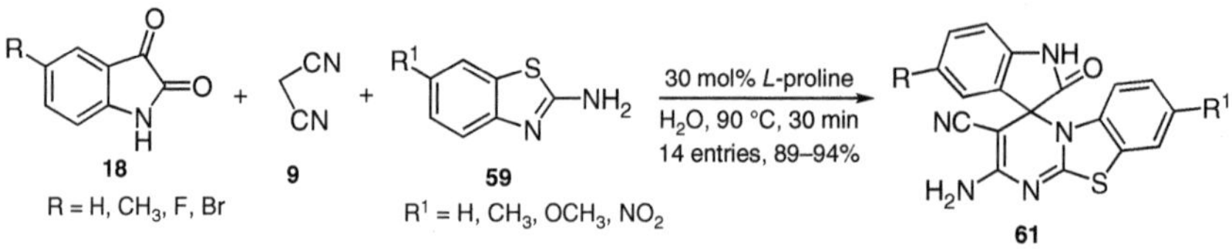

R = H, CH_3, F, Br

R^1 = H, CH_3, OCH_3, NO_2

Scheme 14.24 *L*-proline-catalyzed synthesis of spiro[benzothiazolo[3,2-*a*]pyrimidine-4,3′-indolines in water.

from one-pot three-component reactions of isatins (**18**), malononitrile (**9**), and 2-aminobenzothiazole (**59**) at 90 °C (Scheme 14.24) [48]. All reactions were complete within 30 minutes and afforded the products (**61**) in excellent yields.

14.6 Organocatalyzed Miscellaneous Reactions in Water

14.6.1 Synthesis of Structurally Diverse α,β-Unsaturated Ketones

In 2022 Choudhury et al. [49] synthesized a series of structurally diverse α,β-unsaturated ketones (**62**) *via* the Claisen–Schmidt reaction of substituted benzaldehydes (**7**) and acetone (**22a**) using 1 mol% choline hydroxide as catalyst at 50 °C in water (Scheme 14.25).

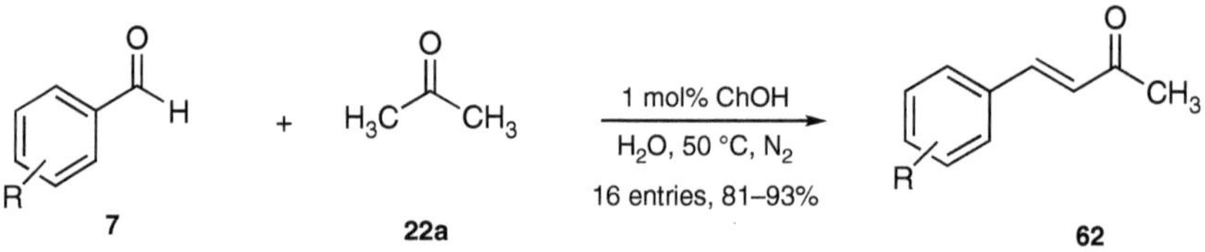

R = H, 2-Me, 3-Me, 4-Me, 2-OMe, 3-OMe, 4-OMe, 4-F, 2-Cl, 4-Cl, 2-Br, 4-Br, 2-NO_2, 4-NO_2, 4-CF_3, 3,5-$(OCH_3)_2$

Scheme 14.25 Choline hydroxide-catalyzed synthesis of structurally diverse α,β unsaturated ketones*via* Claisen–Schmidt reaction in water.

14.6.2 Synthesis of (*E*)-Nitroalkenes

In 2023, Mohanta et al. [50] reported that 0.2 mol% *L*-proline is sufficient to catalyze the Henry condensation reaction of various aldehydes (**7**) and nitroalkanes (**43**) at 100 °C in water affording the (*E*)-nitroalkenes (**37a**) in excellent yields (Scheme 14.26).

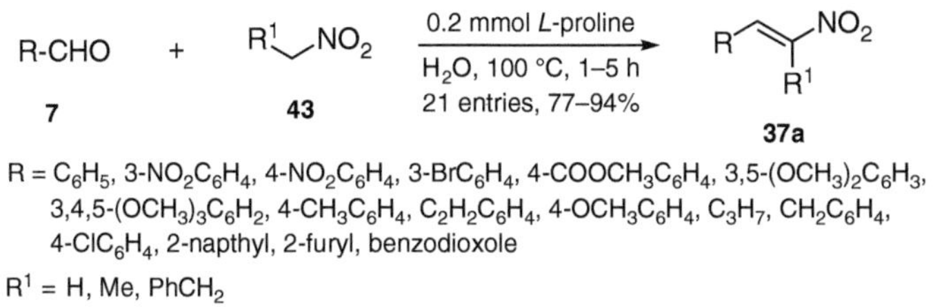

Scheme 14.26 L-proline-catalyzed synthesis of (E)-nitroalkenes in water via Henry condensation reactions.

14.6.3 Synthesis of *Bis*(Pyrazol-4-yl)Arylmethanes

In 2019, Nourzian et al. [51] synthesized a series of *bis*(pyrazol-5-yl)arylmethanes (**65**) *via* the one-pot pseudo five-component tandem Knoevenagel–Michael addition between two molecules of ethyl acetoacetate (**12**), two molecules of hydrazine hydrate (**63**)/phenyl hydrazine (**63a**), and a substituted benzaldehyde (**7**) using 10 mol% guanidine hydrochloride as catalyst in water under reflux conditions (Scheme 14.27). 4,4′,4″,4‴-(1,4-Phenylenebis(methanetriyl))tetrakis(3-methyl-1-phenyl-1*H*-pyrazol-5-ol) (**64a**) was also synthesized in excellent yield under similar conditions *via* a one-pot pseudo nine-component reaction of four molecules of ethyl acetoacetate (**12**), four molecules of phenyl hydrazine (**63a**), and a single molecule of terephthaldehyde (**7b**).

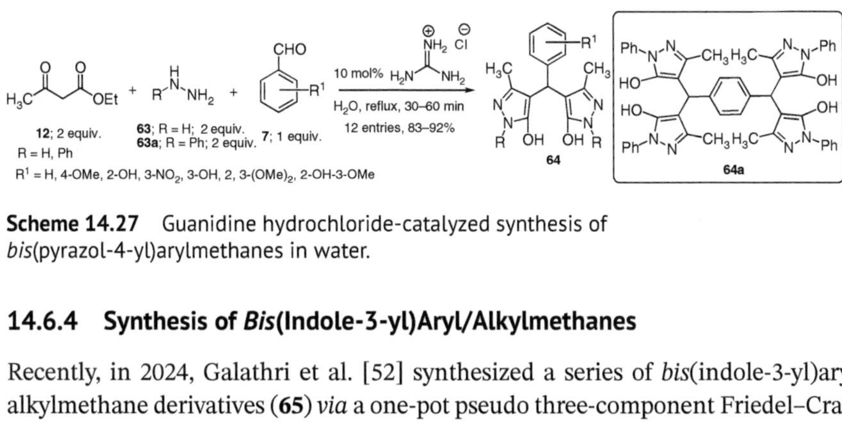

Scheme 14.27 Guanidine hydrochloride-catalyzed synthesis of *bis*(pyrazol-4-yl)arylmethanes in water.

14.6.4 Synthesis of *Bis*(Indole-3-yl)Aryl/Alkylmethanes

Recently, in 2024, Galathri et al. [52] synthesized a series of *bis*(indole-3-yl)aryl/alkylmethane derivatives (**65**) *via* a one-pot pseudo three-component Friedel–Crafts arylation between one molecule of aldehydes (**7**) and two molecules of substituted indoles (**41**) in the presence of a catalytic amount of benzo[*d*]pyrazolo[5,1-*b*][1,3]iodazol-4-ium salt at room temperature in water (Scheme 14.28).

14.6.5 Synthesis of 2-Aminobenzothiazolomethyl Naphthols

In 2019, Lal et al. [53] developed a simple and facile method for the synthesis of a series of 2-benzothiazoloaminomethyl naphthols (**66**) *via* one-pot three-component reactions of 2-aminobenzothiazole (**59**), aromatic aldehydes (**7**), and β-naphthol (**47**) using 10 mol% *L*-valine as an efficient organocatalyst at 70 °C in water (Scheme 14.29). All reactions were completed within three hours.

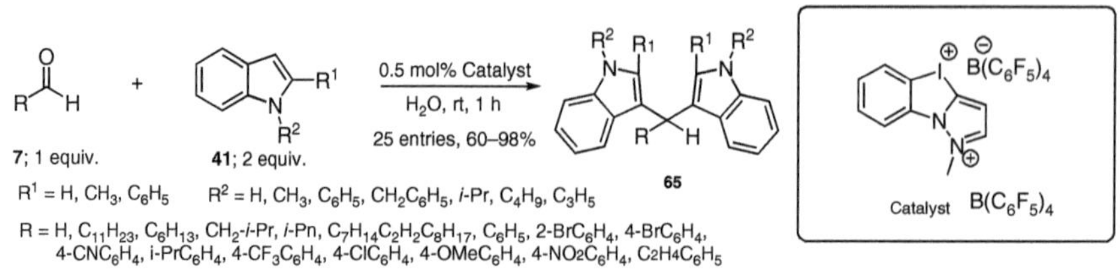

Scheme 14.28 Benzo[*d*]pyrazolo[5,1-*b*][1,3]iodazol-4-ium salt-catalyzed synthesis of *bis*(indole-3-yl)aryl/alkylmethanes in water.

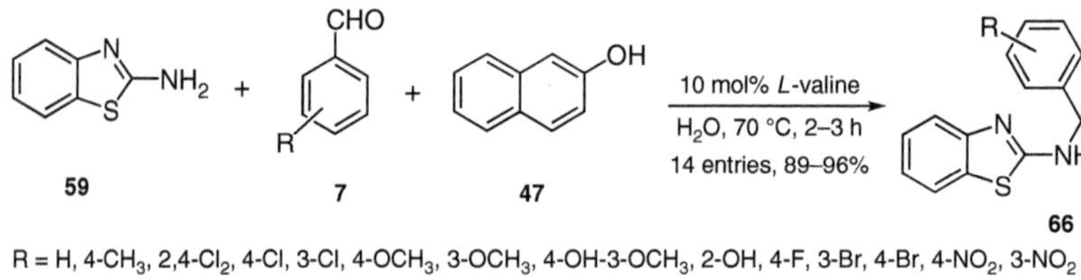

Scheme 14.29 *L*-valine-catalyzed synthesis of 2-aminobenzothiazolomethyl naphthols in water.

14.6.6 Synthesis of *Bis*(Coumarin-3-yl)Arylmethanes

In 2021, Kharrngi et al. [54] synthesized a number of *bis*(coumarin-3-yl)arylmethane derivatives (**67**) *via* one-pot pseudo three-component reactions of one molecule of aromatic aldehyde (**7**) and two molecules of 4-hydroxycoumarin (**22**) using a catalytic amount of proline-derived bifunctional thiourea under reflux conditions in water (Scheme 14.30). Anthelmintic activity of the compounds synthesized was evaluated, and four compounds (**67a–d**) showed a better anthelmintic activity than the standard drug Praziquantel against *Raillietina echinobothrida*.

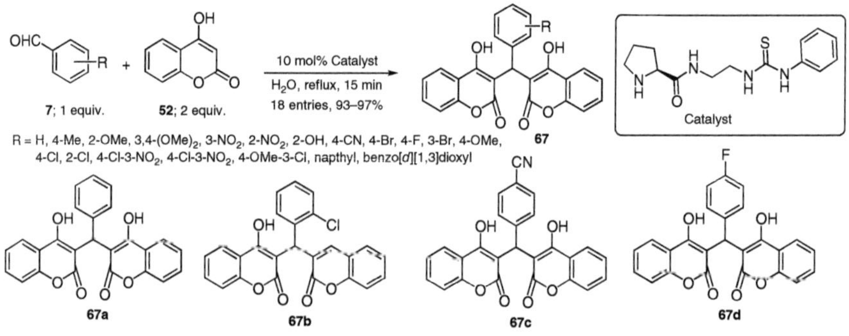

Scheme 14.30 Synthesis of *bis*(coumarin-3-yl)arylmethanes using proline-derived bifunctional thiourea as catalyst in water.

14.6.7 Synthesis of (*E*)-3-(Arylimino)Indolin-2-ones

In 2019, our group reported a facile and water-mediated protocol for the synthesis of a series of biologically promising (*E*)-3-(arylimino)indolin-2-one derivatives (**68**) by the reaction of various substituted anilines (**1**) and isatins (**18**) in the presence of a catalytic amount of mandelic acid or itaconic acid as the catalyst at room temperature (Scheme 14.31) [12]. (*E*)-2-((4-Methoxyphenyl)imino)acenaphthylen-1(2*H*)-one (**69**) was also synthesized under similar conditions by replacing isatin (**18**) with acenaphthylene-1,2-dione (**27d**).

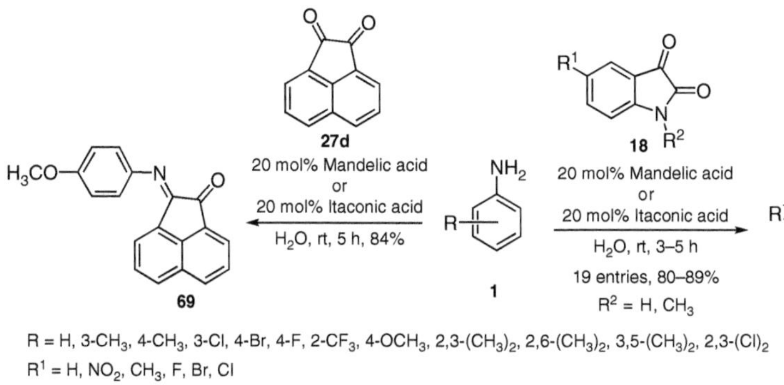

Scheme 14.31 Mandelic acid or itaconic acid-catalyzed synthesis of 2-aminobenzothiazolomethyl naphthols in water.

14.6.8 Synthesis of 4-Hydroxy-4*H*-Chromene-Functionalized Pyrazoles and Barbiturates

A novel series of 4-hydroxy-4*H*-chromene-functionalized pyrazoles and barbiturates was synthesized by the reaction of (*E*)-1-(2-ethoxy-2-methyl-2*H*-chromen-3-yl)-3-phenylprop-2-en-1-one (**70**) and 3-methyl-1-phenyl-1*H*-pyrazol-5(4*H*)-one (**71**) or barbituric acid (**4**) using a catalytic amount of *L*-proline at 80 °C in water (Scheme 14.32) [55].

14.7 Asymmetric Synthesis Using Organocatalysts

14.7.1 Organocatalyzed Asymmetric *Exo*-Selective Diels–Alder Reaction in Water

In 2022, Hayashi and his research group [56] reported an asymmetric Diels–Alder reaction between α-fluoro-α,β-unsaturated aldehydes (**74**) and cyclopentadiene (**75**) using a catalytic amount of perchloric acid salt of diarylprolinol silyl ether at room temperature in water (Scheme 14.33). This method provided the adducts (**76**) with excellent *exo*-selectivity and enantioselectivity.

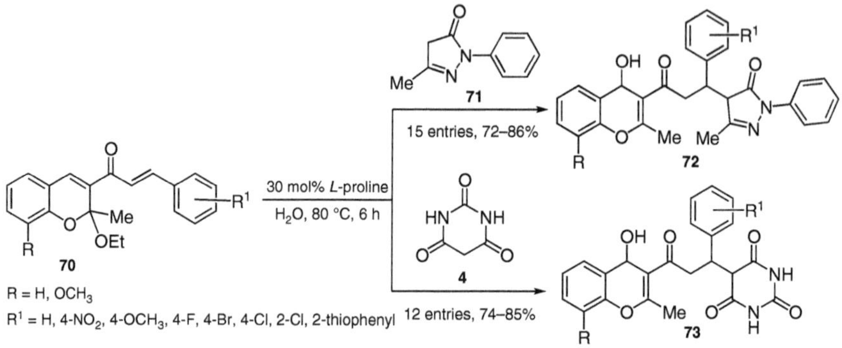

Scheme 14.32 *L*-proline-catalyzed synthesis of 4-hydroxy-4*H*-chromene-functionalized pyrazoles and barbiturates in water.

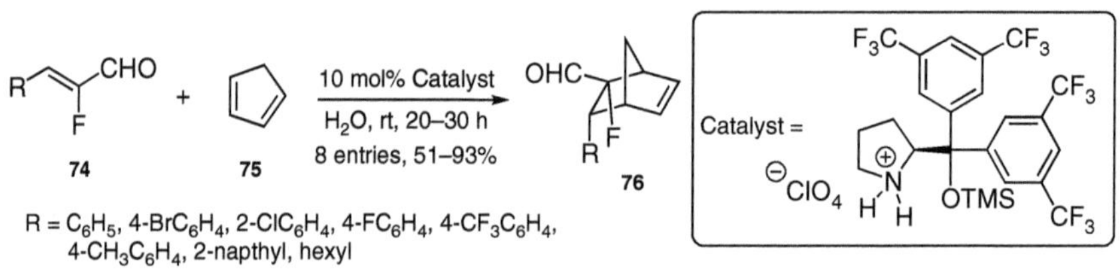

Scheme 14.33 Asymmetric *exo*-selective Diels–Alder reaction in water. OTMS, oxytrimethylsilane.

14.7.2 Organocatalyzed Asymmetric Aldol Reaction in Water

In 2020, Rani et al.[57] synthesized a D-glucosamine-based β-CD inclusion complex under ultrasound-assisted conditions. By using a small amount of this novel species as a heterogeneous catalyst in water, they accomplished the asymmetric Aldol reaction between various ring ketones (**27**) and aromatic aldehydes (**7**) (Scheme 14.34). Interestingly, a 99% *syn* selectivity of the desired product (**77**) was experienced in the reaction of 2-cholrobenzaldehyde and cyclohexanone. Under similar conditions, 3-OH-3-(substituted)indolin-2-ones (**78**) were obtained by the reaction of cyclic ketones **27** and isatin (**18**). In 2020, Keskin et al.[58] applied a novel proline-based

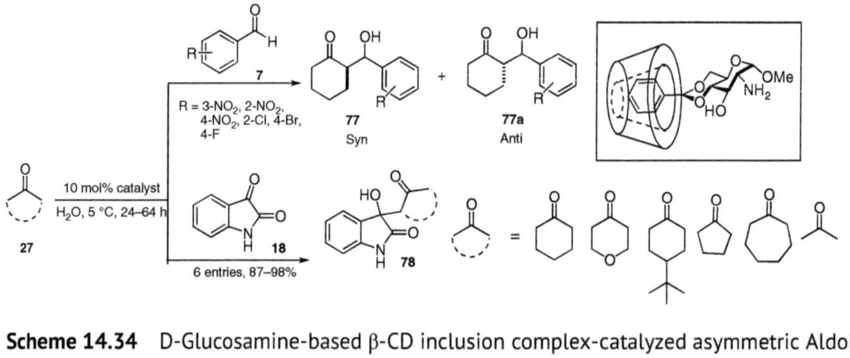

Scheme 14.34 D-Glucosamine-based β-CD inclusion complex-catalyzed asymmetric Aldol reaction in water.

amide *viz.*, (*S*)-methyl-2-((*S*)-3-hydroxy-2-((*S*)-3-pyrrolidine-2-carboxamido)propanamido)-phenylpropanoate as an efficient organocatalyst for a similar type of asymmetric Aldol reaction at 0 °C in water (Scheme 14.35). Next year, Yeo et al. [59] reported another asymmetric Aldol reaction utilizing an isothiouronium salt-based chiral proline-functionalized amide as an efficient bifunctional organocatalyst at 0 °C in aqueous medium (Scheme 14.36).

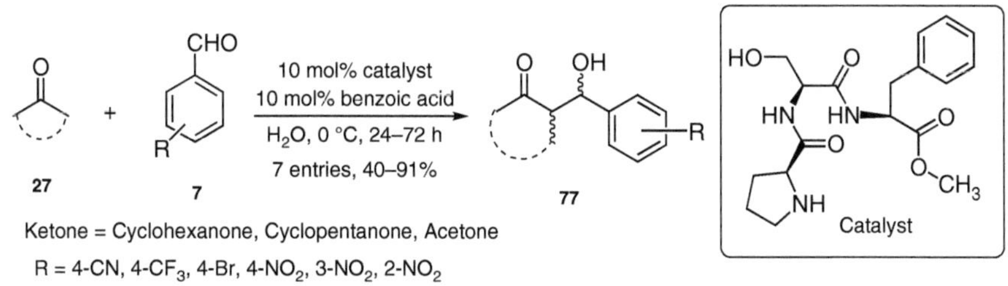

Scheme 14.35 A novel proline-based amide-catalyzed asymmetric Aldol reaction in water.

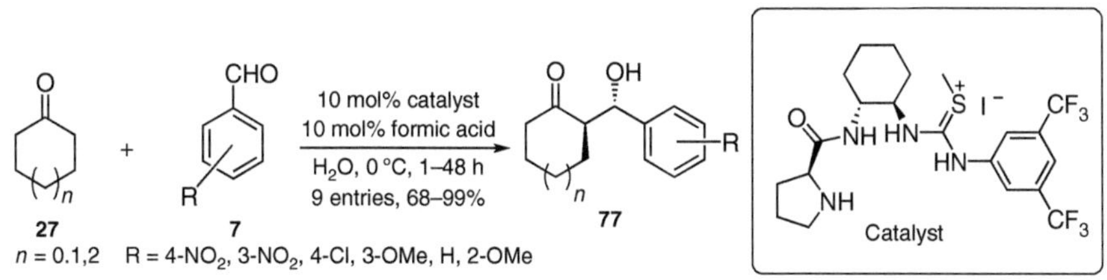

Scheme 14.36 Asymmetric Aldol reaction in water using an isothiouronium salt-based chiral proline-functionalized amide catalyst.

14.7.3 Organocatalyzed Asymmetric Michael Addition in Water

In 2019, Kucherenko et al. [60] used 1 mol% of C_2-symmetric chiral tertiary amine-squaramide as an efficient recyclable organocatalyst in aqueous medium in asymmetric Michael-type reaction of substituted Kojic acids (**79**) and nitroolefins (**37**), which afforded the corresponding adducts (**80**) in good yields and stereoselectivity (Scheme 14.37). In the same year, Lan et al. [61] demonstrated another water-mediated asymmetric Michael reaction between cyclohexanone (**27**) and

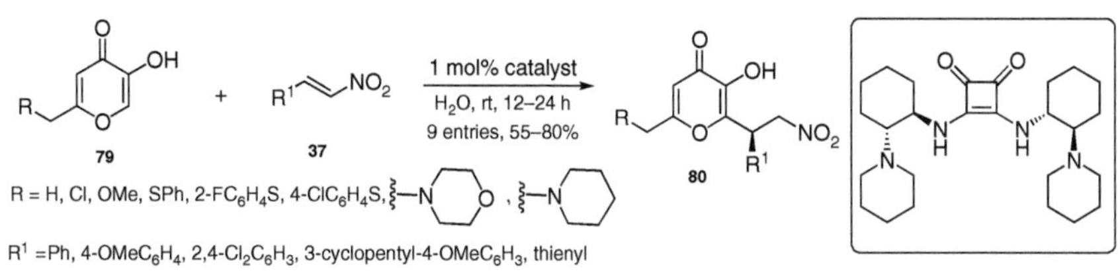

Scheme 14.37 C_2-symmetric tertiary amine-squaramide-catalyzed asymmetric Michael addition in water.

various nitrostyrenes (**37**) utilizing a small amount of pyrrolidine-based triazolo-functionalized chiral porous polymers as an efficient heterogeneous organocatalyst at room temperature (Scheme 14.38).

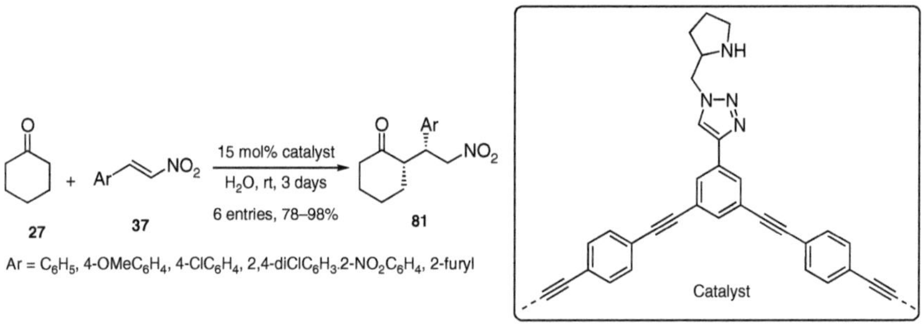

Scheme 14.38 Asymmetric Michael addition in water using pyrrolidine-based triazolo-functionalized chiral porous polymers as efficient heterogeneous organocatalysts.

In 2021, Mahato et al. [62] used another proline-derived hydrophobic organocatalyst in aqueous medium for the samebatch of reaction, which yielded the desired 1,4-conjugated Michael addition products stereoselectively (*dr*: >97 : 3, *ee* up to >99.9%) in excellent (up to 95%) yields at room temperature (Scheme 14.39).

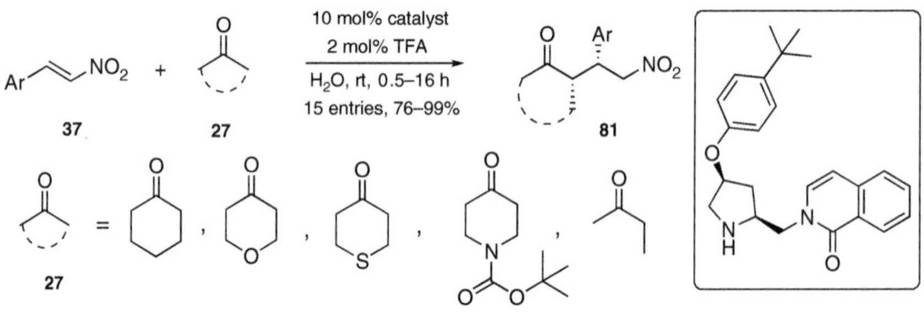

Scheme 14.39 Asymmetric Michael addition reaction using proline-derived hydrophobic organocatalyst in water. TFA, trifluoroacetic acid.

In 2021, Shim et al. [63] prepared a (*R*,*R*)-1,2-diphenylethylenediamine (DPEN)-based thiourea derivative. Using a catalytic amount of this novel compound, they developed another asymmetric Michael addition reaction protocol in aqueous medium with high enantioselectivity of 76–99% *ee* for the*syn* species formed in 90% (as compared to the 10% of thc *anti* form) (Scheme 14.40). Using the same chiral catalyst, they have developed another asymmetric Michael addition reaction between *N*-phenylmaleimide derivatives (**82**) and isobutyraldehyde (**7c**) to afford the corresponding adducts (**83**) in excellent (upto 97%) yields with high enantioselectivity (*synee* upto 99%) (Scheme 14.41) [64]. The same chiral organocatalyst was also found to be efficient in the asymmetric Michael reaction of *trans*-β-nitrostyrene (**37**) and isobutyraldehyde (**7c**)

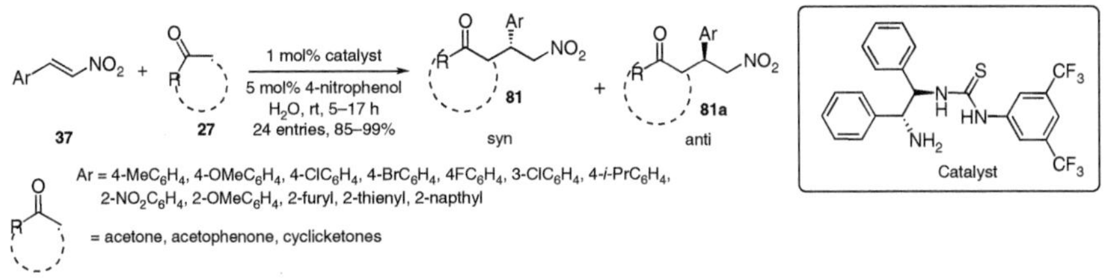

Scheme 14.40 Asymmetric Michael addition reaction using aDPEN-based thiourea organocatalyst in water.

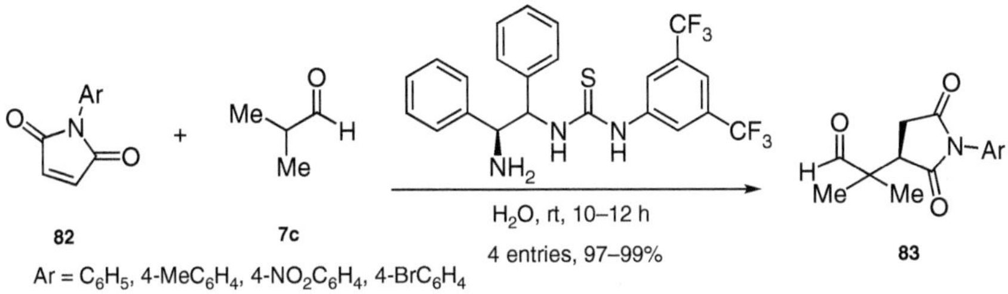

Scheme 14.41 Asymmetric Michael addition reaction between phenylmaleimide derivatives and isobutyraldehyde using aDPEN-based thiourea as the catalyst in water.

at room temperature in water, and that afforded the desired adducts (**84**) in excellent yields with high enantioselectivity (Scheme 14.42) [65]. In 2022, Bae et al. [66] utilized a novel proline-derived bifunctional organocatalyst 1-(3,5-bis(trifluoromethyl)phenyl)-3-((R)-phenyl((S)-pyrrolidin-2-yl)methyl)urea (PHU) for the same asymmetric Michael addition at ambient temperature in aqueous medium (Scheme 14.43).

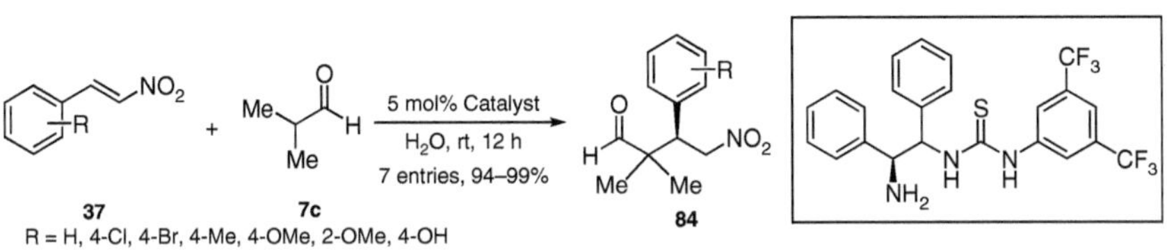

Scheme 14.42 Asymmetric Michael addition reaction between *trans*-β-nitrostyrene and isobutyraldehyde using aDPEN-based thiourea as the catalyst in water.

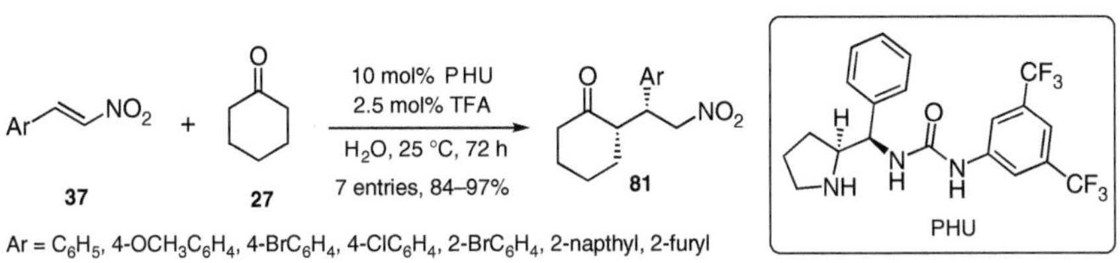

Scheme 14.43 Asymmetric Michael addition reaction using proline-derived bifunctional organocatalyst in water.

14.8 Conclusions

This chapter deals with a large number of literature data reported during the last four years related to various metal-free organocatalyzed transformations in water. The catalytic applicability of various organic substances as catalyst in aqueous medium has been discussed in this chapter. A variety of biologically promising *N*-heterocycles, *O*-heterocycles, *N,O*-heterocycles, *N,S*-heterocycles, and acyclic scaffolds were synthesized by using a variety of amines, amino acids, surfactants, and other organic substances as catalyst under diverse reaction conditions in water. Interestingly, asymmetric Aldol reactions, Michael addition reactions, Diels–Alderreactions also took place enantioselectively by using various organocatalysts in aqueous medium. Mild reaction conditions, the use of nontoxic catalysts and water as solvent are a few of the common advantages of these protocols developed. On a few occasions, the catalysts applied were collected and recycled for further runs. On a few occasions, the plausible mechanism and the probable role of the catalyst were also discussed. We are confident that this chapter will definitely boost the ongoing related research and motivate the readers to carry out diverse organic transformations in aqueous medium by selecting organic substances as catalysts.

Acknowledgments

Dr. B. Banerjee is thankful to the Akal University, Talwandi Sabo, Punjab, India and the Kalgidhar Trust, Baru Sahib, Himachal Pradesh, India for the support.

References

1 Davies, H.M. and Lian, Y. (2012). *Acc. Chem. Res.* 45: 923–935.
2 Shen, Z.L., Wang, S.Y., Chok, Y.K. et al. (2013). *Chem. Rev.* 113: 271–401.
3 Kotha, S. and Khedkar, P. (2012). *Chem. Rev.* 112: 1650–1680.
4 Nugent, W.A. (2012). *Angew.Chem. Int. Ed.* 51: 8936–8949.
5 Kaneda, K. and Mizugaki, T. (2009). *Energy Environ. Sci.* 2: 655–673.
6 Trost, B.M. (2002). *Acc. Chem. Res.* 35: 695–705.
7 Brahmachari, G. and Banerjee, B. (2015). *Curr. Green Chem.* 2: 274–305.
8 Sankar, M., Dimitratos, N., Miedziak, P.J. et al. (2012). *Chem. Soc. Rev.* 41: 8099–8139.
9 Kaur, G., Thakur, S., Kaundal, P. et al. (2018). *ChemistrySelect* 3: 12918–12936.
10 Kaur, G., Singh, A., Kaur, N., and Banerjee, B. (2021). *Synth. Commun.* 51: 1121–1131.
11 Kaur, G., Bala, K., Devi, S., and Banerjee, B. (2018). *Curr. Green. Chem.* 5: 150–167.
12 Kaur, G., Singh, A., Bala, K. et al. (2019). *Curr. Org. Chem.* 23: 1778–1788.
13 Banerjee, B. (2018). *Curr. Org. Chem.* 22: 208–233.

14 Simon, M.-O. and Li, C.-J. (2012). *Chem. Soc. Rev.* 41: 1415–1427.
15 Chanda, A. and Fokin, V.V. (2009). *Chem. Rev.* 109: 725–748.
16 Gawande, M.B., Bonifácio, V.D., Luque, R. et al. (2013). *Chem. Soc. Rev.* 42: 5522–5551.
17 Breslow, R. (1991). *Acc. Chem. Res.* 24: 159–164.
18 Butler, R.N. and Coyne, A.G. (2012). *Chem. Rev.* 110: 6302–6337.
19 Jimeno, C. (2016). *Org. Biomol. Chem.* 14: 6147–6164.
20 Van der Helm, M.P., Klemm, B., and Eelkema, R. (2019). *Nat. Rev. Chem.* 3: 491–508.
21 Bhosle, M.R., Kharote, S.A., Bondle, G.M. et al. (2020). *Chem.Biodivers.* 17: e1900577.
22 Tamaddon, F. and Tadayonfar, S.E. (2020). *J. Mol. Struct.* 1207: 127728–127734.
23 Bhadke, P.K., Gadkari, Y.U., Salim, S.S. et al. (2023). *J. Chem. Sci.* 135: 68–74.
24 Kaur, G., Devi, M., Kumari, A. et al. (2018). *ChemistrySelect* 3: 9892–9910.
25 Dogra, D., Bharti, R., and Sharma, R. (2020). *Orbital: Electron J. Chem.* 12: 232–241.
26 Pariyar, G.C., Mitra, B., Mukherjee, S., and Ghosh, P. (2020). *ChemistrySelect* 5: 104–108.
27 Chate, A.V., Rudrawar, P.P., Bondle, G.M., and Sangeshetti, J.N. (2019). *Synth. Commun.* 50: 1–17.
28 Kumar, P., Amber, N., and Tripathi, V.D. (2023). *Eur.J. Adv. Chem. Res.* 4: 139–147.
29 Khopkar, S. and Shankarling, G. (2020). *J. Chem. Sci.* 132: 31–41.
30 Bhutia, Z.T., Panjikar, P.C., Iyer, S. et al. (2020). *ACS Omega* 5: 13333–13343.
31 Patil, P.J., Salunke, G.D., Deshmukh, M.B. et al. (2019). *ChemistrySelect* 4: 13071–13078.
32 Brahmachari, G. and Banerjee, B. (2014). *ACS Sustain. Chem. Eng.* 2: 411–422.
33 Kaur, M., Priya, A., Singh, A. et al. (2022). *Magnetic Nanocatalysis: Synthetic applications*, vol. 1 (ed. R.S. Varma and B. Banerjee), 411–434. Berlin, Boston: De Gruyter.
34 Banerjee, B., Kaur, M., Sharma, A. et al. (2022). *Curr. Green Chem.* 10: 162–173.
35 Brahmachari, G., Laskar, S., and Banerjee, B. (2014). *J. Heterocyclic Chem.* 51: E303–E308.
36 Auria-Luna, F., Fernández-Moreira, V., Marqués-López, E. et al. (2020). *Sci. Rep.* 10: 11594–11611.
37 Banerjee, B., Priya, A., Kaur, M. et al. (2023). *Catal. Lett.* 153: 3547–3560.
38 Rani, D., Gulati, V., Guleria, M. et al. (2022). *J. Mol. Struct.* 1265: 133341–133348.
39 Thangalipalli, S., Bandalla, S.G., Gavinolla, V. et al. (2023). *ChemistrySelect* 8: e202302065.
40 Bandita, D. and Pasha, M.A. (2012). *Ultrason. Sonochem.* 19: 725–728.
41 Banerjee, B., Priya, A., Sharma, A. et al. (2023). *Arkivoc* 7: 202312116.
42 Chavan, A.S., Kharat, A.S., Bhosle, M.R. et al. (2021). *Synth. Commun.* 51: 1963–1973.
43 Saini, K., Raigar, A.K., Manju et al. (2022). *J. Org. Chem.* 87: 13734–13743.
44 Kaurav, M.S., Sahu, P.K., Sahu, P.K. et al. (2019). *RSC Adv.* 9: 3755–3763.

45 Ghogare, R.S., Patankar-Jain, K., and Momin, S.A.H. (2021). *Lett. Org. Chem.* 18: 83–87.

46 Sayahi, M.H., Shamkhani, F., Mahdavi, M., and Bahadorikhalili, S. (2021). *Res. Chem. Intermed.* 47: 4101–4112.

47 Singh, S. and Lal, J. (2020). *SN Appl. Sci.* 2: 1271–1279.

48 Verma, A., Arya, D.K., Kumar, S. et al. (2023). *Res. Chem. Intermed.* 49: 4865–4877.

49 Choudhury, S.S., Mahapatra, S., Sahu, A.K. et al. (2022). *ACS Sustain. Chem. Eng.* 10: 14271–14279.

50 Mohanta, R., Solo, M., Devi, M., and Bez, G. (2023). *Catal. Lett.* 153: 3003–3008.

51 Noruzian, F., Olyaei, A., Hajinasiri, R., and Sadeghpour, M. (2019). *Synth. Commun.* 49: 2712–2724.

52 Galathri, E.M., Kuczmera, T.J., Nachtsheim, B.J., and Kokotos, C.G. (2024). *Green Chem.* 26: 825–831.

53 Lal, J., Singh, S., and Rani, P. (2019). *Chem. Afr.* 3: 53–60.

54 Kharrngi, B., Dhar, E.D., Basumatary, G. et al. (2021). *J. Chem. Sci.* 133: 1–15.

55 Reddy, C.V.S. and Sumathi, S. (2020). *ChemistrySelect* 5: 8729–8734.

56 Bondzić, B.P., Daskalakis, K., Taniguchi, T. et al. (2022). *Org. Lett.* 24: 7455–7460.

57 Rani, D., Sethi, A., Kaur, K., and Agarwal, J. (2020). *J. Org. Chem.* 85: 9548–9557.

58 Keskin, E., Yolacan, C., and Aydogan, F. (2020). *Acta Chim. Slov.* 67: 1014–1023.

59 Yeo, H.M., Kang, S., and Kim, T.H. (2021). *Tetrahedron Lett.* 80: 153324–153327.

60 Kucherenko, A.S., Kostenko, A.A., Komogortsev, A.N. et al. (2019). *J. Org. Chem.* 84: 4304–4311.

61 Lan, Y., Yang, C., Zhang, Y. et al. (2019). *Polym. Chem.* 10: 3298–3305.

62 Mahato, C.K., Mukherjee, S., Kundu, M. et al. (2021). *J. Org. Chem.* 86: 5213–5226.

63 Shim, J.H., Ahn, B.K., Lee, J.Y. et al. (2021). *Catalysts* 11: 1004–1014.

64 Shim, J.H., Cheun, S.H., Kim, H.S., and Ha, D.C. (2022). *Molecules* 27: 2759–2769.

65 Shim, J.H., Cheun, S.H., Kim, H.S., and Ha, D.C. (2022). *Catalysts* 12: 121–133.

66 Bae, D., Lee, J.W., and Ryu, D.H. (2022). *J. Org. Chem.* 87: 16532–16541.

15

Organic Electrocatalysis in Water

Samina Aslam[1], Sadia Rani[1], Kiran Lal[1], and Nisar Ahmed[2]

[1] *The Women University Multan, Department of Chemistry, LMQ Road, Multan 60000, Pakistan*
[2] *Cardiff University, School of Chemistry, Main Building, Park Place, Cardiff CF10 3AT, UK*

15.1 Introduction

There has been a renewed interest in electrochemical processes in recent times, despite their extensive history [1]. Research and innovative developments have been made in the field of electrochemical organic synthesis [2]. In organic chemistry, the electrochemical technique offers an alternate strategy for efficient transformation [3]. When it comes to synthesis, electrochemical methods are great because they can get rid of waste and reuse metal reagents [4]. The utilization of environmentally friendly electrons, derived from sustainable energy sources, as reactants for the production of high-value compounds has significantly propelled research in the realm of electrochemistry [5]. Electrochemical conversions, in accordance with the 12 principles of green chemistry, enable the replacement of ecologically harmful oxidants or catalysts with electrons [6]. Water is by far the most ideal medium when it comes to environmentally friendly solvents [7]. Because different inorganic salts are more soluble in water, conductivity as a rate-limiting factor can be ignored. Since the discovery of decarboxylation reactions by Faraday and Kolbe, electrochemical reactions in aqueous media have an extended history [8]. In aqueous media, modern electrochemistry is also well established, particularly for the conversion of biomass, reduction of CO_2, water splitting, and waste water disposal [9]. Water has been used as a co-solvent or additive with methanol, dimethyl sulfoxide, and acetonitrile in electrochemical organic synthesis. The solubility of organic chemicals, which would lead to multiphase situations, is one disadvantage of aqueous media. However, this "on-water" immiscibility of reactants and water also has certain special benefits for organic synthesis [10].

The chemical industry aims to develop aqueous solution-based processes. In addition to its inherent eco-friendliness, water possesses several desirable characteristics, such as its availability and the fact that it requires minimal handling of solvents, which make it an extremely attractive choice as a solvent. Developing electroorganic processes that can be conducted in an aqueous solution is highly appealing

Organic Transformations in Water: Principles and Applications, First Edition.
Edited by Gopinathan Anilkumar, Nissy Ann Harry, and Sankuviruthiyil M. Ujwaldev.

in this particular scenario. Typically, electroorganic processes in a water-based solution are constrained by the narrow range of electrical potentials that water can withstand without undergoing direct reduction or oxidation (as represented by Eqs. (15.1) and (15.2)).

$$2H_2O + 2e^- \rightleftharpoons H_2 + 2OH^-, E^0 = -1.1\ \text{V vs. Ag/AgCl (3.5 M KCl)} \quad \text{(alkaline solution)} \tag{15.1}$$

$$2H_2O \rightleftharpoons 4H^+ + O_2 + 4e^-, E^0 = 1.03\ \text{V vs. Ag/AgCl (3.5 M KCl)} \quad \text{(acidic solution)} \tag{15.2}$$

15.2 Direct vs. Mediated Reductive Oxidation and Oxidative Reduction

Low reaction yields and unwanted byproducts are the results of water reduction and oxidation interfering with electroorganic processes. Reductive oxidation and oxidative reduction are synthetic methods that have been used to report many electroorganic processes in water (Scheme 15.1). In reductive oxidation, the peroxydisulfate ($S_2O_8^{2-}$) is uniformly reduced using an electrogenerated $Ru(NH_3)_6^{3+}$ at a potential of −0.2 V vs. Ag/AgCl (3.5 M KCl). This process produces the sulfate radical anion, which is capable of oxidizing species beyond those with a water oxidation potential ($E^{0\prime}(SO_4^{\bullet-}/SO_4^{2-}) = 2.21$ V vs. Ag/AgCl). $SO_4^{\bullet-}$ is produced electrochemically and can effectively remove hydrogen atoms from a range of organic molecules, including toluene and benzyl alcohol, to produce a water-soluble product. Oxidative reduction is the inverse process of reductive oxidation. The powerfully reducing carbon dioxide radical anion ($E^{0\prime}$ ($CO_2^{\bullet-}/CO_2) = -2.1$ V vs. (Ag/AgCl)) is produced by the electrochemically generated $Ru(bpy)_3^{3+}$ homogeneous oxidation of oxalate ($C_2O_4^{2-}$), and it can reduce species at potentials greater than that of water or proton reduction. The C–Br group of benzyl bromide has been uniformly reduced in aqueous solutions using $CO_2^{\bullet-}$ as an oxidizing agent [11].

15.3 Water as a Solvent and an Electrolyte

Water is essential for life. All living organisms rely on it for their chemical processes. Depending on temperature, the properties of water can shift, making it either a good solvent or not so great at dissolving things. The environment in which aqueous media respond can be changed by adjusting temperature in accordance with specified physical–chemical characteristics (Table 15.1) [12].

Electrosynthesis requires a conductive medium for ions to move. Water-based electrolyte solutions are significantly superior at conducting electricity. The electrolytic conductivities of aqueous electrolyte solutions are shown in Table 15.2 [13].

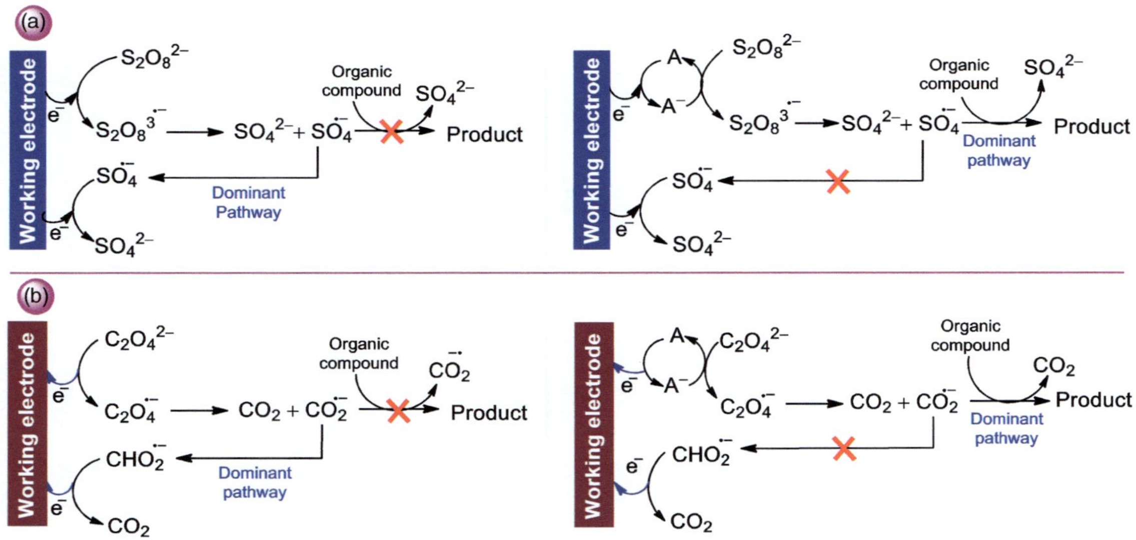

Scheme 15.1 (a) Direct vs. mediated reductive oxidation; (b) direct vs. mediated oxidative reduction processes. Source: Das et al. [11]/Royal Society of Chemistry.

Table 15.1 Physical and chemical properties of water at different temperatures.

	Ambient temp.	Subcritical water	Supercritical water
Temperature	0–100 °C	100–374 °C	>374 °C
Dielectric constant	78.5 (25 °C, 0.1 MPa)	27.1 (250 °C, 5 MPa), 18.2 (330 °C, 30 MPa)	5.9 (400 °C, 25 MPa), 10.5 (400 °C, 50 MPa)
Ionic Product $K_w[mol^2\ L^{-2}]$	10^{-14} (increasing to 10^{-12} at 100 °C	increasing from 10^{-12} (100 °C) to 10^{-11} (300 °C)	Strongly decreasing to below 10^{-20}(400 °C) and below 10^{-23}(550 °C)

Source: Adapted from Möller et al. [12].

Table 15.2 Electrolytic conductivities of aqueous electrolyte solutions, measured at a temperature of 25 °C and derived from Ohmic losses[a)].

Aqueous reaction media	Specific conductivity [mS cm^{-1}]	Ohmic losses [v]	Ohmic loss [Wcm^{-2}]
Water	10^{-2}–10^{-3}		
1 M H_2SO_4	346	0.29	0.029
1 M KOH	223	0.45	0.045
1 M KCl	86	1.16	0.116
1 M $LiClO_4$	71	1.41	0.141

a) Across an electrolyte solution layer of 1 cm thickness and at a current density of 100 mA cm^{-2}.
Source: Schmidt [13]/John Wiley & Sons.

15.4 Electrocatalysis

A catalyst that participates in electrochemical reactions is known as an electrocatalyst. It works on electrode surfaces, or, in certain cases, on the electrode surface itself. It should possess long term stability, elevated activity, a large surface area, and high electrical conductivity. Three functional and multifunctional electrocatalysts for oxygen evolution reactions (OER), oxygen reduction reactions (ORR), and hydrogen evolution reactions (HER) have been discovered [14].

Electrolysis is the process of breaking down and forming bonds in an electrolyte in a "homogeneous" system by transferring electrons to or from an inert electrode. An essential step in fuel cells and electrolysis devices is electrocatalysis. The capacity of electrochemical devices to convert chemicals into electrical forms is enhanced by the availability of renewable power. This allows electrical energy to be stored in electricity-driven conversions or produced as chemical molecules. Environmentally friendly, sustainable, and renewable energy technologies are made possible by electrocatalytic processes that encourage chemical or electrical interconversions. These technologies form part of the long-term energy infrastructure [15]. Reaction kinetics, product transport, reactant and ion transport, and electron conduction all

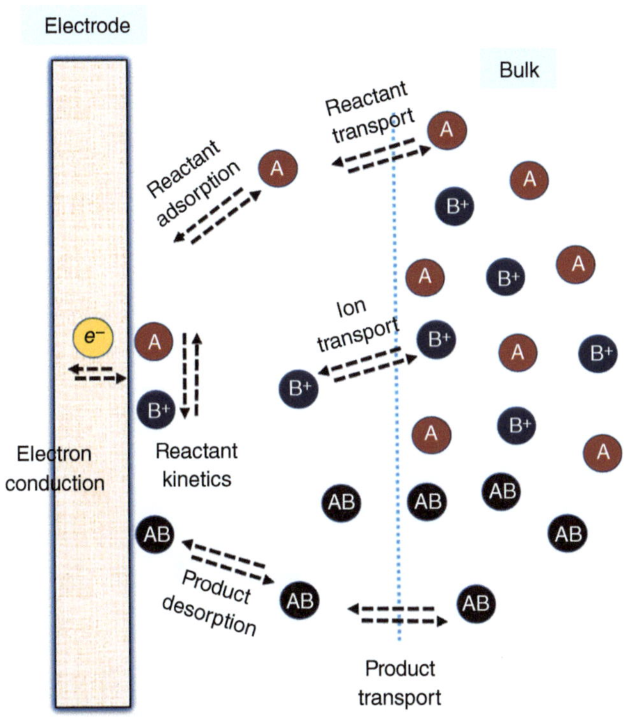

Figure 15.1 Electron conduction, the mobility of species and ions, and electrochemical reactions near the electrode surface. Source: Maheshwari et al. [15]/with permission of Elsevier.

have an impact on an electrochemical device's performance (Figure 15.1). Materials for electrocatalysts must be engineered to enable effective functioning close to equilibrium potential. Electrocatalysis takes place at the interface between the electrode and the electrolyte, where reactants, ions, and products flow between these components.

15.5 Water Electrolysis

Hydrogen and oxygen are created concurrently in conventional water electrolysis because HER and OER are tightly coupled. This can cause explosive H_2/O_2 mixing because of the gas crossover. Reactive oxygen species could be produced by the presence of H_2, O_2, and electrocatalysts, which might decrease the lifespan of an electrolyzer. There have been recent developments in innovative methods to deal with the difficulties of traditional water electrolysis. Among these cutting-edge approaches are decoupled water electrolysis (it uses redox mediators to separate HER from OER both temporally and spatially), overall water electrolysis (it uses bifunctional nonprecious electrocatalysts (or precursors) to drive both HER and OER), hybrid water electrolysis (it combines thermodynamically more favorable

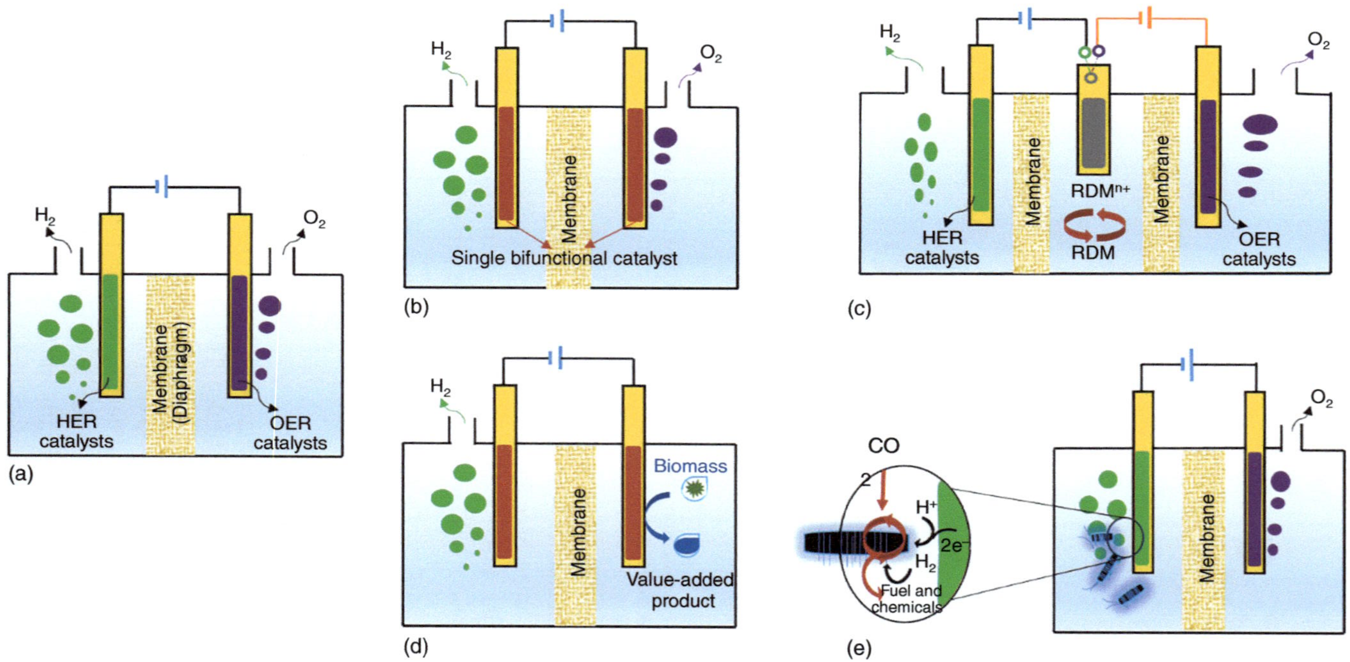

Figure 15.2 (a) Traditional water electrolysis; (b) overall water electrolysis; (c) decoupled water electrolysis; (d) hybrid water electrolysis; (e) tandem water electrolysis. Source: You and Sun [16]/American Chemical Society.

organic upgrading reactions to replace OER), and tandem water electrolysis (it uses biocatalysts to convert the *in situ* produced hydrogen gas with foreign compounds (like N_2 and CO_2) to more valuable products) (Figure 15.2) [16].

15.6 Organic Transformations in Water: Principles and Applications

15.6.1 Kolbe Electrolysis

A promising process that combines the use of electrons as reagents and renewable carboxylic acids as raw materials to produce value-added compounds is called Kolbe electrolysis. The electrolysis was performed using a custom-designed electrochemical microreactor. The focus was on achieving continuous operation and reusing the aqueous electrolyte and unreacted material (Figure 15.3) [6].

There are few reports on electrochemistry in aqueous reaction media at temperatures above 100 °C since such cells must also be pressure and temperature resistant. One example is the electrochemical conversion of cellulose to mixtures of furfurals and levulenic acid in subcritical water or of lactic acid to mixtures of acetic acid, acrylic acid, and acetaldehyde [17]. This explains why the first electro-organic reaction, which involved the electrochemical decarboxylation of acetic acid and dimerization of methyl radicals to produce ethane, as well as a related reaction involving valeric acid **1** that produced octane **2** and is now known as the Kolbe reaction, both occurred in aqueous solutions (Scheme 15.2a) [18].

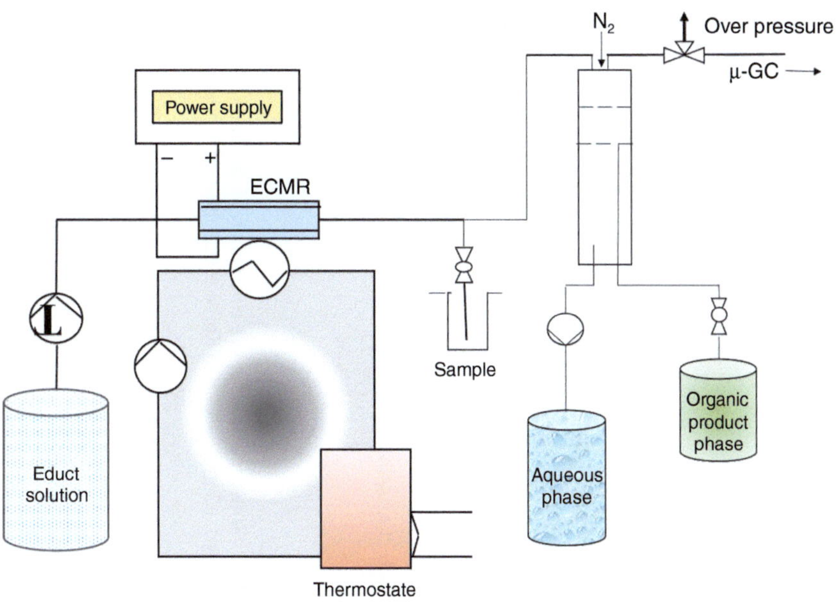

Figure 15.3 Flow chart of single-pass Kolbe electrolysis in water. Source: Baumgarten et al. [6]/John Wiley & Sons/CC BY 4.0.

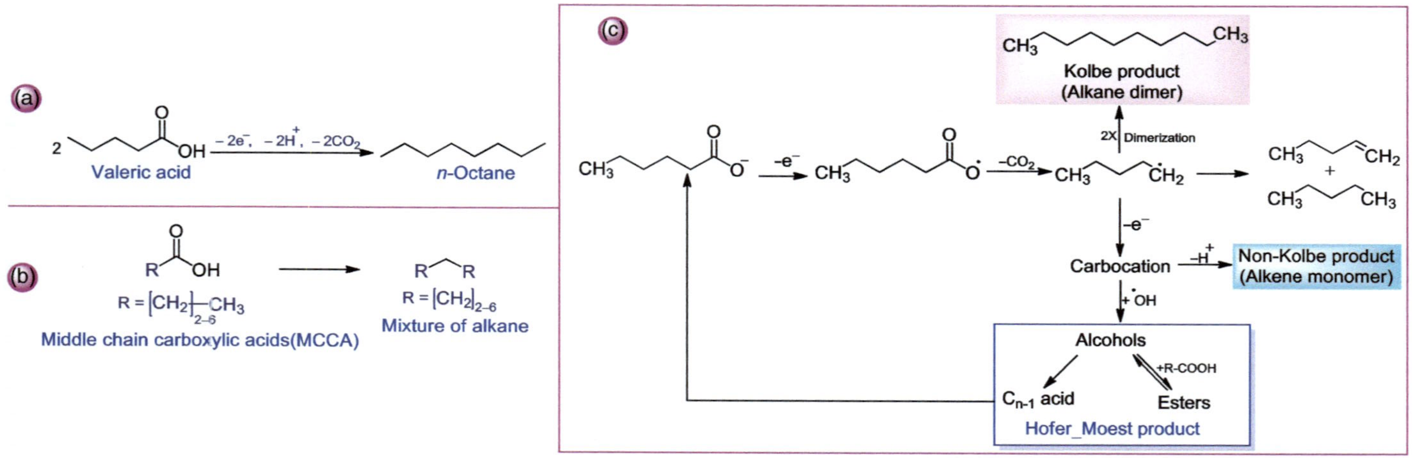

Scheme 15.2 Kolbe reaction for the conversion of: (a) valeric acid to *n*-Octane; (b) medium-chain carboxylic acids (MCCA) to the corresponding mixture of hydrocarbons; (c) mechanism of Kolbe electrolysis of n-hexanoic acid to the Kolbe product n-decane and non-Kolbe products such as alkenes, alcohols, and esters.

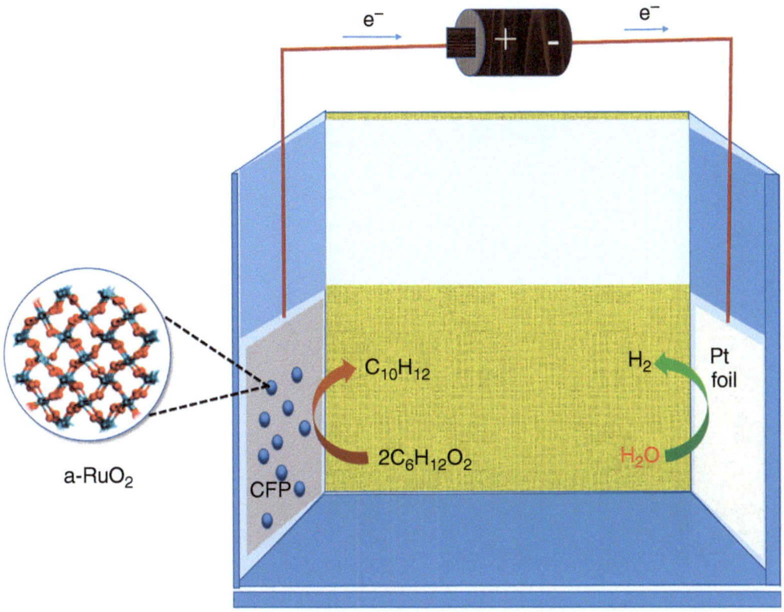

Figure 15.4 Diagram showing the formation of hydrogen during the electrocatalytic decarboxylation of hexanoic acid. Source: Wang et al. [19]/John Wiley & Sons.

Urban et al. have found a way, using the Kolbe reaction, to turn certain chemicals into hydrocarbons when conditions are not too acidic or basic (Scheme 15.2b). The catalytic transformation of volatile carboxylic acids from biomass in an aqueous environment is a key part of creating a biorefinery that can endure. It is still thought that Kolbe electrolysis is the best way to turn aliphatic carboxylic acids (carboxylate) into alkane for making gasoline (Scheme 15.2c).

Amorphous RuO_2 is very good at electrocatalytic oxidative decarboxylation of hexanoic acid and can make decane, which is a Kolbe product, with a yield that is 5.4 times higher than that of regular RuO_2. A fresh design idea for making an effective electrocatalyst for the decarboxylation coupling process has been shown (Figure 15.4) [19]. IrO_2, thin-film Pt, RuO_2, or boron-doped diamonds lack a coulombic efficiency and selectivity similar to pure Pt and are hardly scalable. Platinized titanium that is readily accessible on the market can serve as an anode for the highly effective galvanostatic conversion of n-hexanoic acid to n-decane in a water-based solution [20].

15.6.2 Role of Nanostructures in Electrochemical Conversions

A novel metal nanostructure produced during the electrochemical conversion will be very interesting from the perspective of material chemistry. The reactivity of the nanomaterial adhered to the electrode will be enhanced as a result of its greater surface area. For example, an environmentally friendly method for producing homoallyl alcohols/amines from basic aldehydes/imines and allylic bromides

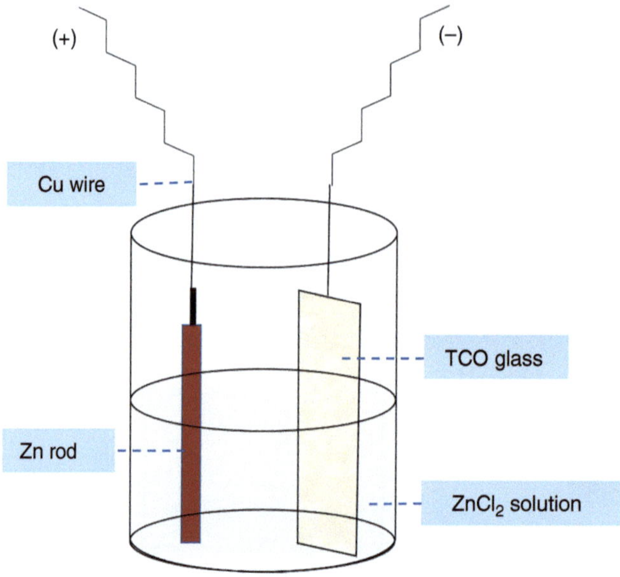

Figure 15.5 Electrochemical set up for the synthesis of thick Zn films. Source: Sinha et al. [21]/Royal Society of Chemistry.

has been devised. This process utilizes $ZnCl_2$ as the reagent. The zinc(0) powder, which was deposited and had a nanowire structure, had a significant impact on the reaction (Figure 15.5) [21]. The organic reaction known as metal-mediated allylation is commonly performed in aqueous solutions [22].

For the fabrication of nanomaterials, electrochemical deposition techniques are appealing due to their ease of use and generally low-cost equipment requirements. An appealing substitute for the deoxygenation of epoxides **1** is the electrolysis process due to its ease of use and low environmental impact. An undivided cell has been used to efficiently synthesize alkenes **2** using epoxides **1**. In this strategy, Zn(0) having a hierarchy-based nanostructure, facilitates the reaction (Scheme 15.3a) [23].

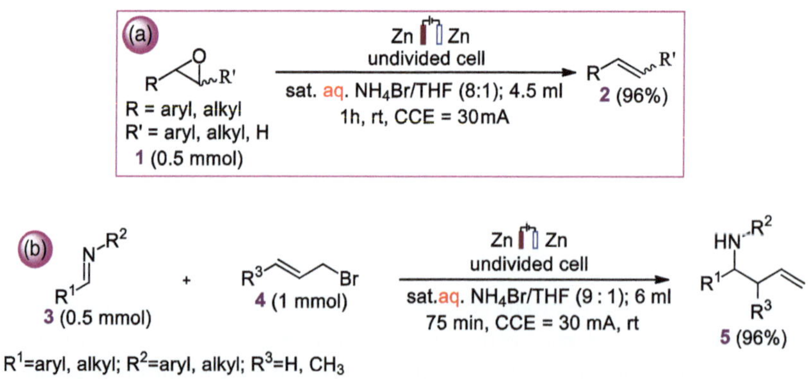

Scheme 15.3 (a) Electrochemical deoxygenation of epoxidesinto alkenes; (b) zinc-mediated electrochemical allylation of iminesin an aqueous medium.

Scheme 15.3b showed that the reactivity of simple imines in an aqueous medium was significantly influenced by the deposited zinc(0) powder with nanoscale hierarchies. This approach became an appealing synthetic substitute for homoallylic amines because of its excellent efficacy and green credentials. In this protocol, allylmetal reagents are added to C=N double bonds. These amines serve as essential structural components for a variety of nitrogen-containing compounds derived from natural sources and biologically active chemicals [24].

15.6.3 Metal-Catalyzed Electrosynthesis in Water

This innovative technique is conducted in water-based solutions in the presence of atmospheric oxygen. Under the influence of $Pd(OAc)_2$, the Zn-mediated allylic alkylations occur without difficulty, involving various alkyl halides (including primary, secondary, and tertiary) **6** and substituted allylic halides **7** (Scheme 15.4a) [25]. An electrochemical C–H/N–H annulation reaction of benzamide **9** with alkyne **10** at room temperature was reported by the Ackermann group in 2018 (Scheme 15.4b) [26]. The Jutand group published a study on transition metal-catalyzed electrochemical oxidative homocoupling in 2008 (Scheme 15.4c). Once one catalytic cycle was completed, the electrons from the anode were transferred via benzoquinone to regenerate Pd^0. Furthermore, it was discovered that benzoquinone was essential in avoiding the precipitation of Pd black, which was thought to result from the complexation of benzoquinone with Pd [27].

15.6.4 Electrochemical Oxidation in Aqueous Media

Oxidized compounds have gained significant value as fundamental components for a diverse range of processes and refined chemistry [28], particularly in the benzylic position. A technique for this transformation that is sustained by C–H functionalization has been shown in Scheme 15.5a. The process relies on electrochemistry for its power supply and draws radicals from *tert*-butyl hydroperoxide for the C–H abstraction using an aqueous-based medium. This approach can be easily adapted and used for large-scale synthesis. Furthermore, medicinal chemists may be able to modify this approach to produce oxidized analogues and enable clinical evaluation after demonstrating its efficacy on bioactive compounds [29]. Electrochemical epoxidation in aqueous media has also been well documented (Scheme 15.5b). Water acts as an oxygen source. The Cantillo group discovered in 2021 that alkenes **18** could be electrochemically epoxidized in THF/water. *In situ* formation of bromohydrin produced epoxides **19**. The reaction mixture produced 1,2-dihydroxyl intermediate **20** at 140 °C. In a single pot, tandem electrolysis produced aldehydes **21** [30]. The lactonization of ketones **22** via a non-Baeyer–Villiger process using MeCN/water as a cosolvent is described in Scheme 15.5c [31]. An essential step in the synthesis of organic materials is the oxidation of alcohol. In order to electrooxidize alcohol **24** to the corresponding product ketone **26** in aqueous conditions, the Tanaka group immobilized 2,2,6,6-tetramethylpiperidine N-oxyl (TEMPO) on a rigid network polymer (PBZTNT-N-Oxyl) **25** (Scheme 15.5d) [32].

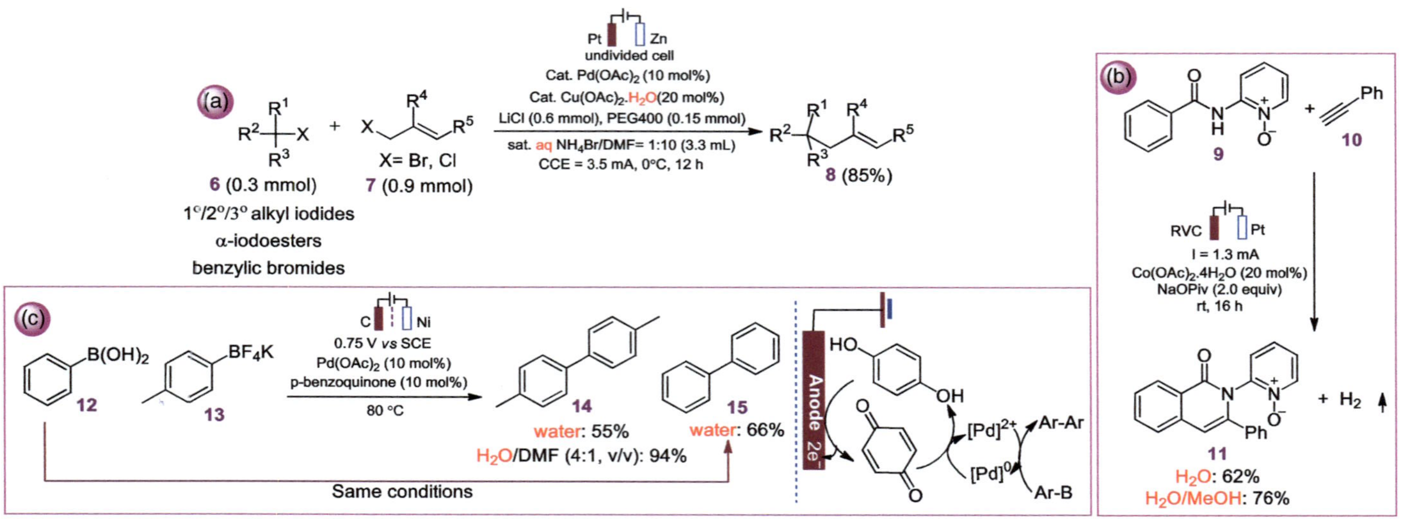

Scheme 15.4 (a) Cu/Pd-catalyzed allylic alkylations; (b) Co-catalyzed C–H/N–H annulation; (c) benzoquinone/Pd-catalyzed electrochemical homocoupling of aryl boron compounds **12** and **13** in aqueous media.

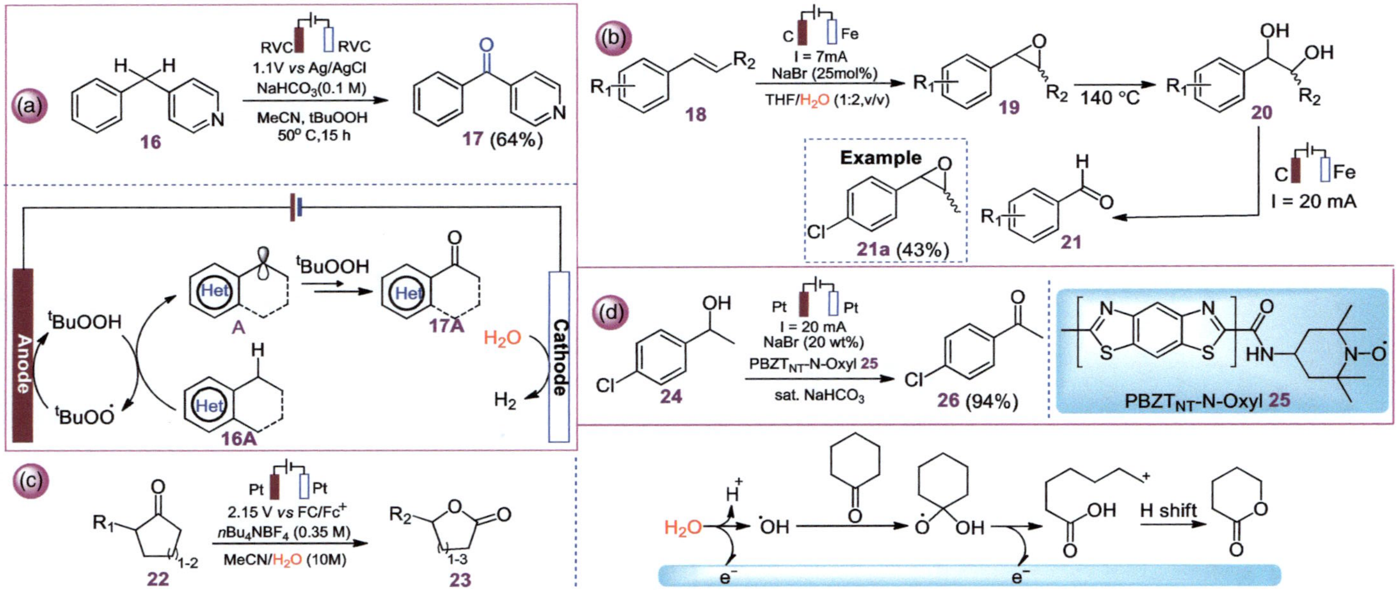

Scheme 15.5 (a) Electrochemical oxidation of benzylic C−H and its proposed mechanism; (b) oxidation of alkene to epoxide; (c) electrochemical production of lactones **23**; (d) electrocatalyzed oxidation of alcohol **24** in aqueous media.

15.6.5 Electrochemical Reduction in Aqueous Media

A moderate yield of Birch reduction of electron-rich arenes such as anisole **27** to the corresponding product **28** was reported in 1983 by the Kariv–Miller group using electrochemistry on aqueous media (Scheme 15.6a) [33]. The electrochemical hydrogenation of dimethyl maleate **29** was reported by the Tajima group in 2013. Amberlite IR-120, an immobilized sulfonic acid, was utilized as the proton source in this protocol (Scheme 15.6b) [34]. Process development, including cathodic reduction in the aqueous phase, demonstrates the benefit of the OER reaction. For the pilot product of 3-exomethylene-7-(*R*)-glutaroylaminocepham-4-carboxylic acid 1-(*S*)-oxide **32** from substrate **31**, Chai et al. set up a 1,3-reductive de-acetylation reaction (Scheme 15.6c) [35].

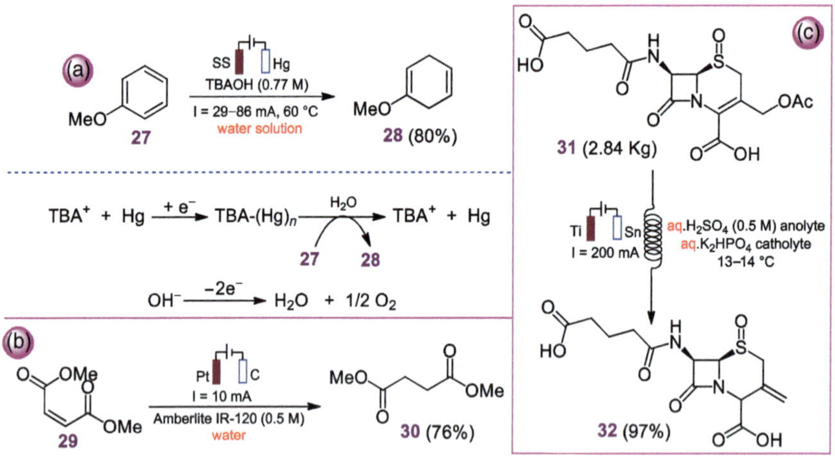

Scheme 15.6 (a) Electrochemical Birch reduction on aqueous media; (b) electro-reduction of maleic ester to succinic ester; (c) pilot scale electrochemical 1,3-reductive deacetylation in aqueous medium.

15.6.6 Emulsion Electrooxidation in Aqueous Media

The Atobe group reported the direct conversion of amine **33** to nitrile **34** by the use of the electrochemical method (Scheme 15.7a) [36]. In 2022, the same group established a flow emulsion electrooxidation process using an undivided flow cell that converted *n*-octylamine **33** to octanenitrile **34** (Scheme 15.7b) [37].

15.6.7 Electrochemical Polymerization in Aqueous Media

Oligo(ethylene oxide) methyl ether methacrylate (OEOMA) polymerization in an undivided cell with an Al sacrificial anode was reported (Scheme 15.8a). The catalyst is destroyed by protonated amine ligands when the pH of the aqueous electrolyte is lowered by the Al^{3+} cation. The ideal solution for the precipitation of $Al(OH)_3$ is a

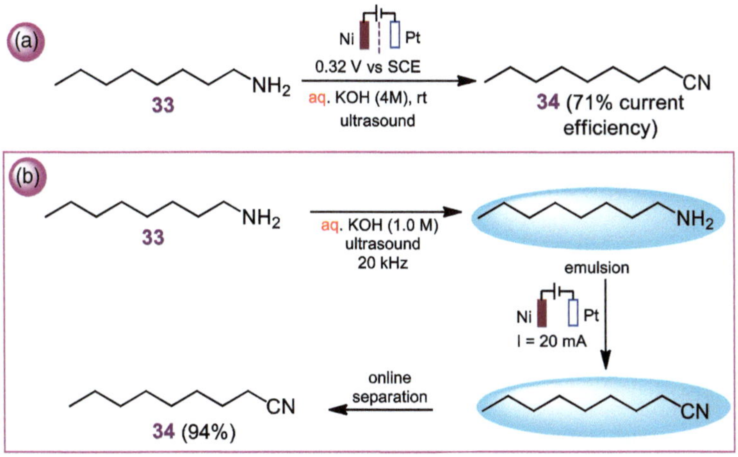

Scheme 15.7 The anodic oxidation of primary amine to nitrile in aqueous media; (a) anodic oxidation in emulsion-formed ultrasonication; (b) integrated flow emulsion electrochemical oxidation of n-octylamine **33**.

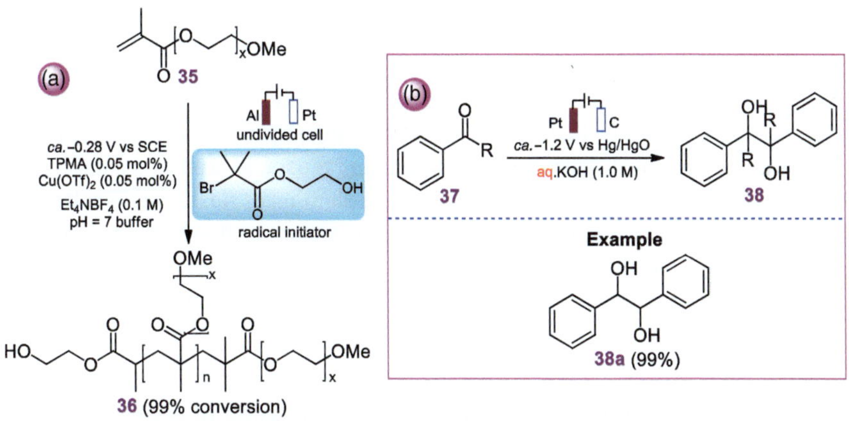

Scheme 15.8 (a) Electrochemical atom transfer radical polymerization (ATRP); (b) electrochemical cathodic dimerization of aldehydes with carbon paper as the cathode in aqueous media.

buffer solution that is neutral [38]. Likewise, the electrochemical synthesis of pinacol product **38** from benzaldehyde **37** in water was reported by the Zhang group in 2021 (Scheme 15.8b) [39].

15.6.8 Electrocatalytic Hydrogenation (ECH)

Water is an effective solvent in the reaction known as ECH of unsaturated compounds. In this strategy, the ECH of platform chemicals produced from biomass, including furfural (Scheme 15.9a), has been highlighted [41]. It has been

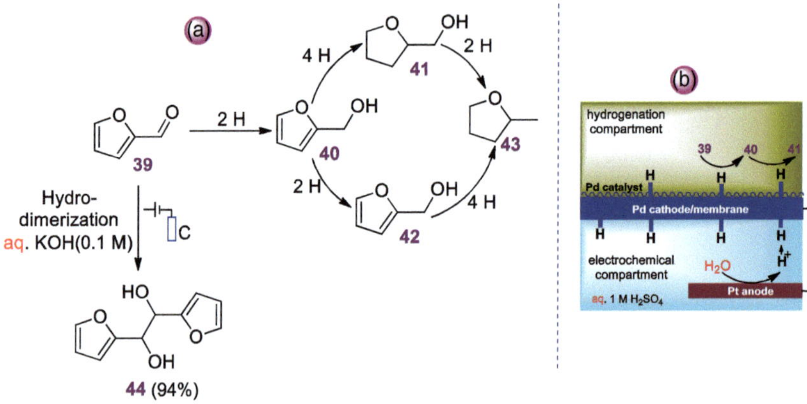

Scheme 15.9 (a) Possible products of the ECH of furfural (**6**). (b) ECH of **39** in a Pd membrane reactor. Source: Delima et al. [40]/Royal Society of Chemistry.

demonstrated that the catalyst material and the electrolysis conditions have a significant impact on the selectivity of the conversion of **39** into the potential hydrogenation products **40–43**. In an aqueous acidic solution (pH = 2.9), **39** can be selectively transformed to 2-methylfuran **42** by using CuPd particles supported on carbon as catalysts. However, tetrahydrofurfuryl alcohol **41** was found to be selectively produced in a Pd membrane reactor (Scheme 15.9b), where a hydrogen-permeable Pd foil acts as a separator between the electrochemical processes in aqueous 1 M H_2SO_4 and the chemical hydrogenation in organic media, as well as a catalyst for ECH [40]. The benefits of this approach include significant flexibility in selecting the solvent for hydrogenation while employing water as the hydrogen source and exceptional selectivity for ECH at high current densities. Hydrodimerization to hydrofuroin is one potential side reaction that could happen under ECH conditions **44** [41].

15.6.9 Examples of Electrochemical Processes in Mixture of Water and Other Solvents

The Terent'ev group claimed in 2021 that they had synthesized bromohydrins using an electrochemical technique from alkenes and inorganic bromide (Scheme 15.10a) [42]. Apart from the widely recognized bromonium pathway, a particular DMSO-Br^+ pathway similar to that documented by Yoshida was suggested as an effective path to the target product [43]. In 2019, Xu et al. revealed that under anodic oxidation conditions, $CF_2HSO_2NHNHBoc$ served as the precursor for the difluoromethylation of α,β-unsaturated compounds **47** (Scheme 15.10b) [44]. Ackermann and his coworkers disclosed the production of aminated heteroarenes **51** through an effective dehydrogenative C–H of **49** and N–H of **50** coupling (Scheme 15.10c) [45].

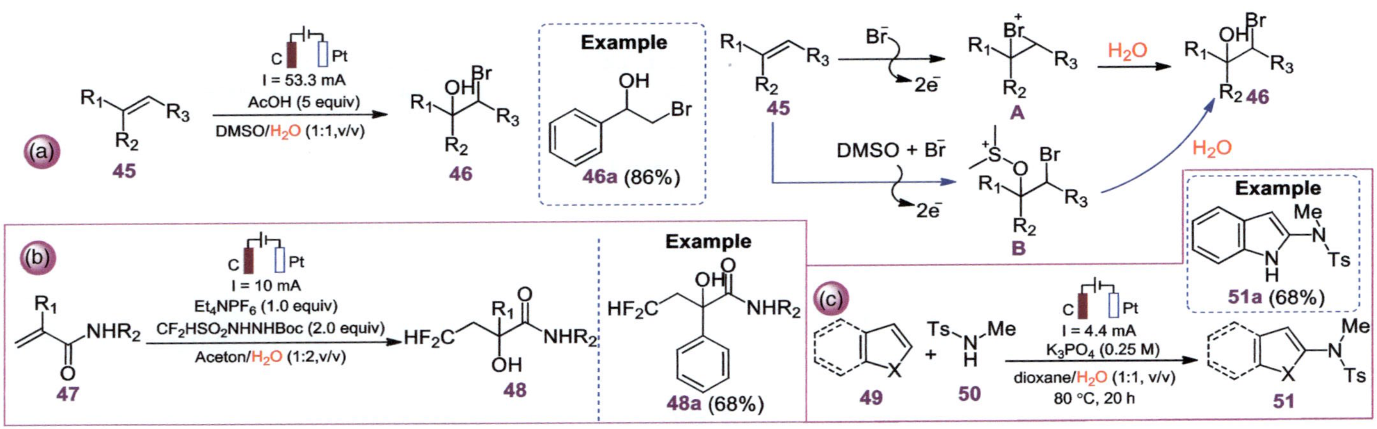

Scheme 15.10 (a) The oxidative electrochemical synthesis of bromohydrin in DMSO/water; (b) electrochemical difluoromethylation of α,β-unsaturated compounds **47** in acetone/water; (c) electrochemical C–H amination of **49** in dioxane/water.

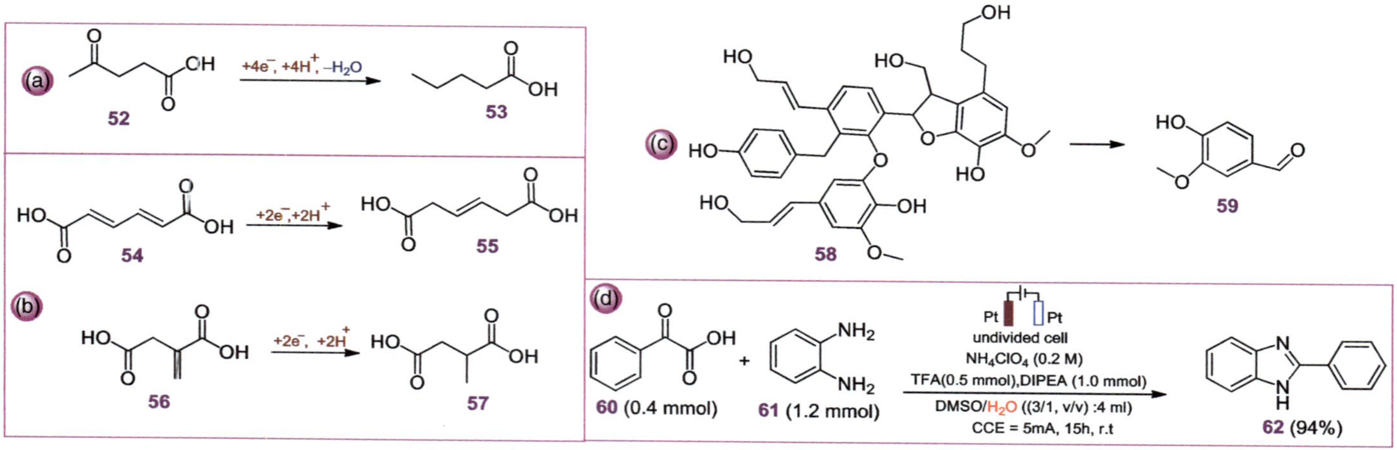

Scheme 15.11 (a) Hydrodeoxygenation of levulenic acid **52** to valeric acid **53**; (b) hydrogenation of muconic acid **54** to trans-3-hexendioic acid **55** and itaconic acid **56** to methylsuccinic acid **57**; (c) production of vanillin **59** from elecrochemical degradation of Kraft-lignin **58**; (d) electrochemical production of benzimidazole **62** in aqueous media.

15.7 Benefits of Electrochemical Conversions

Several benefits can be achieved through the electrochemical conversion of naturally occurring substances in a water environment. Biogenic compounds, which are found in living organisms, typically dissolve well in water because they have a polar nature. This means they can easily undergo electrochemical reactions in watery environments [46]. Valeric acid is soluble in water, but n-octane, the product of the reaction, is not. So, after the reaction, the n-octane forms a separate layer on top of the water, making it easy to collect and separate the desired product. This concept has been known for a long time and is now being applied in the context of electrofuel synthesis. In electrochemical hydrogenation or hydrodeoxygenation reactions in water, various transformations can occur. For example, you can turn levulenic acid **52** into valeric acid **53** or muconic acid **54** into *trans*-3-hexenedioic acid **55** (Scheme 15.11a,b). An example of integrating microbial and electrochemical conversions is making methylsuccinic acid **57** from glucose using a fungus and electricity via an intermediate itaconic acid **56** (Scheme 15.11b) [47]. These products have potential uses in making things like special types of plastics and biofuels. Biomass usually has water in it, and it would take a lot of energy to get rid of this water, for example, by drying it. Using water as part of the process, instead of trying to get rid of it, lowers the amount of energy needed and makes the whole thing cheaper. Many possible sources of biogenic materials are called “waste” because they have diluted, complex, and mixed-up organic matter. Organic electrosynthesis is a way to use these materials and make them more valuable. Waldvogel and coworkers showed how to turn Kraft-lignin **58**, which is a waste product from making paper, into vanillin **59**, which is a valuable taste ingredient (Scheme 15.11c) [48]. This shows how organic electrosynthesis can turn things that used to be thrown away into things that are useful. It has been reported that benzimidazole **62** can be formed in an aqueous solution by a very efficient electrochemical decarboxylative coupling of *o*-phenylenediamine **61** with α-keto acid **60** (Scheme 15.11d). Pharmaceutical agents, natural goods, and functional materials all rely on benzoimidazoles as a fundamental component. This reaction protocol is friendly and easy to operate [49].

15.8 Conclusion

This chapter introduces the role of electrocatalysis in water. In addition, a brief overview of the basic concepts that are necessary to comprehend and explore organic electrosynthesis in aqueous media is provided. Furthermore, it examines a range of applications and explores current research developments in this domain. Aqueous reduction and oxidation reactions are facilitated by water, which is the most convenient safe source of hydrogen and oxygen. An additional force accelerating the reactions is the availability of the HER and OER reactions in water. Furthermore, one potential characteristic should be the high surface tension that organic compounds experience when exposed to water.

15.9 Future Perspectives

Over the coming decades, a crucial and fruitful area of research will be the understanding of the mechanisms behind organic electro-syntheses in aqueous reaction media, as well as the identification of new and previously identified syntheses for a bio-based economy [50]. The utilization of electricity for chemical transformations, which often entail laborious procedures and the usage of a certain amount of reagents, has introduced a novel pathway in the realm of catalysis. The subject of electrosynthesis is advancing through the incorporation of novel concepts and methodologies from other disciplines, including magneto-electrochemistry and ultrasonication. Furthermore, continuous innovation in electrolyte and electrode systems is necessary to overcome the limitations of existing electrode materials, such as mercury and lead. An ideal electrode system in current times should possess corrosion resistance and exhibit a significant overpotential to minimize undesirable side reactions. It is important to thoroughly investigate solvents, not only as a reaction medium, but also as a means to control selectivity in the electrosynthetic route. The viability of electrochemical processes could lead to their widespread adoption as a conventional synthetic technique, rather than a specialized technology [51].

In the last 15 years, density functional theory has been increasingly popular in electrocatalysis research. Therefore, new methods for investigating solvation/electrification effects or probing ion transfer paths in basic electrochemical steps could be obtained by combining DFT, with its special ability to examine bond breaking and forming at the electrocatalyst surface, with force-field approaches that can handle the longer length and time scales of electrolyte dynamics.

References

1 Das, S., Pérez-Ramírez, J., Gong, J. et al. (2020). *Chem. Soc. Rev.* 49: 2937–3004.

2 Leech, M.C. and Lam, K. (2022). *Nat. Rev. Chem.* 6: 275–286.

3 Waldvogel, S.R. and Möhle, S. (2015). *Angew. Chem. Int. Ed.* 54: 6398–6399.

4 Sahloul, K., Sun, L., Requet, A. et al. (2012). *Chem. Eur. J.* 18: 11205–11209.

5 Pollok, D. and Waldvogel, S.R. (2020). *Chem. Sci.* 11: 12386–12400.

6 Baumgarten, N., Etzold, B.J., Magomajew, J., and Ziogas, A. (2022). *ChemistryOpen* 11: e202200171.

7 Yamashita, Y., Noguchi, A., Fushimi, S. et al. (2021). *J. Am. Chem. Soc.* 143: 5598–5604.

8 Kolbe, H. (1847). *J. für Prakt. Chemie* 41: 137–139.

9 Liu, H. and Li, W. (2021). *Curr. Opin. Electrochem.* 30: 100795.

10 Butler, R.N. and Coyne, A.G. (2016). *Org. Biomol. Chem.* 14: 9945–9960.

11 Hosseini, S., Beeler, J.A., Sanford, M.S., and White, H.S. (2023). *Faraday Discuss.* 247: 195–208.

12 Möller, M., Nilges, P., Harnisch, F., and Schröder, U. (2011). *ChemSusChem* 4: 566–579.

13 Schmidt, V.M. (2012). *Elektrochemische Verfahrenstechnik: Grundlagen, Reaktionstechnik, Prozessoptimierung*, 135–229. John Wiley & Sons.
14 Jiang, X., Kuklin, A.V., Baev, A. et al. (2020). *Phys. Rep.* 848: 1–58.
15 Maheshwari, S., Li, Y., Agrawal, N., and Janik, M.J. (2018). *Advances in Catalysis*, vol. 63, 117–167. Elsevier.
16 You, B. and Sun, Y. (2018). *Acc. Chem. Res.* 51: 1571–1580.
17 Akin, O. and Yuksel, A. (2017). *Cellulose* 24: 4729–4741.
18 Faraday, M. (1834). *Ann. Phys.* 109: 481–520.
19 Wang, C., Liu, K., Jin, Y. et al. (2023). *ChemSusChem* 16: e202300222.
20 Neubert, K., Schmidt, M., and Harnisch, F. (2021). *ChemSusChem* 14: 3097–3109.
21 Sinha, A.K., Mondal, B., Kundu, M. et al. (2014). *Org. Chem. Front.* 1: 1270–1275.
22 Roy, U.K. and Roy, S. (2010). *Chem. Rev.* 110: 2472–2535.
23 Huang, J.-M., Lin, Z.-Q., and Chen, D.-S. (2012). *Org. Lett.* 14: 22–25.
24 Huang, J.M., Wang, X.X., and Dong, Y. (2011). *Angew. Chemie* 4: 954–957.
25 Lai, Y.-L. and Huang, J.-M. (2017). *Org. Lett.* 19: 2022–2025.
26 Tian, C., Massignan, L., Meyer, T.H., and Ackermann, L. (2018). *Angew. Chem.* 130: 2407–2411.
27 Amatore, C., Cammoun, C., and Jutand, A. (2008). *Eur. J. Org. Chem.* 2008: 4567–4570.
28 Rani, S., Aslam, S., Lal, K. et al. (2023). *Chem. Rec.* 24: e202300331.
29 Marko, J.A., Durgham, A., Bretz, S.L., and Liu, W. (2019). *Chem. Commun.* 55: 937–940.
30 Jud, W., Kappe, C.O., and Cantillo, D. (2021). *Electrochem. Sci. Adv.* 1: e2100002.
31 Maalouf, J.H., Jin, K., Yang, D. et al. (2020). *ACS Catal.* 10: 5750–5756.
32 Kubota, J., Ido, T., Kuroboshi, M. et al. (2006). *Tetrahedron* 62: 4769–4773.
33 Swenson, K.E., Zemach, D., Nanjundiah, C., and Kariv-Miller, E. (1983). *J. Org. Chem.* 48: 1777–1779.
34 Miura, Y., Tateno, H., and Tajima, T. (2013). *Electrochemistry* 81: 371–373.
35 Chai, D., Genders, D., Weinberg, N. et al. (2002). *Org. Process Res. Dev.* 6: 178–183.
36 Atobe, M., Ikari, S., Nakabayashi, K. et al. (2010). *Langmuir* 26: 9111–9115.
37 Mikami, R., Shida, N., and Atobe, M. (2022). *Org. Process Res. Dev.* 26: 1268–1278.
38 Luo, J., Chavez, M., Durante, C. et al. (2022). *Electrochim. Acta* 432: 141183.
39 Liu, C., Li, R., Zhou, W. et al. (2021). *ACS Catal.* 11: 8958–8967.
40 Delima, R.S., Stankovic, M.D., MacLeod, B.P. et al. (2022). *Energy Environ. Sci.* 15: 215–224.
41 Francke, R. (2022). *Curr. Opin. Electrochem.* 36: 101111.
42 Bityukov, O.V., Vil', V.A., Nikishin, G.I., and Terent'ev, A.O. (2021). *Adv. Synth. Catal.* 363: 3070–3078.
43 Ashikari, Y., Shimizu, A., Nokami, T., and Yoshida, J.-I. (2013). *J. Am. Chem. Soc.* 135: 16070–16073.
44 Xu, H.H., Song, J., and Xu, H.C. (2019). *ChemSusChem* 12: 3060–3063.
45 Zhang, Y., Lin, Z., and Ackermann, L. (2021). *Chem. Eur. J.* 27: 242–246.

46 Nilges, P., dos Santos, T.R., Harnisch, F., and Schröder, U. (2012). *Energy Environ. Sci.* 5: 5231–5235.

47 Holzhäuser, F.J., Artz, J., Palkovits, S. et al. (2017). *Green Chem.* 19: 2390–2397.

48 Schmitt, D., Regenbrecht, C., Hartmer, M. et al. (2015). *Beilstein J. Org. Chem.* 11: 473–480.

49 Wang, H.B. and Huang, J.M. (2016). *Adv. Synth. Catal.* 358: 1975–1981.

50 Möhle, S., Zirbes, M., Rodrigo, E. et al. (2018). *Angew. Chem. Int. Ed.* 57: 6018–6041.

51 Lodh, J., Paul, S., Sun, H. et al. (2023). *Front. Chem.* 10: 1571.

16

Visible Light Photocatalysis in Water

Camilla Russo[1], *Francesca Brunelli*[2], *Gian Cesare Tron*[2], *and Mariateresa Giustiniano*[1]

[1] *University of Naples Federico II, Department of Pharmacy, 80131 Napoli, Italy*
[2] *University of Piemonte Orientale, Department of Drug Science, 28100 Novara, Italy*

16.1 Introduction

The use of visible light to promote organic transformations has become a well-consolidated approach to chemical and material synthesis [1]. In photoredox catalysis, these reactions imply the excitation of a (photo)catalyst able to harvest light energy and reach an electronically excited state, which can further promote either the oxidation or the reduction of a suitable radical precursor (e.g. via single-electron transfer (SET) events). Alternative common mechanistic pathways can involve the formation of electron donor–acceptor (EDA) complexes, energy transfer (EnT), hydrogen atom transfer (HAT) catalysis, and proton-coupled electron transfer (PCET). All these pathways can be more or less influenced by the reaction solvent as the latter not only can determine, as obvious, the solubility of the starting materials, but also can affect the energies between the electronic states of a light-absorbing substrate according to what is defined as solvatochromic effects [2]. In this context, water, the solvent of Life, has attracted increasing interest as a reaction medium for photochemical transformations [3]. It has been shown that water can enhance the reductive power of metal-based photocatalysts, promote reductive proton transfer and the LUMO-lowering effect (lowest unoccupied molecular orbital), and influence chemoselectivity and reaction rates, thus unlocking unique and unconventional reactivities [3]. According to Prof. Kobayashi and coworker [4] and Prof. Lipshutz and coworkers [5], a conceptual classification of the current literature can rely on three main groups: (i) *in-water reactions*, i.e. reactions involving substrates soluble in water (ia: water-soluble catalyst; ib: catalyst not soluble in water); (ii) *in-water reactions involving the use of additives* to achieve solubilization of the substrates such as micelles-forming surfactants (iia and iib depending on the solubility of the catalyst); and (iii) *on-water reactions*, occurring when the substrates are not water-soluble (iiia–c, with iiia referred to reactions involving a soluble catalyst). The second class can also be extended to “artificial enzymes” or “chemzymes”, defined

Organic Transformations in Water: Principles and Applications, First Edition.
Edited by Gopinathan Anilkumar, Nissy Ann Harry, and Sankuviruthiyil M. Ujwaldev.

as water-soluble macromolecules able to host a non-polar guest in a hydrophobic cavity, which include cyclodextrines, antibodies, nanomaterials, dendrimers, and other disperse interface-rich structures [6]. The third class is probably the most interesting one with current ongoing studies focused on understanding how bulk properties of water such as polarity and viscosity, solute-water interactions at the microscopic level, as well as both nonspecific (e.g., Coulombic and van der Waals) and specific (e.g., hydrogen bonding, hydrophobic) interactions, could influence the reaction kinetics and overall outcome. On-water reactions indeed can occur either at interfaces such as catalyst–water, substrate(s)-water, catalyst-substrate(s)-water, or, according to the Marcus model [7], at an oil–water interface, when hydrophobic interactions of the starting materials lead to the formation of suspensions or emulsions. However, the study of the exact phenomena underlying the mechanism of rate acceleration at aqueous interfaces still remains a complex task, also due to the implication of multiple different effects, thus representing a field of intense research. The last paragraph of this chapter will briefly mention photobiocatalytic approaches harnessing the combination of enzymatic reactions with light-promoted transformations.

16.2 In-Water Reactions

As a matter of fact, a strict classification of in-water reactions is not possible due to poor information about the exact solubility of the reactants; such reactions have indeed been reported "in water" even when involving slightly or not soluble starting materials. Accordingly, this section will focus on those transformations involving, at least, water-soluble photocatalysts as well as reactions where the solubilization of the starting materials is achieved by means of additives involving, but not limited to, micelles-forming surfactants.

16.2.1 Soluble Photocatalysts

Most metal-based photoredox catalysts exhibit poor solubility in water (from <1 to 1000 ppm), which can be tuned in some cases by the substitution pattern and the anionic partner [8]. In this context, Roelfes and coworkers [9] and Conrad and coworkers [10] have shown that homogeneous conditions can be achieved by using water-soluble iridium photoredox catalysts where the dative ligands of $[Ir(dF(CF_3)ppy)_2(dtbbpy)]PF_6$ have been replaced by either quaternary ammonium groups or carboxylic acid residues, respectively. The use of $[Ir(dF(CF_3)\text{-}ppy)_2(dNMe_3bpy)]Cl_3$ (Scheme 16.1a) led to a synthetic protocol to modify dehydroalanine (Dha) residues in peptides and proteins, which is performed in water and under physiologically relevant conditions [9], while the introduction of carboxylic acid moieties on the bipyridyl ligand led to catalyst **2** (Scheme 16.1b) applied to the trifluoromethylation of acyl tyrosine amide and an unprotected dipeptide Asp-Tyr **1** in phosphate-buffered saline solvent (DPBS) [10].

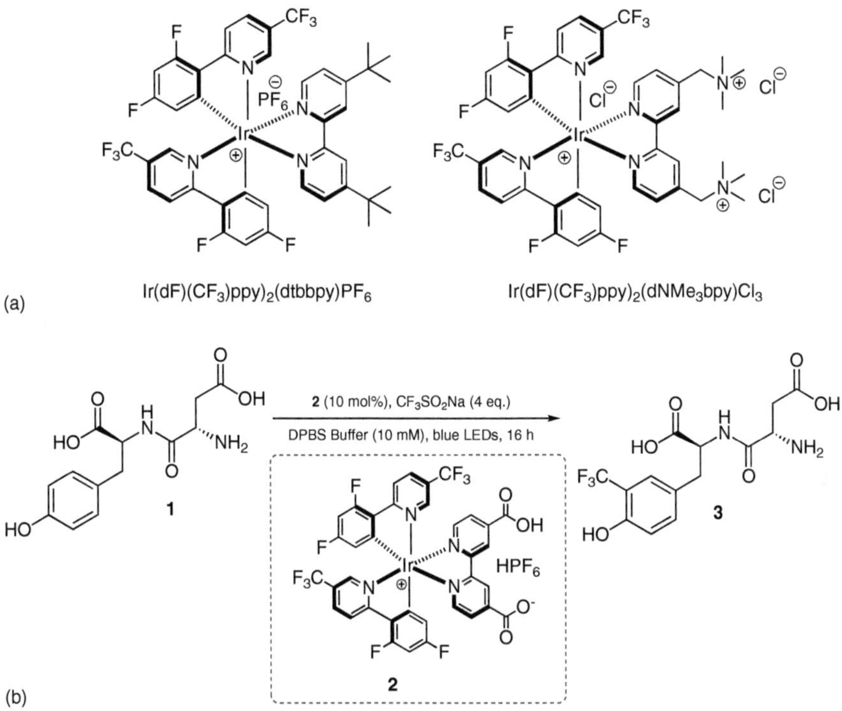

Scheme 16.1 (a) Water-soluble Ir-based photocatalysts; (b) trifluoromethylation of dipeptide **2**.

A relevant example of a water-soluble organophotocatalyst, i.e. alizarin red S, was reported by Wu and coworkers for the bromination and iodination of 8-aminoquinoline amides in water [11]. On the other hand, Romero, Teixidor and coworkers showed how very low loadings of metallacarboranes systems such as $[Co(C_2B_9H_{11})_2]^-$ were effective to promote the oxidation of aromatic and aliphatic alcohols as well as the oxidation of alkenes to epoxides (conversions between 65% and 97% in short reaction times of 15 minutes and excellent chemo-selectivities) [12, 13]. Interestingly, such metallacarborane could also be easily recycled by covalently linking it to magnetic nanoparticles (MNPs), as proved in the photooxidation of alcohols [14]. More recently, the same authors were able to exploit noncovalent interactions—persistent even after water dissolution—between $[Ru^{II}(trpy)(bpy)(H_2O)]$ (trpy = terpyridine and bpy = bipyridine) and the photoredox metallacarborane $[3,3\text{-}Co(1,2\text{-}C_2B_9H_{11})_2]_2$ to develop a catalytic system to photooxidize alcohols to aldehydes and carboxylic acids via a proton-coupled electron-transfer process under UV light irradiation and with a 0.005 mol % of catalyst loading [15].

In this context, the selective oxidation of alcohols to aldehydes was achieved by Wang and coworkers by using mesoporous graphitic carbon nitride (mpg-CN), a metal-free polymeric photocatalyst [16], endowed with high tunability for a wide range of parameters. Albeit requiring a biphasic solvent system to be recycled, one more example worthy of note relies on a cobalt-based phthalocyanine photoredox catalyst [CpPc-$(SO_3Na)_4$], whose efficacy was proved by Baskar and coworkers in the oxidative dehydrogenation of N-heterocycles such as tetrahydro-β-carbolines, indolines, and tetrahydro-(iso)-quinolines [17].

16.2.2 Addition of Solubilizing Agents

16.2.2.1 Photomicellar Catalytic Systems

Micellar catalysis, pioneered by Prof. Lipshutz, represents a consolidated approach enabling the use of water-based reaction media when the reactants are not hydrophilic enough to get homogeneous conditions in such systems [18]. In the context of photoredox catalysis, the same Prof. Lipshutz in 2018 reported the design and the synthesis of an iridium photocatalyst covalently linked to polyethyleneglycol ubiquinol succinate (PQS), a reduced form of dietary supplement CoQ10, able to undergo self-aggregation in water into nanomicelles [19]. The PQS-[Ir] photocatalyst was efficient in a pool of representative transformations, including the difunctionalization of alkenes **4** and the sulfonylation of enol acetates **7** (Scheme 16.2). While the main advantage of this system is the easy recyclability of the catalyst in the same reaction flask (up to ten times without losses in the yield), the need for a preliminary synthesis of photoredox catalysts endowed with different redox potentials and covalently linked to surfactants limits its broad application. Accordingly, most of the literature reports rely on the combination of common both metal-based and organic photoredox catalysts with aqueous micellar solutions such as TPGS-750M, Triton-X, SDS, and CTAC, to cite a few. Representative protocols involve arylation reactions with diazonium ions generated *in situ* by using eosin B in a 2 wt % Triton-X solution [20]; direct arylations of aryl bromides in a microfluidic reactor by exploiting an association colloid formed by castor oil derivative Kolliphor EL (K-EL) and a specifically devised photoredox active surfactant (S-PTh) [21]; the aromatization of 1,4-dihydropyridines to pyridines in an aqueous solution of Triton-X [22] as well as the C–H functionalization of heteroarenes in SDS micellar aqueous solution [23].

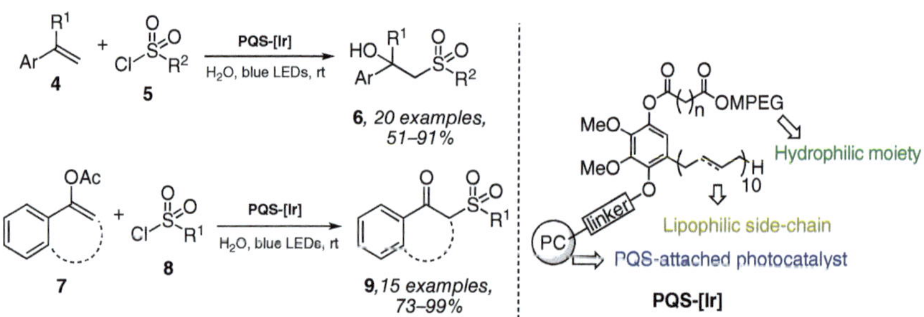

Scheme 16.2 PQS-attached photocatalyst and conversion of alkenes **4** and enol acetates **7** to products **6** and **9**.

Often, the photomicellar conditions proved to afford higher yields while enabling to either avoid or reduce stoichiometric radical promoters, oxidants, and acids as additives. In some other cases, the same reaction performed in acetonitrile does not afford the product, thus highlighting the critical importance of microstructuring the components in the reaction medium. Even more interestingly, the use of an anionic surfactant rather than a cationic one can drive the chemoselectivity toward the formation of different products: this is the case of *o*-chlorobenzamides **10**, which could lead to either intramolecular C–H arylation or *N*-dealkylation depending on the use of a CTAB (cetyl trimethyl ammonium bromide) or an SDS (sodium dodecyl sulfate) micellar solution, respectively [24] (Scheme 16.3). We recently showed that the evaluation of the polarity charge of both the photocatalyst and the surfactant could be key to figuring out optimal photomicellar catalytic conditions [25]. More in detail, solution 1D- and 2D-NMR experiments performed in the presence of paramagnetic probes enabled the investigation of the reaction environment at the atomic level, i.e. the localization of the photocatalyst with respect to the micelles. In our specific case, we developed a synthetic protocol to get amides from isocyanides and tertiary aromatic amines using $[Ir(ppy)_2bpy]PF_6$ in a 2 wt % SDS water solution.

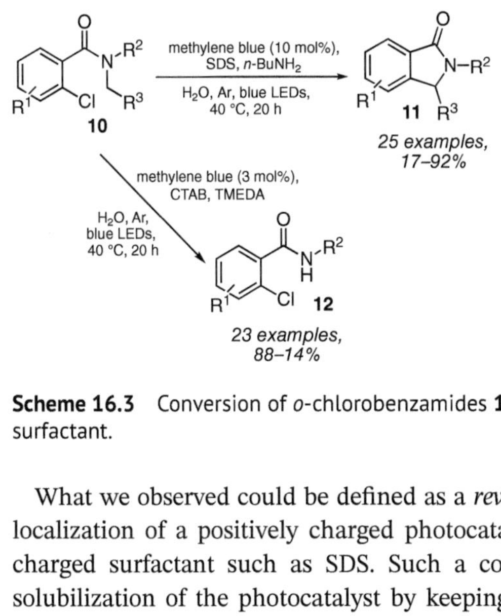

Scheme 16.3 Conversion of *o*-chlorobenzamides **10** to either **11** or **12** depending on the surfactant.

What we observed could be defined as a *reverse polarity principle*, providing the localization of a positively charged photocatalyst on the surface of a negatively charged surfactant such as SDS. Such a condition could indeed enhance the solubilization of the photocatalyst by keeping the interactions with the micelles on their surface, thus exposed to light irradiation. In the case of photocatalytic methods, indeed, the "nanoreactor" model, generally acknowledged to explain the efficiency of the micellar medium, poses some concerns about the ability of the light to penetrate deep inside the micelles and reach the reactants. SDS micelles were also shown to be an optimum solvent system to promote, via 2-aminoanthracene irradiation at 355 nm, the formation of hydrated electrons acting as "super-reductants" in the dimerization of chloro-organic substrates [26]. As an alternative to common surfactants, several organic transformations could harness amphiphilic polymeric nanoparticles incorporating organic photocatalysts such as phenothiazines and acridinium dyes [27, 28] and colloidal platinum nanoparticles dispersed by poly-vinylpyrrolidone [29].

16.2.2.2 Use of Additives Other Than Surfactants

Homogeneous reaction conditions in water can also be achieved by means of chaotropic agents (e.g., urea and guanidinium salts) able to enhance the water solubility of hydrophobic particles, whereas kosmotropic additives, exemplified by sodium chloride, are known to favor the aggregation of such lipophilic substrates by decreasing their water solubility [30]. The origin of the terms, from Greek *chaos* and *kosmo*, would highlight the ability of the additives to induce and stabilize "disorganization" and "organization," respectively, of the hydrogen bond network of pure water. In the first case, such a disorganization would promote the solvation of a hydrophobic substrate, whereas favoring the organization of such polar interactions between water molecules would lead to further repulsion of the non-polar substrate. Albeit such definitions have been indicated as misleading, these agents are commonly exploited for the proteins' purification in salting-in/salting-out methods [31]. In the context of photocatalysis, very recently, Prof. König reported the use of meglumine, a sugar alcohol derived from sorbitol and used as pharmaceutical excipient [32, 33], as an effective solubilizing agent to promote EDA complex formation between haloarenes, as electron-poor species, and indoles, anilines, anisoles, or thiols as electron donors [34] (Scheme 16.4). Importantly, the same transformations did not occur in organic solvents, thus highlighting how the water effects are key to promoting EDA complex formation as well as enhancing its stability. This elegant report was the first to address the study of EDA complexes in water, which, being by definition species arising from aggregation, are expected to exhibit low solubility. The authors reported 91 C–C and C–S cross-coupling products formed upon reductive fragmentation of haloarenes, followed by radical–radical coupling and proton loss (Scheme 16.4).

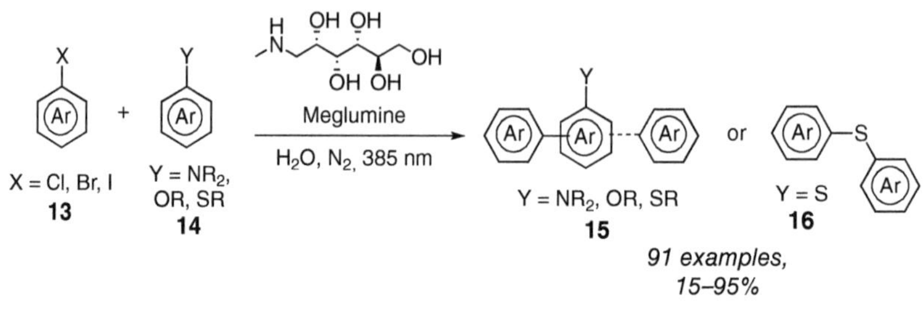

Scheme 16.4 Use of meglumine as an additive for photochemical cross-coupling reactions.

16.3 On-Water Reactions

According to Prof. Kobayashi, heterogeneous mixtures of insoluble reactants undergoing accelerated conversion to products can be defined as on-water reactions [4]. Albeit the original term was used by Prof. Sharpless for catalyst-free reactions [35], the current amount of data available in the literature led to extending this definition to all the reactions occurring at interfaces with water and showing dramatic acceleration with respect to common organic solvents. The Kobayashi model proposes a further classification based on the behavior of the water layer. Water molecules can indeed: (i) be oriented to extrude protons toward the oily phase in the inner layer, (ii) be enwrapped by a secondary layer, (iii) form a bulk water layer. Due to the lack of sufficient data, it is unrealistic to base a definition on the solubility of the reactants. Accordingly, a more convenient approach is based on the achievement of "on-water" like acceleration [36], where it has been shown that only one equivalent of water could be enough to gain such an advantage [37]. Accordingly, this class of reactions is not limited to those involving lipophilic substrates forming aggregates and appearing as aqueous suspensions, but can also be extended to "in-water" reactions exploiting "on-water" effects.

16.3.1 Catalytic Roles of Water as a Solvent for the Acceleration of Reaction Rates

As for photochemical and photoredox catalytic synthetic methods, it is common to find procedures where optimum conditions are achieved by adding water (either a few equivalents or precise volume ratios) to the reaction medium based on an organic solvent (either miscible or not). Unfortunately, most such synthetic protocols do not report sufficient data about the reason(s) for the beneficial effect of water. In some cases, the addition of water can be useful to achieve homogeneous conditions by increasing the solubilization of a complex reaction mixture involving a photocatalyst and one or more inorganic base/additive(s). In this context, the deboronative cyanation reaction to give products **19** starting from trifluoroborate salts **17** (Scheme 16.5) and promoted by a ruthenium-based photocatalyst, a hypervalent iodine oxidant, a cyanide source **18**, and TFA (trifluoroacetic acid) as an additive can be considered a representative example [38].

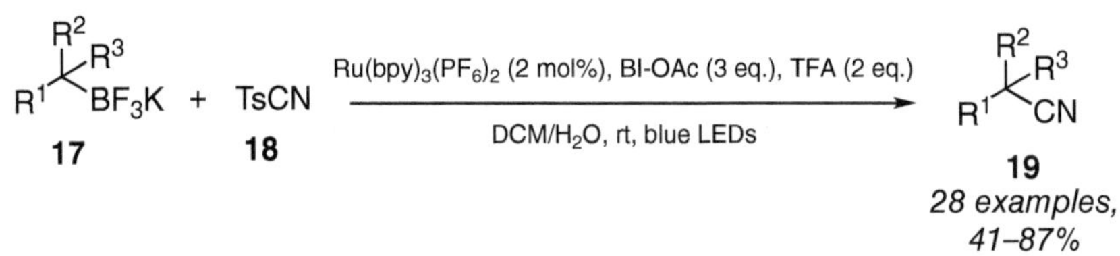

Scheme 16.5 Deboronative cyanation is promoted by a Ru-based photocatalyst.

Here, the dissolution of the different reactants/reagents was achieved with a biphasic 1 : 1 DCM/H_2O solvent system, whereas 2 : 1 mixtures of organic solvents (acetone/DCM and HFIP/DCM) led to poorer outcomes. Some other reports exploit the catalytic role of water and water-based species (such as H_3O^+) in promoting cycloaddition reactions, as shown by Prof. Rueping in the synthesis of isoxazolidines [39]. Homogeneous conditions can be advantageous when the unique features of water in forming hydrogen bonds are effective in the modification of key properties of the reactants, such as their redox potential. Such an effect has been described, for example, by Prof. Zeitler, who showed how the addition of water induces a

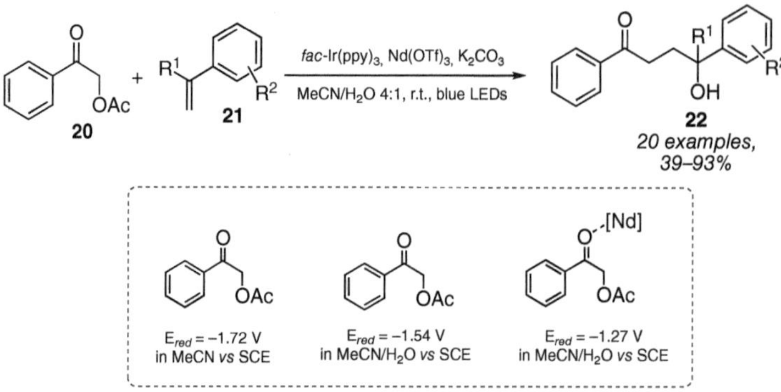

Scheme 16.6 LUMO-lowering effect of water in the photocatalytic cross-coupling of α-carbonyl acetates with styrenes.

LUMO-lowering effect for α-acetylated acetophenone derivatives **20** (Scheme 16.6) (E_{red} = −1.72 V in MeCN *vs.* SCE [saturated calomel electrode]), by increasing their redox potential to −1.54 V *vs.* SCE [40]. Such an effect was further enhanced with a water-compatible Lewis acid such as $Nd(OTf)_3$ (E_{red} = −1.27 V *vs.* SCE for the carbonyl substrate in the presence of the LA).

16.3.2 Exploitation of Non-Solvent Properties of Water

The nonsolvent properties of water, referred to as lipophilic hydrophobic additives and/or reagents, can be a key to achieving a high degree of chemoselectivity, as shown by Jui and coworkers for the radical conjugate addition of nitrogen heterocycles **23** to electron-poor alkenes **24** (Scheme 16.7a) [41]. The authors reported how the addition of increasing amounts of water spanning from 0% to 33% to DMSO (dimethyl sulfoxide) led to decrease the solubility of the Hantzsch ester, herein used

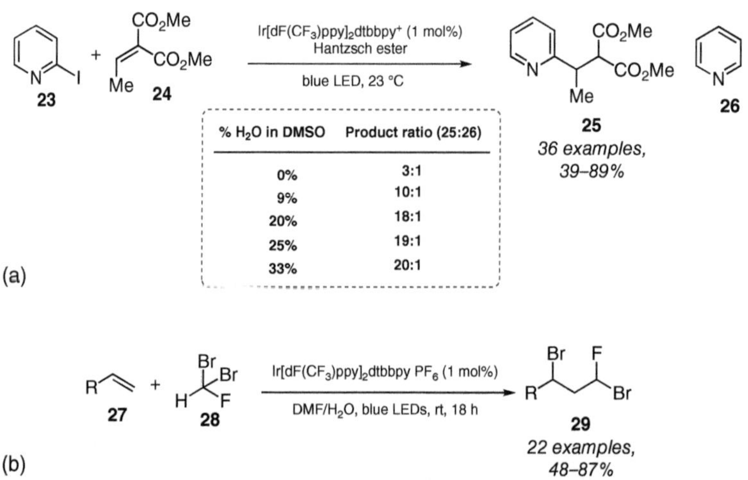

% H_2O in DMSO	Product ratio (25:26)
0%	3:1
9%	10:1
20%	18:1
25%	19:1
33%	20:1

Scheme 16.7 (a) Effect of water in the chemoselective formation of alkylated pyridines **25**; (b) chemoselective synthesis of dibromofluoromethylated adducts **29**.

as a hydrogen atom donor, thus improving the selective formation of the Michael addition products **25** over the formation of the reduced nitrogen heterocycles **26** (Scheme 16.7a). A similar solvent effect, albeit less intuitive, was reported by Qing and coworkers in the chemoselective synthesis of hydro- and dibromofluoromethylated adducts **29** from alkenes **27** and dibromofluoromethane **28** (Scheme 16.7b) [42]. Here, the use of THF (tetrahydrofuran) as both a solvent and a hydrogen atom source led to the exclusive formation of hydro-bromofluoromethylated products, while the switch to a 1 : 4 DMF/H_2O solvent system completely suppressed their formation, affording products **29** in high yields. In such a case, the role of water in promoting the chemoselectivity is unclear, as the use of the sole DMF (*N,N*-dimethyl formamide) as the solvent still afforded a mixture of both hydro- and dibromofluoromethylated products. More or less pronounced solvation effects of reaction intermediates/products could account for the observed chemoselectivity.

Besides these reports about mixtures of water and organic solvents, a breakthrough approach to use pure water as a solvent system has very recently been described by Prof. König in the C(sp^2)–N, C(sp^2)–O, C(sp^2)–S, C(sp^2)–Se (Scheme 16.8) and C(sp^3)–C(sp^2) coupling reactions via EDA complex formation [43]. The authors showed that the melting point depression of solid mixtures can form an oil phase where the hydrophobic interactions with water can further promote the aggregation of the substrates leading to both EDA complex formation and stabilization. The formation of an oily thin film enables maximum engagement of the active hydrogen atoms of the reactants in hydrogen bonds with water at the interface, thus leading to a dramatic acceleration of the photochemical transformations. The latter, being selected as reactions not occurring in common organic solvents, enabled the acknowledgment of the huge potentiality of photochemical transformations at the water interface.

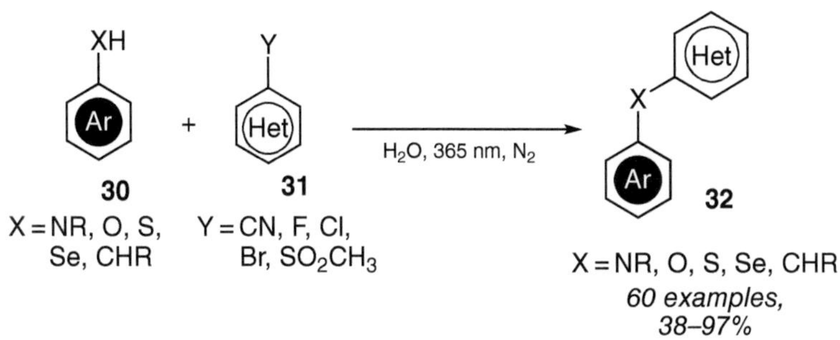

Scheme 16.8 Cross-coupling cleavage of C(sp^2)–Y bonds at the water-oil interface via melting point depression.

16.4 Photobiocatalysis

Photobiocatalysis could be defined as an intense research area aiming at the combination of the unique chemical features of photocatalysis and enzymatic reactions in terms of chemo-, regio-, and stereo-selectivities [44]. Moreover, such a field superbly addresses the challenge of sustainable production of chemicals, as enzymes stand for biodegradable catalysts. Reactions are usually run in water-based media, at room temperature or slightly above, and at atmospheric pressures. Such a section would provide the reader with a general overview of the mechanistic differences between the currently available photobiocatalytic approaches, while the reader should refer to specific review articles for a more comprehensive treatment of the state-of-the-art. According to Prof. Hyster [44], light can be used in biocatalysis by exploiting four main strategies: (i) in photoenzymatic catalysis, a cofactor within the protein active site is photoexcited to promote electron or EnT that triggers the conversion of a substrate into a product, (ii) synergistic photoenzymatic catalysis relies on the excitation of an exogenous cofactor to enable a chemical transformation within a protein active site, (iii) tandem photocatalyst/enzyme reactions, where light is either used to regenerate a cofactor or to provide, via photoexcitation, the conversion of a reactant into a substrate suitable for the biocatalytic process, (iv) the fourth class involves enzymatic reactions coupled to natural photosynthesis. All these reactions are usually performed in aqueous buffers, often (but not always) requiring an organic co-solvent (typically 5–30%) to achieve optimum conversions, yields, selectivities, and turn-over numbers of the enzyme.

16.4.1 Photoenzymatic Catalysis

There are currently only three examples of photoenzymes known in nature: a DNA photolyase able to repair pyrimidine cyclobutane dimers in DNA and lesions formed upon UV light irradiation of DNA [45]. The repair process occurs thanks to the photoexcitation of the reduced flavin adenine dinucleotide (FAD) cofactor (FAD_{hq}^{-}) in the active site of the enzyme. Interestingly, homologous enzymes endowed with deazaflavin or pterin cofactors enable the use of alternative wavelengths [45]. A second class of natural photoenzyme is the protochlorophyllide reductase, where the reduction occurs via photoexcitation of the PChlide substrate rather than the NADPH cofactor [46]. A third and well-exploited class is finally represented by the fatty acid photodecarboxylase (FAP), which promotes the hydrodecarboxylation of fatty acids **33** to alkanes **34** in a redox-neutral reaction (Scheme 16.9) [47].

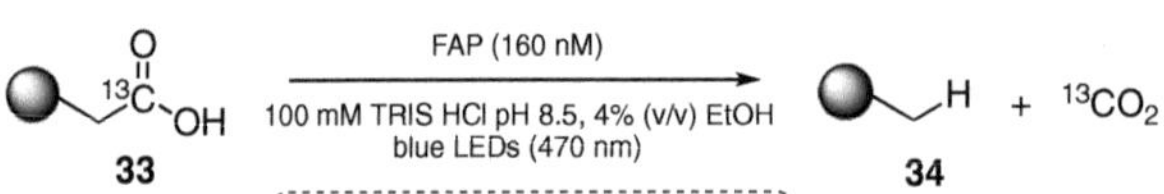

Substrate	Relative Activity (%)
C12:0	8 ± 1
C14:0	40 ± 1
C16:0	99 ± 1
C17:0	96 ± 1
C18:0	36 ± 1
C22:0	< 1

Scheme 16.9 Discovery of fatty acids decarboxylase (FAP).

Mechanistically, the reaction is initiated by the photoexcitation of the FAD cofactor to FAD* with visible light (λ = 325–530 nm with λ_{max} = 467 nm), which promotes the oxidation of the carboxylate. The latter affords an alkyl radical upon decarboxylation, while the termination step can occur either via HAT from Cys432 or PCET from Arg451 as a proton source and $FAD^{\bullet-}$ as the reductant. As for the synthetic scope, FAPs are able to accommodate mono- and di-carboxylic acids from C_{12} to C_{22} to afford C_{n-1} alkanes. Other classes of photoenzymes involve: nicotinamide-dependent ketoreductases (KREDs), flavin-dependent "ene"-reductases (EREDs), flavin-dependent Baeyer–Villiger monooxigenases (BVMs), and flavin-dependent photoenzymes for radical polymerization [44a]. These enzymes are well-known classes of biocatalysts, sometimes with wide and consolidated applications in synthesis, but their combination with light-mediated processes has further expanded their scope. For example, in the case of KREDs the group of Prof. Hyster demonstrated that the reducing power of the NADH/NADPH cofactors could be enhanced, and, as exemplified for the hydrodehalogenation of α-bromolactones, the formation of a charge-transfer (CT) complex with the cofactor inside the catalytic site of the protein could lead to a photoactive substrate (λ_{max} = 390 nm), which is exclusively excited in a chiral environment (i.e. the enzyme catalytic site), thus leading to a stereoselective reaction [48].

16.4.2 Synergistic Photoenzymatic Catalysis

Synergistic photoenzymatic catalysis involves enzymatic transformations relying on electron or EnT from an exogenous photocatalyst to a substrate or cofactor within the protein active site [44]. This family of reactions encloses two subfamilies based on native and non-natural reactivity, respectively. The first class covers cytochromes P450 (the most widely used being $P450_{BM3}$) and Rieske-type non-heme iron enzymes: the bioconjugation with small molecules of photocatalysts, either metal-based polypyridyl complexes or organic ones such as Eosin Y and Rose Bengal, proved to be beneficial for the yield and the total turn-over number, thus improving the overall catalytic activity. In the second class of reactions involving nicotinamide- and flavin-dependent enzymes, an exogenous photocatalyst enables non-native mechanisms, where the electron transfer events are exclusively promoted by the photocatalyst and the enzyme provides a chiral environment for the desired transformation. In this case, the major challenge to address is avoiding background reactions occurring outside the protein; optimized conditions should hence favor the occurrence of the radical formation exclusively within the protein's active site. As a matter of example, Prof. Hyster reported the deacetoxylation of α-acetoxyketones using a nicotinamide-dependent double-bond reductase and Rose Bengal as a photocatalyst (87% yield, 93 : 7 e.r.) [49]: it is worth noting that the latter conferred to the enzyme this catalytic activity, while in its absence the acetoxylation could not occur.

16.4.3 Tandem Photocatalyst/Enzyme Reactions

In these transformations, the photocatalyst acts independently of the enzyme, usually by reducing the cofactor or enabling *in situ* peroxide generation [44]. Alternatively, organic reactions promoted by light can be simply performed in the presence of the enzyme, whose role is to provide a dynamic kinetic resolution. Light-driven cofactor regeneration occurs, for example, for "ene"-reductases, BVMs, FAD-dependent halogenases, ketoreductases, cytochromes P450, and peroxygenases. On the other hand, in photochemical/enzyme cascades, the two reactions (i.e. the light-triggered and the enzymatic ones) can occur either simultaneously or in separate step, resulting in one-pot/one-step and one-pot/two-steps reactions, respectively. In this context, the dynamic kinetic resolution of amines **35** by using lipases, an iridium-based photocatalyst, and thiyl radicals for substrate racemization, very well exemplifies the concept of one-pot/one-step cascades, albeit performed by using pure acetonitrile as a solvent (Scheme 16.10) [50].

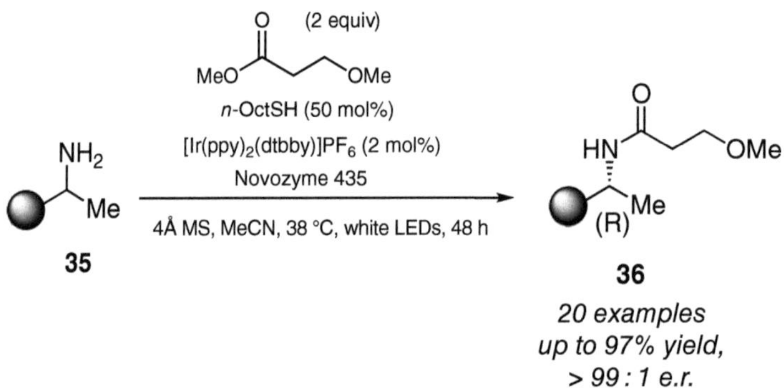

Scheme 16.10 Dynamic kinetic resolution of amines by using lipases.

In a typical one-pot/two-step reaction, instead, the photocatalyst promotes the conversion of a starting material (not elaborated by the enzyme) into a substrate suitable for the enzymatic catalytic step. For example, Schmidt and coworkers achieved an asymmetric C–H oxyfunctionalization of alkanes **37**, **40**, and **43** by photocatalytically oxidizing the latter to aldehydes and ketones **38**, which are then selectively converted to chiral alcohols **39** and **41**, amines **42**, hydroxynitriles **44**, and acyloins **45** in up to 99% e.e. (Scheme 16.11) [51].

16.4.4 Enzymatic Reactions Coupled to Natural Photosynthesis

The development of this class of reactions aims at improving the poor atom efficiency of cofactor regeneration, which affects oxidoreductases, the most widely used enzymes for the synthesis of small organic compounds (alcohol dehydrogenases, enereductases, amine dehydrogenases, and oxygenases). This issue represents a major drawback for the application of biocatalytic methods on scale, i.e. to industrial synthesis. Typically, such reactions require stoichiometric amounts of reductants such as glucose, 2-propanol, and formate, producing equimolar byproducts, which have to be disposed of. A strategy to overcome these limitations, for example, comes from coupling the enzymatic redox process to a photosystem in cyanobacteria that produces NADPH, that can be used as a reductant by the enzyme. It has been shown that such an approach can be harnessed with biocatalysts both native to cyanobacteria and heterologously expressed [44].

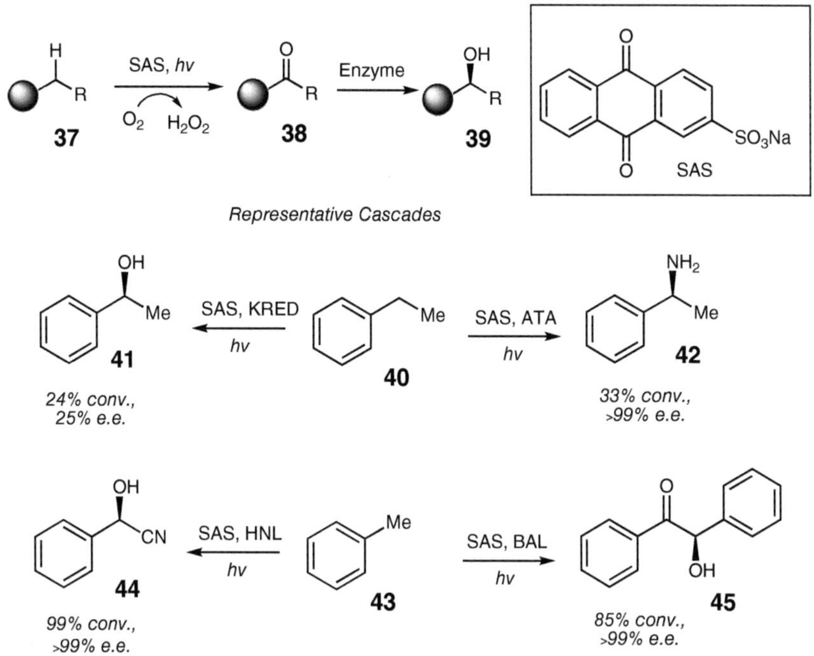

Scheme 16.11 Merging of photoredox and enzyme catalysis for the C–H functionalization of alkanes.

16.5 Summary and Conclusions

In conclusion, the development of photochemical and photoredox catalytic synthetic procedures using water as a reaction medium represents a promising field of research. Water indeed can promote unique reactivity patterns in terms of chemoselectivities and provide significant reaction rate acceleration, while regio- and stereo-specific requirements can be addressed by merging light-activated pathways with enzymatic catalysis. The literature selected in the present chapter highlights how both the solvent and non-solvent properties of water can be valuable tools to drive organic transformations toward the clean and high-yielding formation of the desired products. In particular, the non-solvent effects can promote substrate aggregation into oily phases, which can enable charge transfer complexes' formation and stabilization via hydrogen bonds at the oil–water interface. On the other hand, homogeneous conditions for complex reaction mixtures involving, for example, a photocatalyst, a metal-based co-catalyst, additives (either sacrificial electron donors or acceptors, final oxidants, etc.) can be achieved by exploiting water as a co-solvent. The opportunities that emerged from the current state-of-the-art in the development of light-driven methods in aqueous environments will definitely trigger further studies in the field, contributing to solving the main challenge of a sustainable production of drugs and materials.

References

1 (a) Shaw, M.H., Twilton, J., and MacMillan, D.W.C. (2016). *J. Org. Chem.* 81: 6898–6926. (b) Romero, N.A. and Nicewicz, D.A. (2016). *Chem. Rev.* 116: 10075–10166. (c) Stephenson, C.R.J., Yoon, T.P., and MacMillan, D.W.C. (2018). *Visible Light Photocatalysis in Organic Chemistry*. Germany: Wiley-VCH. (d) Cannalire, R., Pelliccia, S., Sancineto, L. et al. (2021). *Chem. Soc. Rev.* 50: 766–897.

2 McCleverty, J.A. and Meyer, T.J. (2003). *Comprehensive Coordination Chemistry II*. Elsevier Science.

3 (a) Russo, C., Brunelli, F., Tron, G.C., and Giustiniano, M. (2023). *J. Org. Chem.* 88: 6284–6293. (b) Liu, Y., Zuo, L., and Yu, B. (2023). *Curr. Opin. Green Sustain.* 40: 100759. (c) Dou, Q. and Zeng, H. (2023). *Curr. Opin. Green Sustain.* 40: 100766.

4 Kitanosono, T. and Kobayashi, S. (2020). *Chem. Eur. J.* 26: 9408–9429.

5 Cortes-Clerget, M., Yu, J., Kincaid, J.R.A. et al. (2021). *Chem. Sci.* 12: 4237–4266.

6 Ruiz-Lopez, M.F., Francisco, J.S., Martins-Costa, M.T.C., and Anglada, J.M. (2020). *Nat. Rev. Chem.* 4: 459–475.

7 Jung, Y. and Marcus, R.A. (2007). *J. Am. Chem. Soc.* 129: 5492–5502.

8 Jespersen, D., Keen, B., Day, J.I. et al. (2019). *Org. Process Res. Dev.* 23: 1087–1095.

9 van Lier, R.C.W., de Bruijn, A.D., and Roelfes, G. (2021). *Chem. Eur. J.* 27: 1430–1437.

10 Nguyen, T.-T.H., O'Brien, C.J., Tran, M.L.N. et al. (2021). *Org. Lett.* 23: 3823–3827.

11 Qiao, H., Sun, S., Yang, F. et al. (2017). *Adv. Synth. Catal.* 359: 1976–1980.

12 Guerrero, I., Kelemen, Z., Viñas, C. et al. (2020). *Chem. Eur. J.* 26: 5027–5036.

13 Guerrero, I., Viñas, C., Romero, I., and Teixidor, F. (2021). *Green Chem.* 23: 10123–10131.

14 Guerrero, I., Saha, A., Xavier, J.A.M. et al. (2020). *ACS Appl. Mater. Interfaces* 12: 56372–56384.

15 Guerrero, I., Viñas, C., Fontrodona, X. et al. (2021). *Inorg. Chem.* 60: 8898–8907.

16 Long, B., Ding, Z., and Wang, X. (2013). *ChemSusChem* 6: 2074–2078.

17 Srinath, S., Abinaya, R., Prasanth, A. et al. (2020). *Green Chem.* 22: 2575–2587.

18 (a) Lipshutz, B.H., Ghorai, S., and Cortes-Clerget, M. (2018). *Chem. Eur. J.* 24: 6672–6695. (b) La Sorella, G., Strukul, G., and Scarso, A. (2015). *Green Chem.* 17: 644–683. (c) Shen, T., Zhou, S., Ruan, J. et al. (2021). *Adv. Colloid Interface Sci.* 287: 102299.

19 Bu, M.-J., Cai, C., Gallou, F., and Lipshutz, B.H. (2018). *Green Chem.* 20: 1233–1237.

20 Bu, M.-J., Lu, G.-P., Jiang, J., and Cai, C. (2018). *Catal. Sci. Technol.* 8: 3728–3732.

21 Pallini, F., Sangalli, E., Sassi, M. et al. (2021). *Org. Biomol. Chem.* 19: 3016–3023.

22 Ghosh, P.P., Mukherjee, P., and Das, A.R. (2013). *RSC Adv.* 3: 8220–8226.

23 (a) Santos, M.S., Cybularczyk-Cecotka, M., König, B., and Giedyk, M. (2020). *Chem. Eur. J.* 26: 15323–15329. (b) For an example of a MPEG-550 as a phase transfer catalyst, seeFu, Y., Zhu, S., Zhao, X., and Huang, S. (2022). *Green Chem.* 24: 6849–6853.

24 Cybularczyk-Cecotka, M., Predygier, J., Crespi, S. et al. (2022). *ACS Catal.* 12: 3543–3549.

25 Cannalire, R., Santoro, F., Russo, C. et al. (2022). *ACS Org. Inorg. Au* 2: 66–74.

26 Kohlmann, T., Kerzig, C., and Goez, M. (2019). *Chem. Eur. J.* 25: 9991–9996.

27 Eisenreich, F., Meijer, E.W., and Palmans, A.R.A. (2020). *Chem. Eur. J.* 26: 10355–10361.

28 Eisenreich, F., Kuster, T.H.R., van Krimpen, D., and Palmans, A.R.A. (2021). *Molecules* 26: 5882–5893.

29 Kita, Y. and Amao, Y. (2021). *Sustain. Energy Fuels* 5: 6004–6013.

30 Moelbert, S., Normand, B., and De Los Rios, P. (2004). *Biophys. Chem.* 112: 45–57.

31 Zangi, R. (2010). *J. Phys. Chem. B* 114: 643–650.

32 Aloisio, C., de Oliveira, A.G., and Longhi, M. (2016). *J. Pharm. Sci.* 105: 2703–2711.

33 Sravya, G., Suresh, G., Zyryanov, G.V. et al. (2020). *J. Heterocycl. Chem.* 57: 355–369.

34 (a) Tian, Y.-M., Hofmann, E., Silva, W. et al. (2023). *Angew. Chem. Int. Ed.* 62: e202218775. (b) For a further example of EDA complex formation in aqueous medium seeHui, X., Zhang, D., Wu, C. et al. (2023). *Green Chem.* 25: 2274–2278.

35 Narayan, S., Muldoon, J., Finn, M.G. et al. (2005). *Angew. Chem. Int. Ed.* 44: 3275–3279.

36 Schulte, R., Löcker, M., Ihmels, H. et al. (2023). *Chem. Eur. J.* 29: e202203203.

37 Sela, T. and Vigalok, A. (2014). *Org. Lett.* 16: 1964–1967.

38 Dai, J.-J., Zhang, W.-M., Shu, Y.-J. et al. (2016). *Chem. Commun.* 52: 6793–6796.

39 Hou, H., Zhu, S., Pan, F., and Rueping, M. (2014). *Org. Lett.* 16: 2872–2875.

40 Speckmeier, E., Fuchs, P.J.W., and Zeitler, K. (2018). *Chem. Sci.* 9: 7096–7103.

41 Aycock, R.A., Wang, H., and Jui, N.T. (2017). *Chem. Sci.* 8: 3121–3125.

42 (a) Chen, F., Xu, X.-H., and Qing, F.-L. (2021). *Org. Lett.* 23: 2364–2369. (b) For a similar effect seeHu, C.-C., Tong, C.-L., Zhang, Y.-Y. et al. (2023). *Org. Lett.* 25: 1035–1039.

43 Tian, Y.-M., Silva, W., Gschwind, R.M., and König, B. (2024). *Science* 383: 750–756.

44 (a) Emmanuel, M.A., Bender, S.G., Bilodeau, C. et al. (2023). *Chem. Rev.* 123: 5459–5520. (b) Harrison, W., Huang, X., and Zhao, H. (2022). *Acc. Chem. Res.* 55: 1087–1096. (c) Zhang, S., Liu, S., Sun, Y. et al. (2021). *Chem. Soc. Rev.* 50: 13449–13466. (d) Schmermund, L., Jurkas, V., Özgen, F.F. et al. (2019). *ACS Catal.* 9: 4115–4144.

45 Sancar, A. (2003). *Chem. Rev.* 103: 2203–2238.

46 (a) Gabruk, M. and Mysliwa-Kurdziel, B. (2015). *Biochemistry* 54: 5255–5262. (b) Zhang, S., Heyes, D.J., Feng, L. et al. (2019). *Nature* 574: 722–725.

47 (a) Sorigué, D., Légeret, B., Cuiné, S. et al. (2017). *Science* 357: 903–907. (b) Rude, M.A., Baron, T.S., Brubaker, S. et al. (2011). *Appl. Environ. Microbiol.* 77: 1718–1727. (c) Rui, Z., Li, X., Zhu, X. et al. (2014). *Proc. Natl. Acad. Sci. U. S. A.* 111: 18237–18242.

48 (a) Ohnishi, Y., Kagami, M., and Ohno, A. (1975). *Chem. Lett.* 4: 125–128. (b) Fukuzumi, S., Hironaka, K., and Tanaka, T. (1983). *J. Am. Chem. Soc.* 105: 4722–4727.

49 Biegasiewicz, K.F., Cooper, S.J., Emmanuel, M.A. et al. (2018). *Nat. Chem.* 10: 770–775.

50 Yang, Q., Zhao, F., Zhang, N. et al. (2018). *Chem. Commun.* 54: 14065–14068.

51 Zhang, W., Fueyo, E.F., Hollmann, F. et al. (2019). *Eur. J. Org. Chem.* 2019: 80–84.

17

Industrial Applications of Organic Reactions in Water

Chitrakar Ravi and Subbarayappa Adimurthy

CSIR-Central Salt and Marine Chemicals Research Institute, Bhavnagar, Gujarat 364002, India

17.1 Introduction

Water is the nature's solvent. Aqueous media are a crucial interface between chemistry and biology. Even though chemical reactions in aqueous media emphasize the existence of life, water is the worst enemy for synthetic organic chemists, due to the fact that most of the catalysts and reagents have moisture sensitivity which reduces the yield of products and obliterate reproducibility. Such chemical reactions must require organic solvents as medium to enhance their reactivity and stabilize the intermediates formed during the chemical reactions [1]. Despite the fact that organic solvents have toxicity issues such as mutagenicity, teratogenicity, carcinogenicity, and/or nephrotoxicity, still organic solvents dominate the method developments in various aspects. Unlike biomolecules, most of the reactants and catalysts are insoluble in water; some organic reactions require dry organic solvents to stabilize the reaction intermediates as well as to avoid the side product formations by changing the reaction pathway. Such types of reactions require water only in the course of workup procedure to wash the reaction mixture [2]. In general, the operations of organic solvents are also risky due to their flammability, explosivity, and their transportation, and/or storage [3].

An organic reaction in water medium has been known for over a century; indeed, it was preferred by Diels and Alder to perform their well-known [4 + 2] addition of maleic anhydride with furan [4]. Later, many groups developed and organized assemblies such as micelles, cavitands, and organic and inorganic hosts to solubilize organic molecules in water for those reactions which do not occur in water [5, 6]. Further, recent publication from Sharpless and coworkers reported "on-water" reactions between two molecules that are not dissolved but suspended in water [7]. Consequently, the two types of "in-water" and "on-water" reactions expanded to all types of molecules that are insoluble and remain as aggregates in bulk water and to those that are insoluble but remain suspended at the interface involving water as a component. Based on the above observations, numerous synthetically useful reactions in aqueous medium were reported recently [8].

Organic Transformations in Water: Principles and Applications, First Edition.
Edited by Gopinathan Anilkumar, Nissy Ann Harry, and Sankuviruthiyil M. Ujwaldev.

As we know, water has been a preferred medium for biological reactions and very advantageous in green method development due to its nontoxic, cheap, abundant, and nonhazardous nature. Consequently, since last two decades, synthetic chemists have paid much attention to carrying out the reactions in water [9–14]. Given the current emphasis on organic reactions in water, it may be necessary for us to utilize abundantly and freely available, nontoxic, and environmentally friendly water as the medium to perform organic reactions.

Despite a wide variety of organic transformations in aqueous medium, here, we have essentially described a few fundamental organic reactions facilitated in water medium. The details are discussed in the following sections, such as oxidative esterification of alcohols; oxidation of methylarenes; and bromination of arenes, methylarenes, and olefins using green brominating reagent. On the otherhand, "on-water"- and "in-water"-promoted synthesis and functionalization of heterocyclic compounds with an unusual reaction rate acceleration are also represented.

17.2 Oxidative Esterification of Alcohols

The discriminatory oxidation of alcohol is one of the most important transformations in organic synthesis. Furthermore, significance and flexibility of the corresponding carbonyl products are useful in chemical industries. Among all carbonyl compounds, esters are one of the most abundant class of chemicals, frequently utilized as fine chemicals, pharmaceuticals, natural products, agrochemicals, and food additives [15]. Plants and vegetable oils and their essential fatty acids have been used as precursors in the chemical industry for their transformation into high-valued functional chemicals and biofuels [16]. Generally, aliphatic monoesters found in jojoba oil can be easily synthesized from the fatty acids and alcohols derived from triacylglycerol oils [17]. Mainly, long-chain fatty esters were used in the preparation of waxes for industrial applications [18].

Adimurthy and coworkers developed oxidative esterification of aliphatic primary alcohols using environment-friendly reagents under aqueous acidic medium at room temperature (Scheme 17.1). This transformation provided the conversion of various primary alcohols in quantitative yields of the corresponding homocoupled esters. As per the conditions of Scheme 17.1, it doesn't require column chromatography for the separation of products, but simple extraction and removal of the organic solvent provides good yields of pure products. The system has advantages of metal-free conditions, elimination of multistep catalyst preparation (under heterogeneous medium), and used commercially available reagents. The reaction concludes that the oxidative esterification of aliphatic primary alcohols

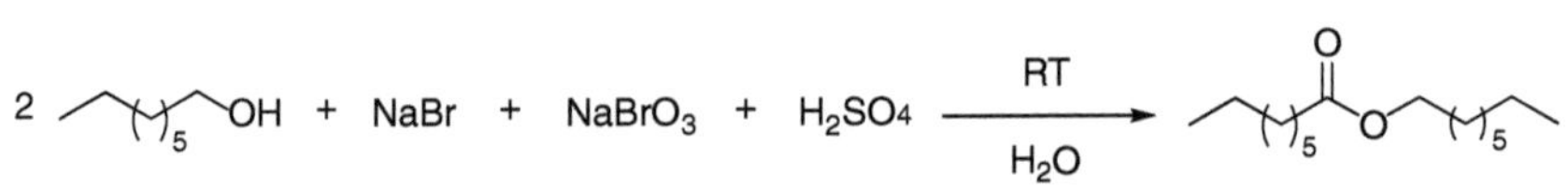

Scheme 17.1 Oxidative esterification of primary alcohols.

with bromide and bromate couple in aqueous acidic medium at room temperature is a modest process with a wide range of substrate scope for both aliphatic and cyclic alcohols and obtained in excellent yields of the desired products [19].

17.3 Green Oxidation of Methylarenes to Benzoic Acids with Bromide/Bromate in Water

The oxidation of methyl aromatics into their corresponding benzoic acids is an important transformation in organic synthesis as these carbonyl derivatives constitute versatile building blocks in pharmaceutical and polymer industries [20]. A variety of oxidizing agents, such as $KMnO_4$, CrO_3, $Na_2Cr_2O_7$, TiO_2, $NaIO_4$, and HNO_3, are available to effect this key transformation [21–25]. However, these metallic reagents are often required in stoichiometric amounts, and these procedures are associated with environmental pollution because of the large amounts of hazardous metallic waste generation. Despite the availability of several methods for the transformation of methyl arenes to aromatic carboxylic acids, the preparation under mild conditions without the use of heavy metal catalysts and reduction of waste is the desirable goal in light of technical, economical, and industrial aspects. As a part of Adimurthy's group activities to develop green procedures that avoid the use of hazardous reagents [26], they developed a novel, heavy-metal-free, and recyclable reagent system for the oxidation of methylarenes to their corresponding benzoic acids using a bromide–bromate couple as reagent under mild conditions in an open atmosphere with water as solvent (Scheme 17.2) [27].

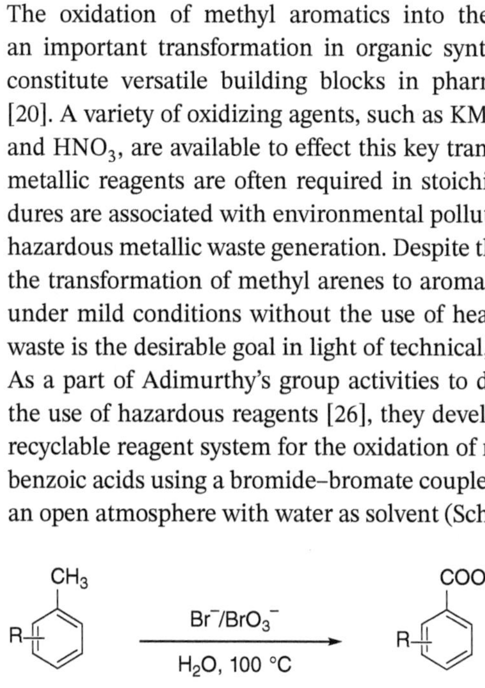

Scheme 17.2 Conversion of methylarenes to aromatic carboxylic acids.

In this transformation, the bromide is not consumed as such in the oxidation process; both bromide–bromate ions ended up in the aqueous effluent as total bromide ions, and they reoxidized a part of the bromide ion to bromate with the oxidizing agents, such as NaOCl, to regenerate the reagent to its original form under ambient conditions (Figure 17.1). Sodium chloride formed in the path of regeneration caused no problem in the oxidation reactions. The regenerated reagent was subsequently used for the oxidation of methylarenes under similar conditions. For a representative example, 4-nitrotoluene was subjected to oxidation by this regenerated reagent system with three consecutive cycles, and the results are the same as with the fresh reagent.

In conclusion, they have developed a very simple and novel procedure for the oxidation of methylarenes to the corresponding benzoic acid derivatives using an economical and easily accessible bromide–bromate couple as oxidizing reagent system. The most significant and green feature of this methodology is the easy

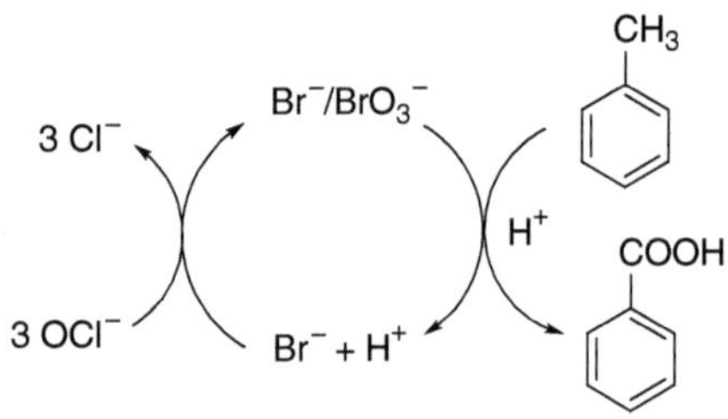

Figure 17.1 Regeneration of bromide/bromate.

operation in water and no use of hazardous organic solvents or toxic reagents. The recyclability of the reagent also adds to its green merit. Thus, this cost-effective and environmentally friendly procedure is a practical alternative to the existing reagents and may find useful applications in academia as well as in industry.

17.4 Environment-Friendly Synthesis of Bromoxynil and Iodoxynil

The majority of the reported methods for bromoxynil synthesis involve the use of liquid bromine in progression [28, 29]. In spite of the industrial significance of bromination process and the continually growing environmental concerns, the use of elemental bromine for bromination reaction is still very common in industry as well as in academia probably due to its easy availability, low cost, and lack of a better alternative. Moreover, purification of the reaction products is often demanding and laborious. By keeping it in mind, Adimurthy et al. developed a high atom-efficient and environment-friendly synthesis of herbicides bromoxynil and iodoxynil using bromide/bromate and iodide/iodate couple as halogenating reagent in water at room temperature (Scheme 17.3) [30].

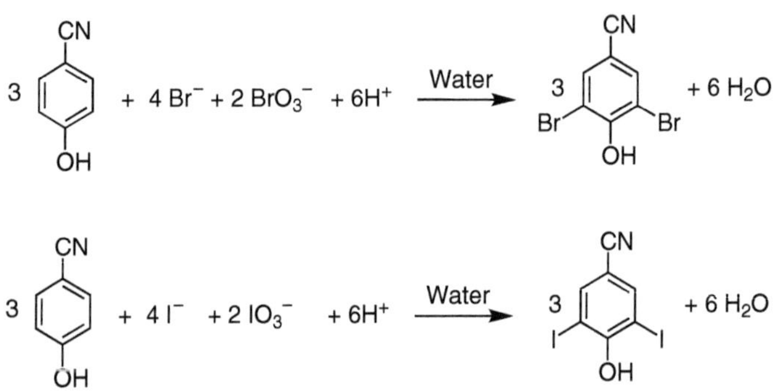

Scheme 17.3 Synthesis of bromoxynil and ioxynil.

The reactions efficiently progressed on water offering high yield and purity of the preferred products. The reagents used for this makeover are reasonably available in a common chemical laboratory. The ease of simplification and high atom-economy of the process make it attractive for the transformation of 4-hydroxybenzonitrile to bromoxynil and ioxynil in small-scale as well as large-scale applications for commercial interest.

17.5 Synthesis of 2,6-Dibromo-4-Nitroaniline from 4-Nitroaniline in an Aqueous Medium

The significance of the bromo-products are well known, which were used as crucial intermediates for the synthesis of various bioactive molecules in pharmaceutical industries [31] and also used as azo-disperse dyes [32], flame retardants [33], pesticides [34], and herbicides [35]. Particularly, 2,6-dibromo-4-nitroaniline (DBPNA) is a very important intermediate used in the azo-disperse dye industry and in the synthesis of star polyfluorenes which are useful in photovoltaic applications [36]. In addition to the above industrial applications, the antifungal behavior of this molecule made it a pharmaceutically important product [37]. DBPNA is a main building block in the synthesis of NanoPutians [38]. Due to the important applications of bromo compounds, a number of brominating reagents have been developed by various groups. Among those reagents, despite its hazardous nature, liquid bromine is a very traditional reagent for industrial manufacture of preferred bromo-compounds. To overcome the hazardous and handling problems associated with liquid bromine at large-scale utilities, Adimurthy and coworkers have introduced a new concept of using liquid bromine intermediate, which consists of an aqueous solution of 5 : 1 bromide–bromate mole ratios during the liquid bromine recovery by cold process [39]. This 5 : 1 mole ratio of Br^-/BrO_3^- could be used for various bromination reactions in combination with mineral acids and suitable oxidants as desirable. Alternately, the 5 : 1 Br^-/BrO_3^- could be converted to 2 : 1 mole ratios using oxidants like sodium hypochlorite and can be used for aromatic substitution reactions under aqueous conditions [40]. In other hand the slow addition of mineral acid to the above aqueous solution containing 2 : 1 Br^-/BrO_3^- salts, which generates hypobromous acid (BrOH) *in situ*, the generated BrOH is responsible for various electrophilic substitution reactions.

Importantly, the method is an organic solvent-free process for the preparation of DBPNA from 4-nitroaniline using 2 : 1 bromide–bromate salts under an aqueous acidic medium at ambient conditions has been developed which is crucial intermediate in azo-disperse dyes synthesis [40]. The 2 : 1 bromide–bromate couple could be obtained by mixing pure $NaBr/NaBrO_3$ salts or by adjusting the 5 : 1 mole ratio of $NaBr/NaBrO_3$ (obtained in an aqueous solution as an intermediate from the bromine manufacture industry by a cold process) to 2 : 1 by the addition of a suitable oxidant. After completion of the reaction, the product was purified by simple filtration and washing with water. The aqueous acidic filtrate was recycled up to five times without loss of purity and yield of the product. The method was extended to other aromatic substrates.

Adimurthy et al. have described a convenient procedure for the quantitative synthesis of 2,6-dibromo 4-nitro aniline from 4-nitroaniline using NaBr and $NaBrO_3$ salts in aqueous acidic medium under ambient conditions without use of organic

$$5NaBr + NaBrO_3 + 3NaOCl \longrightarrow 4NaBr + 2NaBrO_3 + 3NaCl \quad (1)$$

$$2NaBr + NaBrO_3 + 3H^+ \longrightarrow 3BrOH + 3Na^+ \quad (2)$$

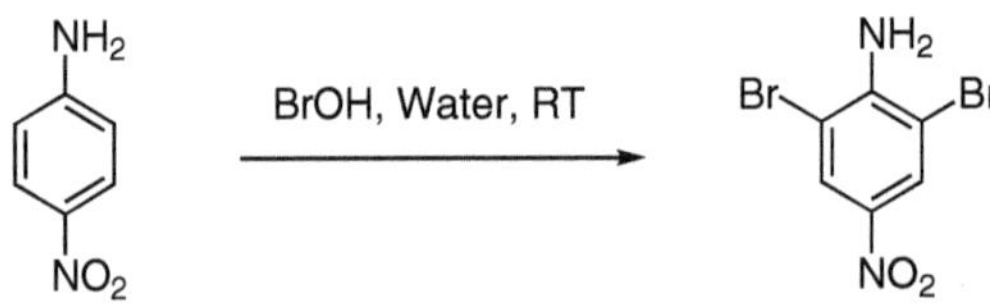

(3)

Scheme 17.4 Synthesis of 2,6-dibromo-4-nitroaniline in an aqueous acidic medium.

solvent in the process (Scheme 17.4). The pure product could be obtained by simple filtration and washing with water without any further purification, which meets the requirement of industrial applications as a dye intermediate. They suggest the mode of addition of brominating reagent to the aqueous acidic 4-nitroaniline solution to get better yield and it helps to recycle the acidic filtrate. During the recycle of filtrate, the dilution of water in the process does not affect either the yield or the purity of the product. They also demonstrated the recyclability of the aqueous effluent in the process.

17.6 HBr–H_2O_2 for Regioselective Synthesis of Bromohydrins and α-Bromoketones and Oxidation of Benzylic/Secondary Alcohols to Carbonyl Compounds in Aqueous Medium

In progress of sustainable, efficient, and selective catalysts for the oxidation of alcohols to aldehydes is a fundamental goal in chemistry. During the recent decades, manifold transition metal-catalyzed reactions have been discovered which significantly improved the organic transformations. Notably, most of the applications are based on complexes of precious metals, such as palladium [41], platinum [42], gold [43], ruthenium [44], and other metal catalysts [45–49]. Also, the (2,2,6,6-tetramethylpiperidin-1-yl)oxyl (TEMPO)-imidazolium salts are efficient metal-free oxidants reported recently for the oxidation of various alcohols [50]. The limited availability of these metals and their high prices makes it highly desirable to search for more economical and environmentally friendly alternatives. In this regard Adimurthy's group developed selective oxidation of various organic substrates using various molor ratios of substrate and $NaBrO_3$ to achieve desired oxidized products (Oxidation of Thiols, Sulfides, and Benzylic/Secondary Alcohols into Disulfides, Sulfoxides, and Aldehydes/Ketones) under ambient conditions [51]. Adimurthy's group further studied and showed that HBr–H_2O_2 can be regioselectively provided bromohydrins and α-bromoketones from olefins and is effective for the selective oxidation of benzylic/secondary alcohols to the corresponding carbonyl compounds under mild conditions [52]. HBr–H_2O_2 is efficiently activated in water under mild conditions, allowing the bromohydroxylation of various styrenes. HBr–H_2O_2 is more economic and easier to handle and offers sufficiently high regioselectivity (100%) for bromohydrin synthesis. Further activation of bromohydrin with a catalytic amount of HBr (50 mol%) and H_2O_2 in water affords α-bromoketones in moderate-to-good yields in a single step. Oxidation of benzylic/secondary alcohols to the corresponding carbonyl compounds has been achieved with the HBr (10 mol%) and H_2O_2 efficiently in aqueous dioxane under ambient conditions (Scheme 17.5).

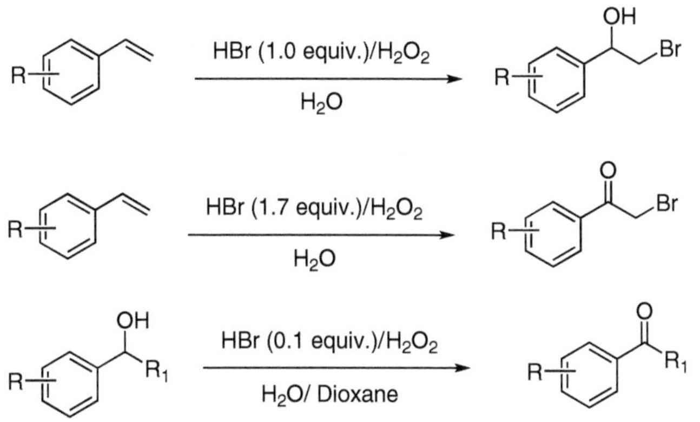

Scheme 17.5 Synthesis of bromohydrins, α-bromoketones, and benzylic/secondary alcohols to carbonyl compounds.

The method described above (Scheme 17.5) is an efficient, selective, and environmentally benign strategy for the synthesis of bromohydrins and α-bromoketones using the HBr–H_2O_2 system in water under mild conditions. The combination of HBr–H_2O_2 generates BrOH *in situ*, which is a good oxidizing as well as a brominating reagent. The addition of BrOH across olefins provide the corresponding bromohydrin derivatives. The dibromo impurity generated during the bromohydrin synthesis is selectively converted to bromohydrin by refluxing in water. This conversion is possible only when the reaction is carried out in water but not in any organic solvent media. Additionally, the HBr (0.7 equiv.) generated during the conversion of bromohydrin to α-bromoketones is utilized to regain bromohydrin by charging a fresh substrate under the conditions specified. Thus, this system generates absolutely zero effluent, representing an environmentally benign process suitable for industrial applications. Catalysis with HBr (10 mol%) is a highly well-organized and selective method for the oxidation of various benzylic alcohols to aldehydes, and secondary alcohols to ketones in aqueous dioxane medium at room temperature. The versatility of this method was applied to a range of styrenes to obtain bromohydrins and α-bromoketones as well as for the selective oxidation of benzylic/secondary alcohols to corresponding carbonyl compounds using simple reaction conditions to get excellent yields of desired products, which makes this method facile, ideal, and attractive synthetic tool.

17.7 Water-Mediated Synthesis of Imidazo[1,2-*a*] Pyridines

Recently, water has been employed as a solvent-cum-catalyst for the construction of various heterocycles [53, 54]. Aqueous reactions have been paying more attention owing to the unique reactivity and selectivity observed that are difficult to achieve in conventional organic solvents [55]. As we know, the organic substrates are generally insoluble in water, but reactions have been reported to proceed "on water" to obtain the desired products [56, 57]. In continuation of Adimurthy's group works on the development of green and sustainable methods [58–60], they also reported "water-mediated" synthesis of methyl imidazo[1,2-*a*]pyridines through intramolecular hydroamination under metal-free conditions (Scheme 17.6).

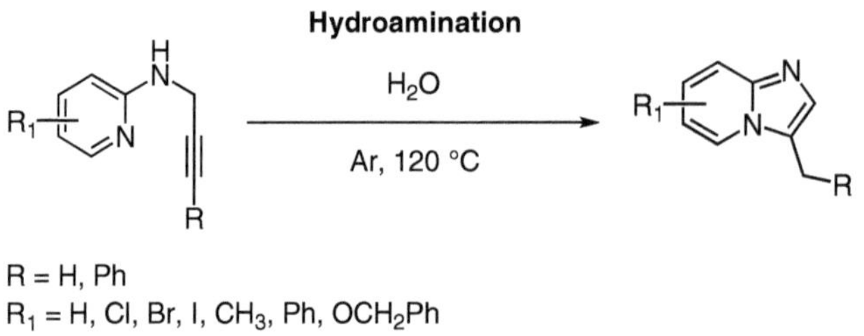

Scheme 17.6 Intramolecular hydroamination in water.

They used substituted *N*-(prop-2-yn-1-yl)pyridin-2-amines [61–63] as starting materials which were readily prepared by known methods [64, 65]. Imidazo[1,2-*a*]pyrazine and imidazo[2,1-*a*]-isoquinoline were also obtained in good yields under similar conditions. They also used the same starting material for Ag-catalyzed intramolecular aminooxygenation for the synthesis of imidazo[1,2-*a*]-pyridine-3-carbaldehydes in moderate-to-good yields in acetonitrile [66, 67]. To gain insight into the reaction mechanism, they performed some control experiments with deuterated water (Scheme 17.7). The starting substrate having free acetylene group delivered terminal–CD3 substituent product, whereas phenyl acetylene substrate

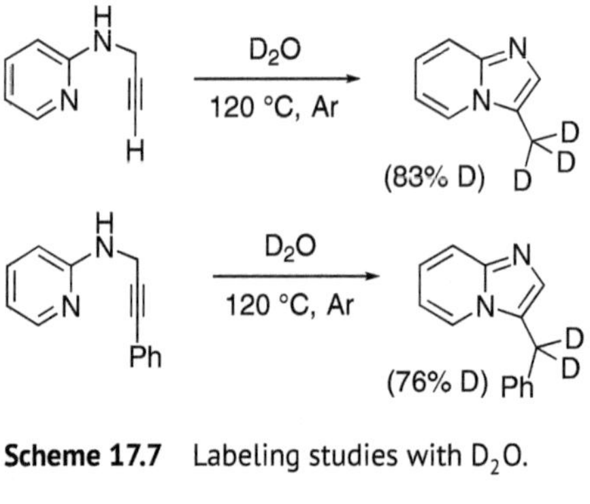

Scheme 17.7 Labeling studies with D_2O.

gave –CD2Ph at terminal position containing product. The observations may be rationalized through the reaction mechanism proposed in Scheme 17.8.

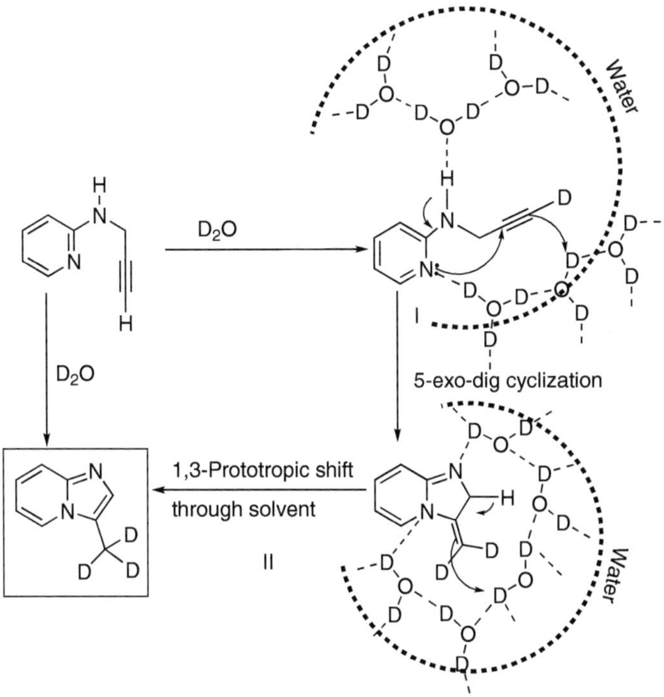

Scheme 17.8 Mechanism for water-mediated intramolecular hydroamination.

In the first step, starting material bearing acidic protons undergoes fast deuterium exchange to yield the intermediate **I** (Scheme 17.8). Then, **I** undergoes intramolecular hydroamination through 5-exo-dig cyclization [53] to produce the vinyl intermediate **II**, incorporating a second deuterium in the process. Generally, **II** being unstable undergoes the rearrangement shown in step 3 during which process $-CD_3$ is formed. It would be evident that, for acetylene substituted at the terminal position, only two deuterium atoms can be incorporated as observed in scheme 17.7.

They summarized as *N*-(prop-2-yn-1-yl)pyridin-2-amines underwent a clean transformation to the corresponding methyl imidazo[1,2-*a*]pyridines without the deliberate addition of any catalyst or reagent with simple water as the medium. The desired products were obtained in 75–97% isolated yield. Studies with D_2O helped to shed light on their proposed mechanism of the reaction system.

17.8 Sulfenylation of N-Heteroarenes in Water (Rice Cooking Method)

Many strategies were successfully engaged for the sulfenylation of imidazo[1,2-*a*]pyridines; it required activators or catalysts, such as copper, iodine, peroxides, and these reactions were only effective in organic solvents [68]. Recently, water has been employed as a solvent-cum-catalyst for the construction of heterocycles [53]. Aqueous reactions have attracted much attention owing to their unique reactivity and selectivity, which are difficult to achieve in conventional organic solvents [55]. Organic substrates are generally insoluble in water; however, reactions proceed "on water" to obtain the desired products.

Adimurthy and coworkers developed a method for the catalyst-free C–H sulfenylation of imidazo[1,2-*a*]pyridines by using sulfonothioates as an odorless thioarylated reagent in aqueous medium (rice cooking method) (Scheme 17.9) [69]. This procedure was employed for the sulfenylation of imidazo[1,2-*a*]pyridines with broad functional-group tolerance and was also extended to the sulfenylation of indoles and imidazothiazoles. The sulfonothioates were activated exclusively in aqueous medium rather than in an organic solvent, and the possibility of the process for scale-up studies was also discussed. In conclusion, they developed an efficient strategy for the sulfenylation of various heterocycles, such as imidazo[1,2-*a*]pyridines, phenylbenzo[*d*]imidazo[2,1-*b*]thiazoles, and indole derivatives, under catalyst-free, aqueous conditions with a high degree of functional group tolerance. The scope of this method was extended to a variety of arene- and alkane sulfonothioates, and the feasibility of performing the reactions at gram-scale was also demonstrated.

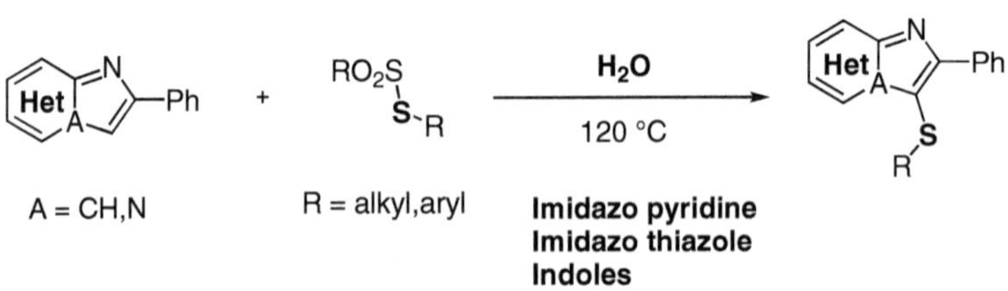

Scheme 17.9 Sulfenylation of imidazo[1,2-*a*]pyridine.

17.9 Summary and Conclusions

In this chapter, we have described the water-mediated organic transformations under different conditions. The chapter mainly focused on the industrially important organic reactions carried out in aqueous medium, such as oxidative esterifications, conversion of methyl arenes to corresponding arene carboxylic acids. The preparations of herbicides and dye intermediate DBPNA under purely aqueous medium were also described. The synthesis of bromhydrins or bromoketones and selective oxidation of benzylic alcohols to carbonyl compounds under metal-free conditions were briefly discussed. Further, water-mediated synthesis of imidazo[1,2-*a*]pyridines and their functionalizations are discussed. An analysis is provided that enfolds both the "on-water" and "in-water" phenomena and water plays a crucial role in each transformation.

Acknowledgments

CSIR-CSMCRI Communication No.177/2023. We are thankful to CSIR-CSMCRI (MLP-0074 and HCP-49) for financial support.

References

1. Cortes-Clerget, M., Yu, J.L., Kincaid, J.R.A. et al. (2021). *Chem. Sci.* 12: 4237–4266.
2. Vaitheesh, J., Ramkumar, V., Giri Babu, V., and Ramamurthy, V. (2021). *J. Photochem. Photobiol. A.* 420: 113492.
3. (a) Romney, D.K., Arnold, F.H., Lipshutz, B.H., and Li, C.-J. (2018). *J. Org. Chem.* 83: 7319–7322; (b) Li, G., Wang, B., and Resasco, D.E. (2020). *ACS Catal.* 10: 1294–1309.
4. Woodward, R.B. and Baer, H. (1948). *J. Am. Chem. Soc.* 70: 1161–1166.
5. Ramamurthy, V. and Inoue, Y. (2011). *Supramolecular Photochemistry: Controlling Photochemical Processes*, 1–51. Hoboken: John Wiley.
6. Turro, N.J., Ramamurthy, V., and Scaiano, J.C. (2010). *Modern Molecular Photochemistry of Organic Molecules*, 1110. Sausalito, CA: University Science Books.
7. Narayan, S., Muldoon, J., Finn, M.G. et al. (2005). *Angew. Chem. Int. Ed.* 44: 3275–3279.
8. Gajewski, J.J. (1997). *Acc. Chem. Res.* 30: 219–225.
9. Kitanosono, T., Masuda, K., Xu, P., and Kobayashi, S. (2018). *Chem. Rev.* 118: 679–746.
10. Gawande, M.B., Bonifacio, V.D.B., Luque, R. et al. (2013). *Chem. Soc. Rev.* 42: 5522–5551.
11. Chanda, A. and Fokin, V.V. (2009). *Chem. Rev.* 109: 725–748.
12. Butler, R.N. and Coyne, A.G. (2010). *Chem. Rev.* 110: 6302–6337.
13. Lindstrom, U.M. (2007). *Organic Reactions in Water: Principles, Strategies and Applications*, 1–28. Oxford: Blackwell Publishing Ltd.
14. Lindstroem, U.M. (2002). *Chem. Rev.* 102: 2751–2771.
15. Otera, J. and Nishikido, J. (2009). *Esterification: Methods, Reactions, and Applications*, 293–320. Hoboken, NJ: Wiley.
16. (a) Biermann, U., Metzger, J.O., and Meier, M.A.R. (2010). *Macromol. Chem. Phys.* 211: 854–862; (b) Biermann, U., Bornscheuer, U., Meier, M.A.R. et al. (2011). *Angew. Chem. Int. Ed.* 50: 3854–3871.
17. (a) Bouzidi, L., Li, S., Di Biase, S. et al. (2012). *Chem. Phys. Lipids* 165: 38–50; (b) Bouzidi, L., Li, S.J., Di Biase, S. et al. (2012). *Ind. Eng. Chem. Res.* 51: 14892–14902; (c) Bouzidi, L., Li, S., Di Biase, S. et al. (2013). *Ind. Eng. Chem. Res.* 52: 2740–2749; (d) Bouzidi, L., Li, S., and Narine, S. (2014). *Symmetry* 6: 655–676.
18. Islam, A. and Ravindra, P. (2017). *Biodiesel Production with Green Technologies*, 1–105. Cham: Springer.

19 Reddy, N.N.K., Ravi, C., and Adimurthy, S. (2018). *Synth. Commun.* 48: 1663–1670.

20 (a) Sheldon, R.A. and Kochi, J.K. (1981). *Metal-Catalyzed Oxidations of Organic Compounds*, 424. New York: Academic Press; (b) Cainelli, G. and Cardillo, G. (1984). *Chromium Oxidations in Organic Chemistry*, 259. Berlin: (Springer; (c) Hudlicky, M. (1990). *Oxidations in Organic Chemistry*. Washington, DC: *American Chemical Society*.

21 (a) Wiberg, K.B. (1965; part A). *Oxidation in Organic Chemistry* (ed. K.B. Wiberg), 69. New York: Academic Press; (b) March, J. (1992). *Advanced Organic Chemistry: Reactions, Mechanisms, and Structure*, 4e, 1328. New York: John Wiley & Sons.

22 Sam, D.J. and Simmons, H.E.C. (1972). *J. Am. Chem. Soc.* 94: 4024–4025.

23 Friedman, L., Fishel, D.L., and Shechter, H. (1965). *J. Org. Chem.* 30: 1453–1457.

24 (a) Muzart, J. (1986). *Tetrahedron Lett.* 27: 3139–3142; (b) Muzart, J. (1987). *Tetrahedron Lett.* 28: 2131–2132; (c) Yamazaki, S. (1999). *Org. Lett.* 1: 2129–2132.

25 (a) Bastock, T.W., Clark, J.H., Matin, K., and Trenbirth, B.W. (2002). *Green Chem.* 4: 615–617; (b) Sasson, Y., Zappi, G.D., and Neumann, R. (1986). *J. Org. Chem.* 51: 2880–2883; (c) Saha, B. and Espenson, J.H. (2005). *J. Mol. Catal. A: Chem.* 241: 33–38.

26 (a) Adimurthy, S., Ramachandraiah, G., Bedekar, A.V. et al. (2006). *Green Chem.* 8: 916–922; (b) Adimurthy, S., Ghosh, S., Patoliya, P.U. et al. (2008). *Green Chem.* 10: 232–237.

27 Patil, R.D., Bhadra, S., Adimurthy, S., and Ranu, B.C. (2010). *Synth. Commun.* 40: 2922–2929.

28 (a) Sheads, R. E., Segal, D. A., *US Patent***1984**, *4*,436,665; (b) Sheads, R. E., *US Patent***1984**, *4*, 439,368.

29 Vidyasagar, A., Nalawala, K.J., and Vershney, A.K. (1993). *Indian J. Chem.* 32B: 872–873.

30 Adimurthy, S., Joshi, G., and Patil, R.D. (2010). *Indian J. Chem.* 49B: 1678–1680.

31 Gordon, G.W. (1999). *Chem. Soc. Rev* 28: 335–346.

32 Peters, A.T. (1988). *J. Soc. Dyers Colour.* 104: 344–348.

33 Ioffe, D. and Kampf, A. (2007). Bromine organic compounds. In: *Kirk-Othmer Encyclopedia of Chemical Technology*, 340–364. New York: John Wiley and Sons, Inc.

34 Dagani, M.J., Barda, H.J., Benya, T.J., and Sanders, D.C. (2002). *Ullmann's Encyclopedia of Industrial Chemistry: Bromine Compounds*, 405–429. Weinheim: Wiley-VCH.

35 Cutulle, M.A., Armel, G.R., Brosnan, J.T. et al. (2014). *J. Agric. Food Chem.* 62: 329–336.

36 Chakraborty, C., Layek, A., Ray, P.P., and Malik, S. (2014). *Eur. Polym. J.* 52: 181–192.

37 Gershon, H., McNeil, M.W., Parmegiani, R., and Godfrey, P.K. (1971). *Appl. Microbiol.* 22: 438–440.

38 Chanteau, S.H. and Tour, J.M. (2003). *J. Org. Chem* 68: 8750–8766.

39 (a) Ramachandraiah, G., Ghosh, P. K., Mehta, A. S., Adimurthy, S., Jethva, A. D., Vaghela, S. S., *US Pat.* 6740253, 2004; (b) Ramachandraiah, G., Ghosh, P. K., Adimurthy, S., BedekarA. V., Shukla, D. B., *US Pat.* 7459139, 2008.

40 Venkatanarayana, P. and Adimurthy, S. (2016). *RSC Adv.* 6: 90184–90187.

41 (a) Wang, H., Deng, S.X., Shen, Z.R. et al. (2009). *Green Chem.* 11: 1499–1502; (b) Feng, B., Hou, Z., Wang, X. et al. (2009). *Green Chem.* 11: 1446–1452; (c) Jenzer, G., Sueur, D., Mallat, T., and Baiker, A. (2000). *Chem. Commun.* 22: 2247–2248.

42 (a) Ng, Y.H., Ikeda, S., Harada, T. et al. (2008). *Chem. Commun.* 44: 3181–3183; (b) Nielsen, R.J. and Goddard, W.A. (2006). *J. Am. Chem. Soc.* 128: 9651–9660; (c) Polshettiwar, V. and Varma, R.S. (2009). *Org. Biomol. Chem.* 7: 37–40; (d) Wang, T., Shou, H., Kou, Y., and Liu, H. (2009). *Green Chem.* 11: 562–568; (e) Maity, P., Gopinath, C.S., Bhaduri, S., and Lahiri, G.K. (2009). *Green Chem.* 11: 554–561.

43 (a) Mitsudome, T., Noujima, A., Mizugaki, T. et al. (2009). *Adv. Synth. Catal.* 351: 1890–1896; (b) Ni, J., Yu, W.J., He, L.H. et al. (2009). *Green Chem.* 11: 756–759.

44 (a) Yu, H., Fu, X., Zhou, C. et al. (2009). *Chem. Commun.* 45: 2408–2410; (b) Kato, C.N., Shinohara, A., Moriya, N., and Nomiya, K. (2006). *Catal. Commun.* 7: 413–416.

45 (a) Dressen, M.H.C.L., Stumpel, J.E., van de Kruijs, B.H.P. et al. (2009). *Green Chem.* 11: 60–64; (b) Lei, Z. and Wang, R. (2008). *Catal. Commun.* 9: 740–742; (c) Yurdakal, S., Palmisano, G., Loddo, V. et al. (2009). *Green Chem.* 11: 510–516.

46 Chen, Y.L. and Chou, T.C. (1996). *Ind. Eng. Chem. Res.* 35: 2172–2176.

47 Sawayama, Y., Shibahara, H., Ichihashi, Y. et al. (2006). *Ind. Eng. Chem. Res.* 45: 8837–8845.

48 Li, Y., Nakashima, D., Ichihashi, Y. et al. (2004). *Ind. Eng. Chem. Res.* 43: 6021–6026.

49 Choudhary, V.R., Dumbre, D.K., and Bhargava, S.K. (2009). *Ind. Eng. Chem. Res.* 48: 9471–9478.

50 Miao, C.X., He, L.N., Wang, J.Q., and Wang, J.L. (2009). *Adv. Synth. Catal.* 351: 2209–2216.

51 Joshi, G., Bhadra, S., Ghosh, S. et al. (2010). *Ind. Eng. Chem. Res.* 49: 1236–1241.

52 Patil, R.D., Joshi, G., and Adimurthy, S. (2010). *Ind. Eng. Chem. Res.* 49: 8100–8105.

53 Zhang, X., Zhou, Y., Wang, H.S. et al. (2011). *Green Chem.* 13: 397–405.

54 Reisch, J. and Scheer, M.J. (1988). *Heterocycl. Chem.* 25: 651–677.

55 Herrerías, C.I., Yao, X., Li, Z., and Li, C.J. (2007). *Chem. Rev.* 107: 2546–2562.

56 Shapiro, N. and Vigalok, A. (2008). *Angew. Chem. Int. Ed.* 47: 2849–2852.

57 Santi, C., Battistelli, B., Testaferri, L., and Tiecco, M. (2012). *Green Chem.* 14: 1277–1280.

58 Chunavala, K.C., Joshi, G., Suresh, E., and Adimurthy, S. (2011). *Synthesis* 4: 635–641.

59 Patil, R.D. and Adimurthy, S. (2011). *Adv. Synth. Catal.* 353: 1695–1700.

60 Donthiri, R.R., Patil, R.D., and Adimurthy, S. (2012). *Eur. J. Org. Chem.* 24: 4457–4460.

61 Thiel, O.R., Achmatowicz, M.M., Reichelt, A., and Larsen, R.D. (2010). *Angew. Chem. Int. Ed.* 49: 8395–8398.

62 Krasnova, L.B., Hein, J.E., and Fokin, V.V. (2010). *J. Org. Chem.* 75: 8662–8665.

63 Cheng, D., Croft, L., Abdi, M. et al. (2007). *Org. Lett.* 9: 5175–5178.

64 Husinec, S., Markovic, R., Petkovic, M. et al. (2011). *Org. Lett.* 13: 2286–2289.

65 Sucunza, D., Samadi, A., Chioua, M. et al. (2011). *Chem. Commun.* 47: 5043–5045.

66 Mohan, D.C., Rao, S.N., and Adimurthy, S. (2013). *J. Org. Chem.* 78: 1266–1272.

67 Mohan, D.C., Sarang, N.B., and Adimurthy, S. (2013). *Tetrahedron Lett* 54: 6077–6080.

68 (a) Ding, Y., Wu, W., Zhao, W. et al. (2016). *Org. Biomol. Chem.* 14: 1428–1431; (b) Ding, Y., Xie, P., Zhu, W. et al. (2016). *RSC Adv.* 6: 81932–81935; (c) Ji, X.-M., Zhou, S.-J., Chen, F. et al. (2015). *Synthesis* 47: 659–671; (d) Ge, W., Zhu, X., and Wei, Y. (2013). *Eur. J. Org. Chem.* 27: 6015–6020; (e) Sang, P., Chen, Z., Zou, J., and Zhang, Y. (2011). *Green Chem.* 13: 397–405; (f) Chen, M., Huang, Z.-T., and Zheng, Q.-Y. (2012). *Chem. Commun.* 48: 11686–11688; (g) Yang, Y., Zhang, S., Tang, L. et al. (2016). *Green Chem.* 18: 2609–2613.

69 Ravi, C., Joshi, A., and Adimurthy, S. (2017). *Eur. J. Org. Chem.* 25: 3646–3651.

Index

Organic Transformations in Water: Principles and Applications, First Edition.
Edited by Gopinathan Anilkumar, Nissy Ann Harry, and Sankuviruthiyil M. Ujwaldev.

b

c

d

e

q

r

S

t

u

v